Numerical Methods for Special Functions

Numerical Methods for Special Functions

Amparo Gil
Universidad de Cantabria
Santander, Cantabria, Spain

Javier Segura
Universidad de Cantabria
Santander, Cantabria, Spain

Nico M. Temme
Centrum voor Wiskunde en Informatica
Amsterdam, The Netherlands

siam. Society for Industrial and Applied Mathematics • Philadelphia

Copyright © 2007 by the Society for Industrial and Applied Mathematics.

10 9 8 7 6 5 4 3 2 1

All rights reserved. Printed in the United States of America. No part of this book may be reproduced, stored, or transmitted in any manner without the written permission of the publisher. For information, write to the Society for Industrial and Applied Mathematics, 3600 University City Science Center, Philadelphia, PA 19104-2688.

Trademarked names may be used in this book without the inclusion of a trademark symbol. These names are used in an editorial context only; no infringement of trademark is intended.

Maple is a registered trademark of Waterloo Maple, Inc.

Mathematica is a registered trademark of Wolfram Research, Inc.

MATLAB is a registered trademark of The MathWorks, Inc. For MATLAB product information, please contact The MathWorks, Inc., 3 Apple Hill Drive, Natick, MA 01760-2098 USA, 508-647-7000, Fax: 508-647-7101, *info@mathworks.com*, *www.mathworks.com*.

Library of Congress Cataloging-in-Publication Data

Gil, Amparo.
   Numerical methods for special functions / Amparo Gil, Javier Segura, Nico M. Temme.
     p. cm.
   Includes bibliographical references and index.
   ISBN 978-0-898716-34-4 (alk. paper)
     1. Functions, Special—Data processing. 2. Numerical analysis. 3. Asymptotic expansions. 4. Apporximation theory. I. Segura, Javier. II. Temme, N. M. III. Title.

QA351.G455 2007
518'.6–dc22

 is a registered trademark.

To our sons Alonso and Javier (A.G. & J.S.)

To my grandsons Ambrus and Fabian (N.M.T.)

Contents

Preface		**xiii**
1	**Introduction**	**1**
I	**Basic Methods**	**13**
2	**Convergent and Divergent Series**	**15**
	2.1 Introduction	15
	2.1.1 Power series: First steps	15
	2.1.2 Further practical aspects	17
	2.2 Differential equations and Frobenius series solutions	18
	2.2.1 Singular points	19
	2.2.2 The solution near a regular point	20
	2.2.3 Power series expansions around a regular singular point	22
	2.2.4 The Liouville transformation	25
	2.3 Hypergeometric series	26
	2.3.1 The Gauss hypergeometric function	28
	2.3.2 Other power series for the Gauss hypergeometric function	30
	2.3.3 Removable singularities	33
	2.4 Asymptotic expansions	34
	2.4.1 Watson's lemma	36
	2.4.2 Estimating the remainders of asymptotic expansions	38
	2.4.3 Exponentially improved asymptotic expansions	39
	2.4.4 Alternatives of asymptotic expansions	40
3	**Chebyshev Expansions**	**51**
	3.1 Introduction	51
	3.2 Basic results on interpolation	52
	3.2.1 The Runge phenomenon and the Chebyshev nodes	54
	3.3 Chebyshev polynomials: Basic properties	56
	3.3.1 Properties of the Chebyshev polynomials $T_n(x)$	56
	3.3.2 Chebyshev polynomials of the second, third, and fourth kinds	60

	3.4	Chebyshev interpolation	62
		3.4.1 Computing the Chebyshev interpolation polynomial . . .	64
	3.5	Expansions in terms of Chebyshev polynomials	66
		3.5.1 Convergence properties of Chebyshev expansions	68
	3.6	Computing the coefficients of a Chebyshev expansion	69
		3.6.1 Clenshaw's method for solutions of linear differential equations with polynomial coefficients	70
	3.7	Evaluation of a Chebyshev sum	75
		3.7.1 Clenshaw's method for the evaluation of a Chebyshev sum	75
	3.8	Economization of power series	80
	3.9	Example: Computation of Airy functions of real variable	80
	3.10	Chebyshev expansions with coefficients in terms of special functions .	83
4	**Linear Recurrence Relations and Associated Continued Fractions**		**87**
	4.1	Introduction	87
	4.2	Condition of three-term recurrence relations	88
		4.2.1 Minimal solutions	89
	4.3	Perron's theorem	92
		4.3.1 Scaled recurrence relations	94
	4.4	Minimal solutions of TTRRs and continued fractions	95
	4.5	Some notable recurrence relations	96
		4.5.1 The confluent hypergeometric family	96
		4.5.2 The Gauss hypergeometric family	102
	4.6	Computing the minimal solution of a TTRR	105
		4.6.1 Miller's algorithm when a function value is known	105
		4.6.2 Miller's algorithm with a normalizing sum	107
		4.6.3 "Anti-Miller" algorithm	110
	4.7	Inhomogeneous linear difference equations	112
		4.7.1 Inhomogeneous first order difference equations. Examples	112
		4.7.2 Inhomogeneous second order difference equations	115
		4.7.3 Olver's method	116
	4.8	Anomalous behavior of some second order homogeneous and first order inhomogeneous recurrences	118
		4.8.1 A canonical example: Modified Bessel function	118
		4.8.2 Other examples: Hypergeometric recursions	120
		4.8.3 A first order inhomogeneous equation	121
		4.8.4 A warning	122
5	**Quadrature Methods**		**123**
	5.1	Introduction	123
	5.2	Newton–Cotes quadrature: The trapezoidal and Simpson's rule . . .	124
		5.2.1 The compound trapezoidal rule	126
		5.2.2 The recurrent trapezoidal rule	129
		5.2.3 Euler's summation formula and the trapezoidal rule . . .	130
	5.3	Gauss quadrature	132

		5.3.1	Basics of the theory of orthogonal polynomials and Gauss quadrature 133
		5.3.2	The Golub–Welsch algorithm 141
		5.3.3	Example: The Airy function in the complex plane 145
		5.3.4	Further practical aspects of Gauss quadrature 146
	5.4	The trapezoidal rule on $\mathbb{R}$ 147	
		5.4.1	Contour integral formulas for the truncation errors 148
		5.4.2	Transforming the variable of integration 153
	5.5	Contour integrals and the saddle point method 157	
		5.5.1	The saddle point method 158
		5.5.2	Other integration contours 163
		5.5.3	Integrating along the saddle point contours and examples 165

II Further Tools and Methods   171

6 Numerical Aspects of Continued Fractions   173

6.1	Introduction 173	
6.2	Definitions and notation 173	
6.3	Equivalence transformations and contractions 175	
6.4	Special forms of continued fractions 178	
	6.4.1	Stieltjes fractions 178
	6.4.2	Jacobi fractions 179
	6.4.3	Relation with Padé approximants 179
6.5	Convergence of continued fractions 179	
6.6	Numerical evaluation of continued fractions 181	
	6.6.1	Steed's algorithm 181
	6.6.2	The modified Lentz algorithm 183
6.7	Special functions and continued fractions 185	
	6.7.1	Incomplete gamma function 186
	6.7.2	Gauss hypergeometric functions 187

7 Computation of the Zeros of Special Functions   191

7.1	Introduction 191	
7.2	Some classical methods 193	
	7.2.1	The bisection method 193
	7.2.2	The fixed point method and the Newton–Raphson method 193
	7.2.3	Complex zeros 197
7.3	Local strategies: Asymptotic and other approximations 197	
	7.3.1	Asymptotic approximations for large zeros 199
	7.3.2	Other approximations 202
7.4	Global strategies I: Matrix methods 205	
	7.4.1	The eigenvalue problem for orthogonal polynomials . . . 206
	7.4.2	The eigenvalue problem for minimal solutions of TTRRs	207
7.5	Global strategies II: Global fixed point methods 213	
	7.5.1	Zeros of Bessel functions 213

		7.5.2	The general case 219
	7.6	Asymptotic methods: Further examples 224
		7.6.1	Airy functions 224
		7.6.2	Scorer functions 227
		7.6.3	The error functions 229
		7.6.4	The parabolic cylinder function 233
		7.6.5	Bessel functions 233
		7.6.6	Orthogonal polynomials 234

8 Uniform Asymptotic Expansions 237

8.1 Asymptotic expansions for the incomplete gamma functions 237
8.2 Uniform asymptotic expansions 239
8.3 Uniform asymptotic expansions for the incomplete gamma functions . 240
    8.3.1 The uniform expansion 242
    8.3.2 Expansions for the coefficients 244
    8.3.3 Numerical algorithm for small values of η 245
    8.3.4 A simpler uniform expansion 247
8.4 Airy-type expansions for Bessel functions 249
    8.4.1 The Airy-type asymptotic expansions 250
    8.4.2 Representations of $a_s(\zeta), b_s(\zeta), c_s(\zeta), d_s(\zeta)$ 253
    8.4.3 Properties of the functions $A_\nu, B_\nu, C_\nu, D_\nu$ 254
    8.4.4 Expansions for $a_s(\zeta), b_s(\zeta), c_s(\zeta), d_s(\zeta)$ 256
    8.4.5 Evaluation of the functions $A_\nu(\zeta), B_\nu(\zeta)$ by iteration . . . 258
8.5 Airy-type asymptotic expansions obtained from integrals 263
    8.5.1 Airy-type asymptotic expansions 264
    8.5.2 How to compute the coefficients α_n, β_n 267
    8.5.3 Application to parabolic cylinder functions 270

9 Other Methods 275

9.1 Introduction 275
9.2 Padé approximations 276
    9.2.1 Padé approximants and continued fractions 278
    9.2.2 How to compute the Padé approximants 278
    9.2.3 Padé approximants to the exponential function 280
    9.2.4 Analytic forms of Padé approximations 283
9.3 Sequence transformations 286
    9.3.1 The principles of sequence transformations 286
    9.3.2 Examples of sequence transformations 287
    9.3.3 The transformation of power series 288
    9.3.4 Numerical examples 288
9.4 Best rational approximations 290
9.5 Numerical solution of ordinary differential equations: Taylor
expansion method 291
    9.5.1 Taylor-series method: Initial value problems 292
    9.5.2 Taylor-series method: Boundary value problem 293
9.6 Other quadrature methods 294

Contents   xi

	9.6.1	Romberg quadrature	294
	9.6.2	Fejér and Clenshaw–Curtis quadratures	296
	9.6.3	Other Gaussian quadratures	298
	9.6.4	Oscillatory integrals	301

III Related Topics and Examples   307

10 Inversion of Cumulative Distribution Functions   309
     10.1 Introduction 309
     10.2 Asymptotic inversion of the complementary error function 309
     10.3 Asymptotic inversion of incomplete gamma functions 312
         10.3.1 The asymptotic inversion method 312
         10.3.2 Determination of the coefficients ε_i 314
         10.3.3 Expansions of the coefficients ε_i 316
         10.3.4 Numerical examples 316
     10.4 Generalizations 317
     10.5 Asymptotic inversion of the incomplete beta function 318
         10.5.1 The nearly symmetric case 319
         10.5.2 The general error function case 322
         10.5.3 The incomplete gamma function case 324
         10.5.4 Numerical aspects 326
     10.6 High order Newton-like methods 327

11 Further Examples   331
     11.1 Introduction 331
     11.2 The Euler summation formula 331
     11.3 Approximations of Stirling numbers 336
         11.3.1 Definitions 337
         11.3.2 Asymptotics for Stirling numbers of the second kind . . . 338
         11.3.3 Stirling numbers of the first kind 343
     11.4 Symmetric elliptic integrals 344
         11.4.1 The standard forms in terms of symmetric integrals . . . 345
         11.4.2 An algorithm 346
         11.4.3 Other elliptic integrals 347
     11.5 Numerical inversion of Laplace transforms 347
         11.5.1 Complex Gauss quadrature 348
         11.5.2 Deforming the contour 349
         11.5.3 Using Padé approximations 352

IV Software   353

12 Associated Algorithms   355
     12.1 Introduction 355
         12.1.1 Errors and stability: Basic terminology 356

		12.1.2	Design and testing of software for computing functions: General philosophy 357
		12.1.3	Scaling the functions 358
	12.2	Airy and Scorer functions of complex arguments 359
		12.2.1	Purpose 359
		12.2.2	Algorithms 359
	12.3	Associated Legendre functions of integer and half-integer degrees . . . 363
		12.3.1	Purpose 363
		12.3.2	Algorithms 364
	12.4	Bessel functions 369
		12.4.1	Modified Bessel functions of integer and half-integer orders 370
		12.4.2	Modified Bessel functions of purely imaginary orders . . 372
	12.5	Parabolic cylinder functions 377
		12.5.1	Purpose 377
		12.5.2	Algorithm 378
	12.6	Zeros of Bessel functions 385
		12.6.1	Purpose 385
		12.6.2	Algorithm 385

List of Algorithms   **387**

Bibliography   **389**

Index   **405**

Preface

Probably, the most extended (pseudo)definition of the set of functions known as "special functions" refers to those mathematical functions which are widely used in scientific and technical applications, and of which many useful properties are known. These functions are typically used in two related contexts:

1. as a way of obtaining simple closed formulas and other analytical properties of solutions of problems from pure and applied mathematics, statistics, physics, and engineering;
2. as a way of understanding the nature of the solutions of these problems, and for obtaining numerical results from the representations of the functions.

Our book is intended to provide assistance when a researcher or a student needs to get the numbers from analytical formulas containing special functions. This book should be useful for those who need to compute a function by their own means, or for those who want to know more about the numerical methods behind the available algorithms. Our main purpose is to provide a guide of available methods for computations and when to use them. Also, because of the large variety of numerical methods that are available for computing special functions, we expect that a broader "numerical audience" will be interested in many of the topics discussed (particularly in the first part of the book). Several levels of reading are possible in this book and most of the chapters start with basic principles. Examples are given to illustrate the use of the methods, pseudoalgorithms are given to describe technical details, and published algorithms for computing a selection of functions are described as practical illustrations for the basic methods of this book.

The presentation of the topics is organized in four parts: Basic Methods, Further Tools and Methods, Related Topics and Examples, and Software. The first part (Basic Methods) describes a set of methods which, in our experience, are the most popular and important ones for computing special functions. This includes convergent and divergent series, Chebyshev expansions, linear recurrence relations, and quadrature methods. These basic chapters are mostly self-contained and start from first principles. We expect that many of the contents are appropriate for advanced numerical analysis courses (parts of the chapters are in fact based on classroom notes); however, because the main focus is on special functions, detailed examples of application are also provided.

The second part of the book (Further Tools and Methods) contains a set of important methods for computing special functions which, however, are probably not so well known as the basic methods (at least for readers who are not very familiar with special functions).

Certainly, this does not mean that these tools are less effective than the selected basic methods; for example, the performance of uniform asymptotic expansions is quite impressive in many instances. The chapters in this second part are: Continued Fractions, Computation of the Zeros of Special Functions, Uniform Asymptotic Expansions, and Other Methods (Padé approximations, sequence transformations, best rational approximations, Taylor's method for ordinary differential equations, and further quadrature methods including the Clenshaw–Curtis and Filon methods).

The third part (Related Topics and Examples) describes some methods that are specific to certain functions. A first chapter is devoted to the (asymptotic) numerical inversion of a class of distribution functions with details for gamma and beta distributions (a topic which researchers in statistics, probability, and econometrics may find useful). A second chapter (Further Examples) describes varied topics such as the Euler summation formula (and applications), the computation of symmetric elliptic integrals (Carlson's method), and the numerical inversion of Laplace transforms.

We thank NIST for the permission to quote part of a section in the DLMF project (from the chapter "Numerical Methods") on solving ordinary differential equations by using Taylor series (our §9.5), and Frank Olver for his assistance in writing this part. We thank the SIAM editorial staff, in particular Louis Primus, for their patience and splendid cooperation.

Finally, the fourth part illustrates the use of the methods by providing descriptions of specific algorithms for computing selected functions: Airy functions, Legendre functions, and parabolic cylinder functions, among others. The corresponding Fortran 90 routines can be downloaded from

$$\text{http://functions.unican.es.}$$

The web page will hold successive actualizations and extensions of the available software.

We would like to thank Dr. Van Snyder for his extensive and useful comments, and Dr. Ernst Joachim Weniger for providing us with notes, and further useful information, on Padé approximations and sequence transformations. Finally, we thank the Spanish Ministry of Education and Science for financial support (projects MTM2004-01367, MTM2006-09050).

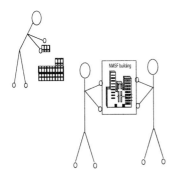

Chapter 1
Introduction

This book deals with numerical methods for computing special functions. This means that we describe numerical methods for computing quantities for which no general definition seems to be available. What makes a function used in applications or in pure mathematics a "special function"? This is perhaps a matter of taste. There is, however, a general practical consensus regarding which functions are special: a special function should be useful for applications and should satisfy certain special properties which allow analytical treatment.

Several centuries ago the astronomers used the trigonometric functions as their basic tools in mathematics. In recent centuries the development of wave theory introduced many other functions of higher level, and many of these functions became the classical functions with which many problems from physics and other applied sciences could be described. In pure mathematics special functions arose that played a role in number theory. Also later, some of the functions that became important in statistics and probability theory were new or related to the classical functions, in particular the gamma and beta distribution functions.

There are many books now that give collections of special functions and/or describe their properties. A classical reference is the *Handbook of Mathematical Functions*, edited by Milton Abramowitz and Irene Stegun. The first edition appeared in 1964 and very soon a complete revision of this important reference, with many new chapters, will be published, together with a free accessible web version. Although the *Handbook* has introductory matter on computing and approximating elementary and special functions, complete algorithms or software are not given in either version of this reference work.

When the *Handbook* was published in 1964, a great number of algorithms and methods for computing special functions were already known, but in later years many new ideas were developed and became available in the form of mathematical software for computing special functions. Software libraries were constructed and several books appeared with collections of software, some of them claiming to cover all the functions considered in the *Handbook*.

In the present book we are not so formidably optimistic that we claim to describe computational methods or algorithms for all functions described in the old or new version of the *Handbook*. However, we describe methods which, according to our own experience, appear most frequently in the computation of special functions; this is so particularly in the first part of the book, devoted to power series, Chebyshev expansions, recurrence relations and continued fractions, and quadrature of integrals, but many other topics are also described in the book. Some methods are illustrated with explicit software examples in the last part of the book.

Airy functions are a good example of functions for which different techniques (convergent and divergent series, Chebyshev expansions, quadrature) can be of interest for computing the functions, depending on the range of the variable. Next we consider this set of functions, solutions of the second order differential equation $y'' - xy = 0$, as an example for introducing basic concepts to be described later to a greater extent. It should come as no surprise that we first discuss the solution of the differential equation using power series.

Convergent power series and differential equations

In many books of mathematical methods for physicists or engineers the words "special functions" appear for the first time when solving certain differential equations (for instance, when solving the Schrödinger equation by separation of variables) and, particularly, when trying to solve the equations by power series. Equations such as the Hermite or Bessel equations appear, which can be solved by using convergent power series.

The fact that we can find a convergent series for a specific solution of a second order ordinary differential equation may seem to indicate that the computation of such a function is of no concern. However, this is not true from a numerical point of view, even when the series does converge for any real or complex value of the argument. It could only be true if we had at our disposal a computer equipped with infinite precision arithmetic, that was infinitely fast, and that was without limitations in the numbers which can be stored. Because this ideal machine does not exist, the use of series will be limited by these three factors (as any other method).

Take, for instance, the elementary example $y'' - y = 0$, with $y_1(z) = e^z$ and $y_2(z) = e^{-z}$ as two independent solutions. Maclaurin series for y_1 and y_2 are convergent for all z. However, when $\Re z > 0$ the range of accurate computation of $y_2(z)$ should be restricted to small $|z|$, because for large $|z|$, the first terms of the series are much larger in modulus than the whole sum, leading to numerical cancellation and severe loss of significant digits. Also, the series for $y_1(z)$ is dangerous for large z because many terms need to be added. Maclaurin series should not be used very far from $z = 0$.

For not-so-elementary functions, the same types of limitations occur when using series (Chapter 2). Take for instance the very important case of the Airy functions, which are solutions of the second order ordinary differential equation

$$y''(z) - zy(z) = 0. \tag{1.1}$$

We can try power series to find the general solution. Substituting $y(z) = \sum_{n=0}^{\infty} a_n z^n$ we readily see that $a_{2+3n} = 0, n \in \mathbb{N}$, and that two linearly independent solutions are

$$y_1(z) = \sum_{k=0}^{\infty} 3^k \left(\frac{1}{3}\right)_k \frac{z^{3k}}{(3k)!}, \quad y_2(z) = \sum_{k=0}^{\infty} 3^k \left(\frac{2}{3}\right)_k \frac{z^{3k+1}}{(3k+1)!}, \tag{1.2}$$

where $(\alpha)_0 = 1, (\alpha)_k = \alpha(\alpha+1)\cdots(\alpha+k-1), k \geq 1$. Elementary methods (the ratio test) can be used to prove that both series are convergent for any complex value of z.

Chapter 1. Introduction

At this moment, let us restrict the problem to real positive $z = x$. We observe that both $y_1(x)$ and $y_2(x)$ are positive and increasing for $x > 0$; all the terms of the series are positive and no cancellations occur. Now, because all the solutions of the differential equation can be written as $y(x) = \alpha y_1(x) + \beta y_2(x)$, and both $y_1(x)$ and $y_2(x)$ tend to $+\infty$ as $x \to \infty$, the equation necessarily has a solution $f(x)$, called a *recessive solution*, such that $f(x)/y(x) \to 0$ for any other solution $y(x)$ not proportional to $f(x)$.

Indeed, as $x \to +\infty$, either $y_1(x)/y_2(x) \to 0$, giving that $y_1(x)$ is recessive, or $y_2(x)/y_1(x) \to 0$, giving that $y_2(x)$ is recessive, or $y_1(x)/y_2(x) \to C \neq 0$, giving that $y_1(x) - Cy_2(x)$ is recessive. The last possibility is what happens for the recessive solutions of (1.1), with $C = 3^{1/3}\Gamma(2/3)/\Gamma(1/3)$. Again, for an algorithm the Maclaurin series can only be considered for not too large x, particularly for computing the recessive solution (such as for e^{-x} in the former example).

Later we will see that the recessive solution is a multiple of Ai(z), which is exponentially small at $+\infty$, and both $y_1(z)$ and $y_2(z)$ given in (1.2) are exponentially large. Hence, the solutions $y_1(z)$ and $y_2(z)$ cannot be used to compute all solutions of (1.1) for all values of z, in particular when z is large, because this may introduce large errors. We say that $y_1(z)$ and $y_2(z)$ do not constitute a *numerically satisfactory pair of solutions* of (1.1) at $+\infty$, because the recessive solution cannot be computed by these two for large positive z. The same situation occurs when considering differential equations for other special functions. For a graph of the Airy function Ai(x), see Figure 1.1.

Obviously, one should never compute an exponentially decreasing function for large values of the variable as a linear combination of two increasing functions for values of the argument for which the computed value is much smaller than the increasing functions. Subtracting two large quantities for obtaining a small quantity is a numerical disaster. For the recessive Airy function Ai(x) we have Ai(1) = 0.135..., and the loss of significant digits becomes noticeable when we use the functions of (1.2) when $x \geq 1$. Therefore, if we need to compute the recessive solution, we must consider an independent method of computation for this function.

In addition, one needs to pay attention to different ranges of the variable in order to select a numerical satisfactory pair. For instance, going back to the elementary case, $y'' - y = 0$, the solutions $y_1(z) = e^z$ and $y_2(z) = e^{-z}$, which constitute a numerically satisfactory pair when $\Re z \gg 0$, but this pair is not a satisfactory pair near $z = 0$. Near the origin, it is better to include $y_3(z) = \sinh z = (y_1(z) - y_2(z))/2$ as a solution when $|z|$ is small; a companion solution could be $y_4(z) = \cosh z$. Of course, $y_3(z)$ and $y_4(z)$ do not constitute a numerically satisfactory pair when $|\Re z|$ is large.

Liouville–Green approximation and dominant asymptotic behavior

More information on the behavior of the solutions of a linear second order differential equation can be obtained by transforming the equation. It is a straightforward matter to check (see also §2.2.4) that, if $y(z)$ is a solution of (1.1), then the function $Y(z) = z^{1/4} y(z)$ satisfies the equation

$$\ddot{Y}(\zeta) + \left(-1 + \frac{5}{36\zeta^2}\right) Y(\zeta) = 0 \tag{1.3}$$

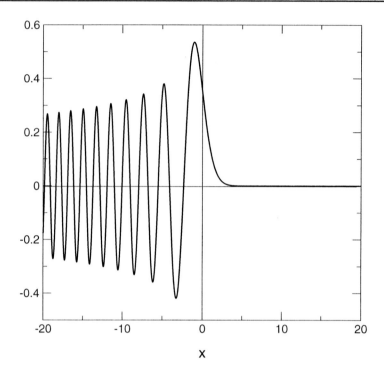

Figure 1.1. *The Airy function* Ai(x) *is oscillating for $x < 0$ and exponentially small for $x \to +\infty$.*

in the variable $\zeta = \frac{2}{3}z^{3/2}$. Looking at (1.3), we can expect that for large ζ the term $5/(36\zeta^2)$ will be negligible and that the solutions will behave exponentially, $Y(\zeta) \sim \exp(\pm\zeta)$. Undoing the changes, we expect that the recessive solution of (1.1) behaves as $z^{-1/4}\exp(-\frac{2}{3}z^{3/2})$. A more detailed analysis [168, Chap. 6] shows that this approximation (the Liouville–Green approximation) makes sense. Therefore, we are certain that a recessive solution exists which decreases exponentially.

Condition of solution of ordinary differential equations

It is not a surprise that the recessive solutions of the defining differential equations are usually the most important in applications; indeed, physical quantities are by nature finite quantities and functions representing these quantities should be bounded. From a numerical point of view, as explained, a special treatment for these functions is needed.

Anyway, isn't the Airy function a solution of a second order differential equation? Isn't it true that most students who have received a course on numerical analysis are familiar with methods for solving second order differential equations? Isn't it true that numerical methods for second order differential equations apply to any solution of the equation? So, can we rely blindly on, let's say, Runge–Kutta? Or do we need more analysis? The answer is, obviously, yes, analysis is necessary and one needs to know if the desired solution is recessive.

Chapter 1. Introduction

Numerical methods for solving initial value problems for ordinary differential equations (like Euler, Taylor, Runge–Kutta, etc.) can be used for computing any solution of the equation. This is good and bad news. The nice part is obvious; the bad part has to do with the condition of the process. If, for instance, we are integrating the Airy equation for computing Ai(x), starting from the known values Ai(0) and Ai$'$(0) and computing values of the function for increasing values of x, we will end up with a completely wrong answer as x increases. The reason is that Ai(x) is recessive as $x \to +\infty$. Precisely because the numerical method can in principle be used for computing any solution, dominant or not, the computation of Ai(x) is badly conditioned because any numerical error will introduce a small component of a dominant solution which, of course, tends to dominate as x increases. Contrarily, the problem of computing a dominant solution would be well conditioned.

A better approach for computing a recessive solution consists of using function values for large x-values, computing initial values by using the asymptotic expansion of the Airy function, and integrating the differential equation with decreasing values of x. In this way, we consider the initial value problem

$$y'' - xy = 0,$$
$$y(b) = \text{Ai}(b), \quad y'(b) = \text{Ai}'(b), \tag{1.4}$$

and compute the solution in the backward direction (from b to 0 with negative step size); when b is chosen large enough we can compute the initial values from asymptotic expansions (that we next discuss). This process is well conditioned.

In §9.5 we describe the use of Taylor's method for solving differential equations defining special functions. We explain that the direction (increasing values or decreasing values of $\Re z$) in which we solve the equation has to be chosen in order to get a stable method, depending on which solution is wanted.

Asymptotic approximations

Earlier we have seen that, as $z \to +\infty$, the Airy function has the asymptotic estimate Ai(z) $\sim C z^{-1/4} \exp(-\zeta)$, where $\zeta = \frac{2}{3} z^{3/2}$. We can improve this estimate by using (1.3) again. Because we are interested in an exponentially decaying solution, we will try a solution of the form

$$Y(\zeta) = \exp(-\zeta) g(\zeta), \quad \zeta = \tfrac{2}{3} z^{3/2}. \tag{1.5}$$

Substitution gives the differential equation

$$\frac{d^2 g}{d\zeta^2} - 2\frac{dg}{d\zeta} + \frac{\lambda}{\zeta^2} g = 0, \quad \lambda = \frac{5}{36}, \tag{1.6}$$

and using a formal series in powers of ζ^{-1}, that is, $g(\zeta) = \sum_{k=0}^{\infty} a_m \zeta^{-m}$, we get, equating term by term,

$$a_{m+1} = -\frac{\lambda + m(m+1)}{2(m+1)} a_m, \quad m = 0, 1, 2, \ldots. \tag{1.7}$$

Therefore, we find the expansion

$$y(z) \sim z^{-1/4} e^{-\zeta} \sum_{k=0}^{\infty} a_m \zeta^{-m}, \quad \zeta = \tfrac{2}{3} z^{3/2}, \tag{1.8}$$

which holds for recessive solutions for any $a_0 \neq 0$. Specifying $a_0 = \frac{1}{2}\pi^{-1/2}$ we obtain the expansion for the recessive Airy function

$$\operatorname{Ai}(z) \sim z^{-1/4} e^{-\zeta} \sum_{k=0}^{\infty} a_m \zeta^{-m}. \tag{1.9}$$

We have not used the equal sign in (1.9), because the series does not converge for any real (or complex) z. It is an asymptotic series (which explains why we use the symbol $\sim$) and it can be used to obtain an interesting approximation for large z, even when the series diverges. In fact, it is possible to prove that this is a valid asymptotic expansion for all large complex z, excluding negative values of z.

When we say that an expansion of the form

$$f(z) \sim \sum_{n=0}^{\infty} a_n z^{-n}, \quad z \to \infty, \tag{1.10}$$

is an *asymptotic expansion*, we assume that (see Chapter 2)

$$z^N \left(f(z) - \sum_{n=0}^{N-1} a_n z^{-n} \right), \quad N = 0, 1, 2, \ldots, \tag{1.11}$$

where the sum is empty when $N = 0$, is a bounded function for large values of z, with limit a_N as $z \to \infty$, for any N. This can also be written as

$$f(z) = \sum_{n=0}^{N-1} a_n z^{-n} + \mathcal{O}\left(z^{-N}\right), \quad z \to \infty. \tag{1.12}$$

The validity is usually restricted to a sector in the z-plane; for the Airy function the expansion (1.9) holds inside the sector $|\operatorname{ph} z| < \pi$.

The notion of asymptotic expansion means that if z is sufficiently large, using a finite number of terms of the series may yield accurate results. In some sense, asymptotic series can be used as if they were convergent series and we can add successive terms in order to get the desired precision. But there is an important difference: we should not add too many terms, because usually the series finally diverges. For fixed and large z, there is a specific number of terms N_z of the series such that if more terms are added, the result worsens. This is not a numerical effect, but rather an intrinsic analytical property of asymptotic series.

Further analysis [168] shows that, under mild conditions, asymptotic expansions of the form (1.9) can be differentiated (when we have an expansion for a function which has a continuous derivative admitting an asymptotic expansion or when the function is analytic in a given sector). Differentiating (1.9), which is the asymptotic expansion for an entire function, gives an asymptotic expansion for $\operatorname{Ai}'(z)$.

Because now we have an approximation for computing $\operatorname{Ai}(x)$ and $\operatorname{Ai}'(x)$ for sufficiently large x, we can also consider using these expansions as the approximation method for large x, and forget about solving the differential equation. In this way we can combine the convergent series expansion for $x \leq x_0$ and the divergent asymptotic series for $x \geq x_0$ in order to build an algorithm for computing the Airy functions for positive x. The only thing to be determined is the value of x_0 and the accuracy attainable. This can be done analytically, by using the available bounds for the remainders of both approximations and setting x_0 so as to optimize the accuracy; also an empirical approach is possible. It turns out that such an algorithm based on convergent and divergent series is indeed possible and that when we take $x_0 = 5.5$ we can get a relative precision better than 10^{-8} for all $x > 0$ when we do the computations with 15 digits.

An asymptotic approximation for dominant solutions can be obtained in the same way by using (see (1.5)) $Y(\zeta) = \exp(\zeta)h(\zeta)$. This gives the asymptotic expansion for the Airy function $\text{Bi}(z)$, and it reads

$$\text{Bi}(z) \sim 2z^{-1/4} e^{\zeta} \sum_{k=0}^{\infty} (-1)^m a_m \zeta^{-m}, \qquad (1.13)$$

where a_m and ζ are the same as in (1.9). This expansion is valid inside the sector $|\text{ph } z| < \frac{1}{3}\pi$.

With respect to the negative real axis, if we proceed as before but with a change of variables, $\zeta(x) = \frac{2}{3}(-x)^{3/2}$, we observe that $Y(x) = x^{1/4}y(x)$ satisfies (1.3), with $-1 + 36/(5\zeta^2)$ replaced by $1 + 36/(5\zeta^2)$, which shows that Airy functions have infinitely many real negative zeros. The corresponding asymptotic expansion shows this behavior explicitly [2, eq. (10.4.60)]; see also (3.133) and Figure 1.1.

Chebyshev methods and Chebyshev expansions

Although we have been relatively successful (8-digit precision using 15–16 digits) for intermediate values of x, the Maclaurin and asymptotic series approximations have limited accuracy and many significant digits are lost. This can be expected because both are approximations around two particular points (0 and ∞). An accurate and efficient method should do better than that.

Another possibility for approximation consists in interpolating the function at a number of points in an interval, that is, finding a polynomial for approximating the function such that the polynomial gives the exact value at some specific points in this interval. If the function to be interpolated has a sufficiently smooth variation inside the interval, this may yield a good approximation for the whole interval. For the sake of having smooth variation, it is better to factor out the dominant exponential term, to write $\text{Ai}(x) = x^{-1/4}\exp(-\zeta)g(\zeta)$, and to interpolate $g(\zeta)$.

The values of x where the functions are interpolated should be conveniently chosen so as to make the error as uniform as possible in the whole interval. For this purpose, the best choices are the Chebyshev nodes (see Chapter 3) and the corresponding interpolation is the so-called *Chebyshev approximation*, which is usually given in the form of a polynomial or of a rational function. For building a Chebyshev approximation one needs, however, to have a numerical method for computing function values at the Chebyshev nodes. Extended precision arithmetic could be considered for computing values from Maclaurin series (or from asymptotics).

A convenient and efficient way for computing Chebyshev approximations is to consider the expansion in terms of Chebyshev polynomials $T_n(x)$. The real finite or infinite interval is transformed into the standard interval $[-1, 1]$ and the transformed function is expanded in the series

$$f(x) = \sum_{n=0}^{\infty} a_n T_n(x), \quad -1 \leq x \leq 1. \tag{1.14}$$

The coefficients can be obtained from certain interpolating formulas, from integrals, or from a differential equation satisfied by the function $f(x)$. For details, also for the Airy function $\text{Ai}(x)$, we refer to Chapter 3.

Quadrature methods

Another way to obtain values of Airy functions for intermediate values (not necessary real) is to consider integral representations of the functions. Also, integral representations are an interesting source for obtaining asymptotic approximations. For instance, it can be proved that the following is a representation for the recessive solution of Airy's equation:

$$\text{Ai}(z) = \frac{1}{2\pi} e^{-\zeta} \int_0^\infty e^{-z^{1/2} t} \cos\left(\frac{1}{3} t^{3/2}\right) t^{-1/2} \, dt, \quad \zeta = \frac{2}{3} z^{3/2}. \tag{1.15}$$

Details on how to obtain this representation can be found, for instance, in [46]. The leading factor clearly shows that this is a recessive solution. Inspection of the integral shows that the main contributions for large z come from t close to 0. Replacing cos by 1, we get a first estimation $\text{Ai}(x) \sim (2\sqrt{\pi})^{-1} z^{-1/4} \exp(\zeta)$, which is the behavior already known from (1.9). In fact, replacing the cosine by a series in powers of $\frac{1}{3} t^{3/2}$ and integrating term by term we get, not surprisingly, the expansion of (1.9). For Laplace-type integrals Watson's lemma (Chapter 2) justifies this procedure and proves its asymptotic nature.

We can also use the above integral for computing $\text{Ai}(z)$, in particular for large positive z. Then the integrand has a rapidly decreasing exponential; this is usually a favorable situation for the trapezoidal rule (see Chapter 5), probably with an additional change of variables for improving convergence. However, for small z we have many oscillations for the integrand, which is always a numerically dangerous situation because cancellations may take place. We will see that by considering alternative representations in terms of path integrals in the complex plane, it is possible to find integral representations with no oscillating terms in the integrand. For instance, we will use in Chapter 5 the integral representation of the Airy function

$$\text{Ai}(z) = \frac{1}{2\pi i} \int_C e^{\phi(w)} \, dw, \quad \phi(w) = \frac{1}{3} w^3 - zw, \tag{1.16}$$

which is defined for any complex number z and where C is a contour in the complex plane that starts at $\infty e^{-\pi i/3}$ and terminates at $\infty e^{+\pi i/3}$. We use this integral for z in the sector $0 \leq \text{ph}\, z \leq \frac{2}{3}\pi$ and will show how we can deform the contour of integration C (using Cauchy's theorem) in such a way that the complex function $\phi(w)$ has a constant imaginary part along the new contour of integration. Along such a path, the integrand is nonoscillating and the contour is a steepest descent contour, which has additional interesting properties. It passes through one or more saddle points of $\phi(w)$, the solution(s) of the equation $\phi'(w) = 0$, giving $w_\pm = \pm\sqrt{z}$. Near the saddle point w_+, the integral gets the main contributions, and

Chapter 1. Introduction

the decay of the integrand away from the saddle points is exponential, which is an interesting situation for using the trapezoidal quadrature rule.

Also, Gaussian quadrature is in some cases an interesting choice, particularly when an integral representation with a classical weight is available (as is the case with Airy functions; see Chapters 5 and 12).

Putting all the ingredients together for the Airy functions

For the particular case of Airy functions it is possible to build efficient and high-accuracy methods for z in the whole complex plane [88, 90] with the three discussed methods:

- Maclaurin series for small $|z|$,
- asymptotic expansions for large $|z|$,
- quadrature of integrals for intermediate values (by using the trapezoidal rule for a steepest descent contour or Gaussian quadrature).

It is interesting to note that the approximations obtained from the second and third methods explicitly give the dominant asymptotic exponential factors $e^{-\frac{2}{3}z^{3/2}}$ and $e^{+\frac{2}{3}z^{3/2}}$, which is important for writing algorithms for the scaled functions in order to avoid underflow or overflow.

In addition, for the particular case of real-variable Airy functions, Chebyshev approximations are also interesting alternatives for the intermediate values once the dominant and rapidly varying exponential factors have been scaled out.

We summarize so far by observing that algorithms for special functions are usually a cocktail of several methods, and the final choice of the ingredients of the cocktail is, in some sense, a matter of taste, particularly in the regions where the most straightforward approximations fail (Maclaurin series for small $|z|$ and asymptotics for large $|z|$). In the intermediate region some researchers like to use numerical methods for solving the differential equation of the Airy function [65], whereas others prefer integrals with the dominant exponential factor in front by using Gaussian quadrature for fixed precision computations [88], or the trapezoidal rule [90] for adjustable precision.

Recurrence relations and continued fractions

The functions considered in this book are (all, or most of them) of hypergeometric type. Examples are Bessel functions, Legendre functions, and Kummer functions. All these functions have extra parameters, and these parameters play a role in recurrence relations for these functions, just as the variable z plays its role in their differential equations. The Airy function is a special case of the Bessel functions.

For example, modified Bessel functions of order ν are solutions of the differential equation

$$z^2 y'' + z y' - (z^2 + \nu^2) y = 0, \qquad (1.17)$$

and the solutions are denoted by $K_\nu(z)$ (the recessive solution at $z = +\infty$) and $I_\nu(z)$. When $\nu = 1/3$ this equation is related to the Airy equation (1.1). In fact we have

$$\mathrm{Ai}(z) = \pi^{-1} \sqrt{z/3} K_{\frac{1}{3}}(\zeta), \quad \zeta = \tfrac{2}{3} z^{3/2}. \qquad (1.18)$$

The three-term recurrence relation for the modified Bessel functions reads

$$y_{\nu+1} + \frac{2\nu}{x} y_\nu - y_{\nu-1} = 0 \qquad (1.19)$$

with $y^{(1)}(x) = I_\nu(x)$ and $y^{(2)}(x) = e^{i\pi\nu} K_\nu(x)$ as linearly independent solutions.

We can use the relation (1.19) for computing function values starting with two initial values. For instance, we have

$$K_{\frac{1}{2}}(z) = \sqrt{\frac{\pi}{2z}} e^{-z}, \quad K_{\frac{3}{2}}(z) = \sqrt{\frac{\pi}{2z}} e^{-z} \left(1 + \frac{1}{z}\right), \qquad (1.20)$$

from which the values $K_{n+1/2}(x)$ can be computed by using

$$K_{\nu+1}(z) = \frac{2\nu}{z} K_\nu(z) + K_{\nu-1}(z). \qquad (1.21)$$

Also, we could consider the same strategy for $I_\nu(z)$, but that problem would be ill conditioned.

The theory of three-term recurrence relations bears many similarities to second order homogeneous ordinary differential equations. For instance, given a pair of linearly independent solutions the following two properties hold: first, the general solution is a linear combination of both solutions; second, when the ratio of two linearly independent solutions has finite limit or tends to ∞ as the recurrence parameter tends to ∞, then the recurrence relation possesses a minimal solution. For recurrence relations, the recessive solution is usually called a minimal solution.

As for ordinary differential equations, before applying a recurrence relation we must know the condition of the process, that is, if the recurrence has a minimal solution. Similarly, as happened with the solution of ordinary differential equations, one should not consider forward computation for a minimal solution, but only backward recurrence (although one should also pay attention to possible transitory behavior; see §4.8). It turns out that $I_\nu(z)$ is minimal and that it should be computed backwards by using the relation

$$I_{\nu-1}(z) = \frac{2\nu}{z} I_\nu(z) + I_{\nu+1}(z). \qquad (1.22)$$

There is an important connection between the existence of minimal solutions of three-term recurrence relations and the convergence of an associated continued fraction, which can be used for computing ratios of the minimal solution $y_\nu/y_{\nu-1}$. The link is provided by Pincherle's theorem. In theory, this sole information suffices for computing the minimal solution up to a constant (not depending on ν) multiplicative factor. This aspect is discussed in detail in Chapter 4.

Asymptotics for large parameters. Uniform asymptotics

For computing functions which have dependence on one or several parameters, asymptotic analysis provides some powerful tools of computation when one of the parameters is large (or small). Take for instance the case of parabolic cylinder functions, which are solutions of the second order ordinary differential equation

$$y''(x) - \left(\frac{x^2}{4} + a\right) y(x) = 0. \tag{1.23}$$

A pair of independent solutions of this ordinary differential equation (numerically satisfactory because it comprises the recessive solution) is denoted as $U(a, x)$, $V(a, x)$ which, up to an a-dependent gamma function factor, are also solutions of a three-term recurrence relation with respect to the parameter a (and constitute also a satisfactory pair with respect to this recurrence relation).

When $x^2/4 + a < 0$, the solutions of (1.23) are oscillatory, and for a proper description of the asymptotic behavior of $U(a, x)$ and $V(a, x)$ when a is a large negative parameter, the transition of the oscillatory behavior (when $x^2/4 + a < 0$) to the nonoscillatory behavior (when $x^2/4 + a > 0$) can be described by Airy functions.

To see this, we observe that when $a < 0$ and x crosses the turning points $\pm 2\sqrt{-a}$ the differential equation (1.23) has the same character as the Airy equation (1.1) at $z = 0$, in the sense that the coefficient $A(z)$ in the defining differential equation $y''(z) + A(z)y(z) = 0$ changes sign. Therefore, it is possible that around the turning points $x = \pm 2\sqrt{-a}$, Airy functions can be used for approximating parabolic cylinder functions. Indeed, we can apply Liouville transformations (see §2.2.4) to (1.23) and obtain a transformed equation

$$\ddot{Y}(\zeta) - \left[\mu^4 \zeta + F(\zeta)\right] Y(\zeta) = 0, \quad \mu = \sqrt{-2a}, \tag{1.24}$$

where ζ and $F(\zeta)$ can be given in explicit forms [166, §7]. When the function $F(\zeta)$ is neglected in (1.24), the reduced equation can be solved in terms of Airy functions. This is the first step in constructing complete asymptotic expansions for large values of μ, in which the Airy functions are the main elements that describe the transition near the turning points of the equation in (1.23). These expansions are uniformly valid with respect to ζ in an interval containing the origin, or with respect to x in an interval containing one of the transition points $\pm 2\sqrt{-a}$. The general theory of this type of uniform asymptotic analysis for solutions of second order linear ordinary differential equations can be found in [168, Chap. 11]. The same uniform Airy-type asymptotic expansions can be obtained by using integral representations of the parabolic cylinder functions.

The computation of the coefficients of the series appearing in the uniform expansions is usually a hard problem, in particular in the neighborhood of the turning points ($\zeta = 0$ in (1.24)), but the reward is very high because with the different types of (uniform) expansions for a set of functions, one can compute these functions for wide ranges of the parameters.

This and that

Many more tools are available for computing special functions. We have only briefly mentioned the methods that are more frequently used in numerical algorithms. For the particular case of parabolic cylinder functions, the algorithms described in Chapter 12 are based on the following five methods:

- Maclaurin series for small x and a,
- Poincaré-type asymptotics for large x,

- recurrence relations with respect to a,
- quadrature methods (trapezoidal rule) for intermediate intervals,
- uniform asymptotic expansions for large a and x.

These methods are essential for many other functions considered in this book, and the first four items are described in the part that we call *Basic Methods*.

But, as mentioned earlier, there are many other numerical methods and problems which are of interest. Methods like rational approximations, sequence transformations, numerical solution of ordinary differential equations, specific quadrature rules for oscillating integrands, among others, are also of interest. On the other hand, the problem of computing zeros of special functions is also an important topic, as well as the numerical inversion of certain statistical distributions, the inversion of Laplace transforms, the computation of elliptic integrals through duplication formulas, and so on. These and other topics will also be treated in this book, not all of them to the same extent or, we confess, with the same devotion (with our sincere apologies).

Part I

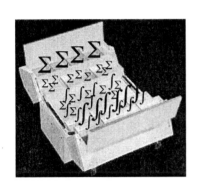

Basic Methods

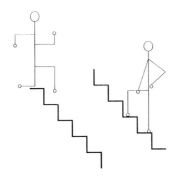

Chapter 2
Convergent and Divergent Series

The divergent series are the invention of the devil.
—Niels Henrik Abel, 1828

2.1 Introduction

For the evaluation of special functions many tools are available, as we will discuss in subsequent chapters, but convergent or divergent series are always to be considered for designing efficient algorithms when one or more parameters assume small or large values. Power series are the main type of series in these circumstances.

To understand the several forms of power series for special functions it is important to know the theory of regular and singular points for second order linear ordinary differential equations. This theory will be considered in §2.2.1.

We discuss the efficiency of power series mainly in connection with the Gauss hypergeometric function. Several connection formulas and transformations for these functions will be given, and we show that all power series for the Gauss functions that follow from these formulas are not sufficient in numerical evaluations for the whole complex z-plane. In these cases, continuation formulas become useful.

For asymptotic expansions we concentrate on Laplace-type integrals, with Watson's lemma as the main tool. We discuss the numerical aspects, including exponentially improved expansions, and give alternatives for standard asymptotic expansions. We describe a few methods for transforming asymptotic expansions into new expansions with better asymptotic convergence. The discussion of uniform asymptotics is postponed for a later chapter (Chapter 8).

2.1.1 Power series: First steps

We first recall the standard forms for Taylor–Maclaurin expansions with remainders. For more details and proofs see [238, Chap. 5].

Let f be a real function of a real variable x, and let it have continuous derivatives of the first n orders ($n \geq 1$) when $x \in [a, b]$, $a = \min\{x, x_0\}$, $b = \max\{x, x_0\}$. Then *Taylor's formula* reads

$$f(x) = f(x_0) + \sum_{m=1}^{n} \frac{f^{(m)}(x_0)}{m!}(x - x_0)^m + R_n, \qquad (2.1)$$

where

$$R_n = \frac{1}{n!} \int_{x_0}^{x} (x - t)^n f^{(n+1)}(t)\, dt. \qquad (2.2)$$

We have *Lagrange's form for the remainder*

$$R_n = \frac{f^{(n+1)}(\zeta)}{(n+1)!}(x - x_0)^{n+1} \quad \text{for some } \zeta \in (a, b), \qquad (2.3)$$

and *Cauchy's form for the remainder*

$$R_n = \frac{f^{(n+1)}(\zeta)}{n!}(x - \zeta)^n (x - x_0) \quad \text{for some } \zeta \in (a, b). \qquad (2.4)$$

When we can take n arbitrarily large and when $\lim_{n \to \infty} R_n = 0$, we obtain the *Taylor series*

$$f(x) = \sum_{m=0}^{\infty} \frac{f^{(m)}(x_0)}{m!}(x - x_0)^m \qquad (2.5)$$

for values of x_0 and x for which $\lim_{n \to \infty} R_n = 0$. In particular, the case $x = 0$ is called *Maclaurin series*.

Bounds for the remainder R_n in (2.1) can be used in estimating the number of terms that are needed to obtain a certain accuracy for computing $f(x)$ on an interval $[a, b]$, but in general the computation of high derivatives of f is not always a simple job. Exceptions are simple functions like e^x, $\sin x$, $\cos x$, and $\log(1+x)$. In practice the check on convergence is done by comparing the size of successive terms in the expansion with the computed partial sums, but care is always needed. A safe strategy for all cases cannot be given.

In the case of elementary and special functions we deal with functions that are analytic in (part of) the complex plane. In that case the remainders can be expressed in terms of a Cauchy integral in which no derivatives of the function occur. That is, if f is analytic on and inside a circle C of radius r centered at the point z_0, then we have

$$f(z) = \sum_{m=0}^{\infty} a_m (z - z_0)^m, \quad a_m = \frac{1}{m!} f^{(m)}(z_0) = \frac{1}{2\pi i} \int_C \frac{f(z)}{(z - z_0)^{m+1}}\, dz. \qquad (2.6)$$

If M_r is an upper bound of $|f(z)|$ on C, we have *Cauchy's inequality*

$$|a_m| = \frac{1}{m!}|f^{(m)}(z_0)| \leq \frac{M_r}{r^m}, \quad m = 0, 1, 2, \ldots. \qquad (2.7)$$

When the series in (2.6) is truncated after the term $a_n(z - z_0)^n$, we can introduce a remainder R_n and write, for $n = 0, 1, 2, \ldots$,

$$f(z) = \sum_{m=0}^{n} a_m(z - z_0)^m + R_n, \quad R_n = \frac{(z - z_0)^{n+1}}{2\pi i} \int_C \frac{f(\zeta)}{(\zeta - z)(\zeta - z_0)^{n+1}}\, d\zeta. \qquad (2.8)$$

2.1. Introduction

Again, when M_r is an upper bound of $|f(z)|$ on C, we have the upper bound for $|R_n|$:

$$|R_n| \leq \frac{|z-z_0|^{n+1}}{r^{n+1}} \frac{rM_r}{r-|z-z_0|}, \quad |z-z_0| < r. \tag{2.9}$$

When we compare the bound in (2.9) with those that can be based on (2.2)–(2.4), we conclude that there is certainly an advantage in the case of analytic functions, because we only need a bound of the function itself on a circle in the complex plane and not one of the higher derivatives. However, in general, for special functions numerical bounds of the functions are not always easy to obtain.

2.1.2 Further practical aspects

When we take $z_0 = 0$ in (2.6) the power series becomes the Maclaurin series

$$f(z) = \sum_{m=0}^{\infty} a_m z^m. \tag{2.10}$$

When f has singular points the series converges for all $|z| < r$, where r is smaller than the smallest modulus of the singularities (in some cases r may be equal to the smallest modulus). For the numerical evaluation of f, the power series can be used efficiently for z properly inside the disk with radius r.

For entire functions, that is, when the radius of convergence of the series in (2.10) is infinite, the domain for using the power series for computations is restricted by two main aspects: efficiency and stability. In addition, when we want to choose n for obtaining a certain accuracy, we also have to choose a realistic value of r.

Example 2.1 (the exponential function, positive argument). For $e^x = \sum x^n/n!$ with $x > 0$, efficiency plays the major role when x becomes large. We want to know for which n the relative error R_n/e^x becomes less than ε. We have $M_r = e^r$. From (2.9) we obtain

$$|R_{n-1}/e^x| \leq \frac{x^n}{r^n} \frac{re^{r-x}}{r-x} < \varepsilon, \quad x < r. \tag{2.11}$$

We can take any $r > x$, say, $r = x + 1$, but this would require a much too large value of n. Observe that $r^{-n} e^r$ is minimal (for positive r) when $r = n$, which is the value we take, and which should be larger than x. The terms $x^n/n!$ of the series initially increase with n, until $n \sim x$, and they decrease when $n > x$. So, the requirement $n > x$ is obvious. With $r = n$ the bound in (2.11) reads

$$|R_{n-1}/e^x| \leq \frac{x^n}{n^n} \frac{ne^{n-x}}{n-x} < \varepsilon, \quad x < n. \tag{2.12}$$

When we compare this with the intuitive approach $x^n/(n!e^x) < \varepsilon$, and replace $n!$ by n^n/e^n (considering Stirling's formula), we see that we have corresponding dominant terms in both inequalities. The value of n that comes from (2.12) is indeed quite optimal. ∎

Example 2.2 (the exponential function, negative argument). When $x < 0$, $|x|$ large, efficiency and stability are both important, because the function value e^x is very small compared with the computed terms in the series.

When we compute the terms with 10 digits, $\sum_{n=0}^{35} (-5)^n/n! = 0.006737954445$, whereas $e^{-5} = 0.006737946999\ldots$. We see a relative error of 10^{-6}. The terms for $n \geq 36$ are all smaller than 10^{-16}. The maximal term is $-26.0\ldots$ and occurs when $n = 5$. ∎

The stability aspect discussed in Example 2.2 occurs not only in this simple example, but may become important in the evaluation of many other special functions. For example, the Airy function Ai(z), a solution of the differential equation $w'' - zw = 0$, is an entire function with power series representation,

$$\begin{aligned}
\text{Ai}(z) &= c_1 f(z) - c_2 g(z), \\
f(z) &= 1 + \tfrac{1}{3!} z^3 + \tfrac{1 \cdot 4}{6!} z^6 + \tfrac{1 \cdot 4 \cdot 7}{9!} z^9 + \cdots, \\
g(z) &= z + \tfrac{2}{4!} z^4 + \tfrac{2 \cdot 5}{7!} z^7 + \tfrac{2 \cdot 5 \cdot 8}{10!} z^{10} + \cdots, \\
c_1 &= 3^{-\frac{2}{3}} \Gamma\left(\tfrac{2}{3}\right), \quad c_2 = 3^{-\frac{1}{3}} \Gamma\left(\tfrac{1}{3}\right)
\end{aligned} \qquad (2.13)$$

(see (2.72) for another notation of f and g). For large complex values with $|\text{ph } z| < \pi$, the Airy function behaves as follows:

$$\text{Ai}(z) \sim \tfrac{1}{2} \pi^{-\frac{1}{2}} z^{-\frac{1}{4}} e^{-\frac{2}{3} z^{\frac{3}{2}}}. \qquad (2.14)$$

We have Ai$(5) = 0.000108344\ldots$. This value cannot be obtained by using the representation of Ai(z) in the first line of (2.13) and the power series for f and g with 10-digit computations. For complex values $z = x + iy$, the exponential term in (2.14) can become large, and in that case the cancellation of leading digits may not occur. In [88] we have used 14-digit precision for the power series for $|y| \leq 3$ and $-2.6 \leq x \leq 1.3$.

2.2 Differential equations and Frobenius series solutions

Power series for special functions are easily derived from the differential equations that define these functions. When we expand the solutions at a particular point, say, the origin, in many important cases the theory of regular and singular points is needed. In this section we give the basic elements of this theory, which is needed to understand the power series expansions of the Gauss hypergeometric functions considered in §2.3.

In the method of separation of variables in a particular coordinate system the well-known equations of mathematical physics, such as the wave equation, the Helmholtz equation, and the Schrödinger equation, may split up into several linear ordinary differential equations of the form

$$f'' + p(z) f' + q(z) f = 0. \qquad (2.15)$$

A small set of equations of the form (2.15) governs the well-known functions of mathematical physics. We have the following important examples:

- the hypergeometric differential equation

$$z(1-z) f'' + [c - (a+b+1)z] f' - ab f = 0, \qquad (2.16)$$

2.2. Differential equations and Frobenius series solutions

- the Bessel differential equation
$$z^2 f'' + zf' + (z^2 - \nu^2) f = 0, \tag{2.17}$$

- the Legendre differential equation
$$(1 - z^2) f'' - 2zf' + \nu(\nu + 1)f = 0, \tag{2.18}$$

- the confluent hypergeometric (Kummer) differential equation
$$zf'' + (c - z)f' - af = 0, \tag{2.19}$$

- the Whittaker differential equation
$$z^2 f'' - \left[\tfrac{1}{4}z^2 - \kappa z + \left(\mu^2 - \tfrac{1}{4} \right) \right] f = 0, \tag{2.20}$$

- the Hermite differential equation
$$f'' - 2zf' + 2\nu f = 0, \tag{2.21}$$

- the Weber or parabolic differential equation
$$f'' - \left(\tfrac{1}{4}z^2 + a \right) f = 0. \tag{2.22}$$

We will meet these equations in the later sections. In this section we consider the classification of these equations with respect to the nature of their singular points, and we consider power series expansions of the solutions of these equations.

2.2.1 Singular points

Consider (2.15), where p and q are given functions which are analytic in a simply connected domain G of the complex plane $\mathbb{C}$, except possibly for a set of isolated points, which are poles for the functions (that is, we assume that p and q are meromorphic in G). We define regular and singular points of the equation in (2.15).

Definition 2.3. *A point $z_0 \in G$ where p and q are analytic is called a regular point of the differential equation. When $z = z_0$ is not a regular point, it is called a singular point. When $z = z_0$ is a singular point but both $(z - z_0)p(z)$ and $(z - z_0)^2 q(z)$ are analytic there, then z_0 is called a regular singular point. If z_0 is neither a regular point nor a regular singular point, then it is said to be an irregular singular point.*

Especially important in the theory of special functions are the expansions in terms of power series in the neighborhood of regular singular points. At regular points the solutions can be represented rather straightforwardly in terms of power series. At regular singular points the situation is more complicated. The general setup for these points is to represent the solution as a power series multiplied by an algebraic term.

Example 2.4 (the hypergeometric equations at $z = 0$ and $z = 1$). The hypergeometric differential equation (2.16) has regular singularities at $z = 0$ and $z = 1$. Kummer's differential equation (2.19) has a regular singularity at $z = 0$. ∎

Transformation of the point at infinity

The transformation $\zeta = 1/z$ yields for (2.15)

$$\frac{d^2 g}{d\zeta^2} + P(\zeta)\frac{dg}{d\zeta} + Q(\zeta)g = 0, \qquad (2.23)$$

where

$$g(\zeta) = f(z) = f(1/\zeta), \quad P(\zeta) = \frac{2}{\zeta} - \frac{1}{\zeta^2}p(1/\zeta), \quad Q(\zeta) = \frac{1}{\zeta^4}q(1/\zeta). \qquad (2.24)$$

A decisive answer about the nature of the point $z = \infty$ of (2.15) is obtained by investigating the functions P and Q in the neighborhood of $\zeta = 0$.

Definition 2.5. *The point $z = \infty$ is called a regular point, or a regular singular point, of the differential equation (2.15) when the point $\zeta = 0$ is a regular point, or a regular singular point, respectively, of the differential equation (2.23).*

To decide about the nature of the point at infinity, let p, q of (2.15) have the expansions

$$zp(z) = p_0 + p_1/z + p_2/z^2 + \cdots, \quad z^2 q(z) = q_0 + q_1/z + q_2/z^2 + \cdots, \qquad (2.25)$$

which converge for sufficiently large values of $|z|$. Then P, Q of (2.23) have the expansions

$$\zeta P(\zeta) = 2 - p_0 - p_1\zeta - p_2\zeta^2 + \cdots, \quad \zeta^2 Q(\zeta) = q_0 + q_1\zeta + q_2\zeta^2 + \cdots, \qquad (2.26)$$

which converge for sufficiently small values of $|\zeta|$. It follows that, when p, q have the above expansions, (2.23) has a regular singular point at zero, and, hence, (2.15) has a regular singular point at infinity.

Example 2.6 (the hypergeometric equations at infinity). The hypergeometric differential equation (2.16) has a regular singularity at infinity. The point at infinity is not a regular singularity of Kummer's differential equation (2.19). ∎

2.2.2 The solution near a regular point

For a regular point of the differential equation (2.15) we have the following result.

Theorem 2.7 (the solution near a regular point). *Let the functions p and q be analytic in the open disk $|z| < R$ and let a_0 and a_1 be arbitrary complex numbers. Then, there exists one and only one function f with the following properties:*

1. *f satisfies the differential equation (2.15);*

2. *f satisfies the initial conditions $f(0) = a_0$, $f'(0) = a_1$; and*

3. *f is analytic inside the disk $|z| < R$.*

Proof. For the proof see [219, Chap. 4]. □

2.2. Differential equations and Frobenius series solutions

The Wronskian of two solutions

We now apply Theorem 2.7 twice:

1. with initial conditions $a_0 = 1$, $a_1 = 0$; we call the unique solution f_1;
2. with initial conditions $a_0 = 0$, $a_1 = 1$; we call the unique solution f_2.

Then the functions f_1 and f_2 are linearly independent solutions of (2.15) and each solution f of (2.15) can be written as a linear combination of f_1 and f_2: $f(z) = f(0)f_1(z) + f'(0)f_2(z)$.

More generally, each pair $\{f_1, f_2\}$ such that any solution of (2.15) can be written in the form
$$f(z) = Af_1(z) + Bf_2(z), \tag{2.27}$$
where A and B are constants, is called a *fundamental system of solutions*; that is, the pair $\{f_1, f_2\}$ is a basis for the linear space of solutions.

Let us consider two arbitrary solutions f_1 and f_2 of the second order differential equation (2.15). We construct with this pair the expression

$$\mathcal{W}[f_1, f_2](z) = \begin{vmatrix} f_1(z) & f_2(z) \\ f_1'(z) & f_2'(z) \end{vmatrix} = f_1(z)f_2'(z) - f_1'(z)f_2(z), \tag{2.28}$$

which is called the *Wronskian* of the pair $\{f_1, f_2\}$. $\mathcal{W}[f_1, f_2]$ plays an important part when investigating the linear independence of a pair of solutions $\{f_1, f_2\}$ of a differential equation. We can consider an analogous expression also for differential equations of higher order (again by using determinants).

It is quite simple to verify, by using (2.15), that the function $\mathcal{W}[f_1, f_2]$ is a solution of the equation $w' = -p(z)w$. It follows that, when $z, z_0 \in G$,

$$\mathcal{W}[f_1, f_2](z) = C \exp\left[-\int_{z_0}^{z} p(\zeta)\, d\zeta\right], \tag{2.29}$$

where C does not depend on z; C equals the value of $\mathcal{W}[f_1, f_2]$ at the point $z = z_0$. Consequently, the Wronskian vanishes identically (when $C = 0$) or it never vanishes in a domain where p is analytic. Another consequence is that the Wronskian of each pair of solutions $\{f_1, f_2\}$ of (2.15) reduces to a constant when the term with the first derivative in (2.15) is missing ($p = 0$).

Now let $\{f_1, f_2\}$ constitute a linearly dependent pair in G. This means there exist two complex numbers A and B with $|A| + |B| \neq 0$, such that
$$Af_1(z) + Bf_2(z) = 0 \quad \forall z \in G. \tag{2.30}$$

Differentiation gives
$$Af_1'(z) + Bf_2'(z) = 0 \quad \forall z \in G. \tag{2.31}$$

These two equations for A and B can be interpreted as a linear system. We have assumed that a solution $\{A, B\} \neq \{0, 0\}$ exists. Consequently, the determinant of the system should vanish. That is, $\mathcal{W}[f_1, f_2](z) = 0$, $z \in G$, when the pair $\{f_1, f_2\}$ constitutes a linearly dependent pair of solutions of (2.15). On the other hand, the Wronskian cannot vanish for a fundamental system $\{f_1, f_2\}$. Summarizing, we have the following theorem.

Theorem 2.8 (the Wronskian of two solutions). *The solutions $\{f_1, f_2\}$ of the differential equation (2.15) are linearly independent if and only if the Wronskian (2.28) does not vanish identically in a domain where the solutions are analytic.*

Power series expansions around a regular point

The method for constructing series expansions around regular and regular singular points is called the Frobenius method. We consider now the case of a regular point.

We consider power series expansions around $z = 0$. Of course, around any other regular point z_0, the method works in the same way. The series

$$p(z) = \sum_{n=0}^{\infty} p_n z^n, \quad q(z) = \sum_{n=0}^{\infty} q_n z^n \qquad (2.32)$$

are convergent in the disc $|z| < R$. We consider a solution of (2.15) with given initial values $f(0) = c_0$, $f'(0) = c_1$. It is not difficult to prove the convergence of the power series $f(z) = \sum_{n=0}^{\infty} c_n z^n$ in which the coefficients must satisfy the relation

$$n(n+1)c_{n+1} + \sum_{k=1}^{n} k c_k p_{n-k} + \sum_{k=0}^{n-1} c_k q_{n-k} = 0, \quad n = 1, 2, \ldots. \qquad (2.33)$$

2.2.3 Power series expansions around a regular singular point

Let $z = 0$ be a regular singular point of (2.15). With a slight change in notation we write the differential equation (2.15) in the form

$$z^2 f'' + z p(z) f' + q(z) f = 0, \qquad (2.34)$$

where we assume that p and q are analytic in $|z| < R$ with power series

$$p(z) = \sum_{n=0}^{\infty} p_n z^n, \quad q(z) = \sum_{n=0}^{\infty} q_n z^n. \qquad (2.35)$$

We assume that at least one of the coefficients p_0, q_0, and q_1 is different from 0. We may expect that, for values of z near the regular singular point $z = 0$, the solutions of (2.34) behave as the solutions of the equation

$$z^2 g'' + z p_0 g' + q_0 g = 0. \qquad (2.36)$$

This is Euler's differential equation, with exact solutions $g(z) = z^\mu$, where μ satisfies the quadratic equation (also called the *indicial equation*)

$$\mu(\mu - 1) + \mu p_0 + q_0 = 0. \qquad (2.37)$$

Actually we try to find a solution of (2.34) of the form

$$f(z) = z^\mu \sum_{n=0}^{\infty} c_n z^n, \qquad (2.38)$$

2.2. Differential equations and Frobenius series solutions

in which the series converges in a neighborhood of the origin, and defines an analytic function there. The result is as follows.

Theorem 2.9 (power series expansions around a regular singular point). *Let the functions p and q be analytic in $|z| < R$.*

1. *If the roots μ_1, μ_2 of (2.37) satisfy $\mu_1 - \mu_2 \notin \mathbb{Z}$, then there exist two solutions*

$$f_1(z) = z^{\mu_1} \sum_{n=0}^{\infty} c_n z^n, \quad f_2(z) = z^{\mu_2} \sum_{n=0}^{\infty} d_n z^n. \tag{2.39}$$

   The power series represent analytic functions in $|z| < R$ and may be differentiated term by term. The functions f_1 and f_2 both satisfy the differential equation (2.34).

2. *If $\mu_1 - \mu_2 \in \mathbb{Z}$, then there exists at least one solution of the form (2.38), where μ satisfies (2.37), and the series converges for all z in $|z| < R$.*

Proof. For the proof see [219, pp. 93–95]. □

We give a few details of the proof. Substitution of (2.38) into (2.34) gives, for $n \geq 0$,

$$(n+\mu)(n+\mu-1)c_n + \sum_{k=0}^{n}(k+\mu)p_{n-k}c_k + \sum_{k=0}^{n}q_{n-k}c_k = 0. \tag{2.40}$$

For $n = 0$ (and $c_0 \neq 0$) this corresponds to the indicial equation (2.37). When μ is a solution of this equation, we can choose c_0 arbitrarily. For special functions c_0 will often be chosen so that f has a convenient normalization.

We assume that the power series in (2.38) converges in a neighborhood of $z = 0$. Collecting the coefficients of c_n in (2.40), we obtain $(n+\mu)(n+\mu-1)+(n+\mu)p_0+q_0$. Using (2.37) and the exponents μ_1, μ_2, we can write

$$\mu(\mu-1) + \mu p_0 + q_0 = (\mu - \mu_1)(\mu - \mu_2). \tag{2.41}$$

It follows that the coefficients of c_n in (2.40) can be written as

$$(n+\mu)(n+\mu-1) + (n+\mu)p_0 + q_0 = (n+\mu-\mu_1)(n+\mu-\mu_2). \tag{2.42}$$

Hence, (2.40) can be written as

$$(n+\mu-\mu_1)(n+\mu-\mu_2)c_n = -\sum_{k=0}^{n-1}(k+\mu)p_{n-k}c_k - \sum_{k=0}^{n-1}q_{n-k}c_k. \tag{2.43}$$

By choosing $\mu = \mu_1$, (2.43) becomes

$$n(n+\mu_1-\mu_2)c_n = -\sum_{k=0}^{n-1}(k+\mu_1)p_{n-k}c_k - \sum_{k=0}^{n-1}q_{n-k}c_k. \tag{2.44}$$

For $n = 1, 2, 3, \ldots$ the coefficients c_n can be computed successively from this relation. Apart from the free choice of c_0 we obtain just one formal series solution of the differential equation. By choosing, on the other hand, $\mu = \mu_2$, (2.43) becomes

$$n(n + \mu_2 - \mu_1)c_n = -\sum_{k=0}^{n-1}(k + \mu_2)p_{n-k}c_k - \sum_{k=0}^{n-1} q_{n-k}c_k. \tag{2.45}$$

In this way we obtain, in general, a second linearly independent solution.

However, the method breaks down when $\mu_1 - \mu_2$ is an element of $\mathbb{Z}$. Let $\mu_1 = \mu_2$. Then both schemes for calculating the coefficients (2.44) and (2.45) are the same. Consequently, in this case at least one of the solutions is not of the form (2.38). Let $m = \mu_1 - \mu_2 = 1, 2, 3, \ldots$; then we find a solution via (2.44). In (2.45), however, the coefficient of c_n has the form $n(n - m)$. We see that at the value $n = m$ the recurrence relation may break down (it may happen that the right-hand side of (2.45) vanishes as well). When indeed the method breaks down, two different solutions of (2.34) cannot have the representation (2.38), and the second solution takes the form

$$f_2(z) = cf_1(z)\ln z + z^{\mu_2}\sum_{n=0}^{\infty} d_n z^n. \tag{2.46}$$

The coefficients d_n may be found by substituting (2.46) into the differential equation (2.34) and comparing equal powers of z. In this method one has a free choice for the coefficient d_0, if $m > 0$. When in the above analysis $m = 0$ we have $d_0 = 0$.

Example 2.10 (Bessel's equation). In Bessel's equation (2.17) $z = 0$ is a regular singular point. The indicial equation (2.37) has the two solutions $\mu_1 = \nu$, $\mu_2 = -\nu$. The method of this section produces for the special choice $c_0 = 2^{-\nu}/\Gamma(\nu + 1)$ the Bessel function

$$J_\nu(z) = \left(\tfrac{1}{2}z\right)^\nu \sum_{n=0}^{\infty} \frac{(-1)^n \left(\tfrac{1}{2}z\right)^{2n}}{\Gamma(n + \nu + 1)\, n!}. \tag{2.47}$$

In general, the pair $\{J_\nu(z), J_{-\nu}(z)\}$ constitutes a fundamental system. However, when $\nu = m \in \mathbb{Z}$ this is no longer true, because of the property

$$J_{-m}(z) = (-1)^m J_m(z), \tag{2.48}$$

which easily follows from (2.47) by using the well-known properties of the gamma function. Hence, the pair $\{J_m(z), J_{-m}(z)\}$ does not constitute a fundamental system when $m \in \mathbb{Z}$.

The scheme of constructing the power series for the second solution breaks down in this case. A second solution of (2.17) has to be obtained in a different way. Also, when $\nu = m + \tfrac{1}{2}$ we have $\mu_2 - \mu_1 = -2m - 1$, which again is a negative integer. The construction of the second solution does not break down in this case.

2.2. Differential equations and Frobenius series solutions

When, in the theory of special functions, the construction by power series of the second solution, as described in this section, is not possible, one often uses a special technique to obtain a second solution. First one assumes that $\mu_1 - \mu_2 \notin \mathbb{Z}$. Then a convenient linear combination of two power series solutions is defined. In the present case of the Bessel functions, the procedure works as follows. Assume that $\nu \notin \mathbb{Z}$ and define as a new solution of Bessel's equation the so-called Neumann function

$$Y_\nu(z) = \frac{\cos \nu\pi \, J_\nu(z) - J_{-\nu}(z)}{\sin \nu\pi}, \qquad (2.49)$$

a linear combination of two functions defined in (2.47). The cosine function lets the numerator vanish if ν approaches an integer value. The sine function takes care of an interesting process then. Other periodic functions can be used here, but the choices in (2.49) give the Neumann function a suitable normalization. The limit of the right-hand side of (2.49), when $\nu \to m$, is properly defined, and can be computed by using l'Hôpital's rule. In this way one obtains $Y_m(z)$, which is of the form

$$Y_m(z) = \frac{2}{\pi} J_m(z) \ln z + z^{-m} \sum_{n=0}^{\infty} c_n z^{2n}. \qquad (2.50)$$

The appearance of the logarithm in the above representation of $Y_m(z)$ is a characteristic feature when the second solution cannot be written in the form (2.38), as we have seen in (2.46). ∎

Other singularities

We have seen that the solutions of differential equations with a regular singular point have an algebraic singularity at this point. Special circumstances generate a solution with a logarithmic term. The equation

$$z^3 f'' + zf' - 2f = 0 \qquad (2.51)$$

has, at $z = 0$, a singularity of a different type. We have the explicit solution $f(z) = \exp(1/z)$ and we see that at least one solution has an essential singularity at $z = 0$. This solution cannot be represented by a power series at the origin, whether or not multiplied by algebraic or logarithmic terms.

Nonlinear differential equations

In this section we have discussed linear equations of the second order. The theory of regular and regular singular points will not work with nonlinear differential equations. Consider, for instance, the equation $y' = 1 + y^2$. No singularities occur in the coefficients of this equation. The solution $y = \tan x$, however, has poles galore.

2.2.4 The Liouville transformation

The differential equations given in (2.16)–(2.22) appear in many other forms in applications. Also, in many occasions transformations of these equations are used. Here we give the *Liouville transformation*.

Consider the equation

$$Y'' + p(x)Y' + q(x)Y = 0, \tag{2.52}$$

and substitute

$$Y(x) = W(x)\exp\left[-\frac{1}{2}\int^x p(\xi)\,d\xi\right]. \tag{2.53}$$

Then the function $W(x)$ satisfies the equation

$$W'' - Q(x)W = 0, \tag{2.54}$$

where

$$Q(x) = \tfrac{1}{4}p^2(x) + \tfrac{1}{2}p'(x) - q(x). \tag{2.55}$$

When x in the final W-equation is transformed by $t(x)$, then, in general, a differential equation is obtained in which the first derivative is present. One has to apply a transformation like (2.53) to the new equation to remove the first derivative. The following transformation (the Liouville transformation) of both the dependent and independent variables directly transforms (2.54) into a form in which the first derivative is missing. Assuming that the third derivative of $t(x)$ exists in the considered x-domain, we write

$$W(x) = \sqrt{\dot{x}}\,w(t), \quad t = t(x), \tag{2.56}$$

where $\dot{x} = dx/dt$. Then (2.54) becomes

$$\ddot{w} - \psi(t)w = 0, \tag{2.57}$$

where

$$\psi(t) = \dot{x}^2 Q(x) + \sqrt{\dot{x}}\,\frac{d^2}{dt^2}\frac{1}{\sqrt{\dot{x}}}. \tag{2.58}$$

The second term in $\psi(t)$ is often expressed in the form

$$\sqrt{\dot{x}}\,\frac{d^2}{dt^2}\frac{1}{\sqrt{\dot{x}}} = -\frac{1}{2}\{x, t\}, \tag{2.59}$$

where $\{x, t\}$ is the *Schwarzian derivative*

$$\{x, t\} = \frac{\dddot{x}}{\dot{x}} - \frac{3}{2}\left(\frac{\ddot{x}}{\dot{x}}\right)^2. \tag{2.60}$$

In [168] this transformation is frequently used for obtaining the *Liouville–Green approximation* (also called the *WKB approximation*) for the solutions of the differential equation.

2.3 Hypergeometric series

Many special functions can be defined by power series that are of hypergeometric type. That is, they can be defined by power series of the form

$$f(z) = \sum_{n=0}^{\infty} c_n z^n, \tag{2.61}$$

2.3. Hypergeometric series

where c_{n+1}/c_n is a rational function of n. Examples are

$$e^z = \sum_{n=0}^{\infty} \frac{z^n}{n!}, \quad (1+z)^a = \sum_{n=0}^{\infty} \binom{a}{n} z^n. \tag{2.62}$$

A useful framework for working with these functions is the class of *generalized hypergeometric functions*. We define

$$_pF_q\left(\begin{matrix} a_1, \ldots, a_p \\ b_1, \ldots, b_q \end{matrix}; z\right) = \sum_{n=0}^{\infty} \frac{(a_1)_n \cdots (a_p)_n}{(b_1)_n \cdots (b_q)_n} \frac{z^n}{n!}, \tag{2.63}$$

where $(a)_n$ is the *Pochhammer symbol*, also called the shifted factorial, defined by

$$(a)_0 = 1, \quad (a)_n = a(a+1)\cdots(a+n-1) \quad (n \geq 1). \tag{2.64}$$

In terms of the gamma function we have

$$(a)_n = \frac{\Gamma(a+n)}{\Gamma(a)}, \quad n = 0, 1, 2, \ldots. \tag{2.65}$$

The series in (2.63) defines an entire function in z if $p \leq q$.

In the case $p = q+1$ the infinite series converges if $|z| < 1$, and defines an analytic function in this disk. This function can be continued analytically outside the disk, with a branch cut from 1 to $+\infty$. Let

$$\gamma_q = (b_1 + \cdots + b_q) - (a_1 + \cdots + a_{q+1}). \tag{2.66}$$

Then on the circle $|z| = 1$, the series (2.63) is absolutely convergent if $\Re \gamma_q > 0$, convergent except at $z = 1$ if $-1 < \Re \gamma_q \leq 0$, and divergent if $\Re \gamma_q \leq -1$.

The binomial coefficient in (2.62) can be written in several forms:

$$\binom{a}{n} = \frac{\Gamma(a+1)}{n!\,\Gamma(a+1-n)} = (-1)^n \frac{\Gamma(n-a)}{n!\,\Gamma(-a)} = (-1)^n \frac{(-a)_n}{n!}, \tag{2.67}$$

and we find

$$e^z = {}_0F_0\left(\begin{matrix} - \\ - \end{matrix}; z\right), \quad (1+z)^a = {}_2F_1\left(\begin{matrix} -a, b \\ b \end{matrix}; -z\right) = {}_1F_0\left(\begin{matrix} -a \\ - \end{matrix}; -z\right). \tag{2.68}$$

The second relation holds for any $b \in \mathbb{C}$ and $|z| < 1$. When $a = m$, a nonnegative integer, the binomial function in (2.62) becomes Newton's binomial formula, with only $m+1$ terms. Also from (2.67) we see that $(-m)_n$ equals 0 when $n \geq m+1$. In general, the power series in (2.63) terminates when one of the a_j equals a nonpositive integer. In that case p and q can be any nonnegative integer.

On the other hand, the $_pF_q$ function of (2.63) is not defined if one of the b_j equals a nonpositive integer, except in the following typical case. Let $a_j = -m$ and $b_j = -m - \ell$, with ℓ, m nonnegative integers. Then we have (cf. (2.67))

$$\frac{(a_j)_n}{(b_j)_n} = \frac{(-m)_n}{(-m-\ell)_n} = \begin{cases} \dfrac{m!}{(m-n)!} \dfrac{(m+\ell-n)!}{(m+\ell)!}, & m \geq n, \\ 0, & m < n. \end{cases} \tag{2.69}$$

Other examples of hypergeometric functions are the Bessel functions, with special case

$$\Gamma(\nu+1)(\tfrac{1}{2}z)^{-\nu}J_\nu(z) = {}_0F_1\left(\genfrac{}{}{0pt}{}{-}{\nu+1}; -\tfrac{1}{4}z^2\right) = e^{-iz}{}_1F_1\left(\genfrac{}{}{0pt}{}{\nu+\tfrac{1}{2}}{2\nu+1}; 2iz\right), \qquad (2.70)$$

where $J_\nu(z)$ denotes the *ordinary Bessel function of the first kind* (see also (2.47)).

The ${}_1F_1$-function is also denoted by

$$M(a,c,z) = {}_1F_1\left(\genfrac{}{}{0pt}{}{a}{c}; z\right), \qquad (2.71)$$

and M is also called the *confluent hypergeometric function*.

The functions f and g used for the Airy function in (2.13) can be written as

$$f(z) = {}_0F_1\left(\genfrac{}{}{0pt}{}{-}{\tfrac{2}{3}}; \tfrac{1}{9}z^3\right), \quad g(z) = z\,{}_0F_1\left(\genfrac{}{}{0pt}{}{-}{\tfrac{4}{3}}; \tfrac{1}{9}z^3\right). \qquad (2.72)$$

2.3.1 The Gauss hypergeometric function

The *Gauss hypergeometric function* is the case $p=2$, $q=1$, that is,

$${}_2F_1\left(\genfrac{}{}{0pt}{}{a,b}{c}; z\right) = \sum_{n=0}^\infty \frac{(a)_n(b)_n}{(c)_n\, n!} z^n = 1 + \frac{ab}{c\,1!}z + \frac{a(a+1)b(b+1)}{c(c+1)\,2!}z^2 + \cdots, \qquad (2.73)$$

where $c \neq 0, -1, -2, \ldots$ and $|z| < 1$. It is a solution of the hypergeometric differential equation (2.16). In Examples 2.4 and 2.6 we have observed that this equation has regular singular points at $z = 0$, $z = 1$, and $z = \infty$. It is not difficult to verify that

$$z^{1-c}\,{}_2F_1\left(\genfrac{}{}{0pt}{}{a-c+1,\, b-c+1}{2-c}; z\right) \qquad (2.74)$$

is a second solution of (2.16). This shows that there exist solutions of (2.16) with different behavior from (2.73) at the singular point $z = 0$. Indeed, for (2.16) the solutions of the indicial equation (2.37) are $\mu_1 = 0$ and $\mu_2 = 1 - c$. When c does not assume integer values, the solutions (2.73) and (2.74) constitute a fundamental pair of solutions of (2.16). When c equals an integer, then the fundamental system has one member in which logarithmic terms occur. This is in full agreement with the theory of §2.2.3.

For the regular singular point at $z = 1$ we have $\mu_1 = 0$ and $\mu_2 = c - a - b$, and a fundamental pair is given by

$${}_2F_1\left(\genfrac{}{}{0pt}{}{a,b}{a+b+1-c}; 1-z\right), \quad (1-z)^{c-a-b}{}_2F_1\left(\genfrac{}{}{0pt}{}{c-b,\, c-a}{c-a-b+1}; 1-z\right), \qquad (2.75)$$

(assuming that $c - a - b$ is not an integer) and at $z = \infty$ we have $\mu_1 = a$ and $\mu_2 = b$, and the fundamental pair

$$z^{-a}\,{}_2F_1\left(\genfrac{}{}{0pt}{}{a,\, a-c+1}{a-b+1}; \tfrac{1}{z}\right), \quad z^{-b}\,{}_2F_1\left(\genfrac{}{}{0pt}{}{b,\, b-c+1}{b-a+1}; \tfrac{1}{z}\right) \qquad (2.76)$$

2.3. Hypergeometric series

(assuming that $a-b$ is not an integer). These functions play a role in the connection formulas given in (2.84) and (2.85).

In applications, nonpositive integer values of c may occur in combination with quantities that remove that singularity. A useful limit is

$$\lim_{c \to -m} \frac{1}{\Gamma(c)} {}_2F_1\left(\begin{matrix} a, b \\ c \end{matrix}; z\right)$$
$$= \frac{(a)_{m+1}(b)_{m+1}}{(m+1)!} z^{m+1} {}_2F_1\left(\begin{matrix} a+m+1, b+m+1 \\ m+2 \end{matrix}; z\right), \quad (2.77)$$

where $m = 0, 1, 2, \ldots$. For $m = 0$ the proof is easy, because

$$\frac{1}{\Gamma(c)} {}_2F_1\left(\begin{matrix} a, b \\ c \end{matrix}; z\right) = \left[\frac{1}{\Gamma(c)} + \frac{ab}{\Gamma(c+1)\,1!}z + \frac{a(a+1)b(b+1)}{\Gamma(c+2)\,2!}z^2 + \cdots\right], \quad (2.78)$$

and in the limit $c = 0$ this indeed reduces to the right-hand side of (2.77) with $m = 0$.

If $\Re(c - a - b) > 0$, the value at $z = 1$ is given by

$$ {}_2F_1\left(\begin{matrix} a, b \\ c \end{matrix}; 1\right) = \frac{\Gamma(c)\Gamma(c-a-b)}{\Gamma(c-a)\Gamma(c-b)}. \quad (2.79)$$

Special cases are

$$ {}_2F_1\left(\begin{matrix} a, b \\ b \end{matrix}; z\right) = (1-z)^{-a}, \quad {}_2F_1\left(\begin{matrix} 1, 1 \\ 2 \end{matrix}; z\right) = -\frac{\ln(1-z)}{z},$$

$$ {}_2F_1\left(\begin{matrix} \frac{1}{2}, 1 \\ \frac{3}{2} \end{matrix}; -z^2\right) = \frac{\arctan z}{z}, \quad {}_2F_1\left(\begin{matrix} \frac{1}{2}, 1 \\ \frac{3}{2} \end{matrix}; z^2\right) = \frac{1}{2z}\ln\frac{1+z}{1-z}, \quad (2.80)$$

$$ {}_2F_1\left(\begin{matrix} \frac{1}{2}, \frac{1}{2} \\ \frac{3}{2} \end{matrix}; z^2\right) = \frac{\arcsin z}{z}, \quad {}_2F_1\left(\begin{matrix} \frac{1}{2}, \frac{1}{2} \\ \frac{3}{2} \end{matrix}; -z^2\right) = \frac{1}{z}\ln\left(z + \sqrt{1+z^2}\right).$$

The power series for the Gauss hypergeometric function provides a simple and efficient means for the computation of this function, when z is properly inside the unit disk. The terms can easily be computed by the recursion in the representation

$$ {}_2F_1\left(\begin{matrix} a, b \\ c \end{matrix}; z\right) = \sum_{n=0}^{\infty} T_n, \quad T_{n+1} = z\frac{(a+n)(b+n)}{(c+n)(n+1)}T_n, \quad n \geq 0, \quad T_0 = 1. \quad (2.81)$$

2.3.2 Other power series for the Gauss hypergeometric function

The power series in (2.73) converges inside the unit disk, and for numerical computations we can use only the disk $|z| \le \rho < 1$, with ρ depending on numerical requirements, such as precision and efficiency. Other power series are available, however, to extend this domain.

The $_2F_1$-function with argument z can be written in terms of one or two other $_2F_1$-functions with argument

$$\frac{1}{z}, \quad 1-z, \quad \frac{1}{1-z}, \quad \frac{z}{z-1}, \quad \frac{z-1}{z}. \tag{2.82}$$

A useful set of relations is (see [2, p. 559] and [219, pp. 110 and 113])

$$\begin{aligned}
{}_2F_1\left(\begin{matrix} a, b \\ c \end{matrix}; z\right) &= (1-z)^{-a} {}_2F_1\left(\begin{matrix} a, c-b \\ c \end{matrix}; \frac{z}{z-1}\right) \\
&= (1-z)^{-b} {}_2F_1\left(\begin{matrix} c-a, b \\ c \end{matrix}; \frac{z}{z-1}\right) \\
&= (1-z)^{c-a-b} {}_2F_1\left(\begin{matrix} c-a, c-b \\ c \end{matrix}; z\right),
\end{aligned} \tag{2.83}$$

and the compound connection formulas

$$\begin{aligned}
{}_2F_1\left(\begin{matrix} a, b \\ c \end{matrix}; z\right) &= \frac{\Gamma(c)\Gamma(c-a-b)}{\Gamma(c-a)\Gamma(c-b)} {}_2F_1\left(\begin{matrix} a, b \\ a+b-c+1 \end{matrix}; 1-z\right) \\
&+ \frac{\Gamma(c)\Gamma(a+b-c)}{\Gamma(a)\Gamma(b)} (1-z)^{c-a-b} {}_2F_1\left(\begin{matrix} c-a, c-b \\ c-a-b+1 \end{matrix}; 1-z\right),
\end{aligned} \tag{2.84}$$

$$\begin{aligned}
{}_2F_1\left(\begin{matrix} a, b \\ c \end{matrix}; z\right) &= \frac{\Gamma(c)\Gamma(b-a)}{\Gamma(b)\Gamma(c-a)} (-z)^{-a} {}_2F_1\left(\begin{matrix} a, 1-c+a \\ 1-b+a \end{matrix}; \frac{1}{z}\right) \\
&+ \frac{\Gamma(c)\Gamma(a-b)}{\Gamma(a)\Gamma(c-b)} (-z)^{-b} {}_2F_1\left(\begin{matrix} b, 1-c+b \\ 1-a+b \end{matrix}; \frac{1}{z}\right),
\end{aligned} \tag{2.85}$$

$$\begin{aligned}
{}_2F_1\left(\begin{matrix} a, b \\ c \end{matrix}; z\right) &= \frac{\Gamma(c)\Gamma(b-a)}{\Gamma(b)\Gamma(c-a)} (1-z)^{-a} {}_2F_1\left(\begin{matrix} a, c-b \\ a-b+1 \end{matrix}; \frac{1}{1-z}\right) \\
&+ \frac{\Gamma(c)\Gamma(a-b)}{\Gamma(a)\Gamma(c-b)} (1-z)^{-b} {}_2F_1\left(\begin{matrix} b, c-a \\ b-a+1 \end{matrix}; \frac{1}{1-z}\right),
\end{aligned} \tag{2.86}$$

2.3. Hypergeometric series

$$
{}_2F_1\left(\begin{matrix}a,\ b\\ c\end{matrix};z\right) = \frac{\Gamma(c)\Gamma(c-a-b)}{\Gamma(c-a)\Gamma(c-b)} z^{-a} {}_2F_1\left(\begin{matrix}a,\ a-c+1\\ a+b-c+1\end{matrix};1-\frac{1}{z}\right)
$$
$$
+ \frac{\Gamma(c)\Gamma(a+b-c)}{\Gamma(a)\Gamma(b)} z^{a-c}(1-z)^{c-a-b} {}_2F_1\left(\begin{matrix}c-a,\ 1-a\\ c-a-b+1\end{matrix};1-\frac{1}{z}\right).
$$
(2.87)

When we restrict the absolute values of the quantities in (2.82) to the bound ρ, $0 < \rho < 1$, we find, writing $z = x + iy$,

$$
\begin{aligned}
|z| \leq \rho &\implies x^2 + y^2 \leq \rho^2; \\
\left|\frac{1}{z}\right| \leq \rho &\implies x^2 + y^2 \geq \frac{1}{\rho^2}; \\
|1 - z| \leq \rho &\implies (x-1)^2 + y^2 \leq \rho^2; \\
\frac{1}{|1-z|} \leq \rho &\implies (x-1)^2 + y^2 \geq \frac{1}{\rho^2}; \\
\left|\frac{z}{1-z}\right| \leq \rho &\implies \left(x - \frac{\rho^2}{1-\rho^2}\right)^2 + y^2 \leq \frac{\rho^2}{(1-\rho^2)^2}; \\
\left|\frac{z}{1-z}\right| \geq \frac{1}{\rho} &\implies \left(x - \frac{1}{1-\rho^2}\right)^2 + y^2 \geq \frac{\rho^2}{(1-\rho^2)^2}.
\end{aligned}
$$
(2.88)

The domains defined by these inequalities do not cover the entire z-plane. The points $z = e^{\pm \pi i/3}$ do not satisfy these six conditions, for any $\rho \in (0, 1)$. When $\rho \to 1$, the domain of points not satisfying the six conditions shrinks to the exceptional points $z = e^{\pm \pi i/3}$. See Figure 2.1, where these points are indicated with black dots for the cases $\rho = \frac{1}{2}$ and $\rho = \frac{3}{4}$. In the light area none of the inequalities of (2.88) holds.

To compute the ${}_2F_1$-functions in a neighborhood of the points $z = e^{\pm \pi i/3}$ many other methods are available, and these will be discussed in later chapters. One very useful method is discussed now.

Bühring's analytic continuation formula

In [24] power series expansions of the Gauss function are derived, which enable computations near these special points. Bühring's expansion reads as follows. If $b - a$ is not an integer, we have for $|\text{ph}(z_0 - z)| < \pi$ the continuation formula

$$
{}_2F_1\left(\begin{matrix}a,\ b\\ c\end{matrix};z\right) = \frac{\Gamma(c)\Gamma(b-a)}{\Gamma(b)\Gamma(c-a)}(z_0 - z)^{-a} \sum_{n=0}^{\infty} d_n(a, z_0)(z - z_0)^{-n}
$$
$$
+ \frac{\Gamma(c)\Gamma(a-b)}{\Gamma(a)\Gamma(c-b)}(z_0 - z)^{-b} \sum_{n=0}^{\infty} d_n(b, z_0)(z - z_0)^{-n},
$$
(2.89)

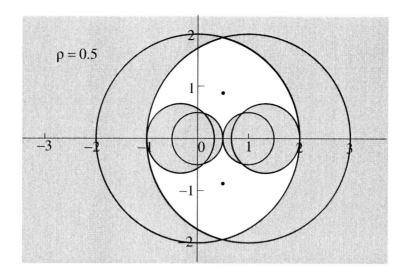

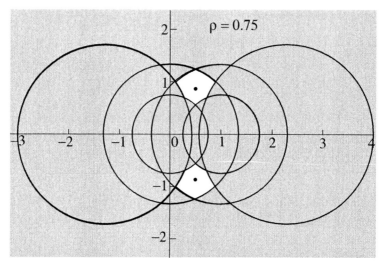

Figure 2.1. *In the light domains none of the inequalities of (2.88) is satisfied. For $\rho \to 1$ these domains shrink to the points $e^{\pm \pi i/3}$, which are indicated by black dots.*

where both series converge outside the circle $|z - z_0| = \max(|z_0|, |z_0 - 1|)$ and the coefficients are given by the three-term recurrence relation

$$d_n(s, z_0) = \frac{n+s-1}{n(n+2s-a-b)}[z_0(1-z_0)(n+s-2)d_{n-2}(s, z_0) \\ + \{(n+s)(1-2z_0) + (a+b+1)z_0 - c\}d_{n-1}(s, z_0)], \quad (2.90)$$

2.3. Hypergeometric series

where $n = 1, 2, 3, \ldots$, with starting values

$$d_{-1}(s, z_0) = 0, \quad d_0(s, z_0) = 1. \tag{2.91}$$

For the case that $b - a$ is an integer a limiting process is needed (as in the next section). Details on the case $a = b$ are given in [24], where also different representations of the coefficients in the series of (2.89) are given.

When we take $z_0 = \frac{1}{2}$, the series in (2.89) converge outside the circle $|z - \frac{1}{2}| = \frac{1}{2}$, and both points $z = e^{\pm\pi i/3}$ discussed earlier are inside the domain of convergence.

For more details on the analytic continuation of the Gauss hypergeometric function and connection formulas, we refer to [12].

2.3.3 Removable singularities

For some combinations of the parameters a, b, and c the relations in (2.84)–(2.87) cannot be used straightforwardly. For example, in (2.84), when $c = a + b$, the gamma functions $\Gamma(c-a-b)$ and $\Gamma(a+b-c)$ are not defined, and the two Gauss functions become the same (as mentioned after (2.75), the two Gauss functions constitute a fundamental pair when $c - a - b$ is not an integer). In agreement with the theory of §2.2.3 logarithmic terms occur in a fundamental pair when $c = a + b$.

To see how this happens in this case we write $c = a + b + \varepsilon$. We expand the Gauss functions in (2.84) in powers of $1 - z$ and obtain for the nth term

$$\frac{\Gamma(a+b+\varepsilon)\Gamma(1+\varepsilon)\Gamma(1-\varepsilon)(1-z)^n}{\Gamma(a)\Gamma(b)\Gamma(a+\varepsilon)\Gamma(b+\varepsilon)\Gamma(1-\varepsilon+n)\Gamma(1+\varepsilon+n)n!} f(\varepsilon), \tag{2.92}$$

where

$$f(\varepsilon) = \frac{1}{\varepsilon}\Big(\Gamma(a+n)\Gamma(b+n)\Gamma(1+\varepsilon+n) \\ -(1-z)^\varepsilon \Gamma(a+\varepsilon+n)\Gamma(b+\varepsilon+n)\Gamma(1-\varepsilon+n)\Big). \tag{2.93}$$

Taking the limit $\varepsilon \to 0$ in $f(\varepsilon)$ we find

$$_2F_1\begin{pmatrix} a, b \\ a+b \end{pmatrix}; z = \sum_{n=0}^{\infty} \frac{\Gamma(a+b)}{\Gamma(a)\Gamma(b)} \\ \times \frac{(a)_n(b)_n}{n!n!}(2\psi(n+1) - \psi(a+n) - \psi(b+n) - \ln(1-z))(1-z)^n, \tag{2.94}$$

where $\psi(z)$ is the logarithmic derivative of the gamma function,

$$\psi(z) = \frac{\Gamma'(z)}{\Gamma(z)}. \tag{2.95}$$

This expansion holds for $|z - 1| < 1$ with $|\mathrm{ph}(1 - z)| < \pi$. That is, there is a branch cut from $z = 1$ to $z = +\infty$, and z is not on this cut. The logarithm $\ln(1 - z)$ assumes its principal branch, which is real for $z < 1$.

Similar modifications for (2.84) are needed when $c = a+b \pm m$, $m = 1, 2, \ldots$. The relations (2.85)–(2.87) also need modifications for certain values of a, b, and c.

We conclude with the observation that the computation of the Gauss hypergeometric function is a nontrivial problem, even for real values of the parameters and argument. With the transformation formulas of (2.83)–(2.87) we cannot reach all points in the complex plane, and these formulas may cause numerical difficulties for certain combinations of the parameters a, b, and c because of removable singularities in the formulas. In [71] many details are discussed for the numerical use of the transformation formulas, and details of a Fortran program are given.

2.4 Asymptotic expansions

In this section we concentrate on Poincaré-type expansions, which are power series with negative powers of the large parameter. In Chapter 8 we consider also uniform expansions.

First we give a definition of an asymptotic expansion.

Definition 2.11. *Let F be function of a real or complex variable z; let $\sum_{n=0}^{\infty} a_n z^{-n}$ denote a (convergent or divergent) formal power series, of which the sum of the first n terms is denoted by $S_n(z)$; let*

$$R_n(z) = F(z) - S_n(z). \tag{2.96}$$

That is,

$$F(z) = a_0 + \frac{a_1}{z} + \frac{a_2}{z^2} + \cdots + \frac{a_{n-1}}{z^{n-1}} + R_n(z), \quad n = 0, 1, 2 \ldots, \tag{2.97}$$

where we assume that when $n = 0$ we have $F(z) = R_0(z)$. Next, assume that for each $n = 0, 1, 2, \ldots$ the following relation holds:

$$R_n(z) = \mathcal{O}\left(z^{-n}\right), \quad \text{as} \quad z \to \infty, \tag{2.98}$$

in some unbounded domain Δ. Then $\sum_{n=0}^{\infty} a_n z^{-n}$ is called an asymptotic expansion of the function F, and we denote this by

$$F(z) \sim \sum_{n=0}^{\infty} a_n z^{-n}, \quad z \to \infty, \quad z \in \Delta. \tag{2.99}$$

This definition is due to Poincaré (1886). Analogous definitions can be given for $z \to 0$, and so on. The relation in (2.98) means that $|z^n R_n(z)|$ is bounded as $z \to \infty$. See also (1.11) and (1.12).

Observe that we do not assume that the infinite series in (2.99) converges for certain z-values. This is not relevant in asymptotics; in the definition only a property of $R_n(z)$ is requested, with n fixed. A common pattern is that the absolute values of the terms in (2.99) decrease initially as n increases, assume a minimal value—for example, when $n \sim |z|$ (as in Example 2.12 and shown in Figure 2.2)—and increase indefinitely as n increases further.

2.4. Asymptotic expansions

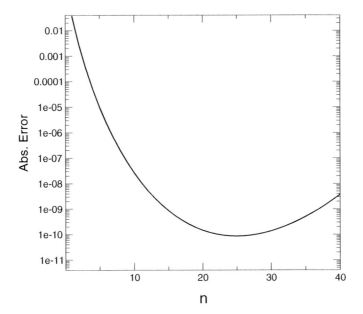

Figure 2.2. *Behavior of the terms in the asymptotic expansion* (2.101). *The absolute error in the approximation of $F(25)$ is plotted as a function of the number of terms (n).*

Example 2.12 (exponential integral $E_1(z)$). The classical example is the so-called exponential integral, that is,

$$F(z) = z \int_z^\infty t^{-1} e^{z-t}\, dt = z\, e^z E_1(z), \quad z \neq 0, \quad |\text{ph } z| < \pi. \tag{2.100}$$

Repeatedly using integration by parts, we obtain

$$F(z) = 1 - \frac{1}{z} + \frac{2!}{z^2} - \cdots + \frac{(-1)^{n-1}(n-1)!}{z^{n-1}} + (-1)^n n!\, z \int_z^\infty \frac{e^{z-t}}{t^{n+1}}\, dt. \tag{2.101}$$

Let us assume that z is positive. Then in this case we have, since $t \geq z$,

$$(-1)^n R_n(z) = n!\, z \int_z^\infty \frac{e^{z-t}}{t^{n+1}}\, dt \leq \frac{n!}{z^n} \int_z^\infty e^{z-t}\, dt = \frac{n!}{z^n}. \tag{2.102}$$

Indeed, $R_n(z) = \mathcal{O}(z^{-n})$ as $z \to +\infty$. Hence

$$z \int_z^\infty t^{-1} e^{z-t}\, dt \sim \sum_{n=0}^\infty (-1)^n \frac{n!}{z^n}, \quad z \to \infty. \tag{2.103}$$

This series is divergent for any finite value of z. However, when z is sufficiently large and n is fixed, the finite part of the series $S_n(z)$, given by

$$S_n(x) = F(z) - R_n(z), \tag{2.104}$$

approximates the function $F(z)$ within any desired accuracy. ∎

2.4.1 Watson's lemma

In Example 2.12 we can derive the asymptotic expansion in a different way. We write $F(z)$ (using the transformation $t = z(1 + u)$) as a Laplace integral

$$F(z) = z \int_0^\infty e^{-zu} f(u)\, du, \quad f(u) = 1/(1 + u). \tag{2.105}$$

We now write

$$f(u) = 1 - u + u^2 - \cdots + (-1)^{n-1} u^{n-1} + (-1)^n u^n/(1+u), \tag{2.106}$$

and we obtain exactly the same expansion, with the same expression and upper bound for $|R_n(z)|$. This approach gives the main ideas for the following result.

Theorem 2.13 (Watson's lemma). *Assume that*

1. *f is analytic at the origin and inside a sector Ω: $|\mathrm{ph}\, t| < \alpha$, where $\alpha > 0$;*

2. *we have*

$$f(t) = \sum_{n=0}^\infty a_n t^n, \quad |t| < R, \tag{2.107}$$

   for some positive number R;

3. *there is a real number σ such that $f(t) = \mathcal{O}(e^{\sigma|t|})$ as $t \to \infty$ inside the sector $|\mathrm{ph}\, t| \leq \alpha - \delta$ for each $\delta \in (0, \alpha)$.*

Then the integral

$$F(z) = \int_0^\infty t^{\lambda-1} f(t) e^{-zt}\, dt, \quad \Re\lambda > 0, \tag{2.108}$$

or its analytic continuation, has the asymptotic expansion

$$F(z) \sim \sum_{n=0}^\infty \Gamma(n + \lambda) \frac{a_n}{z^{n+\lambda}}, \quad z \to \infty, \tag{2.109}$$

in the sector

$$|\mathrm{ph}\, z| \leq \alpha + \tfrac{1}{2}\pi - \delta. \tag{2.110}$$

Proof. For a proof (of a slightly more general theorem), see [168, p. 114]. □

In this result the many-valued functions $t^{\lambda-1}$, $z^{n+\lambda}$ have their principal values on the positive real axis and are defined by continuity elsewhere.

To explain how the bounds in (2.110) arise, we write $\mathrm{ph}\, t = \tau$ and $\mathrm{ph}\, z = \theta$, where $|\tau| < \alpha$. The condition for convergence in (2.108) is $\cos(\tau + \theta) > 0$, that is, $|\tau + \theta| < \tfrac{1}{2}\pi$. Combining this with the bounds for τ we obtain the bounds for θ in (2.110).

In Figure 2.3 we show the sectors where $|\tau| \leq \alpha$ in the t-plane together with the sector where $|\theta| \leq \alpha + \tfrac{1}{2}\pi$ in the z-plane.

2.4. Asymptotic expansions

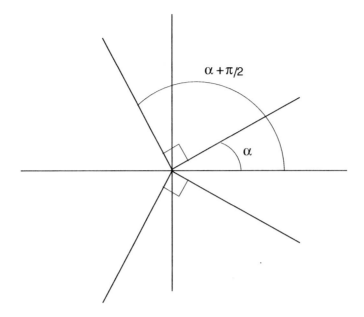

Figure 2.3. *Watson's lemma: sectors in the t-plane and the z-plane.*

Example 2.14 (incomplete gamma function $\Gamma(a, z)$). For the *incomplete gamma function*

$$\Gamma(a, z) = \int_z^\infty e^{-t} t^{a-1} \, dt = z^a e^{-z} \int_0^\infty (t+1)^{a-1} e^{-zt} \, dt, \qquad (2.111)$$

we expand

$$f(t) = (t+1)^{a-1} = \sum_{k=0}^\infty \binom{a-1}{k} t^k. \qquad (2.112)$$

For this function $\alpha = \pi$, and we obtain

$$\Gamma(a, z) \sim z^{a-1} e^{-z} \sum_{k=0}^\infty \binom{a-1}{k} \frac{k!}{z^k}, \quad |\text{ph } z| < \frac{3}{2}\pi. \qquad (2.113)$$

We remark that this asymptotic expansion holds uniformly in the sector $|\text{ph } z| \leq \frac{3}{2}\pi - \delta$. This range is much larger than the usual domain of definition for the incomplete gamma function, which reads $|\text{ph } z| < \pi$. The phrase *or its analytic continuation* is indeed important in Theorem 2.13.

The expansion in (2.113) is usually written in the form

$$\Gamma(a, z) \sim z^{a-1} e^{-z} \sum_{k=0}^\infty \frac{(-1)^k (1-a)_k}{z^k}, \quad |\text{ph } z| < \frac{3}{2}\pi. \qquad (2.114)$$

Observe that the exponential integral is a special case. We have $E_1(z) = \Gamma(0, z)$, which gives the expansion in Example 2.12, again valid uniformly in the sector $|\mathrm{ph}\, z| \le \frac{3}{2}\pi - \delta$.

The expansion (2.114) can be obtained in several other ways, for example, by repeatedly using the recurrence relation

$$\Gamma(a, z) = z^{a-1}e^{-z} + (a-1)\Gamma(a-1, z)$$
$$= z^{a-1}e^{-z}\left(1 + \frac{a-1}{z}\right) + (a-1)(a-2)\Gamma(a-2, z). \tag{2.115}$$

This gives the expansion with explicit remainder

$$\Gamma(a, z) = z^{a-1}e^{-z}\left[\sum_{k=0}^{n-1}\frac{(-1)^k(1-a)_k}{z^k} + \frac{(-1)^n(1-a)_n}{z^n}C_n(a, z)\right], \tag{2.116}$$

where $n = 0, 1, 2, \dots$ and

$$C_n(a, z) = z^{n+1-a}e^z\Gamma(a-n, z). \quad\blacksquare \tag{2.117}$$

2.4.2 Estimating the remainders of asymptotic expansions

As a rule of thumb, when z is large, the absolute values of the terms $a_n z^{-n}$ in expansion (2.99) initially decrease in value, and ultimately they increase in value. Compare Figure 2.2 and the expansion in (2.103) for $z > 0$, where the smallest absolute value of the terms occurs when n is roughly equal to z. We have, with $z = n$, by using Stirling's approximation,

$$\frac{n!}{z^n} \sim \sqrt{2\pi z}\, e^{-z}, \quad z \to \infty. \tag{2.118}$$

Because in Example 2.12 the remainder $R_n(z)$ is less (in absolute value) than $n!/z^n$, we see that at the optimal truncation value $n \sim z$ the error is exponentially small.

It follows that we can determine the interval of z-values for which the asymptotic expansion in (2.103) can be used to obtain a certain precision. For example, if $z = 25$, we have $\sqrt{2\pi z}\, e^{-z} = 1.74\ldots \times 10^{-10}$. Hence, for $z \ge 25$ we can compute the exponential integral with a precision of about 10^{-10}, with maximal 26 terms in the asymptotic expansion. For z-values much larger than 25, fewer terms in the expansion are needed, and we can terminate the summation as soon as $n!/z^n$ is smaller than the required precision.

In our example of the exponential integral we obtained a relation between the remainder $R_n(z)$ and the first neglected term: the remainder is smaller and has the same sign. That is, for $n = 0, 1, \dots$, we have

$$R_n(z) = \theta_n\,(-1)^n \frac{n!}{z^n}, \quad 0 < \theta_n < 1. \tag{2.119}$$

2.4. Asymptotic expansions

For many asymptotic expansions of special functions this type of strict upper bound of the remainder is known. When the large parameter z is complex, information is still available for many cases. In [168] upper bounds for remainders are derived for several kinds of Bessel functions and confluent hypergeometric functions (Whittaker functions) by using the differential equation satisfied by these functions. These bounds are valid also for complex values of parameters and argument. By using these bounds, reliable and efficient algorithms can be designed for the computation of a large class of special functions.

For asymptotic expansions derived from integrals, usually less detailed information on upper bounds of the remainder is available, in particular when the parameters are complex. In [219, §3.6.1] upper bounds are given for the expansion of $\log \Gamma(z)$, also for complex z.

2.4.3 Exponentially improved asymptotic expansions

When information on the remainder in an asymptotic expansion is available, it is possible to improve the accuracy of the expansion by re-expanding the remainder. In that case, not only the large parameter, say z, should be considered as the asymptotic variable but also the truncation number, say n, that gives the smallest remainder.

Consider, for example, expansion (2.116) with remainder given in (2.117). We can write this remainder as an integral

$$C_n(a, z) = z^{n+1-a} e^z \int_z^\infty t^{a-n-1} e^{-t} \, dt = z \int_0^\infty (1+u)^{a-1-n} e^{-zu} \, du. \qquad (2.120)$$

We assume that z is positive and large. As in the previous section, the smallest term in the expansion roughly occurs when $n \sim z$. We write $n = z + \nu$, with $|\nu| \leq \frac{1}{2}$, and obtain

$$C_n(a, z) = z \int_0^\infty (1+u)^{a-\nu-1} e^{-z(u+\ln(1+u))} \, du. \qquad (2.121)$$

We take $v = u + \ln(1+u)$ as the new variable of integration, which gives

$$C_n(a, z) = z \int_0^\infty f(v) e^{-zv} \, dv, \quad f(v) = (1+u)^{\alpha-1} \frac{1+u}{2+u}, \quad \alpha = a - \nu. \qquad (2.122)$$

We can obtain an asymptotic expansion of this integral by applying Watson's lemma. First we need to invert the relation between u and v to obtain the expansion $u = \sum_{k=1}^\infty c_k v^k$. By using the relation $1 + u = (2+u)\frac{du}{dv}$ we easily find

$$c_1 = \tfrac{1}{2}, \quad 2kc_k = c_{k-1} - \sum_{j=1}^{k-1} jc_j c_{k-j}, \quad k = 2, 3, 4, \ldots. \qquad (2.123)$$

This gives the first few coefficients,

$$\begin{aligned} c_1 &= \tfrac{1}{2}, & c_2 &= \tfrac{1}{16}, & c_3 &= -\tfrac{1}{192}, \\ c_4 &= -\tfrac{1}{3072}, & c_5 &= \tfrac{13}{61440}, & c_6 &= -\tfrac{47}{1474560}. \end{aligned} \qquad (2.124)$$

For the function $f(v) = \sum_{k=0}^{\infty} f_k v^k$ we find the following first coefficients:

$$f_0 = \tfrac{1}{2},$$
$$f_1 = \tfrac{1}{8}(-1 + 2\alpha),$$
$$f_2 = \tfrac{1}{64}(1 - 6\alpha + 4\alpha^2),$$
$$f_3 = \tfrac{1}{768}(1 + 14\alpha - 24\alpha^2 + 8\alpha^3), \qquad (2.125)$$
$$f_4 = \tfrac{1}{12288}(-13 - 10\alpha + 100\alpha^2 - 80\alpha^3 + 16\alpha^4),$$
$$f_5 = \tfrac{1}{245760}(47 - 166\alpha - 240\alpha^2 + 520\alpha^3 - 240\alpha^4 + 32\alpha^5),$$
$$f_6 = \tfrac{1}{5898240}(73 + 1274\alpha - 812\alpha^2 - 2240\alpha^3 + 2240\alpha^4 - 672\alpha^5 + 64\alpha^6).$$

Substituting the expansion for f in (2.122) we obtain

$$C_n(a, z) \sim \sum_{k=0}^{\infty} f_k \frac{k!}{z^k}, \qquad (2.126)$$

which holds for large values of z and bounded values of $\alpha = a + z - n$.

To see the benefit of expanding $C_n(a, z)$, we take $a = 0.5$, $z = 6.25$, and $n = 6$. Taking $C_n(a, z) = 1$ in (2.116) gives $\Gamma(a, z) = 0.7223 \, 10^{-3}$, with relative error $0.14 \, 10^{-2}$. When we use the expansion given in (2.126), truncating after $k = 6$, we obtain $\Gamma(a, z) = 0.721303670366 \, 10^{-3}$ with relative error $0.12 \, 10^{-9}$.

In the asymptotic literature the quantity $C_n(a, z)$ (with the optimal choice of n when z is given) is called a *converging factor*. For many special functions converging factors have been developed, usually in a formal way. A rigorous treatment for the exponential integral $E_1(z)$ and the Kummer U-function, also for the case of complex z, is given in [168, pp. 522–536].

The expansion of the converging factor for the expansion (2.116) can be followed by a new expansion of the converging factor for the expansion (2.126), when information on the remainders of this new expansion is available. In this way a further exponential improvement can be obtained, and this leads to what recently in the asymptotic literature has been called *hyperasymptotics*; see [13] for the first ideas and [159] for a detailed analysis of representations of successive remainders in this method, with application to the Kummer U-function. For an application to the Euler gamma function, and for other details on hyperasymptotic methods, we refer the reader to [171, §6.4], where also new insight on the *Stokes phenomenon* is discussed. This topic is important when asymptotic expansions are considered in the complex plane. The development of the theory of hyperasymptotic expansions has also been carried out for the solutions of a class of second order ordinary differential equations; see [160, 161].

2.4.4 Alternatives of asymptotic expansions

In Chapter 9 we discuss several methods for transforming asymptotic sequences into rapidly converging sequences. In this section, we mention alternative expansions that are of a different nature. These expansions are convergent and have asymptotic properties.

2.4. Asymptotic expansions

Hadamard-type expansions

Consider first the integral for the modified Bessel function (for properties of the special functions in this subsection we refer to [2] or [219])

$$I_\nu(z) = \frac{(2z)^\nu e^z}{\sqrt{\pi}\,\Gamma(\nu+\tfrac{1}{2})} \int_0^1 e^{-2zt}[t(1-t)]^{\nu-\tfrac{1}{2}}\,dt. \qquad (2.127)$$

We cannot apply Theorem 2.13 because $[t(1-t)]^{\nu-\tfrac{1}{2}}$ is not analytic in the sector that contains the positive reals (and the integral is over a finite interval, but this is of minor concern). However, proceeding as in Watson's lemma, we substitute the expansion

$$(1-t)^{\nu-\tfrac{1}{2}} = \sum_{k=0}^\infty \binom{\nu-\tfrac{1}{2}}{k}(-t)^k \qquad (2.128)$$

and interchange the order of summation and integration. When we evaluate the resulting integrals not over $[0,1]$ but over $[0,\infty)$, we obtain (cf. (2.109))

$$I_\nu(z) \sim \frac{e^z}{\sqrt{2z\pi}\,\Gamma(\nu+\tfrac{1}{2})} \sum_{k=0}^\infty (-1)^k \binom{\nu-\tfrac{1}{2}}{k} \frac{\Gamma(k+\nu+\tfrac{1}{2})}{(2z)^k}, \quad z\to\infty, \qquad (2.129)$$

which is usually written in the form

$$I_\nu(z) \sim \frac{e^z}{\sqrt{2z\pi}} \sum_{k=0}^\infty \frac{a_k(\nu)}{(2z)^k}, \quad a_k(\nu) = \frac{(\tfrac{1}{2}-\nu)_k(\tfrac{1}{2}+\nu)_k}{k!}. \qquad (2.130)$$

For the Pochhammer symbol $(a)_k$ see (2.64).

The expansion in (2.130) holds for $|\mathrm{ph}\,z| < \tfrac{1}{2}\pi$. But replacing the finite interval by an infinite interval gives a divergent asymptotic expansion. If, after substituting (2.128) into (2.127), we integrate over $[0,1]$, we obtain a convergent expansion of the form (also called a *Hadamard expansion*; see [169])

$$I_\nu(z) = \frac{e^z}{\sqrt{2z\pi}} \sum_{k=0}^\infty \frac{b_k(\nu)}{(2z)^k}, \quad b_k(\nu) = \frac{(\tfrac{1}{2}-\nu)_k(\tfrac{1}{2}+\nu)_k}{k!} P\!\left(\tfrac{1}{2}+\nu+k,\,2z\right), \qquad (2.131)$$

where $P(a,z) = \gamma(a,z)/\Gamma(a)$ is the normalized incomplete gamma function defined by

$$\begin{aligned}P(a,z) &= \frac{1}{\Gamma(a)}\int_0^z e^{-z}t^{a-1}\,dt, \quad \Re a > 0,\\ &= \frac{z^a e^{-z}}{\Gamma(a+1)}\,{}_1F_1\!\left(\begin{matrix}1\\a+1\end{matrix};z\right).\end{aligned} \qquad (2.132)$$

The incomplete gamma functions can be computed by using a backward recursion scheme; see Chapter 4. For a graph of $P(a,x)$ with $a=5$ and $0\le x\le 20$, see Figure 2.4.

For fixed values of z the normalized incomplete gamma function has the asymptotic behavior

$$P(a,z) = \frac{e^{-z}z^a}{\Gamma(a+1)}\left[1+\mathcal{O}(a^{-1})\right], \quad a\to\infty, \qquad (2.133)$$

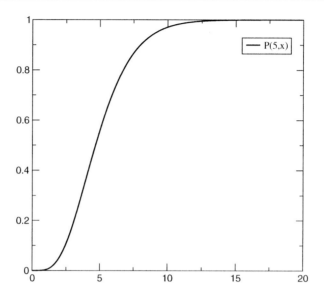

Figure 2.4. *Graph of the incomplete gamma function (2.132) for $a = 5$.*

which follows from the representation in terms of the $_1F_1$-function as given in (2.132). It follows easily that the terms in (2.131) behave like $\mathcal{O}(k^{-\nu-3/2})$. Hence, convergence of (2.131) is guaranteed if $\Re\nu > -\frac{1}{2}$.

The presence of the term $P(\frac{1}{2} + \nu + k, 2z)$ in (2.131) acts as a "smoothing" factor on the coefficients $b_k(\nu)$, since the behavior of $P(a, z)$ when z is large is characterized by a transition point at $z = a$. When the variables are positive, the asymptotic behavior of $P(a, z)$ changes from approximately unity when $a \leq z$ to a rapid decay to zero when a becomes larger than z.

To be more specific, we have the asymptotic behavior as given in (2.133) for $a \gg z$ and

$$P(a, z) = 1 - \frac{e^{-z}z^{a-1}}{\Gamma(a)}\left[1 + \mathcal{O}(z^{-1})\right], \quad z \to \infty, \qquad (2.134)$$

which follows from $P(a, z) = 1 - \Gamma(a, z)/\Gamma(a)$ and the result (2.114) of $\Gamma(a, z)$ in Example 2.14.

So, summing n terms in the Hadamard expansion (2.131), where $n \sim z$ as z is large, gives nearly the same result as summing n terms in the asymptotic expansion (2.130). But including more terms in the Hadamard expansion, trying to benefit from the fact that it is convergent, does not help very much because of the poor convergence of the Hadamard series (recall that convergence of (2.131) is guaranteed if $\Re\nu > -\frac{1}{2}$, and that the rate of convergence is controlled by $k^{-\nu-3/2}$).

In [169, 170] modifications of the Hadamard expansions are discussed from which much faster convergence can be obtained. The method is also used for infinite Laplace-type integrals.

2.4. Asymptotic expansions

An expansion in terms of confluent hypergeometric functions

Next we consider an expansion for the modified Bessel function $K_\nu(z)$ defined by

$$K_\nu(z) = \frac{\sqrt{\pi}(2z)^\nu e^{-z}}{\Gamma(\nu + \frac{1}{2})} \int_0^\infty e^{-2zt}[t(1+t)]^{\nu-\frac{1}{2}} dt. \tag{2.135}$$

We expand this function in terms of the confluent hypergeometric function $U(a, c, z)$, with integral representation

$$U(a, c, z) = \frac{1}{\Gamma(a)} \int_0^\infty e^{-zt} t^{a-1}(1+t)^{c-a-1} dt, \quad \Re z > 0, \quad \Re a > 0. \tag{2.136}$$

By expanding in (2.135)

$$(t+1)^{\nu-\frac{1}{2}} = \sum_{k=0}^\infty \binom{\nu-\frac{1}{2}}{k} t^k, \tag{2.137}$$

the standard asymptotic expansion described by Watson's lemma follows. That is,

$$K_\nu(z) \sim \sqrt{\frac{\pi}{2z}} \frac{e^{-z}}{\Gamma(\nu+\frac{1}{2})} \sum_{k=0}^\infty \binom{\nu-\frac{1}{2}}{k} \frac{\Gamma(k+\nu+\frac{1}{2})}{(2z)^k}, \quad z \to \infty. \tag{2.138}$$

This expansion holds in the sector $|\text{ph } z| < \frac{3}{2}\pi$.

As an alternative, we expand

$$(t+1)^{\nu-\frac{1}{2}} = \sum_{k=0}^\infty c_k \left(\frac{t}{t+1}\right)^k, \quad c_k = (-1)^k \binom{\frac{1}{2}-\nu}{k}. \tag{2.139}$$

This gives a convergent expansion in terms of confluent hypergeometric functions

$$K_\nu(z) = \sqrt{\frac{\pi}{2z}} e^{-z} \sum_{k=0}^\infty (-1)^k \frac{(\nu+\frac{1}{2})_k (\nu-\frac{1}{2})_k}{k!} U\left(k, \frac{1}{2}-\nu, 2z\right). \tag{2.140}$$

We have

$$U\left(0, \tfrac{1}{2}-\nu, 2z\right) = 1, \quad U\left(1, \tfrac{1}{2}-\nu, 2z\right) = (2z)^{\nu+\frac{1}{2}} e^{2z} \Gamma\left(-\tfrac{1}{2}-\nu, 2z\right), \tag{2.141}$$

again an incomplete gamma function. Other U-functions can be obtained by recursion. For large k and $z \neq 0$ the U-function behaves like

$$k! \, U\left(k, \tfrac{1}{2}-\nu, 2z\right) = \mathcal{O}\left(k^\alpha e^{-2\sqrt{2kz}}\right), \tag{2.142}$$

where α is some constant (this follows from [200, eq. (4.6.43)]). We see that the convergence is better than in the previous example. The confluent hypergeometric functions that occur in (2.140) can be computed by backward recursion; see [212] and Chapter 4.

Factorial series

Our third example is an expansion in the form of a *factorial series*, that is, in series of the form

$$F(z) = \sum_{n=0}^{\infty} \frac{a_n \, n!}{z(z+1)\cdots(z+n)}, \qquad (2.143)$$

where a_n are real or complex numbers that do not depend on z. The domain of convergence is usually a half plane $\Re z > z_0$, and coincides with the domain of convergence of the Dirichlet series $\sum_{n=1}^{\infty} a_n n^{-z}$, except for the points $0, -1, -2, \ldots$. For a proof see [124, §258].

There is a formal method to transform an asymptotic expansion

$$F(z) \sim \frac{b_1}{z} + \frac{b_2}{z^2} + \frac{b_3}{z^3} + \cdots, \qquad z \to \infty, \qquad (2.144)$$

into a factorial series. This can be done by using

$$\frac{1}{z^m} = \sum_{n=m}^{\infty} \frac{S_n^{n-m}}{z(z+1)\cdots(z+n)}, \qquad (2.145)$$

where S_n^m is a *Stirling number of the second kind* (see §11.3).

For functions defined in terms of a Laplace integral, that is,

$$F(z) = \int_0^{\infty} f(t) e^{-zt} \, dt, \qquad (2.146)$$

we can use a different method. On the one hand, if f satisfies the conditions of Watson's lemma, we can obtain an asymptotic expansion as in (2.144), and on the other hand we can take the new variable $w = e^{-t}$. This gives

$$F(z) = \int_0^1 w^{z-1} \varphi(w) \, dw, \qquad (2.147)$$

where $\varphi(w) = f(-\ln w)$. Expanding

$$\varphi(w) = \sum_{n=0}^{\infty} a_n (1-w)^n, \qquad (2.148)$$

and substituting this series in (2.147), we again obtain (2.143) in a formal way.

We summarize necessary and sufficient conditions from [157] such that the function $F(z)$ with integral in (2.147) can be written as a factorial series. The function $\varphi(w)$ should be analytic at $w = 1$ and the Maclaurin series at $w = 1$ should be convergent inside the disk $|w - 1| < 1$. Next, if $w = 0$ is a singular point of φ, and $\varphi^{(k)}(w)$ is the first of all derivatives of φ that becomes infinitely large at $w = 0^+$, then there should be a real number λ such that

$$\lim_{w \downarrow 0} \left| w^{z+\lambda} \varphi^{(k)}(w) \right| = \begin{cases} 0 & \text{if } \Re z > \lambda, \\ \infty & \text{if } \Re z < \lambda. \end{cases} \qquad (2.149)$$

In addition, there is a condition for the case that φ has other singularities on the circle $|w - 1| = 1$.

2.4. Asymptotic expansions

Example 2.15 (exponential integral $E_1(z)$). Consider again the exponential integral of Example 2.12 in the form

$$e^z E_1(z) = \int_0^\infty \frac{e^{-zt}}{t+1}\, dt = \int_0^1 w^{z-1} \varphi(w)\, dw, \quad \varphi(w) = \frac{1}{1 - \ln w}. \tag{2.150}$$

The function $\varphi(w)$ is analytic at $w = 1$ and the Maclaurin series at $w = 1$ converges inside the disk $|w - 1| < 1$. Next, $\varphi(0)$ is bounded, and

$$\varphi'(w) = \frac{1}{w(1 - \ln w)^2}, \tag{2.151}$$

which is infinite at $w = 0$. It is clear that the number λ of (2.149) equals 1. The function $\varphi(w)$ has no other singular points on the circle $|w - 1| = 1$, and we conclude that we have the convergent expansion

$$e^z E_1(z) = \sum_{n=0}^\infty \frac{a_n\, n!}{z(z+1)\cdots(z+n)}, \tag{2.152}$$

where the coefficients a_n follow from the expansion (2.148). The first few coefficients are

$$a_0 = 1,\; a_1 = -1,\; a_2 = \tfrac{1}{2},\; a_3 = -\tfrac{1}{3},\; a_4 = \tfrac{1}{6},\; a_5 = -\tfrac{7}{60},\; a_6 = \tfrac{19}{360}. \tag{2.153}$$

Higher coefficients follow easily from a recurrence relation

$$a_n = -\sum_{m=0}^{n-1} \frac{a_m}{n - m}, \quad n = 1, 2, \ldots. \tag{2.154}$$

In Table 2.1 we give the results of computing $ze^z E_1(z)$ for several values of z by using the factorial series in (2.152). We see that convergence for $z = 2$ is too slow for practical purposes. For $z = 10$ we see ten correct digits for $n = 25$. As explained earlier, with the asymptotic expansion for this function we can obtain a precision of about $\exp(-10) = 0.000045\ldots$. ∎

This method can also be used for other special functions, for example, for functions defined by the integral

$$F_\nu(z) = \int_0^\infty \frac{e^{-zt}}{(t+1)^\nu}\, dt, \tag{2.155}$$

which in fact is an incomplete gamma function, $F_\nu(z) = z^{\nu-1} e^z \Gamma(1 - \nu, z)$ (see [219, p. 279]).

In recent research (see, for example, [59]), convergent factorial series for Bessel functions are considered as alternatives for asymptotic expansions.

Table 2.1. *Evaluations of the factorial series* (2.152) *with n terms to compute* $ze^z E_1(z)$, *for several values of n and z.*

n	$z=2$	$z=5$	$z=10$	$z=25$	$z=50$
5	.7222222222	.8520502646	.9156288156	.9628674114	.9807554950
10	.7229745871	.8521142394	.9156333708	.9628674711	.9807554965
15	.7227777794	.8521113347	.9156333405	.9628674711	.9807554965
20	.7227232292	.8521110057	.9156333395	.9628674711	.9807554965
25	.7226979515	.8521109248	.9156333394	.9628674711	.9807554965
30	.7226846022	.8521108995	.9156333394	.9628674711	.9807554965
35	.7226767562	.8521108899	.9156333394	.9628674711	.9807554965
40	.7226717867	.8521108858	.9156333394	.9628674711	.9807554965
45	.7226684565	.8521108834	.9156333394	.9628674711	.9807554965
50	.7226661245	.8521108829	.9156333394	.9628674711	.9807554965
Exact	.7226572338	.8521108815	.9156333394	.9628674711	.9807554965

A convergent asymptotic representation for Laplace-type integrals

We consider the integral

$$F_\lambda(z) = \frac{1}{\Gamma(\lambda)} \int_0^\infty t^{\lambda-1} e^{-zt} f(t)\, dt, \quad \lambda > 0, \tag{2.156}$$

for large values of z with suitable assumptions on f, which will be given later.

In Watson's lemma, the function f is expanded at $t = 0$; in the present method a type of interpolation process is applied to f. The first interpolation point is obtained by considering (we assume that z is positive) the t-value for which $t^\lambda e^{-zt}$ attains its maximum, that is, the point $t_0 = \lambda/z$, and we obtain

$$F_\lambda(z) = f(t_0) z^{-\lambda} + \frac{1}{\Gamma(\lambda)} \int_0^\infty t^{\lambda-1} e^{-zt} [f(t) - f(t_0)]\, dt. \tag{2.157}$$

Next, observing that

$$\frac{d}{dt} t^\lambda e^{-zt} = -\frac{z}{t}(t-t_0) t^\lambda e^{-zt}, \tag{2.158}$$

we obtain

$$F_\lambda(z) = f(t_0) z^{-\lambda} - \frac{1}{z} \frac{1}{\Gamma(\lambda)} \int_0^\infty \frac{f(t) - f(t_0)}{t - t_0} d\left[t^\lambda e^{-zt}\right]. \tag{2.159}$$

Now, integrating by parts, assuming that the integrated terms will vanish, we obtain

$$F_\lambda(z) = f(t_0) z^{-\lambda} + \frac{1}{z} \frac{1}{\Gamma(\lambda)} \int_0^\infty t^\lambda e^{-zt} f_1(t)\, dt, \tag{2.160}$$

where

$$f_1(t) = \frac{d}{dt} \frac{f(t) - f(t_0)}{t - t_0}. \tag{2.161}$$

2.4. Asymptotic expansions

The new integral in (2.160) is of the same form as the integral in (2.156), with λ replaced by $\lambda + 1$ and f by f_1. The new interpolation point is $t_1 = (\lambda + 1)/z$ and we obtain

$$F_\lambda(z) = f(t_0)z^{-\lambda} + \frac{1}{z}\left[f_1(t_1)\lambda z^{-\lambda-1} + \frac{1}{z}\frac{1}{\Gamma(\lambda)}\int_0^\infty t^{\lambda+1}e^{-zt}f_2(t)\,dt \right], \quad (2.162)$$

where

$$f_2(t) = \frac{d}{dt}\frac{f_1(t) - f_1(t_1)}{t - t_1}. \quad (2.163)$$

This gives by further iteration

$$F_\lambda(z) = z^{-\lambda}\sum_{k=0}^{n-1}(\lambda)_k z^{-2k} f_k(t_k) + \frac{1}{z^n}\frac{1}{\Gamma(\lambda)}\int_0^\infty t^{\lambda+n-1}e^{-zt}f_n(t)\,dt, \quad (2.164)$$

where $n = 0, 1, 2, \ldots$, $f_0(t) = f(t)$, and

$$f_{k+1}(t) = \frac{d}{dt}\left[\frac{f_k(t) - f_k(t_k)}{t - t_k}\right], \quad t_k = \frac{\lambda+k}{z}, \quad k = 0, 1, 2, \ldots. \quad (2.165)$$

To obtain the functions f_k and the coefficients in (2.164), we require that f belong to the class C^{2n} for $t > 0$, and that its derivatives admit bounds of the form $|f^{(m)}(t)| \le Me^{\mu t}$, $t > 0$, $m = 0, 1, \ldots, 2n$, for suitable M and μ. With these conditions expansion (2.164) has an asymptotic property and the remainder is of order $\mathcal{O}(z^{-2n-\lambda})$ as $z \to \infty$. Under further mild conditions on f it can be shown that the remainder tends to zero as $n \to \infty$, and that, hence, the expansion is convergent. For proofs, see [72].

Example 2.16 (modified expansion for the incomplete gamma function). For some special functions we can obtain the coefficients $f_k(t_k)$ easily (by using computer algebra). For example, when $f(t) = 1/(1+t)$ we have a special case of the Kummer U-function, that is, the incomplete gamma function (see [219, p. 280]),

$$F_\lambda(z) = \frac{1}{\Gamma(\lambda)}\int_0^\infty t^{\lambda-1}e^{-zt}\frac{dt}{1+t} = U(\lambda, \lambda, z) = e^z\Gamma(1-\lambda, z). \quad (2.166)$$

We write $\zeta = z + \lambda$ and have the expansion

$$e^z\Gamma(1-\lambda, z) \sim z^{-\lambda}\sum_{k=0}^\infty (\lambda)_k z^{-2k} f_k, \quad (2.167)$$

where we write $f_k = f_k(t_k)$, and find the coefficients

$$f_0 = \frac{z}{\zeta},$$

$$f_1 = \frac{z^3}{\zeta(\zeta+1)^2},$$

$$f_2 = \frac{z^5(3\zeta+4)}{\zeta(\zeta+1)^2(\zeta+2)^3}, \quad (2.168)$$

$$f_3 = \frac{z^7[15\zeta^3 + 90\zeta^2 + 175\zeta + 108]}{\zeta(\zeta+1)^2(\zeta+2)^3(\zeta+3)^4},$$

$$f_4 = \frac{z^9[105\zeta^6 + 1680\zeta^5 + 11025\zeta^4 + 37870\zeta^3 + 71540\zeta^2 + 70120\zeta + 27648]}{\zeta(\zeta+1)^2(\zeta+2)^3(\zeta+3)^4(\zeta+4)^5}.$$

Table 2.2. *Verification of the relation in (2.170) for some values of z and λ by using the sum in (2.164) with $n = 5$ and the coefficients given in (2.168).*

z	λ	$z^\lambda F_\lambda(z)$	lhs of (2.170)
10	1	0.9156330438	$-0.38\,10^{-06}$
10	10	0.5121792821	$-0.53\,10^{-05}$
20	1	0.9543709085	$-0.17\,10^{-08}$
20	10	0.6738352273	$-0.14\,10^{-06}$
20	20	0.5061709401	$-0.22\,10^{-06}$
50	1	0.9807554965	$-0.52\,10^{-12}$
50	25	0.6695906479	$-0.16\,10^{-08}$
50	50	0.5024874402	$-0.26\,10^{-08}$

We observe that these coefficients satisfy the order estimate $f_k = \mathcal{O}(1)$ as $z \to \infty$. In addition, the coefficients become small when λ is large. Combining the coefficients with the Pochhammer symbols in (2.164), we conclude that

$$(\lambda)_k f_k(t_k) = \mathcal{O}\left(\lambda^{-k-1}\right), \quad \lambda \to \infty. \qquad (2.169)$$

In fact, the expansion holds for $z \to \infty$, uniformly with respect to $\lambda \geq 0$, and also for complex values of z and λ. In Watson's lemma (see the integral in (2.108)), the asymptotic property of the expansion disappears when λ and z are both large. The same occurs when in (2.113) a and z are of the same size.

For a numerical verification of the method for the case $f(t) = 1/(1+t)$, we have verified the relation

$$z^\lambda F_\lambda(z) + \frac{\lambda}{z} + z^{\lambda+1} F_{\lambda+1}(z) - 1 = 0, \qquad (2.170)$$

which follows from the recurrence relation for the incomplete gamma function

$$\Gamma(a+1, z) = a\Gamma(a, z) + z^a e^{-z}. \qquad (2.171)$$

In Table 2.2 we give the values of $z^\lambda F_\lambda(z)$ together with computed values of the left-hand side of relation (2.170) for a few values of z and λ. We have used expansion (2.167) with five terms and the coefficients given in (2.168). ∎

A modification of the expansion in (2.164) is given in [213], where the coefficients can be computed from the Maclaurin coefficients of f at the fixed interpolation point $t_0 = \lambda/z$. The aim of that method was to obtain an expansion for large z, again holding uniformly with respect to $\lambda \geq 0$. Convergence of the expansion was not discussed. Because of the shifting interpolation points t_k in the method for obtaining (2.164), the computation of the coefficients is much more difficult for that expansion than for the one derived in [213].

Wagner's modification of Watson's lemma

Wagner [227] gives a different method for the integral in (2.156), with a simpler method for obtaining coefficients in a modified expansion.

2.4. Asymptotic expansions

The first step is representing the function f in a proper way. Assume that f is analytic at the origin, with series expansion at the origin $f(t) = \sum_{k=0}^{\infty} a_k t^k$ with $a_0 \neq 0$. Next, write f in the form

$$f(t) = a_0 e^{-rt} + t^m f_1(t), \tag{2.172}$$

such that $m \geq 2$ and f_1 again is analytic at the origin. This can be achieved by taking $r = -a_1/a_0$ and writing $f_1(t) = \sum_{k=0}^{\infty} a_k^{(1)} t^k$. When $a_2 \neq \frac{1}{2} a_0 r^2$ we take $m = 2$ and $a_0^{(1)} = a_2 - \frac{1}{2} a_0 r^2$; otherwise m becomes larger and $a_0^{(1)}$ is chosen accordingly. The integral (2.156) becomes

$$F_\lambda(z) = \frac{a_0}{(z+r)^\lambda} + \frac{\Gamma(\lambda_1)}{\Gamma(\lambda)} F_{\lambda_1}^{(1)}(z),$$

$$F_{\lambda_1}^{(1)}(z) = \frac{1}{\Gamma(\lambda_1)} \int_0^\infty t^{\lambda_1 - 1} e^{-zt} f_1(t)\, dt, \tag{2.173}$$

where $\lambda_1 = \lambda + m$, and the search for new r and m can be continued.

It is rather easy to write an algorithm for this method, and it uses only the Maclaurin coefficients of the original function f. For details and an example for the exponential integral, see [227].

Example 2.17 (modified expansion for the gamma function). We give a modification of the well-known expansion for the gamma function. We start with the integral

$$\Gamma(z+1) = \int_0^\infty u^z e^{-u}\, du, \quad \Re z > -1. \tag{2.174}$$

To write it in the form (2.156) we need a few transformations. First we write $u = zv$, which gives

$$\Gamma(z) = z^z e^{-z} \int_0^\infty e^{-z\phi(v)}\, dv, \quad \phi(v) = v - \ln v - 1. \tag{2.175}$$

Because $\phi'(v) = 0$ if $v = 1$, this integral has a saddle point at $v = 1$, and we substitute

$$\tfrac{1}{2} s^2 = \phi(v), \quad \text{sign}(s) = \text{sign}(v - 1). \tag{2.176}$$

This gives

$$\Gamma(z) = z^z e^{-z} \int_{-\infty}^\infty e^{-\frac{1}{2} z s^2} g(s)\, ds, \quad g(s) = \frac{dv}{ds} = \frac{sv}{v-1}. \tag{2.177}$$

This representation is valid when $\Re z > 0$. We take the even part of $g(s)$ and write

$$\Gamma(z) = \tfrac{1}{2} z^z e^{-z} \int_0^\infty e^{-\frac{1}{2} z s^2} [g(s) + g(-s)]\, ds, \tag{2.178}$$

and substitute $s = \sqrt{2t}$. This gives

$$\Gamma(z) = \sqrt{2\pi}\, z^z e^{-z} F_\lambda(z), \quad F_\lambda(z) = \frac{1}{\Gamma(\lambda)} \int_0^\infty t^{\lambda-1} e^{-zt} f(t)\, dt, \quad \lambda = \frac{1}{2}. \tag{2.179}$$

The function $f(t)$ has an expansion

$$f(t) = 1 + \tfrac{1}{6} t + \tfrac{1}{216} t^2 - \tfrac{139}{97200} t^3 + \cdots, \tag{2.180}$$

Table 2.3. *Comparison of the expansions* (2.182) (*with three terms*) *and* (2.181) (*with three terms*).

z	Expansion (2.182)	Expansion (2.181)
1	1.0015156521	1.0021836242
2	1.0000167806	1.0003143342
3	2.0000025618	2.0001927161
4	6.0000012598	6.0002470703
5	24.000001251	24.000509128
6	120.00000201	120.00147818
7	720.00000465	720.00559676
8	5040.0000143	5040.0262810
9	40320.000055	40320.147800
10	362880.00026	362880.97036

and by substituting this expansion in (2.179) we obtain the first terms of the well-known expansion

$$\Gamma(z) \sim \sqrt{2\pi}\, z^{z-\frac{1}{2}} e^{-z} \left(1 + \frac{1}{12} z^{-1} + \frac{1}{288} z^{-2} - \frac{139}{51840} z^{-3} + \cdots \right), \quad z \to \infty. \qquad (2.181)$$

By using Wagner's method we obtain

$$\Gamma(z) \sim \sqrt{2\pi}\, z^{z-\frac{1}{2}} e^{-z}$$
$$\times \left(\frac{a_0^{(0)}}{(1+r_0/z)^{\frac{1}{2}}} + \frac{3 a_0^{(1)}}{4(1+r_1/z)^{\frac{5}{2}}} z^{-2} + \frac{105 a_0^{(2)}}{16(1+r_2/z)^{\frac{9}{2}}} z^{-4} + \cdots \right), \qquad (2.182)$$

where

$$a_0^{(0)} = 1, \quad a_0^{(1)} = -\frac{1}{108}, \quad a_0^{(2)} = \frac{59593}{306180000},$$
$$r_0 = -\frac{1}{6}, \quad r_1 = -\frac{107}{450}, \quad r_2 = -\frac{133744607}{563153850}. \qquad (2.183)$$

In Table 2.3 we compare the expansions given in (2.181) (with three terms) and (2.182) (also with three terms) for small values of z.

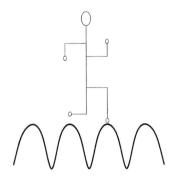

Chapter 3
Chebyshev Expansions

The best is the cheapest.
—Benjamin Franklin

3.1 Introduction

In Chapter 2, approximations were considered consisting of expansions around a specific value of the variable (finite or infinite); both convergent and divergent series were described. These are the preferred approaches when values around these points (either in $\mathbb{R}$ or $\mathbb{C}$) are needed.

In this chapter, approximations in real intervals are considered. The idea is to approximate a function $f(x)$ by a polynomial $p(x)$ that gives a uniform and accurate description in an interval $[a, b]$.

Let us denote by $\mathbb{P}_n$ the set of polynomials of degree at most n and let g be a bounded function defined on $[a, b]$. Then the uniform norm $||g||$ on $[a, b]$ is given by

$$||g|| = \max_{x \in [a,b]} |g(x)|. \tag{3.1}$$

For approximating a continuous function f on an interval $[a, b]$, it is reasonable to consider that the best option consists in finding the *minimax approximation*, defined as follows.

Definition 3.1. *$q \in \mathbb{P}_n$ is the best (or minimax) polynomial approximation to f on $[a, b]$ if*

$$||f - q|| \leq ||f - p|| \quad \forall p \in \mathbb{P}_n. \tag{3.2}$$

Minimax polynomial approximations exist and are unique (see [152]) when f is continuous, although they are not easy to compute in general. Instead, it is a more effective approach to consider near-minimax approximations, based on Chebyshev polynomials.

Chebyshev polynomials form a special class of polynomials especially suited for approximating other functions. They are widely used in many areas of numerical analysis: uniform approximation, least-squares approximation, numerical solution of ordinary and partial differential equations (the so-called spectral or pseudospectral methods), and so on.

In this chapter we describe the approximation of continuous functions by Chebyshev interpolation and Chebyshev series and how to compute efficiently such approximations. For the case of functions which are solutions of linear ordinary differential equations with polynomial coefficients (a typical case for special functions), the problem of computing Chebyshev series is efficiently solved by means of Clenshaw's method, which is also presented in this chapter.

Before this, we give a very concise overview of well-known results in interpolation theory, followed by a brief summary of important properties satisfied by Chebyshev polynomials.

3.2 Basic results on interpolation

Consider a real function f that is continuous on the real interval $[a, b]$. When values of this function are known at a finite number of points x_i, one can consider the approximation by a polynomial P_n such that $f(x_i) = P_n(x_i)$. The next theorem gives an explicit expression for the lowest degree polynomial (the *Lagrange interpolation polynomial*) satisfying these interpolation conditions.

Theorem 3.2 (Lagrange interpolation). *Given a function f that is defined at $n+1$ points $x_0 < x_1 < \cdots < x_n \in [a, b]$, there exists a unique polynomial of degree smaller than or equal to n such that*

$$P_n(x_i) = f(x_i), \quad i = 0, \ldots, n. \tag{3.3}$$

This polynomial is given by

$$P_n(x) = \sum_{i=0}^{n} f(x_i) L_i(x), \tag{3.4}$$

where $L_i(x)$ is defined by

$$L_i(x) = \frac{\pi_{n+1}(x)}{(x - x_i)\pi'_{n+1}(x_i)} = \frac{\prod_{j=0, j\neq i}^{n}(x - x_j)}{\prod_{j=0, j\neq i}^{n}(x_i - x_j)}, \tag{3.5}$$

$\pi_{n+1}(x)$ *being the nodal polynomial,* $\pi_{n+1}(x) = \prod_{j=0}^{n}(x - x_j)$.

Additionally, if f is continuous on $[a, b]$ and $n+1$ times differentiable in (a, b), then for any $x \in [a, b]$ there exists a value $\zeta_x \in (a, b)$, depending on x, such that

$$R_n(x) = f(x) - P_n(x) = \frac{f^{n+1}(\zeta_x)}{(n+1)!} \pi_{n+1}(x). \tag{3.6}$$

Proof. The proof of this theorem can be found elsewhere [45, 48]. □

L_i are called the *fundamental Lagrange interpolation polynomials*.

3.2. Basic results on interpolation

The first part of the theorem is immediate and P_n satisfies the interpolation conditions, because the polynomials L_i are such that $L_i(x_j) = \delta_{ij}$. The formula for the remainder can be proved from repeated application of Rolle's theorem (see, for instance, [48, Thm. 3.3.1]).

For the particular case of Lagrange interpolation over n nodes, a simple expression for the interpolating polynomial can be given in terms of *forward differences* when the nodes are equally spaced, that is, $x_{i+1} - x_i = h$, $i = 0, \ldots, n-1$. In this case, the interpolating polynomials of Theorem 3.2 can be written as

$$P_n(x) = \sum_{i=0}^{n} \binom{s}{i} \Delta^i f_0, \tag{3.7}$$

where

$$s = \frac{x - x_0}{h}, \quad \binom{s}{i} = \frac{1}{i!} \prod_{j=0}^{i-1}(s - j), \quad f_j = f(x_j), \tag{3.8}$$

$$\Delta f_j = f_{j+1} - f_j, \quad \Delta^2 f_j = \Delta(f_{j+1} - f_j) = f_{j+2} - 2f_{j+1} + f_j, \ldots$$

This result is easy to prove by noticing that $f_s = (\Delta + I)^s f_0$, $s = 0, 1, \ldots, n$, and by expanding the binomial of commuting operators Δ and I (I being the identity, $If_i = f_i$).

The formula for the remainder in Theorem 3.2 resembles that for the Taylor formula of degree n (Lagrange form), except that the nodal polynomials in the latter case contain only one node, x_0, which is repeated $n+1$ times (in the sense that the power $(x - x_0)^{n+1}$ appears). This interpretation in terms of *repeated nodes* can be generalized; both the Taylor formula and the Lagrange interpolation formula can be seen as particular cases of a more general interpolation formula, which is *Hermite interpolation*.

Theorem 3.3 (Hermite interpolation). *Let f be n times differentiable with continuity in $[a, b]$ and $n + 1$ times differentiable in (a, b). Let $x_0 < x_1 < \cdots < x_k \in [a, b]$, and let $n_i \in \mathbb{N}$ such that $n_0 + n_1 + \cdots + n_k = n - k$. Then, there exists a unique polynomial P_n of degree not larger than n such that*

$$P_n^{(j)}(x_i) = f^{(j)}(x_i), \quad j = 0, \ldots, n_i, \quad i = 0, \ldots, k. \tag{3.9}$$

Furthermore, given $x \in [a, b]$, there exists a value $\zeta_x \in (a, b)$ such that

$$f(x) = P_n(x) + \frac{f^{n+1}(\zeta_x)}{(n+1)!} \pi_{n+1}(x), \tag{3.10}$$

where $\pi_{n+1}(x)$ is the nodal polynomial

$$\pi_{n+1}(x) = (x - x_0)^{n_0+1} \cdots (x - x_k)^{n_k+1} \tag{3.11}$$

in which each node x_i is repeated $n_i + 1$ times.

Proof. For the proof we refer to [45]. □

An explicit expression for the interpolating polynomial is, however, not so easy as for Lagrange's case. A convenient formalism is that of *Newton's divided difference formula*, also for Lagrange interpolation (see [45] for further details).

For the case of a single interpolation node x_0 which is repeated n times, the corresponding interpolating polynomial is just the Taylor polynomial of degree n at x_0. It is very common that successive derivatives of special functions are known at a certain point $x = x_0$ (Taylor's theorem, (2.1)), but it is not common that derivatives are known at several points. Therefore, in practical evaluation of special functions, Hermite interpolation different from the Taylor case is seldom used.

Lagrange interpolation is, however, a very frequently used method of approximation and, in addition, will be behind the quadrature methods to be discussed in Chapter 5. For interpolating a function in a number of nodes, we need, however, to know the values which the function takes at these points. Therefore, in general we will need to rely on an alternative (high-accuracy) method of evaluation.

However, for functions which are solutions of a differential equation, Clenshaw's method (see §3.6.1) provides a way to compute expansions in terms of Chebyshev polynomials. Such infinite expansions are related to a particular and useful type of Lagrange interpolation that we discuss in detail in §3.6.1 and introduce in the next section.

3.2.1 The Runge phenomenon and the Chebyshev nodes

Given a function f which is continuous on $[a, b]$, we may try to approximate the function by a Lagrange interpolating polynomial.

We could naively think that as more nodes are considered, the approximation will always be more accurate, but this is not always true. The main question to be addressed is whether the polynomials P_n that interpolate a continuous function f in $n+1$ equally spaced points are such that

$$\lim_{n\to\infty} ||f - P_n|| = \lim_{n\to\infty} ||R_n|| = 0, \qquad (3.12)$$

where, if f is sufficiently differentiable, the error can be estimated through (3.6).

A pathological example for which the Lagrange interpolation does not converge is provided by $f(x) = |x|$ in the interval $[-1, 1]$, for which equidistant interpolation diverges for $0 < |x| < 1$ (see [189, Thm. 4.7]), as has been proved by Bernstein.

A less pathological example, studied by Runge, showing the *Runge phenomenon*, gives a clear warning on the problems of equally spaced nodes. Considering the problem of interpolation of

$$f(x) = \frac{1}{1 + x^2} \qquad (3.13)$$

on $[-5, 5]$, Runge observed that $\lim_{n\to\infty} ||f - P_n|| = \infty$, but that convergence takes place in a smaller interval $[-a, a]$ with $a \simeq 3.63$.

This bad behavior in Runge's example is due to the values of the nodal polynomial $\pi_{n+1}(x)$, which tends to present very strong oscillations near the endpoints of the interval (see Figure 3.1).

3.2. Basic results on interpolation

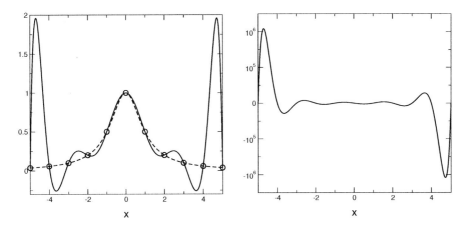

Figure 3.1. *Left: the function $f(x) = 1/(1+x^2)$ is plotted in $[-5, 5]$ together with the polynomial of degree 10 which interpolates f at $x = 0, \pm 1, \pm 2, \pm 3, \pm 4, \pm 5$. Right: the nodal polynomial $\pi(x) = x(x-1)(x-4)(x-9)(x-16)(x-25)$.*

The uniformity of the error in the interval of interpolation can be considerably improved by choosing the interpolation nodes x_i in a different way. Without loss of generality, we will restrict our study to interpolation on the interval $[-1, 1]$; the problem of interpolating f with nodes x_i in the finite interval $[a, b]$ is equivalent to the problem of interpolating $g(t) = f(x(t))$, where

$$x(t) = \frac{a+b}{2} + \frac{b-a}{2}t \tag{3.14}$$

with nodes $t_i \in [-1, 1]$.

Theorem 3.4 explains how to choose the nodes in $[-1, 1]$ in order to minimize uniformly the error due to the nodal polynomial and to quantify this error. The nodes are given by the zeros of a Chebyshev polynomial.

Theorem 3.4. *Let $x_k = \cos((k+1/2)\pi/(n+1))$, $k = 0, 1, \ldots, n$. Then the monic polynomial $\hat{T}_{n+1}(x) = \prod_{k=0}^{n}(x - x_k)$ is the polynomial of degree $n+1$ with the smallest possible uniform norm (3.1) in $[-1, 1]$ in the sense that*

$$\|\hat{T}_{n+1}\| \leq \|q_{n+1}\| \tag{3.15}$$

for any other monic polynomial q_{n+1} of degree $n+1$. Furthermore,

$$\|\hat{T}_{n+1}\| = 2^{-n}. \tag{3.16}$$

The selection of these nodes will not guarantee convergence as the number of nodes tends to infinity, because it also depends on how the derivatives of the function f behave, but certainly enlarges the range of functions for which convergence takes place and eliminates the problem for the example provided by Runge. Indeed, taking as nodes

$$x_k = 5\cos((k+1/2)\pi/11), \quad k = 0, 1, \ldots, 10, \tag{3.17}$$

instead of the 11 equispaced points, the behavior is much better, as illustrated in Figure 3.2.

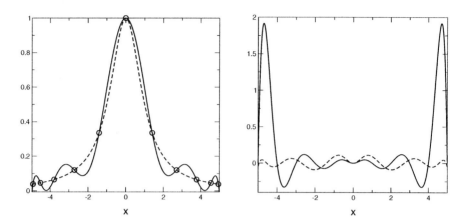

Figure 3.2. *Left: the function $f(x) = 1/(1 + x^2)$ is plotted together with the interpolation polynomial for the 11 Chebyshev points (see (3.17)). Right: the interpolation errors for equispaced points and Chebyshev points is shown.*

Before proving Theorem 3.4 and further results, we summarize the basic properties of Chebyshev polynomials, the zeros of which are the nodes in Theorem 3.4.

3.3 Chebyshev polynomials: Basic properties

Let us first consider a definition and some properties of the Chebyshev polynomials of the first kind.

Definition 3.5 (Chebyshev polynomial of the first kind $T_n(x)$). *The Chebyshev polynomial of the first kind of order n is defined as follows*:

$$T_n(x) = \cos\left[n \cos^{-1}(x)\right], \quad x \in [-1, 1], \quad n = 0, 1, 2, \ldots. \tag{3.18}$$

From this definition the following property is evident:

$$T_n(\cos \theta) = \cos(n\theta), \quad \theta \in [0, \pi], \quad n = 0, 1, 2, \ldots. \tag{3.19}$$

3.3.1 Properties of the Chebyshev polynomials $T_n(x)$

The polynomials $T_n(x)$, $n \geq 1$, satisfy the following properties, which follow straightforwardly from (3.19).

(i) The Chebyshev polynomials $T_n(x)$ satisfy the following three-term recurrence relation:
$$T_{n+1}(x) = 2x T_n(x) - T_{n-1}(x), \quad n = 1, 2, 3, \ldots, \tag{3.20}$$
with starting values $T_0(x) = 1$, $T_1(x) = x$.

3.3. Chebyshev polynomials: Basic properties

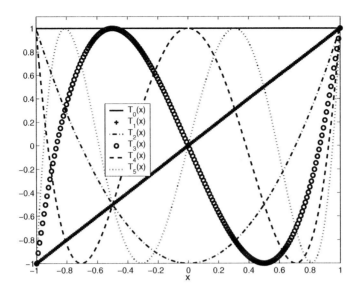

Figure 3.3. *Chebyshev polynomials of the first kind $T_n(x)$, $n = 0, 1, 2, 3, 4, 5$.*

Explicit expressions for the first six Chebyshev polynomials are

$$T_0(x) = 1, \qquad T_1(x) = x,$$
$$T_2(x) = 2x^2 - 1, \qquad T_3(x) = 4x^3 - 3x, \qquad (3.21)$$
$$T_4(x) = 8x^4 - 8x^2 + 1, \quad T_5(x) = 16x^5 - 20x^3 + 5x.$$

The graphs of these Chebyshev polynomials are plotted in Figure 3.3.

(ii) The leading coefficient (of x^n) in $T_n(x)$ is 2^{n-1} and $T_n(-x) = (-1)^n T_n(x)$.

(iii) $T_n(x)$ has n zeros which lie in the interval $(-1, 1)$. They are given by

$$x_k = \cos\left(\frac{2k+1}{2n}\pi\right), \quad k = 0, 1, \ldots, n-1. \qquad (3.22)$$

$T_n(x)$ has $n + 1$ extrema in the interval $[-1, 1]$ and they are given by

$$x'_k = \cos\frac{k\pi}{n}, \quad k = 0, 1, \ldots, n. \qquad (3.23)$$

At these points, the values of the polynomials are $T_n(x'_k) = (-1)^k$.

With these properties, it is easy to prove Theorem 3.4, which can also be expressed in the following way.

Theorem 3.6. *The polynomial $\hat{T}_n(x) = 2^{1-n} T_n(x)$ is the minimax approximation on $[-1, 1]$ to the zero function by a monic polynomial of degree n and*

$$\|\hat{T}_n\| = 2^{1-n}. \tag{3.24}$$

Proof. Let us suppose that there exists a monic polynomial p_n of degree n such that $|p_n(x)| \leq 2^{1-n}$ for all $x \in [-1, 1]$, and we will arrive at a contradiction.

Let x'_k, $k = 0, \ldots, n$, be the abscissas of the extreme values of the Chebyshev polynomial of degree n. Because of property (ii) of this section we have

$$p_n(x'_0) < 2^{1-n} T_n(x'_0), \quad p_n(x'_1) > 2^{1-n} T_n(x'_1), \quad p_n(x'_2) > 2^{1-n} T_n(x'_2), \ldots.$$

Therefore, the polynomial

$$Q(x) = p_n(x) - 2^{1-n} T_n(x)$$

changes sign between each two consecutive extrema of $T_n(x)$. Thus, it changes sign n times. But this is not possible because $Q(x)$ is a polynomial of degree smaller than n (it is a subtraction of two monic polynomials of degree n). □

Remark 1. The monic Chebyshev polynomial $\hat{T}_n(x)$ is not the minimax approximation in $\mathbb{P}_n$ (Definition 3.1) of the zero function. The minimax approximation in $\mathbb{P}_n$ of the zero function is the zero polynomial.

Further properties

Next we summarize additional properties of the Chebyshev polynomials of the first kind that will be useful later. For further properties and proofs of these results see, for instance, [148, Chaps. 1–2].

(a) **Relations with derivatives.**

$$\begin{cases} T_0(x) = T'_1(x), \\ T_1(x) = \frac{1}{4} T'_2(x), \\ T_n(x) = \frac{1}{2} \left(\frac{T'_{n+1}(x)}{n+1} - \frac{T'_{n-1}(x)}{n-1} \right), \quad n \geq 2, \end{cases} \tag{3.25}$$

$$(1 - x^2) T'_n(x) = n \left[x T_n(x) - T_{n+1}(x) \right] = n \left[T_{n-1}(x) - x T_n(x) \right]. \tag{3.26}$$

(b) **Multiplication relation.**

$$2 T_r(x) T_q(x) = T_{r+q}(x) + T_{|r-q|}(x), \tag{3.27}$$

with the particular case $q = 1$,

$$2x T_r(x) = T_{r+1}(x) + T_{|r-1|}(x). \tag{3.28}$$

3.3. Chebyshev polynomials: Basic properties

(c) Orthogonality relation.

$$\int_{-1}^{1} T_r(x) T_s(x) (1-x^2)^{-1/2} \, dx = N_r \delta_{rs}, \tag{3.29}$$

with $N_0 = \pi$ and $N_r = \frac{1}{2}\pi$ if $r \neq 0$.

(d) Discrete orthogonality relation.

1. With the zeros of $T_{n+1}(x)$ as nodes: Let $n > 0$, $r, s \leq n$, and let $x_j = \cos((j+1/2)\pi/(n+1))$. Then

$$\sum_{j=0}^{n} T_r(x_j) T_s(x_j) = K_r \delta_{rs}, \tag{3.30}$$

where $K_0 = n+1$ and $K_r = \frac{1}{2}(n+1)$ when $1 \leq r \leq n$.

2. With the extrema of $T_n(x)$ as nodes: Let $n > 0$, $r, s \leq n$, and $x_j = \cos(\pi j/n)$. Then

$$\sum_{j=0}^{n}{}'' T_r(x_j) T_s(x_j) = K_r \delta_{rs}, \tag{3.31}$$

where $K_0 = K_n = n$ and $K_r = \frac{1}{2}n$ when $1 \leq r \leq n-1$.

The double prime indicates that the terms with suffixes $j=0$ and $j=n$ are to be halved.

(e) Polynomial representation.

The expression of $T_n(x)$ in terms of powers of x is given by (see [38, 201])

$$T_n(x) = \sum_{k=0}^{\lfloor n/2 \rfloor} d_k^{(n)} x^{n-2k}, \tag{3.32}$$

where

$$d_k^{(n)} = (-1)^k 2^{n-2k-1} \frac{n}{n-k} \binom{n-k}{k}, \quad 2k < n, \tag{3.33}$$

and

$$d_k^{(2k)} = (-1)^k, \quad k \geq 0. \tag{3.34}$$

(f) Power representation.

The power x^n can be expressed in terms of Chebyshev polynomials as follows:

$$x^n = 2^{1-n} \sum_{k=0}^{\lfloor n/2 \rfloor}{}' \binom{n}{k} T_{n-2k}(x), \tag{3.35}$$

where the prime indicates that the term for $k=0$ is to be halved.

The first three properties are immediately obtained from the definition of Chebyshev polynomials.

Property (c) means that the set of Chebyshev polynomials $\{T_n(x)\}$ is an orthogonal set with respect to the weight function $w(x) = (1 - x^2)^{-1/2}$ in the interval $(-1, 1)$. This concept is developed in Chapter 5, and it is shown that this orthogonality implies the first discrete orthogonality of property (d); see (5.86). This property, as well as the second discrete orthogonality, can also be easily proved using trigonometry (see [148, Chap. 4]). See also [148, Chap. 2] for a proof of the last two properties.

Shifted Chebyshev polynomials

Shifted Chebyshev polynomials are also of interest when the range of the independent variable is $[0, 1]$ instead of $[-1, 1]$. The *shifted Chebyshev polynomials of the first kind* are defined as

$$T_n^*(x) = T_n(2x - 1), \quad 0 \leq x \leq 1. \tag{3.36}$$

Similarly, one can also build shifted polynomials for a generic interval $[a, b]$.

Explicit expressions for the first six shifted Chebyshev polynomials are

$$\begin{aligned} T_0^*(x) &= 1, \\ T_1^*(x) &= 2x - 1, \\ T_2^*(x) &= 8x^2 - 8x + 1, \\ T_3^*(x) &= 32x^3 - 48x^2 + 18x - 1, \\ T_4^*(x) &= 128x^4 - 256x^3 + 160x^2 - 32x + 1, \\ T_5^*(x) &= 512x^5 - 1280x^4 + 1120x^3 - 400x^2 + 50x - 1. \end{aligned} \tag{3.37}$$

3.3.2 Chebyshev polynomials of the second, third, and fourth kinds

Chebyshev polynomials of the first kind are a particular case of Jacobi polynomials $P_n^{(\alpha,\beta)}(x)$ (up to a normalization factor). Jacobi polynomials, which can be defined through the Gauss hypergeometric function (see §2.3) as

$$P_n^{(\alpha,\beta)}(x) = \binom{n+\alpha}{n} {}_2F_1\left(\begin{array}{c} -n, n+\alpha+\beta+1 \\ \alpha+1 \end{array}; \frac{1-x}{2}\right), \tag{3.38}$$

are orthogonal polynomials on the interval $[-1, 1]$ with respect to the weight function $w(x) = (1 - x)^\alpha (1 + x)^\beta$, $\alpha, \beta > -1$, that is,

$$\int_{-1}^{1} P_r^{(\alpha,\beta)}(x) P_s^{(\alpha,\beta)}(x) w(x)\, dx = M_r \delta_{rs}. \tag{3.39}$$

In particular, for the case $\alpha = \beta = -1/2$ we recover the orthogonality relation (3.29). Furthermore,

$$T_n(x) = {}_2F_1\left(\begin{array}{c} -n, n \\ 1/2 \end{array}; \frac{1-x}{2}\right). \tag{3.40}$$

3.3. Chebyshev polynomials: Basic properties

As we have seen, an important property satisfied by the polynomials $T_n(x)$ is that, with the change $x = \cos\theta$, the zeros and extrema are equally spaced in the θ variable. The zeros of $T_n(x)$ (see (3.18)) satisfy

$$\theta_k - \theta_{k-1} = |\cos^{-1}(x_k) - \cos^{-1}(x_{k-1})| = \pi/n, \qquad (3.41)$$

and similarly for the extrema.

This is not the only case of Jacobi polynomials with equally spaced zeros (in the θ variable), but it is the only case with both zeros and extrema equispaced. Indeed, considering the Liouville–Green transformation (see §2.2.4) with the change of variable $x = \cos\theta$, we can prove that

$$u_n^{(\alpha,\beta)}(\theta) = \left(\sin\tfrac{1}{2}\theta\right)^{\alpha+1/2} \left(\cos\tfrac{1}{2}\theta\right)^{\beta+1/2} P_n^{(\alpha,\beta)}(\cos\theta), \quad 0 \leq \theta \leq \pi, \qquad (3.42)$$

satisfies the differential equation

$$\frac{d^2 u_n^{(\alpha,\beta)}(\theta)}{d\theta^2} + \Omega(\theta) u_n^{(\alpha,\beta)}(\theta) = 0,$$
$$\Omega(\theta) = \frac{1}{4}\left[(2n+\alpha+\beta+1)^2 + \frac{\tfrac{1}{4}-\alpha^2}{\sin^2\tfrac{1}{2}\theta} + \frac{\tfrac{1}{4}-\beta^2}{\cos^2\tfrac{1}{2}\theta}\right]. \qquad (3.43)$$

From this, we observe that for the values $|\alpha| = |\beta| = \tfrac{1}{2}$, and only for these values, $\Omega(\theta)$ is constant and therefore the solutions are trigonometric functions

$$u_n^{(\alpha,\beta)} = C^{(\alpha,\beta)} \cos(\theta w_n^{(\alpha,\beta)} + \phi^{(\alpha,\beta)}), \quad w_n^{(\alpha,\beta)} = n + (\alpha+\beta+1)/2 \qquad (3.44)$$

with $C^{(\alpha,\beta)}$ and $\phi^{(\alpha,\beta)}$ values not depending on θ. The solutions $u_n^{(\alpha,\beta)}$, $|\alpha| = |\beta| = \tfrac{1}{2}$ have therefore equidistant zeros and extrema. The distance between zeros is

$$\theta_k - \theta_{k-1} = \frac{\pi}{n + (\alpha+\beta+1)/2}. \qquad (3.45)$$

Jacobi polynomials have the same zeros as the solutions $u_n^{(\alpha,\beta)}$ (except that $\theta = 0, \pi$ may also be zeros for the latter). Therefore, Jacobi polynomials have equidistant zeros for $|\alpha| = |\beta| = \tfrac{1}{2}$. However, due to the sine and cosine factors in (3.42), the extrema of Jacobi polynomials are only equispaced when $\alpha = \beta = -\tfrac{1}{2}$.

The four types of Chebyshev polynomials are the only classical orthogonal (hypergeometric) polynomials for which the elementary change of variables $x = \cos\theta$ makes all zeros equidistant. Furthermore, these are the only possible cases for which equidistance takes place, not only in the θ variable but also under more general changes of variable (also including confluent cases) [52, 53].

Chebyshev polynomials are proportional to the Jacobi polynomials with equispaced θ zeros. From (3.42) such Chebyshev polynomials can be written as

$$T_n^{\alpha,\beta}(\theta) = C^{(\alpha,\beta)} \frac{\cos(\theta w_n^{(\alpha,\beta)} + \phi^{(\alpha,\beta)})}{\left(\sin\tfrac{1}{2}\theta\right)^{\alpha+1/2} \left(\cos\tfrac{1}{2}\theta\right)^{\beta+1/2}}. \qquad (3.46)$$

$C^{(\alpha,\beta)}$ can be arbitrarily chosen and it is customary to take $C^{(\alpha,\beta)} = 1$, except when $\alpha = \beta = \frac{1}{2}$, in which case $C^{(\alpha,\beta)} = \frac{1}{2}$. On the other hand, for each selection of α and β (with $|\alpha| = |\beta| = \frac{1}{2}$) there is only one possible selection of $\phi^{(\alpha,\beta)}$ in $[0, \pi)$ which gives a polynomial solution. This phase is easily selected by requiring that $T_n^{\alpha,\beta}(\theta)$ be finite as $\theta \to 0, \pi$. With the standard normalization considered, the four families of polynomials $T_n^{\alpha,\beta}(\theta)$ (proportional to $P_n^{(\alpha,\beta)}$) can be written as

$$T_n^{(-1/2,-1/2)}(\theta) = \cos(n\theta) = T_n(x),$$

$$T_n^{(1/2,1/2)}(\theta) = \frac{\sin((n+1)\theta)}{\sin\theta} = U_n(x),$$

$$T_n^{(-1/2,1/2)}(\theta) = \frac{\cos((n+\frac{1}{2})\theta)}{\cos(\frac{1}{2}\theta)} = V_n(x), \qquad (3.47)$$

$$T_n^{(1/2,-1/2)}(\theta) = \frac{\sin((n+\frac{1}{2})\theta)}{\sin(\frac{1}{2}\theta)} = W_n(x).$$

These are the Chebyshev polynomials of first (T), second (U), third (V), and fourth (W) kinds. The third- and fourth-kind polynomials are trivially related because $P_n^{(\alpha,\beta)}(x) = (-1)^n P_n^{(\beta,\alpha)}(-x)$.

Particularly useful for some applications are Chebyshev polynomials of the second kind. The zeros of $U_n(x)$ plus the nodes $x = -1, 1$ (that is, the x zeros of $u_n^{(1/2,1/2)}(\theta(x))$) are the nodes of the Clenshaw–Curtis quadrature rule (see §9.6.2). All Chebyshev polynomials satisfy three-term recurrence relations, as is the case for any family of orthogonal polynomials; in particular, the Chebyshev polynomials of the second kind satisfy the same recurrence as the polynomials of the first kind. See [2] or [148] for further properties.

3.4 Chebyshev interpolation

Because the scaled Chebyshev polynomial $\hat{T}_{n+1}(x) = 2^{-n} T_{n+1}(x)$ is the monic polynomial of degree $n + 1$ with the smallest maximum absolute value in $[-1, 1]$ (Theorem 3.6), the selection of its n zeros for Lagrange interpolation leads to interpolating polynomials for which the Runge phenomenon is absent.

By considering the estimation for the Lagrange interpolation error (3.6) under the condition of Theorem 3.2, taking as interpolation nodes the zeros of $T_{n+1}(x)$,

$$x_k = \cos\left(\left(k + \frac{1}{2}\right)\frac{\pi}{n+1}\right), \quad k = 0, \ldots, n, \qquad (3.48)$$

and considering the minimax property of the nodal polynomial $\hat{T}_{n+1}(x)$ (Theorem 3.6), the following error bound can be obtained:

$$|R_n(x)| = \frac{|f^{(n+1)}(\zeta_x)|}{(n+1)!}|\hat{T}_{n+1}(x)| \le 2^{-n}\frac{|f^{(n+1)}(\zeta_x)|}{(n+1)!} \le \frac{1}{2^n(n+1)!}\|f^{(n+1)}\|, \qquad (3.49)$$

3.4. Chebyshev interpolation

where $\|f^{(n+1)}\| = \max_{x \in [-1,1]} |f^{(n+1)}(x)|$. By considering a linear change of variables (3.14), an analogous result can be given for Chebyshev interpolation in an interval $[a, b]$.

Interpolation with Chebyshev nodes is not as good as the best approximation (Definition 3.1), but usually it is the best practical possibility for interpolation and certainly much better than equispaced interpolation. The best polynomial approximation is characterized by the Chebyshev equioscillation theorem.

Theorem 3.7 (Chebyshev equioscillation theorem). *For any continuous function f in $[a, b]$, a unique minimax polynomial approximation in $\mathbb{P}_n$ (the space of the polynomials of degree n at most) exists and is uniquely characterized by the alternating or equioscillation property that there are at least $n + 2$ points at which $f(x) - P_n(x)$ attains its maximum absolute value, with alternating signs.*

Proof. Proofs of this theorem can be found, for instance, in [48, 189]. $\square$

Because the function $f(x) - P_n(x)$ alternates signs between each two consecutive extrema, it has at least $n + 1$ zeros; therefore P_n is a Lagrange interpolating polynomial, interpolating f at $n + 1$ points in $[a, b]$. The specific location of these points depends on the particular function f, which makes the computation of best approximations difficult in general.

Chebyshev interpolation by a polynomial in $\mathbb{P}_n$, interpolating the function f at the $n + 1$ zeros of $T_{n+1}(x)$, can be a reasonable approximation and can be computed in an effective and stable way. Given the properties of the error for Chebyshev interpolation on $[-1, 1]$ and the uniformity in the deviation of Chebyshev polynomials with respect to zero (Theorem 3.6), one can expect that Chebyshev interpolation gives a fair approximation to the minimax approximation when the variation of f is soft. In addition, the Runge phenomenon does not occur.

Uniform convergence (in the sense of (3.12)) does not necessarily hold but, in fact, there is no system of preassigned nodes that can guarantee uniform convergence for any continuous function f (see [189, Thm. 4.3]). The sequence of best uniform approximations p_n for a given continuous function f does uniformly converge. For the Chebyshev interpolation we need to consider some additional "level of continuity" in the form of the *modulus of continuity*.

Definition 3.8. *Let f be a function defined in an interval $[a, b]$. We define the modulus of continuity as*

$$\omega(\delta) = \sup_{\substack{x_1, x_2 \in [a,b] \\ |x_1 - x_2| < \delta}} |f(x_1) - f(x_2)|. \quad (3.50)$$

With this definition, it is easy to see that continuity is equivalent to $\omega(\delta) \to 0$ as $\delta \to 0$, while differentiability is equivalent to $\omega(\delta) = \mathcal{O}(\delta)$.

Theorem 3.9 (Jackson's theorem). *The sequence of best polynomial approximations $B_n(f) \in \mathbb{P}_n$ to a function f, continuous on $[-1, 1]$, satisfies*

$$\|f - B_n(f)\| \leq K\omega(1/n), \quad (3.51)$$

K being a constant.

Proof. For the proof see [189, Chap. 1]. □

This result means that the sequence of best approximations converges uniformly for continuous functions. The situation is not so favorable for Chebyshev interpolation.

Theorem 3.10. *Let $P_n \in \mathbb{P}_n$ be the Chebyshev interpolation polynomial for f at $n+1$ points. Then*

$$\|f - P_n\| \leq M(n), \quad \text{with} \quad M(n) \sim C\omega(1/n)\log n, \tag{3.52}$$

as $n \to \infty$, C being a constant.

Proof. For the proof see [189, Chap. 4]. □

The previous theorem shows that continuity is not enough and that the condition $\log(\delta)\omega(\delta) \to 0$ as $\delta \to 0$ is required. This is more demanding than continuity but less demanding than differentiability. When such a condition is satisfied for a function f it is said that the function is *Dini–Lipschitz continuous*.

3.4.1 Computing the Chebyshev interpolation polynomial

Using the orthogonality properties of Chebyshev polynomials, one can compute the Chebyshev interpolation polynomials in an efficient way.

First, we note that, because of the orthogonality relation (3.29), which we abbreviate as $\langle T_r, T_s \rangle = N_r \delta_{rs}$, the set $\{T_k\}_{k=0}^n$ is a set of linearly independent polynomials; therefore, $\{T_k\}_{k=0}^n$ is a base of the linear vector space $\mathbb{P}_n$.

Now, given the polynomial $P_n \in \mathbb{P}_n$ that interpolates f at the $n+1$ zeros of $T_{n+1}(x)$, because $\{T_k\}_{k=0}^n$ is a base we can write P_n as a combination of this base, that is,

$$P_n(x) = \sum_{k=0}^n {}'c_k T_k(x), \tag{3.53}$$

where the prime indicates that the first term is to be halved (which is convenient for obtaining a simple formula for all the coefficients c_k). For computing the coefficients, we use the discrete orthogonality relation (3.30). Because P_n interpolates f at the $n+1$ Chebyshev nodes, we have at these nodes $f(x_k) = P_n(x_k)$. Hence,

$$\sum_{j=0}^n f(x_j)T_k(x_j) = \sum_{i=0}^n {}'c_i \sum_{j=0}^n T_i(x_j)T_k(x_j) = \sum_{i=0}^n {}'c_i K_i \delta_{ik} = \tfrac{1}{2}(n+1)c_k. \tag{3.54}$$

Therefore, the coefficients in (3.53) can be computed by means of the formula

$$c_k = \frac{2}{n+1}\sum_{j=0}^n f(x_j)T_k(x_j), \quad x_j = \cos\left(\left(j+\tfrac{1}{2}\right)\pi/(n+1)\right). \tag{3.55}$$

3.4. Chebyshev interpolation

This type of Chebyshev sum can be efficiently computed in a numerically stable way by means of Clenshaw's method discussed in §3.7. The coefficients can also be written in the form

$$c_k = \frac{2}{n+1} \sum_{j=0}^{n} f(\cos\theta_j) \cos(k\theta_j), \quad \theta_j = \left(j + \tfrac{1}{2}\right) \pi/(n+1), \tag{3.56}$$

which, apart from the factor $2/(n+1)$, is a discrete cosine transform (named DCT-II or simply DCT) of the vector $f(\cos\theta_j)$, $j = 0, \ldots, n$.

Interpolation by orthogonal polynomials

The method used previously for computing interpolation polynomials can be used for building other interpolation formulas. All that is required is that we use a set of orthogonal polynomials $\{p_n\}$, satisfying

$$\int_a^b p_n(x) p_m(x) w(x) \, dx = M_n \delta_{nm}, \tag{3.57}$$

where $M_n \neq 0$ for all n, for a suitable weight function $w(x)$ on $[a, b]$ (nonnegative and continuous on (a, b)) and satisfying a discrete orthogonality relation over the interpolation nodes x_k of the form

$$\sum_{j=0}^{n} w_{j,r} p_r(x_j) p_s(x_j) = \delta_{rs}, \quad r, s \leq n. \tag{3.58}$$

When this is satisfied,[1] it is easy to check, by proceeding as before, that the polynomial interpolating a function f at the nodes x_k, $k = 0, \ldots, n$, can be written as

$$P_n(x) = \sum_{j=0}^{n} a_j p_j(x), \quad a_j = \sum_{k=0}^{n} w_{k,j} f(x_k) p_j(x_k). \tag{3.59}$$

In addition, the coefficients can be computed by using a Clenshaw scheme, similar to Algorithm 3.1.

Chebyshev interpolation of the second kind

For later use, we consider a different type of interpolation, based on the nodes $x_k = \cos(k\pi/n)$, $k = 0, \ldots, n$. These are the zeros of $U_{n-1}(x)$ complemented with $x_0 = 1$, $x_n = -1$ (that is, the zeros of $u_n^{(1/2,1/2)}(\cos^{-1} x)$; see (3.42)). Also, these zeros are the extrema of $T_n(x)$.

We write this interpolation polynomial as

$$P_n(x) = \sum_{k=0}^{n} {}'' c_k T_k(x), \tag{3.60}$$

---

[1] In Chapter 5, it is shown that this type of relation always exists when the x_k are chosen to be the zeros of p_{n+1} and $w_{k,j} = w_k$ are the weights of the corresponding Gaussian quadrature rule.

and considering the second discrete orthogonality property (3.31), we have

$$c_k = \frac{2}{n} \sum_{j=0}^{n}{}'' f(x_j) T_k(x_j), \quad x_j = \cos(j\pi/n), \quad j = 0, \ldots, n. \quad (3.61)$$

This can also be written as

$$c_k = \frac{2}{n} \sum_{j=0}^{n}{}'' f(\cos(j\pi/n)) \cos(kj\pi/n), \quad (3.62)$$

which is a discrete cosine transform (named DCT-I) of the vector $f(\cos(j\pi/n))$, $j = 0, \ldots, n$.

3.5 Expansions in terms of Chebyshev polynomials

Under certain conditions of the interpolated function f (Dini–Lipschitz continuity), Chebyshev interpolation converges when the number of nodes tends to infinity. This leads to a representation of f in terms of an infinite series of Chebyshev polynomials.

More generally, considering a set of orthogonal polynomials $\{p_n\}$ (see (3.57)) and a continuous function in the interval of orthogonality $[a, b]$, one can consider series of orthogonal polynomials

$$f(x) = \sum_{k=0}^{\infty} c_k p_k(x). \quad (3.63)$$

Taking into account the orthogonality relation (3.57), we have

$$c_k = \frac{1}{M_k} \int_a^b f(x) p_k(x) w(x) \, dx. \quad (3.64)$$

Proofs of the convergence for this type of expansion for some classical cases (Legendre, Hermite, Laguerre) can be found in [134]. Apart from continuity and differentiability conditions, it is required that

$$\int_a^b f(x)^2 w(x) \, dx \quad (3.65)$$

be finite. Expansions of this type are called *generalized Fourier series*. The base functions $\{p_n\}$ can be polynomials or other suitable orthogonal functions.

Many examples exist of the use of this type of expansion in the solution of problems of mathematical physics (see, for instance, [134]). For the sake of uniform approximation, Chebyshev series based on the Chebyshev polynomials of the first kind are the most useful ones and have faster uniform convergence [5]. For convenience, we write the Chebyshev series as

$$f(x) = \sum_{k=0}^{\infty}{}' c_k T_k(x) = \tfrac{1}{2} c_0 + \sum_{k=1}^{\infty} c_k T_k(x), \quad -1 \leq x \leq 1. \quad (3.66)$$

With this, and taking into account the orthogonality relation (3.29),

$$c_k = \frac{2}{\pi} \int_{-1}^{1} \frac{f(x) T_n(x)}{\sqrt{1-x^2}} dx = \frac{2}{\pi} \int_0^{\pi} f(\cos\theta) \cos(k\theta) \, d\theta. \quad (3.67)$$

3.5. Expansions in terms of Chebyshev polynomials

For computing the coefficients, one needs to compute the cosine transform of (3.67). For this purpose, fast algorithms can be used for computing fast cosine transforms. A discretization of (3.67) using the trapezoidal rule (Chapter 5) in $[0, \pi]$ yields

$$c_k \approx \frac{2}{n} \sum_{j=0}^{n}{}'' f\left(\cos \frac{\pi j}{n}\right) \cos \frac{\pi k j}{n}, \tag{3.68}$$

which is a discrete cosine transform. Notice that, when considering this approximation, and truncating the series at $k = n$ but halving the last term, we have the interpolation polynomial of the second kind of (3.55).

Another possible discretization of the coefficients c_k is given by (3.60). With this discretization, and truncating the series at $k = n$, we obtain the interpolation polynomial of the first kind of degree n.

Chebyshev interpolation can be interpreted as an approximation to Chebyshev series (or vice versa), provided that the coefficients decay fast and the discretization is accurate. In other words, Chebyshev series can be a good approximation to near minimax approximations (Chebyshev), which in turn are close to minimax approximations.

On the other hand, provided that the coefficients c_k decrease in magnitude sufficiently rapidly, the error made by truncating the Chebyshev expansion after the terms $k = n$, that is,

$$E_n(x) = \sum_{k=n+1}^{\infty} c_k T_k(x), \tag{3.69}$$

will be given approximately by

$$E_n(x) \approx c_{n+1} T_{n+1}(x), \tag{3.70}$$

that is, the error approximately satisfies the equioscillation property (Theorem 3.7).

How fast the coefficients c_k decrease depends on continuity and differentiability properties of the function to be expanded. The more regular these are, the faster the coefficients decrease (see the next section).

Example 3.11 (the Chebyshev expansion of arccos x). Let us consider the Chebyshev expansion of $f(x) = \arccos x$; $f(x)$ is continuous in $[-1, 1]$ but is not differentiable at $x = \pm 1$. Observing this, we can expect a noticeable departure from the equioscillation property, as we will see.

For this case, the coefficients can be given in explicit form. From (3.67) we obtain $c_0 = \pi$ and for $k \geq 1$,

$$\begin{aligned}
c_k &= \frac{2}{\pi} \int_0^\pi \theta \cos k\theta \, d\theta \\
&= \frac{2}{\pi} \left\{ \left[\frac{\theta \sin k\theta}{k}\right]_0^\pi - \int_0^\pi \frac{\sin k\theta}{k} d\theta \right\} \\
&= \frac{2}{\pi} \left\{ \left[\frac{\theta \sin k\theta}{k} + \frac{\cos k\theta}{k^2}\right]_0^\pi \right\} \\
&= \frac{2}{\pi} \frac{(-1)^k - 1}{k^2},
\end{aligned} \tag{3.71}$$

from which it follows that

$$c_{2k} = 0, \quad c_{2k-1} = -\frac{2}{\pi}\frac{2}{(2k-1)^2}. \tag{3.72}$$

We conclude that the resulting Chebyshev expansion of $f(x) = \arccos x$ is

$$\arccos x = \frac{\pi}{2}T_0(x) - \frac{4}{\pi}\sum_{k=1}^{\infty}\frac{T_{2k-1}(x)}{(2k-1)^2}. \tag{3.73}$$

This corresponds with the Fourier expansion

$$|t| - \frac{\pi}{2} = -\frac{4}{\pi}\sum_{k=1}^{\infty}\frac{\cos(2k-1)t}{(2k-1)^2}, \quad t \in [-\pi, \pi]. \tag{3.74}$$

The absolute error when the series is truncated after the term $k = 5$ is shown in Figure 3.4. Notice the departure from equioscillation close to the endpoints $x = \pm 1$, where the function is not differentiable. ∎

In the preceding example the coefficients c_k of the Chebyshev expansion can be obtained analytically. Unfortunately, this situation represents an exception and numerical methods have to be applied in order to obtain the coefficients c_k (see §3.6). In a later section we give examples of Chebyshev expansions with explicit coefficients for some special functions (see §3.10).

3.5.1 Convergence properties of Chebyshev expansions

The rate of convergence of the series in (3.74) is comparable with that of the series $\sum_{k=1}^{\infty} 1/k^2$, which is not very impressive. The bad convergence is caused by the analytic property of this function: $\arccos x$ is not differentiable at the endpoints ± 1 of the interval.

The useful applications of Chebyshev expansions arise when the expansion converges much faster. We give two theorems, the proof of which can be found in [148, §5.7]. We consider expansions of the form (3.66) with partial sum denoted by

$$S_n(x) = \frac{1}{2}c_0 + \sum_{k=1}^{n} c_k T_k(x). \tag{3.75}$$

Theorem 3.12 (functions with continuous derivatives). *When a function f has $m + 1$ continuous derivatives on $[-1, 1]$, where m is a finite number, then $|f(x) - S_n(x)| = \mathcal{O}(n^{-m})$ as $n \to \infty$ for all $x \in [-1, 1]$.*

Theorem 3.13 (analytic functions inside an ellipse). *When a function f on $x \in [-1, 1]$ can be extended to a function that is analytic inside an ellipse E_r defined by*

$$E_r = \left\{z : \left|z + \sqrt{z^2 - 1}\right| = r\right\}, \quad r > 1, \tag{3.76}$$

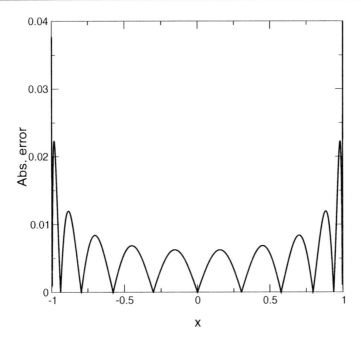

Figure 3.4. *Error after truncating the series in* (3.73) *after the term* $k = 5$.

then $|f(x) - S_n(x)| = \mathcal{O}(r^{-n})$ as $n \to \infty$ for all $x \in [-1, 1]$.

The ellipse E_r has semiaxis of length $(r + 1/r)/2$ on the real z-axis and of length $(r - 1/r)/2$ on the imaginary axis.

For entire functions f we can take any number r in this theorem, and in fact the rate of convergence can be of order $\mathcal{O}(1/n!)$. For example, we have the generating function for the modified Bessel coefficients $I_n(z)$ given by

$$e^{zx} = I_0(z)T_0(x) + 2\sum_{n=1}^{\infty} I_n(z)T_n(x), \quad -1 \le x \le 1, \tag{3.77}$$

where z can be any complex number. The Bessel functions behave like $I_n(z) = \mathcal{O}((z/2)^n/n!)$ as $n \to \infty$ with z fixed, and the error $|e^{zx} - S_n(x)|$ has a similar behavior. The absolute error when the series is truncated after the $n = 5$ term is shown in Figure 3.5.

3.6 Computing the coefficients of a Chebyshev expansion

In general, the Chebyshev coefficients of the Chebyshev expansion of a function f can be approximately obtained by the numerical computation of the integral of (3.67). To improve the speed of computation, fast Fourier cosine transform algorithms for evaluating the sums in (3.68) can be considered. For numerical aspects of the fast Fourier transform we refer the reader to [226].

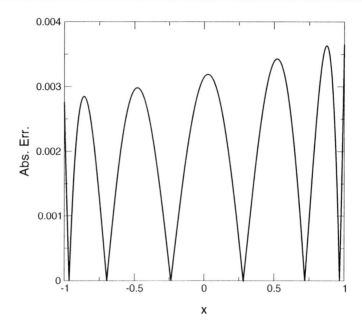

Figure 3.5. *Error after truncating the series for e^{2x} in (3.77) after the term $n = 5$. Compare with Figure 3.4.*

In the particular case when the function f is a solution of an ordinary linear differential equation with polynomial coefficients, Clenshaw [37] proposed an alternative method, which we will now discuss.

3.6.1 Clenshaw's method for solutions of linear differential equations with polynomial coefficients

The method works as follows.

Let us assume that f satisfies a linear differential equation in the variable x with polynomial coefficients $p_k(x)$,

$$\sum_{k=0}^{m} p_k(x) f^{(k)}(x) = h(x), \qquad (3.78)$$

and where the coefficients of the Chebyshev expansion of the function h are known. In general, conditions on the solution f will be given at $x = 0$ or $x = \pm 1$.

Let us express formally the sth derivative of f as follows:

$$f^{(s)}(x) = \tfrac{1}{2} c_0^{(s)} + c_1^{(s)} T_1(x) + c_2^{(s)} T_2(x) + \cdots. \qquad (3.79)$$

Then the following expression can be obtained for the coefficients:

$$2r c_r^{(s)} = c_{r-1}^{(s+1)} - c_{r+1}^{(s+1)}, \quad r \geq 1. \qquad (3.80)$$

3.6. Computing the coefficients of a Chebyshev expansion

To see how to arrive to this equation, let us start with

$$f'(x) = \tfrac{1}{2}c_0^{(1)} + c_1^{(1)}T_1(x) + c_2^{(1)}T_2(x) + \cdots + c_{n-1}^{(1)}T_{n-1}(x)$$
$$+ c_n^{(1)}T_n(x) + c_{n+1}^{(1)}T_{n+1}(x) + \cdots \quad (3.81)$$

and integrate this expression. Using the relations in (3.25), we obtain

$$\begin{aligned}
f(x) = &\frac{1}{2}c_0 + \frac{1}{2}c_0^{(1)}T_1(x) + \frac{1}{4}c_1^{(1)}T_2(x) + \cdots \\
&+ \frac{1}{2}c_{n-1}^{(1)}\left(\frac{T_n(x)}{n} - \frac{T_{n-2}(x)}{n-2}\right) \\
&+ \frac{1}{2}c_n^{(1)}\left(\frac{T_{n+1}(x)}{n+1} - \frac{T_{n-1}(x)}{n-1}\right) \\
&+ \frac{1}{2}c_{n+1}^{(1)}\left(\frac{T_{n+2}(x)}{n+2} - \frac{T_n(x)}{n}\right) + \cdots.
\end{aligned} \quad (3.82)$$

Comparing the coefficients of the Chebyshev polynomials in this expression and the Chebyshev expansion of f, we arrive at (3.80) for $s = 1$. Observe that a relation for c_0 is not obtained in this way. Substituting in (3.82) given values of f at, say, $x = 0$ gives a relation between c_0 and an infinite number of coefficients $c_n^{(1)}$.

A next element in Clenshaw's method is using (3.28) to handle the powers of x occurring in the differential equation satisfied by f. Denoting the coefficients of $T_r(x)$ in the expansion of $g(x)$ by $C_r(g)$ when $r > 0$ and twice this coefficient when $r = 0$, and using (3.28), we infer that

$$C_r\left(xf^{(s)}\right) = \tfrac{1}{2}\left(c_{r+1}^{(s)} + c_{|r-1|}^{(s)}\right). \quad (3.83)$$

This expression can be generalized as follows:

$$C_r\left(x^p f^{(s)}\right) = \frac{1}{2^p}\sum_{j=0}^{p}\binom{p}{j}c_{|r-p+2j|}^{(s)}. \quad (3.84)$$

When the expansion (3.79) is substituted into the differential equation (3.78) together with (3.80), (3.84), and the associated boundary conditions, it is possible to obtain an infinite set of linear equations for the coefficients $c_r^{(s)}$. Two strategies can be used for solving these equations.

Recurrence method. The equations can be solved by recurrence for $r = N-1, N-2, \ldots, 0$, where N is an arbitrary (large) positive integer, by assuming that $c_r^{(s)} = 0$ for $r > N$ and by assigning arbitrary values to $c_N^{(s)}$. This is done as follows.

Consider $r = N$ in (3.80) and compute $c_{N-1}^{(s)}$, $s = 1, \ldots, m$. Then, considering (3.84) and the differential equation (3.78), select r appropriately in order to compute $c_{N-1}^{(0)} = c_{N-1}$. We repeat the process by considering $r = N-1$ in (3.80) and computing $c_{N-2}^{(s)}$, etc. Obviously and unfortunately, the computed coefficients c_r will not satisfy, in general, the boundary conditions, and we will have to take care of these in each particular case.

Iterative method. The starting point in this case is an initial guess for c_r which satisfies the boundary conditions. Using these values, we use (3.80) to obtain the values of $c_r^{(s)}$, $s = 1, \ldots, m$, and then the relation (3.84) and the differential equation (3.78) to compute corrected values of c_r.

The method based on recursions is, quite often, more rapidly convergent than the iterative method; therefore, and in general, the iterative method could be useful for correcting the rounding errors arising in the application of the method based on recursions.

Example 3.14 (Clenshaw's method for the J-Bessel function). Let us consider, as a simple example (due to Clenshaw), the computation of the Bessel function $J_0(t)$ in the range $0 \leq t \leq 4$. This corresponds to solving the differential equation for $J_0(4x)$, that is,

$$xy'' + y' + 16xy = 0 \tag{3.85}$$

in the range $0 \leq x \leq 1$ with conditions $y(0) = 1$, $y'(0) = 0$. This is equivalent to solving the differential equation in $[-1, 1]$, because $J_0(x) = J_0(-x)$, $x \in \mathbb{R}$.

Because $J_0(4x)$ is an even function of x, the $T_r(x)$ of odd order do not appear in its Chebyshev expansion. By substituting the Chebyshev expansion into the differential equation, we obtain

$$C_r(xy'') + C_r(y') + 16 C_r(xy) = 0, \quad r = 1, 3, 5, \ldots, \tag{3.86}$$

and considering (3.84),

$$\tfrac{1}{2}\left(c''_{r-1} + c''_{r+1}\right) + c'_r + 8\left(c_{r-1} + c_{r+1}\right) = 0, \quad r = 1, 3, 5, \ldots. \tag{3.87}$$

This equation can be simplified. First, we see that by replacing $r \to r-1$ and $r \to r+1$ in (3.87) and subtracting both expressions, we get

$$\begin{aligned}
&\tfrac{1}{2}\left(c''_{r-2} + c''_r - c''_r - c''_{r+2}\right) + \left(c'_{r-1} - c'_{r+1}\right) \\
&+ 8\left(c_{r-2} + c_r - c_r - c_{r+2}\right) = 0, \quad r = 2, 4, 6, \ldots.
\end{aligned} \tag{3.88}$$

It is convenient to eliminate the terms with the second derivatives. This can be done by using (3.80). In this way,

$$r\left(c'_{r-1} + c'_{r+1}\right) + 8\left(c_{r-2} - c_{r+2}\right) = 0, \quad r = 2, 4, 6, \ldots. \tag{3.89}$$

Now, expressions (3.80) and (3.89) can be used alternatively in the recurrence process, as follows:

$$\left.\begin{aligned}
c'_{r-1} &= c'_{r+1} + 2rc_r \\
c_{r-2} &= c_{r+2} - \tfrac{1}{8}r\left(c'_{r-1} + c'_{r+1}\right)
\end{aligned}\right\} \quad r = N, N-2, N-4, \ldots, 2. \tag{3.90}$$

As an illustration, let us take as first trial coefficient $\tilde{c}_{20} = 1$ and all higher order coefficients zero. Applying the recurrences (3.90) and considering the calculation with 15 significant digits, we obtain the values of the trial coefficients given in Table 3.1.

Using the coefficients in Table 3.1, the trial solution of (3.85) at $x = 0$ is given by

$$\tilde{y}(0) = \tfrac{1}{2}\tilde{c}_0 - \tilde{c}_2 + \tilde{c}_4 - \tilde{c}_6 + \tilde{c}_8 - \cdots = 8050924923505.5, \tag{3.91}$$

3.6. Computing the coefficients of a Chebyshev expansion

Table 3.1. *Computed coefficients in the recurrence processes* (3.90). *We take as starting values* $\tilde{c}_{20} = 1$ *and 15 significant digits in the calculations.*

r	$\tilde{c}_r$	$\tilde{c}'_{r+1}$
0	807138731281	−8316500240280
2	−5355660492900	13106141731320
4	2004549104041	−2930251101008
6	−267715177744	282331031920
8	18609052225	−15413803680
10	−797949504	545186400
12	23280625	−13548600
14	−492804	249912
16	7921	−3560
18	−100	40
20	1	

Table 3.2. *Computed coefficients of the Chebyshev expansion of the solution of* (3.85).

r	c_r
0	$0.1002541619689529 \ 10^{-0}$
2	$-0.6652230077644372 \ 10^{-0}$
4	$0.2489837034982793 \ 10^{-0}$
6	$-0.3325272317002710 \ 10^{-1}$
8	$0.2311417930462743 \ 10^{-2}$
10	$-0.9911277419446611 \ 10^{-4}$
12	$0.2891670860329331 \ 10^{-5}$
14	$-0.6121085523493186 \ 10^{-7}$
16	$0.9838621121498511 \ 10^{-9}$
18	$-0.1242093311639757 \ 10^{-10}$
20	$0.1242093311639757 \ 10^{-12}$

and the final values for the coefficients c_r of the solution $y(x)$ of (3.85) will be obtained by dividing the trial coefficients by $\tilde{y}(0)$. This gives the requested values shown in Table 3.2.

The value of $y(1)$ will then be given by

$$y(1) = \tfrac{1}{2}c_0 + c_2 + c_4 + c_6 + c_8 + \cdots = -0.3971498098638699, \qquad (3.92)$$

the relative error being $0.57 \ 10^{-13}$ when compared with the value of $J_0(4)$. ∎

Remark 2. Several questions arise in this successful method. The recursion given in (3.90) is rather simple, and we can find its exact solution; cf. the expansion of the J_0 in (3.139). In Chapter 4 we explain that in this case the backward recursion scheme for computing the Bessel coefficients is stable. In more complicated recursion schemes this information is not available. The scheme may be of large order and may have several solutions of which the asymptotic behavior is unknown. So, in general, we don't know if Clenshaw's method for differential equations computes the solution that we want, and if for the wanted solution the scheme is stable in the backward direction.

Example 3.15 (Clenshaw's method for the Abramowitz function). Another but not so easy example of the application of Clenshaw's method is provided by MacLeod [146] for the computation of the Abramowitz functions [1],

$$\mathcal{J}_n(x) = \int_0^\infty t^n e^{-t^2 - x/t}\, dt, \quad n \text{ integer}. \tag{3.93}$$

Chebyshev expansions for $\mathcal{J}_1(x)$ for $x \geq 0$ can be obtained by considering the following two cases depending on the range of the argument x.

If $0 \leq x \leq a$,

$$\mathcal{J}_1(x) = f_1(x) - \sqrt{\pi} x g_1(x) - x^2 h_1(x) \log x, \tag{3.94}$$

where f_1, g_1, and h_1 satisfy the system of equations

$$\begin{aligned} x g_1''' + 3 g_1'' + 2 g_1 &= 0, \\ x^2 h_1''' + 6 x h_1'' + 6 h_1' + 2 x h_1 &= 0, \\ x f_1''' + 2 f_1 &= 3 x^2 h_1'' + 9 x h_1' + 2 h_1, \end{aligned} \tag{3.95}$$

with appropriate initial conditions at $x = 0$. The functions f_1, g_1, and h_1 are expanded in a series of the form $\sum_{k=0}^\infty c_k T_k(t)$, where $t = (2x^2/a^2) - 1$.

If $x > a$,

$$\mathcal{J}_1(x) \sim \sqrt{\frac{\pi}{3}} \sqrt{\frac{\nu}{3}} e^{-\nu} q_1(\nu) \tag{3.96}$$

with $\nu = 3(x/2)^{2/3}$. The function $q_1(\nu)$ can be expanded in a Chebyshev series of the variable

$$t = \frac{2B}{\nu} - 1, \quad B = 3\left(\frac{a}{2}\right)^{2/3}, \tag{3.97}$$

and q_1 satisfies the differential equation

$$4\nu^3 q_1''' - 12\nu^3 q_1'' + (12\nu^3 - 5\nu) q_1' + (5\nu + 5) q_1 = 0, \tag{3.98}$$

where the derivatives are taken with respect to ν. The function q_1 is expanded in a series of the form $\sum_{k=0}^\infty c_k T_k(t)$, where t is given in (3.97).

The transition point a is selected in such a way that a and B are exactly represented. Also, the number of terms needed for the evaluation of the Chebyshev expansions for a prescribed accuracy is taken into account.

The differential equations (3.95) and (3.98) are solved by using Clenshaw's method. ∎

3.7 Evaluation of a Chebyshev sum

Frequently one has to evaluate a partial sum of a Chebyshev expansion, that is, a finite series of the form

$$S_N(x) = \tfrac{1}{2}c_0 + \sum_{k=1}^{N} c_k T_k(x). \tag{3.99}$$

Assuming we have already computed the coefficients c_k, $k = 0, \ldots, N$, of the expansion, it would be nice to avoid the explicit computation of the Chebyshev polynomials appearing in (3.99), although they easily follow from the relations (3.18) and (3.19). A first possibility for the computation of this sum is to rewrite the Chebyshev polynomials $T_k(x)$ in terms of powers of x and then use the Horner scheme for the evaluation of the resulting polynomial expression. However, one has to be careful when doing this because for some expansions there is a considerable loss of accuracy due to cancellation effects.

3.7.1 Clenshaw's method for the evaluation of a Chebyshev sum

An alternative and efficient method for evaluating this sum is due to Clenshaw [36]. This scheme of computation, which can also be used for computing partial sums involving other types of polynomials, corresponds to the following algorithm.

ALGORITHM 3.1. Clenshaw's method for a Chebyshev sum.
Input: x; $c_0, c_1, \ldots, c_N$.
Output: $\widetilde{S}_N(x) = \sum_{k=0}^{N} c_k T_k(x)$.

- $b_{N+1} = 0$; $b_N = c_N$.
- DO $r = N - 1, N - 2, \ldots, 1$:

  $b_r = 2x b_{r+1} - b_{r+2} + c_r$.

- $\widetilde{S}_N(x) = x b_1 - b_2 + c_0$.

Let us explain how we arrived at this algorithm. For simplicity, let us first consider the evaluation of

$$\widetilde{S}_N(x) = \sum_{k=0}^{N} c_k T_k(x) = \tfrac{1}{2}c_0 + S_N(x). \tag{3.100}$$

This expression can be written in vector form as follows:

$$\widetilde{S}_N(x) = \mathbf{c}^T \mathbf{t} = (c_0, c_1, \ldots, c_N) \begin{pmatrix} T_0(x) \\ T_1(x) \\ \vdots \\ T_N(x) \end{pmatrix}. \tag{3.101}$$

On the other hand, the three-term recurrence relation satisfied by the Chebyshev polynomials (3.20) can also be written in matrix form,

$$\begin{pmatrix} 1 & & & & & \\ -2x & 1 & & & & \\ 1 & -2x & 1 & & & \\ & 1 & -2x & 1 & & \\ & & \ddots & \ddots & \ddots & \\ & & & 1 & -2x & 1 \end{pmatrix} \begin{pmatrix} T_0(x) \\ T_1(x) \\ T_2(x) \\ T_3(x) \\ \vdots \\ T_N(x) \end{pmatrix} = \begin{pmatrix} 1 \\ -x \\ 0 \\ 0 \\ \vdots \\ 0 \end{pmatrix}, \quad (3.102)$$

or

$$\mathbf{At} = \mathbf{d}, \quad (3.103)$$

where $\mathbf{A}$ is the $(N+1) \times (N+1)$ matrix of the coefficients of the recurrence relation and $\mathbf{d}$ is the right-hand side vector of (3.102).

Let us now consider a vector $\mathbf{b}^T = (b_0, b_1, \ldots, b_N)$ such that

$$\mathbf{b}^T \mathbf{A} = \mathbf{c}^T. \quad (3.104)$$

Then,

$$\widetilde{S}_n = \mathbf{c}^T \mathbf{t} = \mathbf{b}^T \mathbf{A} \mathbf{t} = \mathbf{b}^T \mathbf{d} = b_0 - b_1 x. \quad (3.105)$$

For S_N, we have

$$S_N = \widetilde{S}_N - \frac{1}{2} c_0 = (b_0 - b_1 x) - \frac{1}{2}(b_0 - 2xb_1 + b_2) = \frac{1}{2}(b_0 - b_2). \quad (3.106)$$

The coefficients b_r can be computed using a recurrence relation if (3.104) is interpreted as the corresponding matrix equation for the recurrence relation (and considering $b_{N+1} = b_{N+2} = 0$). In this way,

$$b_r - 2xb_{r+1} + b_{r+2} = c_r, \quad r = 0, 1, \ldots, N. \quad (3.107)$$

The three-term recurrence relation is computed in the backward direction, starting from $r = N$. With this, we arrive at

$$\widetilde{S}_N = xb_1 - b_2 + c_0, \quad (3.108)$$

$$S_N = xb_1 - b_2 + \frac{c_0}{2}. \quad (3.109)$$

Error analysis

The expressions provided by (3.105) or (3.106), together with the use of (3.107), are simple and avoid the explicit computation of the Chebyshev polynomials $T_n(x)$ (with the exception of $T_0(x) = 1$ and $T_1(x) = x$). However, these relations will be really useful if one can be sure that the influence of error propagation when using (3.107) is small. Let us try to quantify this influence by following the error analysis due to Elliott [62].

3.7. Evaluation of a Chebyshev sum

Let us denote by

$$\widehat{Q} = Q + \delta Q \qquad (3.110)$$

an exact quantity Q computed approximately (δQ represents the absolute error in the computation).

From (3.107), we obtain

$$\hat{b}_n = \left[\hat{c}_n + 2\hat{x}\hat{b}_{n+1} - \hat{b}_{n+2} \right] + r_n, \qquad (3.111)$$

where r_n is the roundoff error arising from rounding the quantity inside the brackets. We can rewrite this expression as

$$\hat{b}_n = \left[\hat{c}_n + 2x\hat{b}_{n+1} - \hat{b}_{n+2} \right] + \eta_n + r_n, \qquad (3.112)$$

where, neglecting the terms of order higher than 1 for the errors,

$$\eta_n = 2(\delta x)\hat{b}_{n+1} \approx 2(\delta x)b_{n+1}. \qquad (3.113)$$

From (3.107), it is clear that δb_n satisfies a recurrence relation of the form

$$Y_n - 2xY_{n+1} + Y_{n+2} = \delta c_n + \eta_n + r_n, \qquad (3.114)$$

which is the same recurrence relation (with a different right-hand side) as that satisfied by b_n. It follows that

$$\delta b_0 - \delta b_1 x = \sum_{n=0}^{N} (\delta c_n + \eta_n + r_n) T_n(x). \qquad (3.115)$$

On the other hand, because of (3.105), the computed $\widetilde{S}_N$ will be given by

$$\widehat{\widetilde{S}}_N = \left[\hat{b}_0 - \hat{b}_1 \hat{x} \right] + s, \qquad (3.116)$$

where s is the roundoff error arising from computing the expression inside the brackets. Hence,

$$\widehat{\widetilde{S}}_N = (b_0 - xb_1) + (\delta b_0 - x(\delta b_1)) - b_1(\delta x) + s, \qquad (3.117)$$

and using (3.115) it follows that

$$\delta \widetilde{S}_N = (\delta b_0 - x\delta b_1) - b_1 \delta x + s = \sum_{n=0}^{N} (\delta c_n + \eta_n + r_n) T_n(x) - b_1 \delta x + s. \qquad (3.118)$$

Let us rewrite this expression as

$$\delta \widetilde{S}_N = \sum_{n=0}^{N} (\delta c_n + r_n) T_n(x) + 2\delta x \sum_{n=0}^{N} b_{n+1} T_n(x) - b_1 \delta x + s. \qquad (3.119)$$

At this point, we can use the fact that the b_n coefficients can be written in terms of the Chebyshev polynomials of the second kind $U_n(x)$ as follows:

$$b_n = \sum_{k=n}^{N} c_k U_{k-n}(x). \qquad (3.120)$$

We see that the term $\sum_{n=0}^{N} b_{n+1} T_n(x)$ in (3.119) can be expressed as

$$\begin{aligned}\sum_{n=0}^{N} b_{n+1} T_n(x) &= \sum_{n=1}^{N} b_n T_{n-1}(x) = \sum_{n=1}^{N} \left(\sum_{k=n}^{N} c_k U_{k-n}(x) \right) T_{n-1}(x) \\ &= \sum_{k=1}^{N} c_k \left(\sum_{n=1}^{k} U_{k-n}(x) T_{n-1}(x) \right) \\ &= \tfrac{1}{2} \sum_{k=1}^{N} c_k (k+1) U_{k-1}(x),\end{aligned} \qquad (3.121)$$

where, in the last step, we have used

$$\sum_{n=1}^{k} \sin(k-n+1)\theta \cos(n-1)\theta = \tfrac{1}{2}(k+1)\sin k\theta. \qquad (3.122)$$

Substituting (3.121) in (3.119) it follows that

$$\delta \widetilde{S}_N = \sum_{n=0}^{N} (\delta c_n + r_n) T_n(x) + \delta x \sum_{n=1}^{N} n c_n U_{n-1}(x) + s. \qquad (3.123)$$

Since $|T_n(x)| \leq 1$, $|U_{n-1}(x)| \leq n$, and assuming that the local errors $|\delta c_n|$, $|r_n|$, $|\delta x|$, $|s|$ are quantities that are smaller than a given $\epsilon' > 0$, we have

$$\left|\delta \widetilde{S}_N\right| \leq \epsilon' \left((2N+3) + \sum_{n=1}^{N} n^2 |c_n| \right). \qquad (3.124)$$

In the case of a Chebyshev series where the coefficients are slowly convergent, the second term on the right-hand side of (3.124) can provide a significant contribution to the error.

On the other hand, when x is close to ± 1, there is a risk of a growth of rounding errors, and, in this case, a modification of Clenshaw's method [164] seems to be more appropriate. We describe this modification in the following algorithm.

ALGORITHM 3.2. Modified Clenshaw's method for a Chebyshev sum.
Input: $x; c_0, c_1, \ldots, c_N$.
Output: $\widetilde{S}_N(x) = \sum_{k=0}^{N} c_k T_k(x)$.

- IF ($x \approx \pm 1$) THEN

3.7. Evaluation of a Chebyshev sum

- $b_N = c_N$; $d_N = b_N$.
- IF $(x \approx 1)$ THEN
  DO $r = N - 1, N - 2, \ldots, 1$:
  $d_r = 2(x - 1)b_{r+1} + d_{r+1} + c_r$.
  $b_r = d_r + b_{r+1}$.
- ELSEIF $(x \approx -1)$ THEN
  DO $r = N - 1, N - 2, \ldots, 1$:
  $d_r = 2(x + 1)b_{r+1} - d_{r+1} + c_r$.
  $b_r = d_r - b_{r+1}$.
- ENDIF

- ELSE

  Use Algorithm 3.1.

- $\widetilde{S}_N(x) = xb_1 - b_2 + c_0$.

Oliver [165] has given a detailed analysis of Clenshaw's method for evaluating a Chebyshev sum, also by comparing it with other polynomial evaluation schemes for the evaluation of (3.100); in addition, error bounds are derived. Let us consider two expressions for (3.100),

$$\widetilde{S}_N(x) = \sum_{n=0}^{N} c_n T_n(x), \qquad (3.125)$$

$$\widetilde{S}_N(x) = \sum_{n=0}^{N} d_n x^n, \qquad (3.126)$$

and let $\hat{S}_N(x)$ be the actually computed quantity, assuming that errors are introduced at each stage of the computation process of $\widetilde{S}_N(x)$ using (3.125) (considering Clenshaw's method) or (3.126) (considering Horner's scheme). Then,

$$\left|\hat{S}_N(x) - S_N(x)\right| \leq \epsilon \sum_{n=0}^{N} \rho_n(x)|c_n| \qquad (3.127)$$

or

$$\left|\hat{S}_N(x) - S_N(x)\right| \leq \epsilon \sum_{n=0}^{N} \sigma_n(x)|d_n|, \qquad (3.128)$$

depending on the choice of the polynomial expression; ϵ is the accuracy parameter and ρ_n and σ_n are error amplification factors. Reference [165] analyzed the variation of these factors with x. Two conclusions of this study were that

- the accuracy of the methods of Clenshaw and Horner are sensitive to values of x and the errors tend to reach their extreme at the endpoints of the interval;
- when a polynomial has coefficients of constant sign or strictly alternating sign, converting into the Chebyshev form does not improve upon the accuracy of the evaluation.

3.8 Economization of power series

Chebyshev polynomials also play a key role in the so-called economization of power series. Suppose we have at our disposal a convergent Maclaurin series expansion for the evaluation of a function $f(x)$ in the interval $[-1, 1]$. Then, a plausible approximation to $f(x)$ may be the polynomial $p_n(x)$ of degree n, which is obtained by truncating the power series after $n + 1$ terms. It may be possible, however, to obtain a "better" nth-degree polynomial approximation. This is the idea of economization: it involves finding an alternative representation for the function containing $n + 1$ parameters that possesses the same functional form as the initial approximant. This alternative representation also incorporates information present in the higher orders of the original power series to minimize the maximum error of the new approximant over the range of x.

Let $\mathcal{S}_{N+1}$ denote

$$\mathcal{S}_{N+1} = \sum_{i=0}^{N+1} a_i x^i, \tag{3.129}$$

the original Maclaurin series for $f(x)$ truncated at order $N + 1$. Then, one can obtain an "economic" representation by subtracting from $\mathcal{S}_{N+1}$ a suitable polynomial $\mathcal{P}_{N+1}$ of the same order such that the leading orders cancel. That is,

$$\mathcal{C}_N = \mathcal{S}_{N+1} - \mathcal{P}_{N+1} = \sum_{i=0}^{N} a'_i x^i, \tag{3.130}$$

where a'_i denotes the resulting expansion coefficient of x^i. Obviously, the idea is to choose $\mathcal{P}_{N+1}$ in such a way that the maximum error of the new Nth order series representation is significantly reduced. Then, an optimal candidate is

$$\mathcal{P}_{N+1} = a_{N+1} \frac{T_{N+1}}{2^N}. \tag{3.131}$$

The maximum error of this new Nth order polynomial $\mathcal{C}_N$ is nearly the same as the maximum error of the $(N + 1)$th order polynomial $\mathcal{S}_{N+1}$ and considerably less than $\mathcal{S}_N$.

Of course, this procedure may be adapted to ranges different from $[-1, 1]$ by using Chebyshev polynomials adjusted to the required range. For example, the Chebyshev polynomials $T_k(x/c)$ (or shifted Chebyshev polynomials $T_k^*(x/c)$) should be used in the range $[-c, c]$ (or $[0, c]$).

3.9 Example: Computation of Airy functions of real variable

The computation of the Airy functions of real variables $\text{Ai}(x)$, $\text{Bi}(x)$ and their derivatives [186] for real values of the argument x is a useful example of application of Chebyshev

3.9. Example: Computation of Airy functions of real variable

expansions for computing special functions. As is usual, the real line is divided into a number of intervals and we consider expansions on intervals containing the points $-\infty$ or $+\infty$. An important aspect is selecting the quantity that has to be expanded in terms of Chebyshev polynomials. The Airy functions have oscillatory behavior at $-\infty$, and exponential behavior at $+\infty$. It is important to expand quantities that are slowly varying in the interval of interest.

When the argument of the Airy function is large, the asymptotic expansions given in (10.4.59)–(10.4.64), (10.4.66), and (10.4.67) of [2] can be considered. The coefficients c_k and d_k used in these expansions are given by

$$c_0 = 1, \quad c_k = \frac{\Gamma(3k + \tfrac{1}{2})}{54^k \, k! \, \Gamma(k + \tfrac{1}{2})}, \quad k = 0, 1, 2, \ldots,$$

$$d_0 = 1, \quad d_k = -\frac{6k+1}{6k-1} c_k, \quad k = 1, 2, 3, \ldots. \tag{3.132}$$

Asymptotic expansions including the point $-\infty$

We have the representations

$$\mathrm{Ai}(-z) = \pi^{-1/2} z^{-1/4} \left\{ \sin(\zeta + \tfrac{1}{4}\pi) \, f(z) - \cos(\zeta + \tfrac{1}{4}\pi) \, g(z) \right\},$$

$$\mathrm{Ai}''(-z) = -\pi^{-1/2} z^{1/4} \left\{ \cos(\zeta + \tfrac{1}{4}\pi) \, p(z) + \sin(\zeta + \tfrac{1}{4}\pi) \, q(z) \right\},$$

$$\mathrm{Bi}(-z) = \pi^{-1/2} z^{-1/4} \left\{ \cos(\zeta + \tfrac{1}{4}\pi) \, f(z) + \sin(\zeta + \tfrac{1}{4}\pi) \, g(z) \right\},$$

$$\mathrm{Bi}'(-z) = \pi^{-1/2} z^{1/4} \left\{ \sin(\zeta + \tfrac{1}{4}\pi) \, p(z) - \cos(\zeta + \tfrac{1}{4}\pi) \, q(z) \right\}, \tag{3.133}$$

where $\zeta = \tfrac{2}{3} z^{3/2}$. The asymptotic expansions for the functions $f(z), g(z), p(z), q(z)$ are

$$f(z) \sim \sum_{k=0}^{\infty} (-1)^k c_{2k} \zeta^{-2k}, \quad g(z) \sim \sum_{k=0}^{\infty} (-1)^k c_{2k+1} \zeta^{-2k-1},$$

$$p(z) \sim \sum_{k=0}^{\infty} (-1)^k d_{2k} \zeta^{-2k}, \quad q(z) \sim \sum_{k=0}^{\infty} (-1)^k d_{2k+1} \zeta^{-2k-1}, \tag{3.134}$$

as $z \to \infty$, $|\mathrm{ph}\, z| < \tfrac{2}{3}\pi$.

Asymptotic expansions including the point $+\infty$

Now we use the representations

$$\mathrm{A}_i(z) = \tfrac{1}{2} \pi^{-1/2} z^{-1/4} e^{-\zeta} \tilde{f}(z), \quad \mathrm{A}'_i(z) = -\tfrac{1}{2} \pi^{-1/2} z^{1/4} e^{-\zeta} \tilde{p}(z),$$

$$\mathrm{B}_i(z) = \tfrac{1}{2} \pi^{-1/2} z^{-1/4} e^{\zeta} \tilde{g}(z), \quad \mathrm{B}'_i(z) = \tfrac{1}{2} \pi^{-1/2} z^{1/4} e^{\zeta} \tilde{q}(z), \tag{3.135}$$

where the asymptotic expansions for the functions $\tilde{f}(z)$, $\tilde{g}(z)$, $\tilde{p}(z)$, $\tilde{q}(z)$ are

$$\tilde{f}(z) \sim \sum_{k=0}^{\infty}(-1)^k c_k \zeta^{-k}, \quad \tilde{p}(z) \sim \sum_{k=0}^{\infty}(-1)^k d_k \zeta^{-k},$$
$$\tilde{g}(z) \sim \sum_{k=0}^{\infty} c_k \zeta^{-k}, \quad \tilde{q}(z) \sim \sum_{k=0}^{\infty} d_k \zeta^{-k}, \quad (3.136)$$

as $z \to \infty$, with $|\mathrm{ph}\, z| < \pi$ (for $\tilde{f}(z)$ and $\tilde{p}(z)$) and $|\mathrm{ph}\, z| < \frac{1}{3}\pi$ (for $\tilde{g}(z)$ and $\tilde{q}(z)$).

Chebyshev expansions including the point $-\infty$

The functions $f(z)$, $g(z)$, $p(z)$, $q(z)$ are the slowly varying quantities in the representations in (3.133), and these functions are computed approximately by using Chebyshev expansions. We write $z = x$. A possible selection is the x-interval $[7, +\infty)$ and for the shifted Chebyshev polynomials we take the argument $t = (7/x)^3$ (the third power also arises in the expansions in (3.134)). We have to obtain the coefficients in the approximation

$$f(x) \approx \sum_{r=0}^{m_1} a_r T_r^*(t), \quad g(x) \approx \frac{1}{\zeta}\sum_{r=0}^{m_2} b_r T_r^*(t),$$
$$p(x) \approx \sum_{r=0}^{m_3} c_r T_r^*(t), \quad q(x) \approx \frac{1}{\zeta}\sum_{r=0}^{m_4} d_r T_r^*(t). \quad (3.137)$$

Chebyshev expansions including the point $+\infty$

Now we consider Chebyshev expansions for the functions $\tilde{f}(x)$, $\tilde{g}(x)$, $\tilde{p}(x)$, $\tilde{q}(x)$, and we have

$$\tilde{f}(x) \approx \sum_{r=0}^{n_1} \tilde{a}_r T_r^*(\tilde{t}), \quad \tilde{g}(x) \approx \sum_{r=0}^{n_2}(-1)^r \tilde{a}_r T_r^*(\tilde{t}),$$
$$\tilde{p}(x) \approx \sum_{r=0}^{n_3} \tilde{c}_r T_r^*(\tilde{t}), \quad \tilde{q}(x) \approx \sum_{r=0}^{n_4}(-1)^r \tilde{c}_r T_r^*(\tilde{t}), \quad (3.138)$$

where $\tilde{t} = (7/x)^{3/2}$.

The number of terms in the Chebyshev expansions m_j, n_j, $j = 1, 2, 3, 4$, are determined for a prescribed accuracy of the functions. The Chebyshev coefficients of functions defined by convergent power series can be computed by rearrangement of the corresponding power series expansions using (3.35). For power series that represent asymptotic expansions this method is not available. In the present case we have used the Maple program `chebyshev` with default accuracy of 10^{-10}. The first three coefficients of the Chebyshev expansions (3.137) and (3.138) are given in Table 3.3. For $\tilde{a}_r$ and $\tilde{c}_r$, see Table 3.4 for more—and more precise—values.

Table 3.3. *The first coefficients of the Chebyshev expansions* (3.137) *and* (3.138).

Coef.	$r=0$	$r=1$	$r=2$
a_r	1.0001227513	0.0001230753	0.0000003270
b_r	0.0695710154	0.0001272386	0.0000006769
c_r	0.9998550220	−0.0001453297	−0.0000003548
d_r	−0.0973635868	−0.0001420772	−0.0000007225
$\tilde{a}_r$	0.9972733954	−0.0026989587	0.0000271274
$\tilde{c}_r$	1.0038355798	0.0038027377	0.0000322597

3.10 Chebyshev expansions with coefficients in terms of special functions

As we have seen in §3.6.1, the coefficients in a Chebyshev expansion can be obtained from recurrence relations when the function satisfies a linear differential equation with polynomial coefficients. All special functions of hypergeometric type satisfy such a differential equation, and in §3.6.1 we have explained how the method works for the Bessel function $J_0(4x)$ in the range $-1 \leq x \leq 1$. However, for this particular function we can obtain expansions in which the coefficients can be expressed in terms of known special functions, which in fact are again Bessel functions. We have (see [143, p. 37])

$$J_0(ax) = \sum_{n=0}^{\infty} \epsilon_n (-1)^n J_n^2(a/2) T_{2n}(x),$$

$$J_1(ax) = 2 \sum_{n=0}^{\infty} (-1)^n J_n(a/2) J_{n+1}(a/2) T_{2n+1}(x),$$

(3.139)

where $-1 \leq x \leq 1$ and $\epsilon_0 = 1$, $\epsilon_n = 2$ if $n > 0$. The parameter a can be any complex number. Similar expansions are available for J-Bessel functions of any complex order, in which the coefficients are $_1F_2$-hypergeometric functions, and explicit recursion relations are available for computing the coefficients. For general integer order, the coefficients are products of two J-Bessel functions, as in (3.139). See again [143].

Another example is the expansion for the error function,

$$e^{a^2 x^2} \operatorname{erf}(ax) = \sqrt{\pi} e^{\frac{1}{2} a^2} \sum_{n=0}^{\infty} I_{n+\frac{1}{2}} \left(\tfrac{1}{2} a^2 \right) T_{2n+1}(x), \quad -1 \leq x \leq 1, \quad (3.140)$$

in which the modified Bessel function is used. Again, a can be any complex number.

The expansions in (3.139) and (3.140) can be viewed as expansions near the origin. Other expansions are available that can be viewed as expansions at infinity, and these may be considered as alternatives for asymptotic expansions of special functions. For example, for the confluent hypergeometric U-functions we have the convergent expansion in terms of shifted Chebyshev polynomials (see (3.36)):

$$(\omega z)^a U(a, c, \omega z) = \sum_{n=0}^{\infty} C_n(z) T_n^*(1/\omega), \quad z \neq 0, \ |\text{ph}\, z| < \tfrac{3}{2}\pi, \ 1 \leq \omega \leq \infty. \quad (3.141)$$

Furthermore, $a, 1+a-c \neq 0, -1, -2, \ldots$. When equalities hold for these values of a and c, the Kummer U-function reduces to a Laguerre polynomial. This follows from

$$U(a, c, z) = z^{1-c} U(1 + a - c, 2 - c, z) \tag{3.142}$$

and

$$U(-n, \alpha + 1, z) = (-1)^n n! L_n^\alpha(z), \quad n = 0, 1, 2, \ldots. \tag{3.143}$$

The expansion (3.141) is given in [143, p. 25]. The coefficients can be represented in terms of generalized hypergeometric functions, in fact, Meijer G-functions, and they can be computed from the recurrence relation

$$\frac{2C_n(z)}{\epsilon_n} = 2(n+1)A_1 C_{n+1}(z) + A_2 C_{n+2}(z) + A_3 C_{n+3}(z), \tag{3.144}$$

where $b = a + 1 - c$, $\epsilon_0 = \frac{1}{2}$, $\epsilon_n = 1$ $(n \geq 1)$, and

$$\begin{aligned}
A_1 &= 1 - \frac{(2n+3)(n+a+1)(n+b+1)}{2(n+2)(n+a)(n+b)} - \frac{2z}{(n+a)(n+b)}, \\
A_2 &= 1 - \frac{2(n+1)(2n+3-z)}{(n+a)(n+b)}, \\
A_3 &= -\frac{(n+1)(n+3-a)(n+3-b)}{(n+2)(n+a)(n+b)}.
\end{aligned} \tag{3.145}$$

For applying the backward recursion algorithm it is important to know that

$$\sum_{n=0}^{\infty} (-1)^n C_n(z) = 1, \quad |\mathrm{ph}\, z| < \tfrac{3}{2}\pi. \tag{3.146}$$

This follows from

$$\lim_{\omega \to \infty} (\omega z)^a U(a, c, \omega z) = 1 \quad \text{and} \quad T_n^*(0) = (-1)^n. \tag{3.147}$$

The standard backward recursion scheme (see Chapter 4) for computing the coefficients $C_n(z)$ works only for $|\mathrm{ph}\, z| < \pi$, and for $\mathrm{ph}\, z = \pm\pi$ a modification seems to be possible; see [143, p. 26].

Although the expansion in (3.141) converges for all $z \neq 0$ in the indicated sector, it is better to avoid small values of the argument of the U-function. Luke gives an estimate of the coefficients $C_n(z)$ of which the dominant factor that determines the speed of convergence is given by

$$C_n(z) = \mathcal{O}\left(n^{2(2a-c-1)/3} e^{-3n^{\frac{2}{3}} z^{\frac{1}{3}}}\right), \quad n \to \infty, \tag{3.148}$$

and we see that large values of $\Re z^{1/3}$ improve the convergence. For example, if we denote $\zeta = \omega z$ and we want to use the expansion for $|\zeta| \geq R(> 0)$, we should choose z and ω ($\omega \geq 1$) such that z and ζ have the same phase, say θ. We cannot choose z with modulus larger than R, and an appropriate choice is $z = Re^{i\theta}$. Then the expansion gives an approximation of the U-function on the half-ray with $|\zeta| \geq R$ with phase θ, and the coefficients $C_n(z)$ can be used for all ζ on this half-ray. For a single evaluation we can take $\omega = 1$.

3.10. Chebyshev expansions with coefficients in terms of special functions

Table 3.4. *Coefficients of the Chebyshev expansion* (3.153).

n	$C_n(z)$	$D_n(z)$
0	0.99727 33955 01425	1.00383 55796 57251
1	−0.00269 89587 07030	0.00380 27374 06686
2	0.00002 71274 84648	−0.00003 22598 78104
3	−0.00000 05043 54523	0.00000 05671 25559
4	0.00000 00134 68935	−0.00000 00147 27362
5	−0.00000 00004 63150	0.00000 00004 97977
6	0.00000 00000 19298	−0.00000 00000 20517
7	−0.00000 00000 00938	0.00000 00000 00989
8	0.00000 00000 00052	−0.00000 00000 00054
9	−0.00000 00000 00003	0.00000 00000 00003

The expansion in (3.141) can be used for all special cases of the Kummer U-function, that is, for Bessel functions (Hankel functions and K-modified Bessel function), for the incomplete gamma function $\Gamma(a, z)$, with special cases the complementary error function and exponential integrals.

Example 3.16 (Airy function). For the Airy function $\text{Ai}(x)$ we have the relations

$$\xi^{\frac{1}{6}} U(\tfrac{1}{6}, \tfrac{1}{3}, \xi) = 2\sqrt{\pi} x^{\frac{1}{4}} e^{\frac{1}{2}\xi} \text{Ai}(x),$$
$$\xi^{-\frac{1}{6}} U(-\tfrac{1}{6}, -\tfrac{1}{3}, \xi) = -2\sqrt{\pi} x^{-\frac{1}{4}} e^{\frac{1}{2}\xi} \text{Ai}'(x), \qquad (3.149)$$

where $\xi = \tfrac{4}{3} x^{\frac{3}{2}}$. For the expansions of the functions $\tilde{f}(x)$ and $\tilde{p}(x)$ in (3.138) we take $\omega = (x/7)^{3/2}$ and $z = \tfrac{4}{3} 7^{3/2} = 24.69\ldots$. To generate the coefficients with this value of z we determine the smallest value of n for which the exponential factor in (3.148) is smaller than 10^{-15}. This gives $n = 8$.

Next we generate for both U-functions in (3.149) the coefficients $C_n(z)$ by using (3.144) in the backward direction, with starting values

$$\widetilde{C}_{19}(z) = 1, \quad \widetilde{C}_{20}(z) = 0, \quad \widetilde{C}_{21}(z) = 0. \qquad (3.150)$$

We also compute for normalization

$$S = \sum_{n=0}^{18} (-1)^n \widetilde{C}_n(z) = -0.902363242772764 \, 10^{25}, \qquad (3.151)$$

where the numerical value is for the Ai case. Finally we compute

$$C_n(z) = \widetilde{C}_n(z)/S, \quad n = 0, 1, 2, \ldots, 9. \qquad (3.152)$$

This gives the coefficients $C_n(z)$ of the expansions

$$2\sqrt{\pi}x^{\frac{1}{4}}e^{\frac{2}{3}x^{3/2}}\mathrm{Ai}(x) \approx \sum_{n=0}^{9} C_n T_n^*\left((7/x)^{3/2}\right),$$

$$-2\sqrt{\pi}x^{\frac{1}{4}}e^{\frac{2}{3}x^{3/2}}\mathrm{Ai}'(x) \approx \sum_{n=0}^{9} D_n T_n^*\left((7/x)^{3/2}\right),$$

(3.153)

$x \geq 7$, of which the coefficients are given in Table 3.4.

The values of the coefficients in the first three rows of Table 3.4 correspond approximately with the same as those of the coefficients $\tilde{a}_r$ and $\tilde{c}_r$ in Table 3.3. ∎

Chapter 4

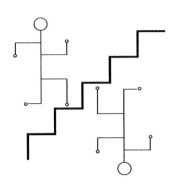

Linear Recurrence Relations and Associated Continued Fractions

"Obvious" is the most dangerous word in mathematics.
—E. T. Bell

If you don't know where you are going, you'll probably end up somewhere else.
—L. Carroll, from *Alice in Wonderland*

4.1 Introduction

Many families of special functions satisfy discrete relations among functions with different parameter values. For example, the Bessel functions $J_\nu(z)$ and $Y_\nu(z)$ both satisfy the following identity between functions of consecutive orders:

$$y_{\nu+1} - \frac{2\nu}{z} y_\nu + y_{\nu-1} = 0. \tag{4.1}$$

A relation of the type

$$C_n y_{n+1} + B_n y_n + A_n y_{n-1} = 0 \tag{4.2}$$

is called a *linear homogeneous three-term recurrence relation* or a *second order homogeneous linear difference equation*.

In this chapter we describe methods for computing special functions that satisfy linear homogeneous three-term recurrence relations (TTRRs from now on) and linear inhomogeneous first order difference equations $y_n = \alpha_n y_{n-1} + \beta_n$. With just these two types of difference equations, algorithms for the computation of many special functions become available. Also second order inhomogeneous equations are briefly discussed.

87

For computation using recurrence relations, the analysis of the condition of the recursive process is compulsory for a sensible use of the recurrence; the first section deals with this fundamental idea and introduces the concept of minimal (or recessive) solutions: solutions which become much smaller than other solutions when recursion is applied. Perron's theorem, in its various versions, is a useful result for analyzing the existence of minimal solutions; this is described in §4.3. Another useful result for numerical computations is Pincherle's theorem, which states the equivalence between the existence of a minimal solution of a TTRR, f_n, and the convergence of the associated continued fraction for the ratio f_n/f_{n-1} (in Chapter 6 methods for the computation of continued fractions are discussed); a proof of Pincherle's theorem is sketched in §4.4, where also estimates for the accuracy of the continued fraction are provided. Section 4.5 provides examples of notable recurrence relations, focusing on confluent and Gauss hypergeometric recursions, and describes the determination of the minimal solution.

Section 4.6 deals with methods of computation of solutions of TTRRs, particularly of minimal solutions. In all these methods, a basic principle which must be respected is that if a solution is recessive as n increases, the recursion should be applied in the direction of decreasing n (backward recursion). Similar precautions need to be considered when computing solutions of inhomogeneous first order recursions, and it is necessary to analyze whether backward or forward recursion is advisable, depending on the asymptotic behavior of the wanted solution (§4.7.1). For second order inhomogeneous equations the situation may worsen, because solutions may appear which cannot be computed either by forward or by backward recursion; the alternative is to solve a boundary value problem instead of an initial value problem, which is the main idea behind Olver's method, which is briefly described.

Not even for the homogeneous second order equations is it always true that either the forward or the backward recurrences are well conditioned. Knowing the asymptotic behavior of the solutions is not always enough and there exist striking examples (both for second order homogeneous and first order inhomogeneous equations) for which some solutions change their behavior drastically, ruining the simple recipe of computing a solution by forward or backward recursion according only to its asymptotic behavior. In §4.8 we provide examples of this type of behavior (modified Bessel functions, confluent hypergeometric functions, and exponential integrals).

4.2 Condition of three-term recurrence relations

TTRRs, particularly for the homogeneous case, most frequently appear in the computation of special functions.

Linear homogeneous TTRRs are functional relations of the form (4.2) which allow us to compute, from a pair of starting values y_0 and y_1, successive values y_n, $n > 1$.

We assume that A_n and C_n do not vanish for any n and that, therefore, the relation can be applied in both the direction of increasing and of decreasing n. In the first case, we speak of forward recursion, while in the second case we speak of backward recursion. With these conditions, we can also write the recurrence as

$$y_{n+1} + b_n y_n + a_n y_{n-1} = 0. \tag{4.3}$$

4.2. Condition of three-term recurrence relations

The theory of linear TTRRs is very similar to the theory of second order linear ordinary differential equations. To begin with, any solution can be written as a linear combination of any two linearly independent solutions of the recurrence relation. A solution is specified when two values, y_0 and y_1, are provided.

Two solutions determined by the values $y_k^{(1)}$ and $y_k^{(2)}$, $k = n-1, n$, are independent if their *Casorati determinant* is different from zero, that is, if

$$D_n = \begin{vmatrix} y_n^{(1)} & y_n^{(2)} \\ y_{n-1}^{(1)} & y_{n-1}^{(2)} \end{vmatrix} \neq 0. \tag{4.4}$$

To see this, observe that from (4.3) we find

$$D_{n+1} = a_n D_n, \tag{4.5}$$

and that we have assumed that $a_n \neq 0$ for any n. So, when two solutions have a Casorati determinant different from zero for a given n, the same holds for any n. In the opposite case, if the Casorati determinant (4.4) is zero, then there exists $\lambda \neq 0$ such that $y_k^{(2)} = \lambda y_k^{(1)}$, $k = n-1, n$, and then, by linearity of the TTRR and because $a_n \neq 0$ for all n, this would be true for all k.

For a reliable computation of special functions via recurrence relations, it is necessary to analyze the condition of the recursive process. We need to analyze the impact of a small perturbation of the initial values y_0 and y_1 on the final result for y_N, $N > 1$, with N possibly large.

4.2.1 Minimal solutions

Problems may arise when we try to compute a solution which becomes much smaller than other solutions as n becomes large.

Let us denote by $\{f_n, g_n\}$ a pair of independent solutions of a TTRR (4.3). Let us suppose that we are interested in computing the solution defined by the starting values f_0, f_1 and that, due to the limited accuracy of the computer number storage, a small error takes place in the initial values. Instead of f_0 and f_1, we have the numerical starting values y_0, y_1

$$\begin{pmatrix} y_0 \\ y_1 \end{pmatrix} = \mu \begin{pmatrix} f_0 \\ f_1 \end{pmatrix} + \gamma r_0 \begin{pmatrix} g_0 \\ g_1 \end{pmatrix}. \tag{4.6}$$

The quantity $r_0 = f_0/g_0$ is included to ensure that f_i and $r_0 g_i$, $i = 0, 1$, are of comparable size (if f_0 or g_0 are zero we should consider a different factor f_i/g_j, $i, j = 0, 1$). Because the second component in (4.6) is due to a numerical error, we will have that $\mu = 1 + \epsilon_1$, $\gamma = \epsilon_2$, where ϵ_1, ϵ_2 are two small numbers (of order ϵ-machine). After the application of the recurrence relation and assuming that no further rounding errors take place we have, by linearity of the TTRR,

$$y_k = \mu f_k + \gamma r_0 g_k = f_k(\mu + \gamma r_0 g_k/f_k), \tag{4.7}$$

and the relative error with respect to the wanted solution, f_k, is

$$\epsilon_r = y_k/f_k - 1 = (\mu - 1) + \gamma r_0 g_k/f_k = \epsilon_1 + \epsilon_2 r_0/r_k, \qquad (4.8)$$

where

$$r_k = f_k/g_k. \qquad (4.9)$$

The error may become large when the ratio r_0/r_k becomes large. The initial deviation introduces a component of the second solution g_n, which increases with respect to f_n and tends to dominate. Furthermore, when this tendency is maintained for arbitrary large k, the errors may grow indefinitely and the ratios of numerical solutions y_k/y_{k-1} will approach those of the unwanted solution, g_k/g_{k-1}, indeed:

$$\frac{y_k}{y_{k-1}} \frac{g_{k-1}}{g_k} - 1 = \frac{\mu(r_k - r_{k-1})}{\gamma r_0 + \mu r_{k-1}}, \qquad (4.10)$$

which goes to zero when $r_k \to 0$, $k \to +\infty$. In fact, this will be true no matter what the initial values y_0 and y_1 are except when $\gamma = 0$; however, numerically γ cannot be exactly zero for all steps of the recursion.

This situation takes place when the TTRR admits a minimal solution, which can be defined as follows.

Definition 4.1. *A solution of a TTRR f_n is said to be a* minimal solution *if there exists a linearly independent solution g_n such that*

$$\lim_{n \to +\infty} \frac{f_n}{g_n} = 0. \qquad (4.11)$$

The solution g_n is called a dominant solution.

Obviously, a minimal solution cannot contain a nonzero component of a dominant solution. Therefore, the minimal solutions of a TTRR are unique up to constant (not depending on n) multiplicative factors.

It is also immediate to check the following lemma.

Lemma 4.2. *A recurrence relation admits a minimal solution if and only if, given any two linearly independent solutions of the recurrence relation, $y_n^{(1)}$ and $y_n^{(2)}$, one of the following limits is finite:* $\lim_{n \to +\infty}(y_n^{(1)}/y_n^{(2)})$ *or* $\lim_{n \to +\infty}(y_n^{(2)}/y_n^{(1)})$.

Proof. If $\lim_{n \to +\infty}(y_n^{(1)}/y_n^{(2)}) = 0$, then $y_n^{(1)}$ is minimal. If $\lim_{n \to +\infty}(y_n^{(2)}/y_n^{(1)}) = 0$, then $y_n^{(2)}$ is minimal. If $\lim_{n \to +\infty}(y_n^{(1)}/y_n^{(2)}) = C \neq 0$, then $f_n = y_n^{(1)} - C y_n^{(2)}$ is minimal; indeed

$$\lim_{n \to +\infty} \frac{f_n}{y_n^{(2)}} = 0. \qquad \square \qquad (4.12)$$

4.2. Condition of three-term recurrence relations

Example 4.3. Simple examples of minimal solutions are provided by the recurrence relations with constant coefficients. For instance, the recursion $y_{n+1} - 2\cosh x\, y_n + y_{n-1} = 0$ has $\{f_n, g_n\} = \{e^{-nx}, e^{nx}\}$ as an independent pair of solutions. For positive x, f_n is minimal and g_n is dominant. Because f_n is minimal, one should never try to compute $f_{40} = e^{-40}$ starting from $f_0 = 1$, $f_1 = e^{-1}$ for $x = 1$ and applying forward recursion: the numerical solution will become contaminated with a dominant component from the very beginning, which will dominate long before $n = 40$ is reached.

Indeed, we obtain the completely wrong answer

$$y_{39} = -0.388796451906197, \quad y_{40} = -1.05685833018597,$$

and we notice that $y_{40}/y_{39} = 2.71828182845905$, which is the number e (with all digits correct), as corresponds to a dominant solution like $y_n = e^{nx}$.

On the other hand, the pair $\{p_n, q_n\} = \{\sinh(nx), \cosh(nx)\}$ also constitute a linearly independent pair which can be computed in a stable way in the forward direction. Both p_n and q_n are dominant. Computing the minimal solution $f_n = e^{-nx}$ from these dominant solutions gives a large error when $x > 0$ and n is large. ∎

From the previous discussion, we conclude that when there is a minimal solution, forward recursion is possible for dominant solutions, but it should never be used for a minimal solution. Furthermore, the ratios y_k/y_{k-1} approach the ratios of dominant solutions as k becomes large.

Reversing the argument, the minimal solutions can be safely computed by backward recursion, but not the dominant solutions. Furthermore, if the recursion for computing a solution y_k is applied starting from high enough values of $n = N \gg k$, the computed ratios y_k/y_{k-1} will approach those of the minimal solution. Indeed, if we start from $y_n = \mu f_n + \gamma r_N g_n$, $n = N, N - 1$, and we compute by backward recursion, we have

$$\frac{y_k/y_{k-1}}{f_k/f_{k-1}} - 1 = \gamma \frac{r_N}{r_{k-1}} \frac{r_{k-1}/r_k - 1}{\mu + \gamma r_N/r_{k-1}}, \quad (4.13)$$

which goes to zero as higher values N for starting the recurrence are considered. This is true regardless of the values of μ and γ, except when μ is exactly zero, which can be excluded from a numerical point of view.

Therefore, before applying a recurrence relation for computing a function, it is necessary to determine whether it has a minimal solution and, if so, if the function under consideration is minimal. If there is a minimal solution but our function is dominant, we should try forward recursion. However, if it is minimal, we have to apply backward recursion. If there are no minimal solutions, both directions of recursion are in principle possible.

Of course, we would need to compute two starting values for the recurrence in each case. However, when the recurrence admits a minimal solution and it is applied in the appropriate direction, the ratios of functions tend to converge to the right value and, therefore, we might only need a normalization condition. This is the basic idea behind Miller's method for computing minimal solutions, which we discuss in §4.6.

4.3 Perron's theorem

Perron's theorem (also known as the Perron–Kreuser theorem [241]) is a useful result for investigating the existence of minimal solutions of a recurrence relation. This theorem is a generalization of Poincaré's theorem (also known as the Poincaré–Perron theorem [60]).

Theorem 4.4 (Poincaré). *Let $y_{n+1} + b_n y_n + a_n y_{n-1} = 0$ with $\lim_{n \to +\infty} b_n = b$ and $\lim_{n \to +\infty} a_n = a$. Let t_1 and t_2 denote the zeros of the characteristic equation $t^2 + bt + a = 0$. Then, if $|t_1| \neq |t_2|$, the difference equation has two linearly independent solutions f_n and g_n satisfying*

$$\lim_{n \to +\infty} \frac{f_n}{f_{n-1}} = t_1, \quad \lim_{n \to +\infty} \frac{g_n}{g_{n-1}} = t_2. \tag{4.14}$$

If $|t_1| = |t_2|$, then

$$\limsup_{n \to \infty} |y_n|^{\frac{1}{n}} = |t_1| \tag{4.15}$$

for any nontrivial solution y_n of the recurrence.

When $|t_1| \neq |t_2|$ the minimal solution is the one whose ratio (f_n/f_{n-1} or g_n/g_{n-1}) converges to the root with the smallest modulus. When $|t_1| = |t_2|$ the theorem is inconclusive with respect to the existence of a minimal solution.

We will not prove this result; see [183]. We only remark the resemblance to the case of a recurrence relation with constant coefficients $y_{n+1} + by_n + ay_{n-1} = 0$. This recurrence relation has the general solution $y_n = c_1 t_1^n + c_2 t_2^n$ when the roots t_1, t_2 of the characteristic equation $t^2 + bt + a = 0$ are different, while the general solution is $y_n = (c_1 + c_2 n) t_1^n$ when the roots are equal.

From Poincaré's theorem the following result can be proved.

Theorem 4.5 ("intuitive" Perron theorem). *Let $y_{n+1} + b_n y_n + a_n y_{n-1} = 0$, with b_n and a_n such that $b_n^2 - 4a_n > 0$ for $n > N$ and $a_n \sim an^\alpha, b_n \sim bn^\beta$ as $n \to +\infty$ ($a, b \neq 0$). Then there exists a pair of independent solutions $\{f_n, g_n\}$ such that*

$$\lim_{n \to +\infty} \frac{1}{t_1(n)} \frac{f_n}{f_{n-1}} = 1, \quad \lim_{n \to +\infty} \frac{1}{t_2(n)} \frac{g_n}{g_{n-1}} = 1, \tag{4.16}$$

where $t_i(n)$ are the solutions of $t^2 + b_n t + a_n = 0$.

The minimal solution is the one corresponding to the smallest $|t_i|$.

This result is saying, in some sense, that when the condition $b_n^2 - 4a_n > 0$ is met we can treat the difference equation as if it were a relation with constant coefficients. When this simple recipe fails, things become more difficult and we must obtain additional asymptotic information in order to decide whether the recurrence has a minimal solution.

Theorem 4.5 is also valid for complex coefficients when we replace the condition $b_n^2 - 4a_n > 0$ with $\lim_{n \to \infty} |t_1(n)/t_2(n)| \neq 1$.

We are not proving the previous theorem. We have named it the "intuitive" Perron theorem because of its parallelism with Poincaré's theorem. Using Poincaré's theorem we will prove the more general version of this result appearing in most texts [75, 241, 219], that is, the complete version of Perron's theorem from where the verification of Theorem 4.5 is immediate.

4.3. Perron's theorem

Theorem 4.6 (Perron). *We consider the TTRR*

$$y_{n+1} + b_n y_n + a_n y_{n-1} = 0, \qquad (4.17)$$

and we suppose that the coefficients satisfy

$$a_n \sim an^\alpha, \qquad b_n \sim bn^\beta, \qquad n \to \infty, \qquad (4.18)$$

$a \neq 0, b \neq 0$.

1. *If $\beta > \alpha/2$, then there are two independent solutions f_n and g_n of the recurrence relation such that*

    $$\frac{f_n}{f_{n-1}} \sim -\frac{a}{b} n^{\alpha-\beta}, \qquad \frac{g_n}{g_{n-1}} \sim -bn^\beta. \qquad (4.19)$$

2. *If $\beta = \alpha/2$, let t_1 and t_2 be the roots of the characteristic polynomial $\phi(t) = t^2 + bt + a$.*

    (a) *If $|t_1| < |t_2|$, then there are two independent solutions f_n and g_n of the recurrence relation such that*

    $$\frac{f_n}{f_{n-1}} \sim t_1 n^\beta, \qquad \frac{g_n}{g_{n-1}} \sim t_2 n^\beta. \qquad (4.20)$$

    (b) *If $|t_1| = |t_2|$, then*

    $$\limsup_{n \to \infty} \left[\frac{|y_n|}{n!^\beta} \right]^{1/n} = |t_1| \qquad (4.21)$$

    for any nontrivial solution of the recurrence relation.

3. *If $\beta < \alpha/2$, then*

    $$\limsup_{n \to \infty} \left[\frac{|y_n|}{n!^{\alpha/2}} \right]^{1/n} = \sqrt{|a|} \qquad (4.22)$$

    for any nontrivial solution of the recurrence relation.

In cases 1 and 2(a), the solution f_n is minimal. In the remaining cases, the theorem is inconclusive with respect to the existence of a minimal solution.

Proof. This theorem is a consequence of Poincaré's theorem. We can prove it by simply rescaling the functions by simple functions of n. The second and third cases are directly obtained from the first case by considering the recurrence relation for the associated functions $y_n = (n!)^{\alpha/2} \hat{y}_n$. Only the first case needs further explanation.

For case 1, we define a new set of functions, $\hat{y}_n$, $y_n = (n!)^\beta \hat{y}_n$ which satisfy a recurrence relation

$$\hat{y}_{n+1} + \hat{b}_n \hat{y}_n + \hat{a}_n \hat{y}_{n-1} = 0, \qquad (4.23)$$

with $\hat{b}_n \to b$ and $\hat{a}_n \sim a_n/n^{2\beta} \sim an^{\alpha-2\beta}$ (and therefore $\hat{a}_n \to 0$). By Poincaré's theorem, there exists a pair of independent solutions of this new recurrence, $\{\hat{f}_n, \hat{g}_n\}$, such that

$$\frac{\hat{f}_n}{\hat{f}_{n-1}} \to 0, \qquad \frac{\hat{g}_n}{\hat{g}_{n-1}} \to -b. \qquad (4.24)$$

On the other hand, denoting by $H_n^{(i)} = \hat{y}_n/\hat{y}_{n-1}$, with $i = 1, 2$, respectively, for $\hat{y}_n = f_n$ and $\hat{y}_n = g_n$, we can write the recurrence relation as

$$H_{n+1}^{(i)} H_n^{(i)} + \hat{b}_n H_n^{(i)} + \hat{a}_n = 0. \tag{4.25}$$

The relation (4.25) holds for large enough n and with finite values of $H_k^{(i)}$, $k = n, n-1$. Indeed, for large enough n, neither $\hat{f}_n$ nor $\hat{g}_n$ can be equal to zero because $\hat{g}_n/\hat{g}_{n-1} \to -b \neq 0$ and, if $\hat{f}_n = 0$, we have, using the recurrence relation, $\hat{f}_{n+2}/\hat{f}_{n+1} = -\hat{b}_{n+1} \simeq -b \neq 0$, which contradicts the fact that $\hat{f}_n/\hat{f}_{n-1} \to 0$.

Multiplying (4.25) for $i = 1$ by $H_n^{(2)}$ and the relation for $i = 2$ by $H_n^{(1)}$, we have upon subtracting both expressions

$$H_n^{(1)} H_n^{(2)} (H_{n+1}^{(1)} - H_{n+1}^{(2)}) = \hat{a}_n (H_n^{(1)} - H_n^{(2)}), \tag{4.26}$$

and then

$$H_n^{(1)} = \frac{\hat{a}_n}{H_n^{(2)}} \frac{H_n^{(1)} - H_n^{(2)}}{H_{n+1}^{(1)} - H_{n+1}^{(2)}} \sim \frac{\hat{a}_n}{H_n^{(2)}}, \tag{4.27}$$

because $H_n^{(1)} \to 0$ and $H_n^{(2)} \to -b$.

Therefore,

$$\frac{\hat{f}_n}{\hat{f}_{n-1}} \sim -\frac{a}{b} n^{\alpha - 2\beta}, \tag{4.28}$$

and, returning to the original solutions, we have the asymptotic behavior, as claimed in the theorem. □

4.3.1 Scaled recurrence relations

The proof of Perron's theorem illustrates the construction of scaled functions. This may be useful for enlarging the range of functions computable from the recurrence relations.

When Theorem 4.5 gives information regarding the existence of minimal solutions, then, given the characteristic roots $t_1(n)$ and $t_2(n)$, we have a pair of solutions such that $f_n/f_{n-1} \sim t_1$ and $g_n/g_{n-1} \sim t_2$. Then we have the rough estimations $f_n \sim t_1^n$ and $g_n \sim t_2^n$. When both minimal and dominant solutions need to be computed for the wider possible range of the order n, it may be convenient to renormalize the functions, and therefore the recurrence, with n-dependent factors. In this way, the range of computation can be enlarged, avoiding overflows for the dominant solution and underflows for the minimal solution.

Then, because considering that the characteristic roots $t_1(n)$ and $t_2(n)$ satisfy

$$t_1(n) t_2(n) \sim 1 \iff a_n \sim 1, \tag{4.29}$$

when $a_n \sim a n^\alpha$, the best choice is to renormalize the functions in the following (or similar) way $y_n = \sqrt{|a|} n^{\alpha/2} w_n$, where the new set w_n satisfies a new recurrence

$$w_{n+1} + \hat{b}_n w_n + \hat{a}_n w_{n-1} = 0. \tag{4.30}$$

such that $\hat{a}_n \to \text{sign}(a)$. Therefore the corresponding roots of the characteristic equation $\lambda^2 + \hat{b}_n \lambda + \hat{a}_n = 0$ are such that $|\lambda_1(n) \lambda_2(n)| \to 1$ and the increasing of the dominant solution is expected to have the same speed as the decreasing of the minimal solution for the new recurrence. This is used in Chapter 12 for some hypergeometric recursions.

4.4. Minimal solutions of TTRRs and continued fractions

This is analogous to the way in which we can proceed for enlarging the range of computation of functions which are solutions of a differential equation $y'' + A(x)y = 0$. Under certain conditions, the Liouville transformation with change of variable $z'(x) = \sqrt{|A(x)|}$ can be used for extracting the dominant exponential factors (see Chapter 1, particularly (1.3), and §2.2.4).

4.4 Minimal solutions of TTRRs and continued fractions

For computing minimal solutions of TTRRs, the continued fraction for the ratios of consecutive solutions is also a useful tool. The basic result to be considered is Pincherle's theorem, which we show next. Further information on continued fractions, including methods of computation, can be found in Chapter 6.

Theorem 4.7 (Pincherle). *Given a TTRR $y_{n+1} + b_n y_n + a_n y_{n-1} = 0$, the continued fraction*

$$\frac{-a_k}{b_k +} \frac{-a_{k+1}}{b_{k+1} +} \cdots \tag{4.31}$$

converges if and only if the recurrence relation possesses a minimal solution. Furthermore, if f_n is a minimal solution, then the continued fraction converges to f_k/f_{k-1}.

Proof. The continued fraction is obtained from the recurrence relation by iterating

$$\frac{y_k}{y_{k-1}} = \frac{-a_k}{b_k + \dfrac{y_{k+1}}{y_k}}. \tag{4.32}$$

By starting backward recursion with initial values $y_{k+m} = 0$, $y_{k+m-1} = 1$, $m > 0$, we would obtain values y_k, y_{k-1} such that

$$\frac{y_k}{y_{k-1}} = \frac{-a_k}{b_k +} \frac{-a_{k+1}}{b_{k+1} +} \cdots \frac{-a_{k+m-1}}{b_{k+m-1} + \dfrac{y_{k+m}}{y_{k+m-1}}}. \tag{4.33}$$

But because $y_{k+m} = 0$ this is just the mth approximant of the continued fraction for $\frac{y_k}{y_{k-1}}$ (see §6.2). The argument is still valid if there is a $y_p = 0$, $k \leq p \leq k+m-2$.

Let $\{f_n, g_n\}$ be a pair of linearly independent solutions. We can write $y_n = \mu f_n + \gamma g_n$. Let us denote $N = k+m$; because $y_N = 0$ and $y_{N-1} = 1$, solving the corresponding system of equations we get

$$\mu = \frac{g_N}{D_N}, \quad \gamma = -\frac{f_N}{D_N}, \quad D_N = \begin{vmatrix} g_N & f_N \\ g_{N-1} & f_{N-1} \end{vmatrix} \neq 0. \tag{4.34}$$

The Casorati determinant D_N is different from zero because $\{f_n, g_n\}$ is a pair of linearly independent solutions.

Then, similarly to how it is done in (4.13),

$$H_{k,N} = \frac{y_k}{y_{k-1}} = \frac{\mu f_k + \gamma g_k}{\mu f_{k-1} + \gamma g_{k-1}} = \frac{f_k - r_N g_k}{f_{k-1} - r_N g_{k-1}}, \qquad (4.35)$$

where $r_N = f_N/g_N = -\gamma/\mu$.

Thus, because g_k and g_{k-1} cannot simultaneously be 0 (because then $g_n = 0$ for all n) and the same is true for f_k and f_{k-1}, we see that $\lim_{m \to \infty} H_{k,N} = 0$ ($N = k + m$) exists if and only if the limits $\lim_{N \to +\infty} r_N$ or $\lim_{N \to +\infty} 1/r_N$ exist. Then, by Theorem 4.7, the continued fraction converges if and only if the recurrence relation admits a minimal solution.

When f_n is the minimal solution then $r_N \to 0$ and the continued fraction converges to the ratio f_k/f_{k-1}. This completes the proof of the theorem. $\square$

A different proof of the theorem can be found, for instance, in [75].

From the proof given here, we can obtain estimations of the error of the mth approximant assuming that $r_N \to 0$ (and therefore that f_n is the minimal solution). We have for the relative error (of course assuming in this case that $f_k \neq 0$ and $f_{k-1} \neq 0$)

$$\epsilon_r(k, N) = H_{k,N} \frac{f_{k-1}}{f_k} - 1 = \frac{r_N}{r_{k-1}} \frac{1 - r_{k-1}/r_k}{1 - r_N/r_{k-1}}. \qquad (4.36)$$

The faster $|r_N|$ decreases as N becomes large, the faster the successive approximants converge (apparently or not; see §4.8) to the ratio f_k/f_{k-1}.

4.5 Some notable recurrence relations

In this section we provide examples of notable recurrence relations and identify their minimal solutions with the aid of Perron's theorem and additional asymptotic information. Necessarily, the number of cases treated is not exhaustive but includes a varied range of functions of the hypergeometric family.

4.5.1 The confluent hypergeometric family

Example 4.8 (Bessel and Coulomb wave functions).

Bessel functions

The recurrence relation for Bessel functions reads

$$y_{\nu+1} - \frac{2\nu}{z} y_\nu + y_{\nu-1} = 0, \quad z \neq 0. \qquad (4.37)$$

The claim of the theorem is that there exist two independent solutions f_n and g_n such that

$$\frac{f_\nu}{f_{\nu-1}} \sim \frac{z}{2\nu}, \quad \frac{g_\nu}{g_{\nu-1}} \sim \frac{2\nu}{z}, \qquad (4.38)$$

4.5. Some notable recurrence relations

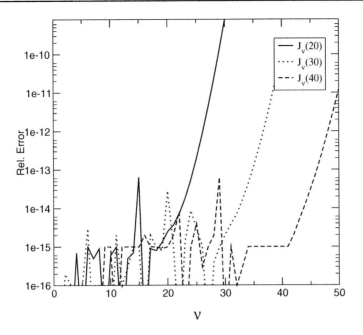

Figure 4.1. *Relative errors for the computation of $J_\nu(20)$, $J_\nu(30)$, $J_\nu(40)$ from forward recursion of* (4.37).

and f_ν is minimal. From known asymptotic behavior, one can identify the Bessel function of the first class as the minimal solution. Indeed $f_\nu = J_\nu(z)$ because (see [2, Chap. 9] for information on the Bessel functions)

$$J_\nu(z) \sim \frac{1}{\sqrt{2\pi\nu}} \left(\frac{ez}{2\nu}\right)^\nu. \tag{4.39}$$

A linearly independent solution is the Bessel function of the second kind $Y_\nu(z)$, with asymptotic behavior as $\nu \to +\infty$,

$$Y_\nu(z) \sim -\sqrt{\frac{2}{\pi\nu}} \left(\frac{ez}{2\nu}\right)^{-\nu}. \tag{4.40}$$

Therefore $\{J_\nu(z), Y_\nu(z)\}$ is a numerically satisfactory pair of solutions of the recurrence relation, because it comprises the minimal solution. We note that the Casorati determinant is

$$J_\nu(z)Y_{\nu-1}(z) - Y_\nu(z)J_{\nu-1}(z) = \frac{2}{\pi z}. \tag{4.41}$$

In Figure 4.1 we show the relative errors when we compute $J_\nu(x)$ for $x = 20, 30, 40$ by using (4.37) in the forward direction. We observe that the errors grow rapidly when ν becomes larger than x. As long as $\nu < x$ no serious instabilities occur.

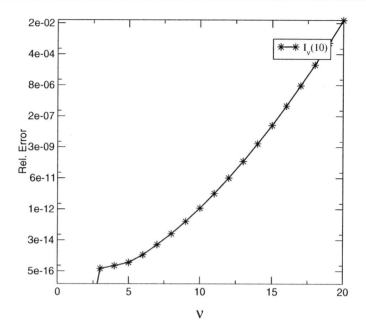

Figure 4.2. *Relative error for the computation of $I_\nu(10)$ from forward recursion of* (4.42).

Modified Bessel functions

The modified Bessel function recurrence is similar to this case. The relation is

$$y_{\nu+1} + \frac{2\nu}{z} y_\nu - y_{\nu-1} = 0, \quad z \neq 0. \tag{4.42}$$

Pincherle's theorem guarantees the existence of a minimal solution, and the asymptotic analysis of known solutions serves to identify the $I_\nu(z)$ as minimal and $e^{i\pi\nu} K_\nu(z)$ as a dominant solution.

In Figure 4.2 we show the relative errors when we compute $I_\nu(10)$ by using (4.42) in the forward direction. We observe that the errors grow rapidly from the start, whereas for $J_\nu(x)$ (see Figure 4.1) the recursion becomes unstable when $\nu > x$.

Coulomb wave functions

Coulomb wave functions can be seen as generalizations of the Bessel functions. The recurrence relation

$$y_{L+1} + b_L y_L + a_L y_{L-1} = 0, \tag{4.43}$$

where

$$b_L = -\frac{2L+1}{c_L}\left[\eta + \frac{L(L+1)}{\rho}\right], \quad a_L = \frac{L+1}{c_L}\sqrt{L^2 + \eta^2},$$
$$c_L = L\sqrt{(L+1)^2 + \eta^2}, \tag{4.44}$$

4.5. Some notable recurrence relations

is satisfied by the Coulomb functions. When $\eta = 0$, $\nu = L + 1/2$, and $\rho = x$ the recurrence relation (4.43) transforms into the Bessel recurrence relation. Perron's theorem gives for Coulomb functions similar results as for Bessel functions. $F_L(\nu, \rho)$ and $G_L(\nu, \rho)$ are called the regular and irregular Coulomb functions, respectively. For the standard minimal and dominant solutions $F_L(\nu, \rho)$ and $G_L(\nu, \rho)$ (see [2, Chap. 14]), we have

$$\frac{F_L}{F_{L-1}} \sim \frac{\rho}{2L}, \quad \frac{G_L}{G_{L-1}} \sim \frac{2L}{\rho}. \qquad \blacksquare \qquad (4.45)$$

Example 4.9 (confluent hypergeometric functions in various directions). Next to the special cases of the previous example we consider some recurrences satisfied by the confluent hypergeometric functions themselves. We refer to [2, Chap. 13] for general information on these functions. We recall the definition in the form of the hypergeometric series,

$$_1F_1\left(\begin{matrix}a\\c\end{matrix}; x\right) = \sum_{n=0}^{\infty} \frac{(a)_n}{n!(c)_n} x^n. \qquad (4.46)$$

This function is also denoted by $M(a, c, x)$. Recursion can be considered with respect to a, to c, and to a and c together, all in forward and backward directions. In fact we can consider recursion with respect to n, $M(a + \epsilon_1 n, c + \epsilon_2 n, x)$, where $\epsilon_j = -1, 0, 1$, $j = 1, 2$ (ϵ_j not both zero). We see that we have eight possible directions of recursions, which we denote by

$$(++), \quad (+0), \quad (+-), \quad (0+), \quad (0-), \quad (-+), \quad (-0), \quad (--). \qquad (4.47)$$

Because of the relation

$$M(a, c, x) = e^z M(b - c, c, -x), \qquad (4.48)$$

not all eight recursions need to be considered. Case 4 is related to case 1, case 7 to case 2, and case 8 to case 5, although there is a change of sign in x, which may be important enough to consider each case separately. In this example we give information on the three cases in which only the $+$ directions occur.

Recursion in the $(+0)$ direction

The recurrence relation $y_{n+1} + b_n y_n + a_n y_{n-1} = 0$ satisfied by $M(a+n, c, x)$ has coefficients

$$b_n = -\frac{2n + 2a + x - c}{n + a}, \quad a_n = \frac{a + n - c}{n + a}. \qquad (4.49)$$

A second solution is given by

$$f_n = \Gamma(1 + a + n - c) U(a + n, c, x), \qquad (4.50)$$

where $U(a, c, x)$ is the standard second solution of the hypergeometric differential equation (see (2.136)).

Perron's theorem does not provide information regarding the existence of minimal solutions. Using the known asymptotic behavior [219], and writing $g_n = M(a + n, c, x)$, we have

$$\frac{f_n}{g_n} \sim e^{-4\sqrt{nx}}. \qquad (4.51)$$

Therefore, f_n is minimal for $x > 0$ when $n \to +\infty$.

This information is useful for computing modified Bessel functions $K_\nu(x)$, particularly when x is large. Indeed, defining $z_n(x) = U(\nu + \frac{1}{2} + n, 2\nu + 1, 2x)$, then $K_\nu(x)$ can be written as

$$K_\nu(x) = \pi^{1/2}(2x)^\nu e^{-x} z_0(x). \tag{4.52}$$

Therefore $K_\nu(x)$ can be related to the recursion of a minimal solution, in particular, to the minimal solution of the $(+0)$ confluent recurrence. In §4.6, we analyze in detail methods for the computation of minimal solutions and in particular Miller's method. This confluent recurrence relation is used for computing $K_\nu(x)$ for real ν [185, 209] and for computing $K_{ia}(x)$ for real a [91, 94, 95].

Recursion in the $(0+)$ direction

For simplicity in the coefficients, it is preferable to consider the recurrence relation $y_{n+1} + b_n y_n + a_n y_{n-1} = 0$ satisfied by $g_n = U(a, c + n, x)$, which has as a second independent solution $f_n = \Gamma(c + n - a)M(a, c + n, x)/\Gamma(c + n)$. The coefficients are

$$b_n = \frac{1 - c - n}{x} - 1, \quad a_n = \frac{c + n - a - 1}{x}. \tag{4.53}$$

Perron's theorem gives a positive answer with respect to the existence of a minimal solution and provides the estimates

$$\frac{f_n}{f_{n-1}} \sim 1, \quad \frac{g_{n+1}}{g_n} \sim \frac{n}{x}, \quad n \to +\infty, \tag{4.54}$$

which indeed correspond to the known asymptotic behavior for the functions defined above (see [219]) and show that f_n is minimal.

Recursion in the $(++)$ direction

For simplicity in the coefficients, we consider the recurrence relation $y_{n+1} + b_n y_n + a_n y_{n-1} = 0$ satisfied by

$$f_n = \frac{1}{\Gamma(c+n)} M(a + n, c + n, x). \tag{4.55}$$

The coefficients are

$$b_n = \frac{c + n - x - 1}{(a+n)x}, \quad a_n = -\frac{1}{(a+n)x}, \tag{4.56}$$

and a second solution is provided by $g_n = (-1)^n U(a + n, c + n, x)$.

Perron's theorem guarantees the existence of a minimal solution. The asymptotic behavior of the minimal (f_n) and the dominant (g_n) solutions obtained from Perron's theorem is

$$\frac{f_n}{f_{n-1}} \sim \frac{1}{n}, \quad \frac{g_n}{g_{n-1}} \sim -\frac{1}{x}, \quad n \to +\infty. \tag{4.57}$$

Comparing this with the known asymptotic behavior (see [213]), one can confirm that f_n, as defined above, is minimal and g_n is dominant. ∎

4.5. Some notable recurrence relations

Example 4.10 (parabolic cylinder functions). The *parabolic cylinder functions* $U(a, z)$ and $V(a, z)$ are linearly independent solutions of the differential equation

$$\frac{d^2w}{dz^2} - \left(\frac{1}{4}z^2 + a\right) w = 0, \tag{4.58}$$

with Wronskian $\mathcal{W}[U, V] = \sqrt{2/\pi}$. We give a few properties of these functions. For more information we refer to §12.5 or [2, Chap. 19].

Parabolic cylinder functions are particular cases of confluent hypergeometric functions but, due to their wide applicability, deserve a separate analysis. It is easy to verify that the following functions are even and odd solutions of (4.58):

$$y_1(z) = e^{-\frac{1}{4}z^2} {}_1F_1\left(\begin{array}{c}\frac{1}{2}a + \frac{1}{4} \\ \frac{1}{2}\end{array}; \frac{1}{2}z^2\right), \quad y_2(z) = ze^{-\frac{1}{4}z^2} {}_1F_1\left(\begin{array}{c}\frac{1}{2}a + \frac{3}{4} \\ \frac{3}{2}\end{array}; \frac{1}{2}z^2\right), \tag{4.59}$$

which shows the relations with the confluent hypergeometric functions. The functions $U(a, z)$ and $V(a, z)$ are defined as follows:

$$U(a, z) = \sqrt{\pi} 2^{-\frac{1}{4} - \frac{1}{2}a} \left[\frac{y_1(z)}{\Gamma(\frac{1}{4} + \frac{1}{2}a)} - \frac{\sqrt{2}\, y_2(z)}{\Gamma(\frac{3}{4} + \frac{1}{2}a)}\right],$$

$$V(a, z) = \frac{1}{\pi}\Gamma\left(\frac{1}{2} + a\right) [\sin(\pi a)U(a, z) + U(a, -z)]. \tag{4.60}$$

These functions satisfy the following different recurrence relations (see [2, §19.6]):

$$U(a - 1, x) = xU(a, x) + \left(a + \tfrac{1}{2}\right) U(a + 1, x) \tag{4.61}$$

and

$$V(a + 1, x) = xV(a, x) + \left(a - \tfrac{1}{2}\right) V(a - 1, x), \tag{4.62}$$

where we have denoted z by x because we will consider real variables.

The recurrence for the U-function in (4.61) has as a second solution $\Gamma(\frac{1}{2} - a)V(a, x)$. A similar situation happens for the recurrence for the V-function, (4.62), which is satisfied by $U(a, x)/\Gamma(\frac{1}{2} - a)$. Exceptions are the negative semi-integer values for the U-recursions and positive semi-integer values for the V-recursion.

Recursion in the positive a direction

For positive a we can take as a linearly independent pair of solutions of (4.61) the pair

$$\{y_{1,a}, y_{2,a}\} = \left\{U(a, x), e^{i\pi a}V(a, x)/\Gamma\left(a + \tfrac{1}{2}\right)\right\}, \tag{4.63}$$

which can also be used in the negative direction, but not when $a = -k + 1/2$, $k \in \mathbb{N}$, because for these values we have $y_{2,a} = 0$.

For the recurrence relation (4.61) Perron's theorem provides no information regarding the existence of minimal solutions. Hence, asymptotic information is needed in order to reveal the behavior of the solutions with respect to the parameter a. For this information we refer to [221].

For this recurrence and for $a \to +\infty$ it follows that $U(a, x)$ is minimal. Indeed, from [221, eqs. (2.29), (2.34)] and the second line in (4.60) we obtain

$$|y_1(a)/y_2(a)| = \left|\Gamma\left(a + \tfrac{1}{2}\right) U(a, x)/V(a, x)\right| \sim \pi \exp(-2\sqrt{ax}). \quad (4.64)$$

Therefore, $U(a, x)$ is minimal for $x > 0$ as $a \to +\infty$.

Recursion in the negative a direction

In this case the situation is quite different. Now we consider the recurrence relation (4.62), which is satisfied by the following pair of linearly independent solutions:

$$\{w_1(a), w_2(a)\} = \left\{V(a, x), U(a, x)/\Gamma\left(\tfrac{1}{2} - a\right)\right\}. \quad (4.65)$$

This pair can also be used for positive a but not for $a = -1/2 + k, k \in \mathbb{N}$.

Considering [221, eqs. (2.23), (2.27)] we see that, as $a \to -\infty$,

$$\begin{aligned}
w_2(a) &= \frac{U(a, x)}{\Gamma(\tfrac{1}{2} - a)} = \frac{2h(a)}{\Gamma(\tfrac{1}{2} - a)} \left[\cos \phi(a, x) + \mathcal{O}\left(a^{-1}\right)\right], \\
w_1(a) &= V(a, x) = \frac{2h(a)}{\Gamma(\tfrac{1}{2} - a)} \left[\sin \phi(a, x) + \mathcal{O}\left(a^{-1}\right)\right],
\end{aligned} \quad (4.66)$$

where $\phi(a, x) = \tfrac{1}{2}a\pi - \tfrac{1}{4}\pi + \mathcal{O}(a^{-1})$ and $h(a) = 2^{-\tfrac{1}{2}} e^{\tfrac{1}{2}a}(-a)^{-\tfrac{1}{2}a - \tfrac{1}{4}}$.

Therefore, any solution of (4.62) has the asymptotic behavior

$$y(a) = C \frac{2h(a)}{\Gamma(\tfrac{1}{2} - a)} \left[\sin(\phi(a, x) + \Phi) + \mathcal{O}\left(a^{-1}\right)\right] \quad (4.67)$$

for some constants C and Φ, and no minimal solution exists as $a \to -\infty$.

From more detailed information [221] on the asymptotic behavior of $U(a, x)$ and $V(a, x)$, it follows that the transition from monotonic to oscillatory behavior, as a function of a, takes place when the parabola $\tfrac{1}{4}x^2 + a = 0$ is crossed, and that the behavior of the solutions for negative a in the nonoscillatory region $x^2/4 + a > 0$ is similar to the behavior for positive a, particularly for not too small x [86]. See also Figure 4.3.

When $a = -n - \tfrac{1}{2}$, the function $U(a, x)$ becomes a Hermite polynomial. We have

$$U\left(-n - \tfrac{1}{2}, x\right) = 2^{-\tfrac{1}{2}n} e^{-\tfrac{1}{4}x^2} H_n(x/\sqrt{2}), \quad n = 0, 1, 2, \ldots. \quad (4.68)$$

All n zeros of the Hermite polynomial $H_n(x/\sqrt{2})$ are within the range $\tfrac{1}{4}x^2 < n + \tfrac{1}{2}$. ■

4.5.2 The Gauss hypergeometric family

Legendre functions are an important special case of the Gauss hypergeometric functions, and we consider two types of recursion. Two recursions for the Gauss functions themselves will be considered in Example 4.12.

4.5. Some notable recurrence relations

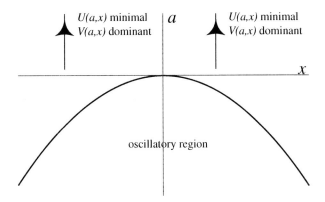

Figure 4.3. *The parabola $a + \frac{1}{4}x^2 = 0$. Inside the parabola, that is, when $a + \frac{1}{4}x^2 < 0$, the functions $U(a,x)$ and $V(a,x)$ are oscillatory. The recurrence relations (4.61) and (4.62) have no dominant or minimal solutions inside the parabola.*

Example 4.11 (Legendre functions with respect to order and degree). We consider the Legendre functions of the first and second kind, $P_\nu^m(z)$ and $Q_\nu^m(z)$, for $z \in \mathbb{C}$ in the half-plane $\Re z > 0$, $z \notin (0, 1]$. The degree ν is a complex number, $\nu \neq -1, -2, \ldots$, and the order m is a nonnegative integer. The recurrence relations with respect to the order m and the degree ν have coefficients which are simple rational functions of m and ν.

Recursion with respect to the order m

Both $P_\nu^m(z)$ and $Q_\nu^m(z)$ satisfy the following recurrence relation:

$$y_{m+1} + \frac{2mz}{\sqrt{z^2-1}} y_m + (m+\nu)(m-\nu-1) y_{m-1} = 0. \tag{4.69}$$

Perron's theorem is positive regarding the existence of a minimal solution. According to this theorem, there exist two linearly independent solutions such that

$$\frac{f_m}{f_{m-1}} \sim mt_1, \quad \frac{g_m}{g_{m-1}} \sim mt_2, \tag{4.70}$$

where

$$t_1 = -\sqrt{\frac{z-1}{z+1}}, \quad t_2 = 1/t_1. \tag{4.71}$$

We see that (because $\Re z > 0$, $z \notin (0, 1]$) $|t_1| < 1 < |t_2|$ and f_m is minimal. From known analytical information (see [75, §6]), it is observed that P_ν^m is the minimal solution as $m \to +\infty$.

Particular cases are prolate and oblate spheroidal harmonics and toroidal functions, discussed in Chapter 12. Also, the *conical functions* $P_{-1/2+i\tau}^m$ and $Q_{-1/2+i\tau}^m$, $\tau \in \mathbb{R}$, are important examples, particularly the P-function, which is real for real $z > -1$ and for real $z > 1$ satisfies (4.69), that is,

$$y_{m+1} + \frac{2mz}{\sqrt{z^2-1}} y_m + ((m-1/2)^2 + \tau^2) y_{m-1} = 0. \tag{4.72}$$

Recursion with respect to the degree ν

Both $P_\nu^m(z)$ and $Q_\nu^m(z)$ satisfy the following recurrence relation:

$$y_{\nu+1} - z\frac{2\nu+1}{\nu-m+1}y_\nu + \frac{\nu+m}{\nu-m+1}y_{\nu-1} = 0. \tag{4.73}$$

Perron's theorem is positive regarding the existence of a minimal solution. According to this theorem, there exist two independent solutions such that

$$\frac{f_\nu}{f_{\nu-1}} \sim t_1, \quad \frac{g_\nu}{g_{\nu-1}} \sim t_2, \tag{4.74}$$

where

$$t_1 = z + \sqrt{z^2-1}, \quad t_2 = 1/t_1. \tag{4.75}$$

We see that (because $\Re z > 0$, $z \notin (0,1]$) $|t_1| > 1 > |t_2|$ and f_ν is minimal. From known analytical information (see [75, §6]), it is observed that Q_ν^m is the minimal solution as $\nu \to +\infty$. ∎

Example 4.12 (Gauss hypergeometric functions). The condition of the recurrence relations satisfied by families of Gauss hypergeometric functions

$$y_n(z) = {}_2F_1\left(\begin{matrix}a+\epsilon_1 n, \; b+\epsilon_2 n \\ c+\epsilon_3 n\end{matrix}; z\right) \tag{4.76}$$

was investigated in [96] for the case $\epsilon_j = -1, 0, 1$ and not all ϵ_j equal to 0. All these recurrence relations $y_{n+1} + b_n y_n + a_n y_{n-1} = 0$ are such that the coefficients a_n and b_n have finite limits as $n \to +\infty$ except, possibly, at one or several of the singular points $z = 0, 1, \infty$. Therefore, the first case of Theorem 4.6 applies and the characteristic roots t_1 and t_2 are finite except, possibly, at the singular points.

For $|\epsilon_j| \leq 1$, the following result was obtained.

Theorem 4.13. *Around $z = 0$ the functions*

$$y_n = {}_2F_1\left(\begin{matrix}a+\epsilon_1 n, \; b+\epsilon_2 n \\ c+\epsilon_3 n\end{matrix}; z\right) \tag{4.77}$$

are minimal solutions as $n \to +\infty$ of the corresponding TTRR if and only if $\epsilon_3 > 0$. The minimality holds in the open connected region including $z = 0$, where the characteristic roots have different moduli.

It appears that this result remains true for all $\epsilon_j \in \mathbb{Z}$, $\epsilon_3 > 0$. This result is a consequence of the fact that $y_n/y_{n-1} \sim \mathcal{O}(1)$ around $z = 0$ and the fact that $t_1(z)t_2(z)$ appears to be singular at $z = 0$ when $\epsilon_3 > 0$.

From this result, minimal solutions for recurrence relations for the functions (4.76) can be built for all z in the complex plane, except on the critical curves $|t_1| = |t_2|$. In [96] it is shown that the 26 recurrence relations for $|\epsilon_i| \leq 1$ reduce to only four independent recursions. An important tool to achieve this is the set of the linear transformation formulas given in (2.83).

The representative recurrence relations chosen in [96] are $(++0), (00+), (++-)$, and $(+0-)$. The first two cases are related to *Norlünd's continued fraction* [158], which is also described by Ince [110, 111]. ∎

4.6 Computing the minimal solution of a TTRR

The solutions of a TTRR are uniquely determined by providing two initial conditions.

For a minimal solution, because only the backward recursion is well conditioned, we need to compute two starting values f_N and f_{N-1} for obtaining f_k, $k < N$, using backward recursion.

When such values f_N and f_{N-1} can be accurately computed by some means and we can apply the recurrence relation without overflow/underflow limitations, this is certainly a possibility for computing minimal solutions. This approach has been considered in Algorithm 12.9.

It is also possible to look for an alternative way to compute the two conditions for backward recurrence, taking into account the minimality of the solution.

4.6.1 Miller's algorithm when a function value is known

In *Miller's algorithm* the two prerequisites for determining the minimal solution are, first, the fact that $\lim_{n \to +\infty} f_n/g_n = 0$ for any other linearly independent solution g_n and, second, a normalization condition.

Assume that we want to compute the sequence

$$f_0, f_1, f_2, \ldots, f_k, \quad (4.78)$$

where k is a given (possibly large) integer, and f_n is the minimal solution.

The idea is to start the recurrence from a sufficiently large $N \gg k$ and to generate a sequence of values y_n, $n = N, N-1, \ldots, 1, 0$. Because f_n is minimal the ratios y_k/y_{k-1} will approach the ratios of the minimal solution f_k/f_{k-1} when $k \ll N$ (see (4.13)). This is true for any set of starting values $\{y_N, y_{N-1}\}$ (not both equal to zero).

A possible choice is $y_N = 0$, $y_{N-1} = 1$. With this choice, the deviation of y_k/y_{k-1} from f_k/f_{k-1} is given by (4.36), which, since f_n is minimal, goes to zero as $N \to +\infty$. Therefore, by using backward recursion, we can compute accurately the ratios f_k/f_{k-1}, $k \ll N$, and hence the minimal solution is determined up to a normalization constant. This constant can be determined in different ways.

If the function f_0 is not zero and a stable and efficient method of numerical computation is known for f_0, then we can generate the approximations

$$\hat{f}_n = \frac{f_0}{y_0} y_n \simeq f_n, \quad n = 0, 1, \ldots, k \ll N. \quad (4.79)$$

If the value of f_0 is accurate, we only need to estimate N for starting the backward recurrence. This can be done by using (4.36); the following estimate will be enough in most occasions:

$$\epsilon_r(k, m) = \frac{y_k/y_{k-1}}{f_k/f_{k-1}} - 1 \simeq \frac{r_N}{r_k}. \quad (4.80)$$

ALGORITHM 4.1. Miller's method for nonvanishing minimal solutions.
Input: ϵ (tolerance), f_0, k

Output: $\hat{f}_n$, $n = 0, \ldots, k$ (numerical approximations to f_n)

- Choose N such that $\left|\frac{r_N}{r_k}\right| < \epsilon$
- Set $y_N = 0$, $y_{N-1} = 1$.
- DO $n = N - 1, \ldots, 1$

$$y_{n-1} = -\frac{1}{a_n}(y_{n+1} + b_n y_n).$$

- DO $n = 0, \ldots, k$

$$\hat{f}_n = \frac{f_0}{y_0} y_n.$$

For a numerical verification we can repeat this algorithm with a larger value of N, and compare the results with those obtained earlier.

Example 4.14. Let us apply Algorithm 4.1 for computing modified Bessel functions $y_n = I_{n+1/2}(x)$. We have

$$y_0 = \sqrt{\frac{2}{\pi x}} \sinh x \qquad (4.81)$$

and

$$y_{n-1} = y_{n+1} + \frac{2n+1}{x} y_n. \qquad (4.82)$$

From Perron's theorem we know that, if k is large,

$$r_k/r_{k-1} \sim \left(\frac{x}{2k+1}\right)^2. \qquad (4.83)$$

With this we can estimate $r_N/r_k \sim (x/(2k))^{2(N-k)}$. Then we have, for $x = 5$ and $k = 20$, $|r_N/r_k| < 10^{-16}$ if $N > 29$.

Starting the recursion from $y_{30} = 0$, $y_{29} = 1$ and proceeding as indicated in Algorithm 4.1 we find that the numerical values $\hat{f}_n$ are correct with 16-digit accuracy for $0 \le n \le 22$. ∎

Remark 3. In order to minimize the risk of overflow in the computation of the function values $y_N, \ldots, y_0$, one can consider the recurrence relation for ratios of functions. We have

$$H_n = -\frac{a_n}{b_n + H_{n+1}}, \quad n = N - 1, \ldots, 1, \qquad (4.84)$$

where $H_n = y_n/y_{n-1}$, assuming that the y_n do not vanish. Then, the numerical estimations can be computed as follows:

$$\hat{f}_0 = f_0, \quad \hat{f}_n = H_n \hat{f}_{n-1}, \quad n = 1, \ldots, k. \qquad (4.85)$$

Also, the ideas described in §4.3.1 can be used to enlarge the computable range.

4.6.2 Miller's algorithm with a normalizing sum

When the function y_0 is difficult to compute or may become zero, Algorithm 4.1 or the variant discussed in Remark 3 should not be used. Zeros occur, for instance, in the case of the computation of the Bessel functions $J_\nu(x)$, which are oscillatory for positive x. For these cases, alternative normalizing conditions should be used. When a sum for the minimal solutions

$$S = \sum_{n=0}^{\infty} \lambda_n f_n \qquad (4.86)$$

is known, we can use this as a normalizing condition. By computing the series for the numerically generated functions y_n and comparing with S, the normalization can be fixed. That is, assuming that

$$f_n \simeq C y_n \qquad (4.87)$$

is true within the required accuracy, which is possible if $n \ll N$ (with C not depending on n), we have

$$S = \sum_{n=0}^{\infty} \lambda_n f_n \approx C \sum_{n=0}^{p} \lambda_n y_n = S_p, \qquad (4.88)$$

where $p \leq N$ should be chosen considering the speed of convergence of the series and the requested accuracy.

Then, using the estimate $C \approx S/S_p$ we can compute the numerical approximations $\hat{f}_n \approx C y_n$.

The series S_p can be evaluated together with the application of backward recursion. Defining

$$s_n = \frac{1}{y_{n-1}} \sum_{m=n}^{p} \lambda_m y_m, \qquad (4.89)$$

and assuming that the y_n do not vanish, we have

$$s_n = H_n(\lambda_n + s_{n+1}), \qquad (4.90)$$

and we can compute this recursively until we arrive at s_1, with $S_p = (\lambda_0 + s_1)y_0$. The numerical approximations provided by the algorithm would be

$$\hat{f}_n = \frac{S}{S_p} y_n, \qquad (4.91)$$

which, in terms of the ratios H_n can also be generated as $\hat{f}_n = H_n \hat{f}_{n-1}$, $n = 1, \ldots, k$, starting from $\hat{f}_0 = S y_0/S_p = S/(\lambda_0 + s_1)$.

In summary, combining this with the backward recurrence with starting value $H_N = 0$ (as given in Remark 3) and taking $p = N$ in (4.88), we can write the following algorithm. Again, we can repeat this algorithm with a larger N for a numerical verification.

ALGORITHM 4.2. Miller's algorithm with a normalizing sum.
Input: ϵ (tolerance), $S = \sum_{n=0}^{\infty} \lambda_n f_n$, k

Output: $\hat{f}_n$, $n = 0, \ldots, k$ (numerical approximations to f_n)

- Choose N "large enough"
- Set $H_N = 0$, $s_N = 0$.
- DO $n = N - 1, \ldots, 1$
$$H_n = -\frac{a_n}{b_n + H_{n+1}},$$
$$s_n = H_n(\lambda_n + s_{n+1}).$$
- $\hat{f}_0 = \dfrac{S}{\lambda_0 + s_1}.$
- DO $n = 1, \ldots, k$
$$\hat{f}_n = H_n \hat{f}_{n-1}.$$

To prove that this is indeed a useful algorithm, several questions should be addressed. First, we will prove that under certain conditions for the dominant solutions the results tend to improve as larger N are chosen; in other words, we will obtain the conditions for which the method converges to the minimal solution as $N \to +\infty$. Second, because the method is based on ratios of functions $H_n = y_n/y_{n-1}$ and some of the values y_n, $n = 0, \ldots, k$, may vanish numerically, we need to prove the stability of the method in this case. In the third place, one has to consider the numerical stability of the evaluation of the normalizing conditions. Finally, it is necessary to specify what is meant by "choose N 'large enough'" in the first line of the algorithm.

Condition for convergence

Proof of convergence. To prove the convergence of the algorithm as $N \to +\infty$, we compare the numerical approximations $\hat{f}_n = \dfrac{S}{S_N} y_n$, which are obtained from the starting values $y_N = 0$, $y_{N-1} = 1$, with the values of the minimal solution f_n.

Let g_n be a dominant solution; we have $y_n = \mu(f_n - r_N g_n)$. Now

$$S_N = \sum_{n=0}^{N} \lambda_n y_n = \mu \left(S - \sum_{n=N+1}^{\infty} \lambda_n f_n - r_N \sum_{n=0}^{N} \lambda_n g_n \right) \quad (4.92)$$
$$= \mu(S - \bar{S}_N^f - r_N S_N^g),$$

where, because the series $S = \sum_{m=0}^{\infty} \lambda_m y_m$ is assumed to be convergent, we have

$$\bar{S}_N^f = \sum_{n=N+1}^{\infty} \lambda_n f_n \to 0 \quad \text{as } N \to +\infty. \quad (4.93)$$

4.6. Computing the minimal solution of a TTRR

The relative error of the numerical approximation when $n < N$ can be evaluated as follows:

$$\epsilon_{n,N} = \frac{\hat{f}_n - f_n}{f_n} = \frac{S}{S_N} \frac{y_n}{f_n} - 1 = \mu \frac{S}{S_N}(1 - r_N/r_n) - 1 = \frac{S(1 - r_N/r_n)}{S - \bar{S}_N^f - r_N S_N^g} - 1. \quad (4.94)$$

Because $r_N \to 0$ and $\bar{S}_N^f \to 0$ as $N \to +\infty$ we infer that the relative error $\epsilon_{n,N}$ tends to 0 (i.e., the algorithm converges) if and only if

$$\lim_{N \to +\infty} r_N S_N^g = \lim_{N \to +\infty} r_N \sum_{n=0}^{N} \lambda_n g_n = 0. \quad \square \quad (4.95)$$

Stability of the recursion for H_n

The numerical stability of the recurrence relation for the ratios H_n, (4.84), when a function value y_{n-1} is nearly vanishing is not a problem. Indeed, if it happens that $b_n + H_{n+1} = \epsilon$ with ϵ a small number, then $H_n = -a_n/\epsilon$ will carry a loss of significant digits, as corresponds to the fact that $y_{n-1} \simeq 0$. Then $H_{n-1} = -a_{n-1}/(b_{n-1} + a_n/\epsilon) \simeq -\epsilon a_{n-1}/a_n$, which, as expected, is small and carries loss of digits, again because $y_{n-1} \simeq 0$. But then $H_{n-2} = -a_{n-2}/(b_{n-1} + H_{n-1})$ and the loss of digits will be unnoticeable for H_{n-2}, as we require. A more detailed analysis reveals that the condition number of H_{n-2} with respect to the variation of ϵ is

$$\frac{\epsilon}{H_{n-2}} \frac{\partial H_{n-2}}{\partial \epsilon} = \frac{-a_{n-1}}{b_{n-2} a_n} \epsilon + \mathcal{O}(\epsilon^2). \quad (4.96)$$

Therefore, the computation is safe unless $a_n b_{n-2}/a_{n-1}$ is very small.

In the extremely improbable situation in which $b_n + H_{n+1} = 0$ (or very close to zero) such that H_n would overflow, we should take $b_n + H_{n+1} = \epsilon$ with ϵ small but such that $1/\epsilon$ does not overflow, and then continue the process.

Stability in the computation of the normalizing sum

The third condition that must be satisfied is the stability in the computation of the normalizing condition. This is a far more important condition to verify, because it may happen that the computation based on the recursion becomes strongly unstable, spoiling all successive values. The ideal situation is to use series for which all the terms have the same sign and they decrease fast enough. A canonical example of a sum rule that should never be used is (see [2, eq. (9.6.38)])

$$e^{-z} = I_0(z) + 2 \sum_{n=1}^{\infty} (-1)^n I_n(z), \quad (4.97)$$

which carries huge cancellations (because $I_n(z) \sim e^z/\sqrt{2\pi z}$ for $\Re z \to +\infty$).

A much more adequate possibility would be, of course,

$$e^z = I_0(z) + 2 \sum_{n=1}^{\infty} I_n(z), \quad (4.98)$$

whose sum has the correct asymptotic behavior for large positive values of $\Re z$.

Estimation of N for the starting value

Finally, the value N for starting the recurrence process should be chosen by estimating the relative error (4.94) using some asymptotic estimates. We have

$$\epsilon_{n,N} \approx \frac{1}{S}(\bar{S}_N^f + r_N S_n^g) - \frac{r_N}{r_n} \approx \frac{1}{S}(\lambda_{N+1} f_{N+1} + r_N \lambda_N g_N) - \frac{r_N}{r_n}, \qquad (4.99)$$

where we have estimated $\bar{S}_N^f$ with its first term and S_n^g with its last term.

The second term in (4.99) is equal to $\lambda_N f_N$ and, if the series converges fast enough, we can neglect it with respect to the first term. In addition, if we are computing functions in the range $0 \leq n \leq k$, because in most occasions r_n is decreasing (note that $r_n \to 0$ as $n \to \infty$), we can set $n = k$ in the last term and estimate

$$\epsilon_N \approx \frac{1}{S}\lambda_{N+1} f_{N+1} - \frac{f_k}{g_k}\frac{g_N}{f_N}. \qquad (4.100)$$

Hence, the relative error has two main contributions: one corresponding to the error in the normalization condition, which is estimated by considering the first neglected term in the sum S, and one corresponding to the condition of backward recursion, which can be approximated by (4.80). For the second term, estimations can be obtained from Perron's theorem, as done in (4.14).

Possible improvements of Algorithm 4.2 can be considered in order to reduce the error and to avoid the a priori selection of a given value. For instance, the recurrence for the ratios H_n, which starts with $H_N = 0$, can instead be started with the actual ratio $H_n = f_n/f_{n-1}$ computed in the form of a continued fraction (Theorem 4.6) with prescribed accuracy. Also, strategies can be considered for a recurrent evaluation of the sum $\hat{S}_N^f$ (see [75]).

Many examples of application of Miller's method are given in Gautschi's article [75]: Bessel and modified Bessel functions, Legendre functions (including conical functions), Coulomb wave functions, and incomplete beta and gamma functions.

4.6.3 "Anti-Miller" algorithm

When initial values for a dominant solution are known, it is possible to compute values for the minimal solution using the Casorati determinant.

Let us assume that the pair $\{f_n, g_n\}$ is a linearly independent pair of solutions of a TTRR, with f_n minimal and g_n dominant. Further, we assume that g_0 and g_1 are accurately known and that, therefore, the solution g_n can be accurately generated by forward recursion.

Then, if the Casorati determinant, (4.34), D_N is known, because g_{N-1} and g_N can be computed by recursion and $H_N^f = f_N/f_{N-1}$ can be computed by means of a continued fraction, we can use these values to compute f_N and f_{N-1}. Indeed,

$$f_{N-1} = \frac{1}{g_{N-1}}\frac{D_N}{H_N^g - H_N^f}, \quad f_N = H_N f_{N-1} = \frac{1}{g_N}\frac{D_N}{(H_N^g)^{-1} - (H_N^f)^{-1}}, \qquad (4.101)$$

where $H_N^g = g_N/g_{N-1}$.

4.6. Computing the minimal solution of a TTRR

This appears to be a numerically stable step when, as is the case, one of the solutions involved is minimal and N is large enough. Indeed, $H_N^g - H_N^f \neq 0$ because g_n and f_n are linearly independent. Furthermore, in the cases for which Theorem 4.6 provides a positive answer regarding the existence of minimal solutions, H_n^f and H_n^g have different asymptotic behavior as $n \to +\infty$.

Only when one of the functions $f_n, g_n, n = N, N-1$, vanishes may we encounter instabilities. However, it is easy to eliminate this problem by considering the following result.

Theorem 4.15. *Let y_n be a solution of $y_{n+1} + b_n y_n + a_n y_{n-1} = 0$. If there exists an integer k such that $b_n b_{n-1} - 4a_n > 0$ and $b_n b_{n-1} > 0$ for all $n \geq k$, then there exists at most one value $m \geq k$ for which $y_m = 0$.*

Proof. Let us define a new set of functions $\tilde{y}_n$ such that $y_n = \gamma_n \tilde{y}_n$, with $\gamma_{n+1}/\gamma_n = -b_n/2$ (for instance, taking $\gamma_n = (-1)^n 2^{-n} b_0 \cdots b_{n-1}$). Then

$$\tilde{y}_{n+1} - 2\tilde{y}_n + \frac{4a_n}{b_n b_{n-1}} \tilde{y}_{n-1} = 0. \tag{4.102}$$

Using the difference operator $\Delta \tilde{y}_j = \tilde{y}_{j+1} - \tilde{y}_j$ we can write

$$\Delta^2 \tilde{y}_{n-1} + \left[-1 + \frac{4a_n}{b_n b_{n-1}} \right] \tilde{y}_{n-1} = 0. \tag{4.103}$$

From this equation, it is easy to check that, assuming $\tilde{y}_m = 0$, $m \geq k$, then the values $\tilde{y}_n$, $n \geq m + 2$, have the same sign as $\tilde{y}_{m+1}$, which cannot be zero if $\tilde{y}_n$ is not a trivial solution. □

We conclude that if the condition of Theorem 4.15 is fulfilled, then g_n or f_n can vanish for only one value of $n \geq k$, and by choosing k high enough the method is free of cancellation problems.

Notice that in the first case in Perron's theorem, $\Omega_n \to -1$, and in the second case $\Omega_n \to -1 + 4a/b^2$, which is indeed negative when $|t_1| \neq |t_2|$. Therefore, in these cases there exists a value k such that the condition $b_n b_{n-1} - 4a_n > 0$ for all $n \geq k$ of Theorem 4.15 is fulfilled.

A particular case of Theorem 4.15 is the following corollary.

Corollary 4.16. *If $a_n < 0$ and b_n has constant sign, then, given a solution y_n of the TTRR $y_{n+1} + b_n y_n + a_n y_{n-1} = 0$, there is at most one value of n for which $y_n = 0$.*

Furthermore, in the case of the previous theorem we have $H_n^f H_n^g < 0$ for large enough n (see Theorem 4.6), which makes the computation of (4.101) definitively stable.

When Theorem 4.15 applies, in the worst-case scenario in which the only possible zero of g_N (or g_{N-1}) is hit (exactly or approximately) when applying (4.101), only by considering a larger value of N will the problem be solved. A way to test for this possible problem is to compare H_n^g with the estimate from Perron's theorem. Also, one can keep track of possible cancellations when applying the recurrence to compute g_N and g_{N-1}. With respect to possible accidental cancellations of f_N or f_{N-1} appearing in the evaluation of the continued fraction, similar precautions could be considered.

The following algorithm can therefore be safely used. We call it the anti-Miller algorithm because it starts in the opposite direction compared to Miller's method, using forward recursion for a dominant solution.

ALGORITHM 4.3. Anti-Miller algorithm.
Input: ϵ (tolerance), $g_0, g_1, k, N \geq k$, with k such that $b_n b_{n-1} - 4a_n > 0$ for $n \geq k$.
Output: (Numerical approximations of) $\{f_n, g_n\}, n = 0, \ldots, N$.

- Compute g_0, g_1.

- DO $n = 1, \ldots, N-1$

  $g_{n+1} = -b_n g_n - a_n g_{n-1}$.

- Compute $H_N^g = g_N/g_{N-1}$; compute $H_N^f = f_N/f_{N-1}$ (use continued fractions).

- $f_{N-1} = \dfrac{D_N}{g_{N-1}(H_N^g - H_N^f)}, \ f_N = H_N^f f_{N-1}$.

- DO $n = N-1, \ldots, 1$

  $f_{n-1} = -(b_n f_n + f_{n+1})/a_n$.

In Chapter 12, examples of the application of this method are provided (Bessel and modified Bessel functions of half-integer order, prolate and oblate spheroidal, and toroidal functions).

4.7 Inhomogeneous linear difference equations

4.7.1 Inhomogeneous first order difference equations. Examples

Let us consider the numerical computation of solutions of linear inhomogeneous first order equations

$$y_n = \alpha_n y_{n-1} + \beta_n. \tag{4.104}$$

Examples of functions which satisfy these types of recurrences are, for example, the gamma function, the incomplete gamma function, and the exponential integrals. We will suppose that $\alpha_n \neq 0$ for all n (so that the recursion can be applied in both directions) and that the recurrence is genuinely nonhomogeneous, that is, we take $\beta_n \neq 0$.

The study of the numerical condition of first order inhomogeneous equations resembles very much the case of second order homogeneous equations, and we will encounter cases in which the computation of the solution by forward recursion is a badly conditioned process.

4.7. Inhomogeneous linear difference equations

The general solution of a first order equation can be written as $y_n = f_n + Cg_n$, where f_n is a particular solution of the inhomogeneous equation and g_n is the solution of the homogeneous equation $y_n = a_n y_{n-1}$ (unique up to a constant factor). Once we fix an initial value y_0, a particular solution of the inhomogeneous equation is fixed.

The process of forward recursion will be badly conditioned for a solution f_n when it happens that the ratio $r_n = f_n/g_n$ becomes very small for large n. Indeed, the initial numerical value will never be exactly f_0, but it will be contaminated by some amount, so that instead we will have

$$y_0 = f_0(1+\epsilon) = f_0 + \epsilon \frac{f_0}{g_0} g_0, \qquad (4.105)$$

and hence

$$\frac{y_n}{f_n} - 1 = \epsilon r_0/r_n, \qquad (4.106)$$

which shows that the relative error may become arbitrarily large.

The similarity with the TTRR is clear; furthermore, both cases are equivalent. Indeed, it is a simple matter to check that all the solutions of (4.104) are also solutions of

$$y_{n+1} - \left(\alpha_{n+1} + \frac{\beta_{n+1}}{\beta_n}\right) y_n + \frac{\beta_{n+1}}{\beta_n} \alpha_n y_{n-1} = 0, \qquad (4.107)$$

and that also the solution of the associated homogeneous equation (4.104) satisfies (4.107).

Not all the solutions of (4.107) are solutions of (4.104), but if y_n is a solution of (4.107), then there exists a number C, not depending on n, such that Cy_n is a solution of (4.104), except when y_n is precisely a solution of the homogeneous equation.

What is true for the TTRR regarding the existence of minimal solutions remains true for first order inhomogeneous equations, but with two important new features. First, Definition 4.1 also applies for the solutions of (4.104), but if there exists a minimal solution for (4.104), it is unique and no multiplicative factors are allowed; the minimal solution is truly unique for a first order inhomogeneous equation. Second, the minimal solution of (4.107) may be the solution of the homogeneous equation; this means that (4.107) may have a minimal solution which is not a solution of (4.104).

We can easily adapt Perron's theorem for this case by using (4.107).

Theorem 4.17. *Let $y_n = \alpha_n y_{n-1} + \beta_n$ and let us suppose that $\beta_{n+1}/\beta_n \sim bn^\beta$ and $\alpha_n \sim an^\alpha$ as $n \to +\infty$. Then the following hold:*

1. *If $\beta < \alpha$ or if $\beta = \alpha$ and $b < a$, the recurrence relation has a minimal solution f_n such that*

$$\frac{f_n}{f_{n-1}} \sim bn^\beta, \quad \frac{g_n}{g_{n-1}} \sim an^\alpha, \qquad (4.108)$$

   where g_n is a dominant solution.

2. If $\beta > \alpha$ or if $\beta = \alpha$ and $b > a$, then the recurrence relation does not possess a minimal solution and all solutions satisfy

$$\frac{y_n}{y_{n-1}} \sim bn^\beta. \tag{4.109}$$

3. If $\alpha = \beta$ and $a = b$, the theorem is inconclusive with respect to the existence of minimal solutions. All solutions satisfy

$$\limsup_{n\to+\infty} \left|\frac{y_n}{n!^\beta}\right|^{1/n} = |b|. \tag{4.110}$$

Proof. This result follows easily applying Theorem 4.6 to (4.107).

It is a straightforward matter to check that the third case in Theorem 4.6 cannot take place.

In the second case of Theorem 4.17, the first order inhomogeneous equation has no minimal solution, because the minimal solution of (4.107) is the solution of the homogeneous equation. □

Example 4.18. The incomplete gamma function $\gamma(n, x) = \int_0^x e^{-t} t^{n-1} dt$ satisfies the first order inhomogeneous equation

$$\gamma(n+1, x) = n\gamma(n, x) - x^n e^{-x}, \quad n > 0, \tag{4.111}$$

as follows from integrating by parts.

Because $\beta_{n+1}/\beta_n \sim x$ and $\alpha_n \sim n$ we conclude that there is a minimal solution satisfying

$$\frac{f_n}{f_{n-1}} \sim x, \tag{4.112}$$

and this limit corresponds to the incomplete gamma function (see §8.1).

Therefore, the incomplete gamma function is the minimal solution and should not be computed by forward recursion. For the dominant solutions we have

$$\frac{g_n}{g_{n-1}} \sim n, \tag{4.113}$$

which corresponds with the solution of the homogeneous equation $y_n^H = n!$ and which is the source of the instabilities for f_n.

It can be expected that the condition of the recurrence relation will worsen as x gets smaller, as can be checked by plotting the ratio

$$\rho_n = \frac{y_n^H}{y_1^H} \frac{f_1}{f_n} = \frac{(1-e^{-x})n!}{\gamma(n+1, x)}. \quad \blacksquare \tag{4.114}$$

A second example, in principle with opposite behavior, is provided by the exponential integrals.

4.7. Inhomogeneous linear difference equations

Example 4.19. The exponential integrals

$$E_n(x) = \int_1^{+\infty} e^{-xt} t^{-n} dt, \quad n = 0, 1, 2, \ldots, \tag{4.115}$$

are considered for $x > 0$. Integrating by parts, we see that the functions $y_n = E_{n+1}(x)$ satisfy the recurrence relation

$$y_n = \frac{1}{n}(-x y_{n-1} + e^{-x}). \tag{4.116}$$

We are in the second case of the theorem and the inhomogeneous first order equation does not have minimal solutions.

However, the fact that the α_n-coefficient is negative can cause bad conditioning, particularly when x is large and α_n is a large negative number, whereas the solution is positive.

The potential danger of the forward recurrence can also be observed by considering the associated TTRR $y_{n+1} + b_n y_n + a_n y_{n-1} = 0$ with

$$b_n = (x-n)/(n+1), \quad a_n = -x/(n+1). \tag{4.117}$$

The fact that the coefficient b_n changes sign for $n = [x] = n_0$ and that a_n is negative is, in fact, an indication that the condition of the recurrence process may suffer an inversion when the value n_0 is reached in a forward or backward direction. This type of phenomenon is described in §4.8. ∎

4.7.2 Inhomogeneous second order difference equations

For second order inhomogeneous equations, a new type of asymptotic behavior can be expected. Let us consider the numerical solution of an equation

$$C_n y_{n+1} + B_n y_n + A_n y_{n-1} = D_n. \tag{4.118}$$

Let us suppose that C_n never vanishes. Therefore, we can write

$$y_{n+1} + b_n y_n + a_n y_{n-1} = d_n. \tag{4.119}$$

It is easy to verify that all the solutions of (4.119) are also solutions of

$$d_n y_{n+2} + (d_n b_{n+1} - d_{n+1}) y_{n+1} + (d_n a_{n+1} - d_{n+1} b_n) y_n - a_n d_{n+1} y_{n-1} = 0, \tag{4.120}$$

and that all solutions of (4.120) are solutions of (4.119) when multiplied by an adequate constant factor with the exception of the solution of the homogeneous equation, which is a solution of (4.120) but not of (4.119) (we assume $d_n \neq 0$).

Although the problem is essentially equivalent to solving a third order difference equation, with three independent solutions, the simple scheme which worked so well for second order equations breaks down. Namely, when the recurrence relations possesses a minimal solution, not all the other solutions are dominant. The subdominant but superminimal solution should not be computed either with forward or backward recursion.

Example 4.20. Let $y_{n+1} - 4y_n + 3y_{n-1} = 2^n$. The corresponding third order equation is $y_{n+2} - 6y_{n+1} + 11y_n - 6y_{n-1} = 0$; the roots of the characteristic equation are $t = 1, 2, 3$, and therefore a triplet of independent solutions is given by $\{f_n, h_n, g_n\} = \{1, 2^n, 3^n\}$ and the general solution of the homogeneous equation is $y_n = A + B2^n + C3^n$. Substituting this in the second order nonhomogeneous equation, we find $B = -2$.

The solution f_n is minimal because $\lim_{n \to +\infty} f_n/g_n = \lim_{n \to +\infty} f_n/h_n = 0$, and g_n is dominant because it is also dominant over h_n, that is, $\lim_{n \to +\infty} h_n/g_n = 0$. The solution h_n is neither dominant nor minimal; it is subdominant because $\lim_{n \to +\infty} h_n/g_n = 0$, but it is superminimal because $\lim_{n \to +\infty} h_n/f_n = +\infty$. ∎

Let now the pair $\{f_n \ g_n\}$ be a numerically satisfactory pair of the homogeneous equation

$$y_{n+1} + b_n y_n + a_n y_{n-1} = 0. \tag{4.121}$$

By numerically satisfactory pair, we mean a pair of independent solutions which includes the minimal solution when the recurrence relation admits one; in such a case, we denote by f_n the minimal solution.

Any solution of the inhomogeneous equation (4.119) can be written as

$$y_n = k_1 f_n + k_2 g_n + h_n, \tag{4.122}$$

h_n being any particular solution.

When we want to compute h_n numerically, different strategies can be considered depending on the asymptotic growth of h_n with respect to f_n and g_n. If h_n is a dominant solution of the inhomogeneous equation and $h_n/g_n \to \infty$ and $h_n/f_n \to \infty$ as $n \to +\infty$, then forward recursion from starting values $h_k, h_{k-1}, k < n$, is well conditioned, because if the solution becomes contaminated with f_n and g_n solutions, these components are damped out with respect to h_n for high enough values of n. On the other hand, if h_n is minimal and $h_n/g_n \to 0$ and $h_n/f_n \to 0$ as $n \to +\infty$, then backward recursion starting from h_N, h_{N-1} is the right choice. When, for any of the two previous cases, one or two of the limits are finite, then generally the processes will also be stable.

However, as in Example 4.20, when h_n is subdominant but superminimal because $h_n/f_n \to \infty$ and $h_n/g_n \to 0$ (or $h_n/f_n \to 0$ and $h_n/g_n \to \infty$), neither forward nor backward recursion should be considered. Fortunately, for this type of problem we can use Olver's method [167].

4.7.3 Olver's method

To describe this method, assume that $f_n/g_n \to 0$ and that we want to compute a solution of the inhomogeneous equation which is subdominant ($h_n/g_n \to 0$) but superminimal ($h_n/f_n \to \infty$).

The idea is to pose the problem as a boundary value problem for computing the numerical solution y_n. The boundary values are a known value $y_0 = h_0 = k$ (or a more general normalizing condition), and the asymptotic value $h_n/g_n \to 0$. These two conditions determine univocally a solution.

4.7. Inhomogeneous linear difference equations

The first condition is exactly taken into account while the second one is approximated by assuming $y_N = 0$ for large enough N, and solving the resulting system of equations:

$$\left. \begin{array}{l} b_1 y_1 + c_1 y_2 = d_1 - a_1 k, \\ a_2 y_1 + b_2 y_2 + c_2 y_3 = d_2, \\ \vdots \\ a_{N-2} y_{N-3} + b_{N-2} y_{N-2} + c_{N-1} y_{N-1} = d_{N-2}, \\ a_{N-1} y_{N-2} + b_{N-1} y_{N-1} = d_{N-1}. \end{array} \right\} \quad (4.123)$$

It is easy to prove that, for fixed r, the solution y_r of the system (4.123) approaches the solution h_r as $N \to +\infty$.

Indeed, because the numerical solution y_n, for a fixed N, is a particular solution of the inhomogeneous equation, then

$$y_n = A_N f_n + B_N g_n + h_n, \quad 0 \le n \le N. \quad (4.124)$$

Taking the special values $n = 0, N$ and because $y_0 = h_0 = k$, we have

$$A_N = \frac{g_0 h_N}{f_0 g_N - g_0 f_N}, \quad B_N = -\frac{f_0 h_N}{f_0 g_N - g_0 f_N}. \quad (4.125)$$

The denominators will be different from zero if N is large enough because $f_n/g_n \to 0$. Hence, for large N,

$$y_r = A_N f_r + B_N g_r + h_r \sim \frac{g_0 h_N}{f_0 g_N} f_r - \frac{h_N}{g_N} g_r + h_r \sim h_r. \quad (4.126)$$

Therefore, when there exists a stable method for solving the system of equations and N is large enough, we can solve the problem accurately. There are several possible ways to solve this system [241]. Here, we describe briefly Olver's approach.

The idea is to solve the system by *forward elimination* followed by *backsubstitution*. Forward elimination starts by eliminating y_1 from the first and the second equations and then y_2 is eliminated in the third equation, and so on. The resulting process can be written in the following compact form:

$$p_{n+1} y_n - p_n y_{n+1} = e_n, \quad n = 1, \ldots, N-1, \quad (4.127)$$

where

$$\begin{array}{l} p_{n+1} + b_n p_n + a_n = 0, \\ e_n = a_n e_{n-1} - d_n p_n, \end{array} \quad n = 1, \ldots, N-1, \quad (4.128)$$

where $e_0 = k$, $p_0 = 1$, $p_1 = 1$.

Assume that we computed the values p_n and e_n from the recurrence relations in (4.128). Then we can compute the numerical solutions in the following way. The recurrence relation for p_n is well conditioned (at least for large n). Indeed, because p_n is a solution of the homogeneous recurrence relation, it is a linear combination of f_n and g_n. Using $p_0 = 1$ and $p_1 = 1$, we have

$$p_n = \frac{1}{f_0 g_1 - g_0 f_1} (f_0 g_n - g_0 f_n). \quad (4.129)$$

Therefore p_n has a component of g_n, which is dominant, and therefore p_n is dominant and forward recursion is expected to be a stable process. It follows that if the computation of e_n is stable (which seems to be the case in most occasions [167]), the use of (4.128) will be numerically safe. Theorem 4.17 may be used in analyzing the stability of the process.

After this, backsubstitution starts from (4.127) for $n = N$ (with $y_N = 0$). Using this relation, the successive values $y_{N-1}, \ldots, y_1$ can be generated. Again, the condition can be analyzed, at least asymptotically, using Theorem 4.17.

Further details and examples can be found in Olver's original paper [167] and in Wimp's book [241].

4.8 Anomalous behavior of some second order homogeneous and first order inhomogeneous recurrences

It is usually assumed that asymptotic information is enough for predicting stable directions for recursion, at least for second order homogeneous equations (for the inhomogeneous case or higher order cases, some solutions are badly conditioned in any direction). This, however, is not always true and there exist examples for which, for finite orders n, a minimal solution interchanges its role with certain dominant solutions. This, as a consequence, implies the anomalous convergence of the associated continued fraction to a value different from the ratio of consecutive minimal solutions (a phenomenon first observed by Gautschi [77] in connection with the $(++)$ confluent recurrence).

4.8.1 A canonical example: Modified Bessel function

Let us consider real solutions of real recurrence relations. Simple examples of such recurrences displaying anomalous behavior are provided by those of the form

$$y_{n+1} - y_{n-1} = f(n) y_n, \quad n \in \mathbb{Z}, \quad f(n) = -f(-n). \tag{4.130}$$

We will assume that the recurrence admits a minimal solution as $n \to +\infty, n \in \mathbb{Z}$. Without loss of generality, we can assume that $f(n) > 0$ when $n > 0$; otherwise, the recurrence satisfied by $(-1)^n y_n$ will satisfy this property.

The solutions of these recurrences satisfy the following properties:

1. The solutions are symmetric, that is, $y_n = y_{-n}$.

2. The minimal solution f_n has alternating sign and $|f_n| > |f_{n'}|$ if $|n| < |n'|$; f_n is also minimal as $n \to -\infty$.

3. The dominant solutions g_n with $g_0 g_1 > 0$ have constant sign and satisfy $|g_n| < |g_{n'}|$ if $|n| < |n'|$.

4.8. Anomalous behavior of some recurrences

Indeed, the solutions are symmetric as a consequence of the symmetry of the recurrence. Due to this symmetry, the other two properties only need to be satisfied for $n \geq 0$. The minimal solution as $n \to +\infty$ (due to symmetry, also minimal as $n \to -\infty$) has, on account of Theorem 4.5, alternating sign for large enough $n > 0$; considering backward recursion $y_{n-1} = y_{n+1} - f(n)y_n$ it is clear that it is always alternating and with increasing modulus as $n \geq 0$. Finally, when $g_0 g_1 > 0$, forward recursion $y_{n+1} = y_{n-1} + f(n)y_n$ shows that $g_n > 0$ for all n and it increases as n increases.

Consider now the quantity $R_n = |r_n| = |f_n/g_n|$ with f_n and g_n as described above; as n increases, R_n gives an increasing sequence for negative n and decreasing for positive n. Therefore, the roles of f_n and g_n are reversed when crossing the value $n = 0$, both for forward and backward recursion. For forward recursion with starting negative values of n, g_n decreases while f_n increases, contrary to the situation when $n > 0$.

The simplest example of well-known functions satisfying these properties are modified Bessel functions of integer order. The recurrence relation

$$y_{\nu+1}(x) - \frac{2\nu}{x} y_\nu(x) - y_{\nu-1}(x) = 0 \qquad (4.131)$$

has as a pair of independent solutions $K_\nu(x)$ and $(-1)^{|\nu|} I_\nu(x)$ (or $e^{i\pi\nu} I_\nu(x)$ if we don't mind abandoning the real notation), where the first one is dominant and the second one is minimal.

For integer orders, $K_n = K_{-n}$ and $I_n = I_{-n}$ and the change of behavior at $n = 0$ follows from the properties previously described. For real orders, the change of behavior for the solutions of (4.131) around $\nu = 0$ still takes place, although, differently from integer orders, $I_{\alpha+n}$, $\alpha \notin \mathbb{Z}$, is not minimal as $n \to -\infty$ ($n \in \mathbb{N}$), and the behavior is not completely symmetric around $\nu = 0$. See Figure 4.4, where different regions are clearly distinguished. When n is negative enough, backward recursion is well conditioned both for $K_{n+0.1}$ and $I_{n+0.1}$ because both are dominant as $n \to -\infty$ ($n \in \mathbb{Z}$). Forward recursion is badly conditioned for this range of values for both I and K. For not so large negative n the condition of the I-function changes and it starts behaving as dominant for increasing n (this is certainly true when $-40 < n < 0$; see Figure 4.4); forward recursion for I is well conditioned in this range and badly conditioned for K (the opposite happens for backward recursion). For $n > 0$ the situation is the reverse of the behavior for moderate negative values of n.

This type of behavior is accompanied, when n is negative, with the apparent convergence of the associated continued fraction to the ratio of solutions which behave as minimal for negative n as n increases, that is, to ratios of K-functions instead of ratios of I-functions (see [51] for further details). This can be understood from (4.36) since, as n increases, we have an initial increasing of $|r_n|$ followed by a steep decrease when $n > 0$. Figure 4.4 illustrates this fact.

This reversal in the behavior of the solutions of a TTRR whose central coefficient changes sign for a given $n = n_0$, as shown in [51], takes place when there is a range of values around n_0 where the minimal solution has an invariable pattern of signs (alternating or with constant sign, depending on the sign of b_n for large n). For the modified Bessel function case, this property holds without restriction for integer orders, but it also holds for real orders when ν is not very negative.

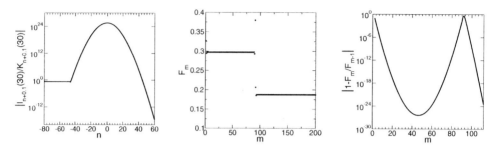

Figure 4.4. *Left: the ratio $I_{n+0.1}(30)/K_{n+0.1}(30)$ as a function of $n \in \mathbb{Z}$. Several regions can be distinguished. Center: successive approximants of the continued fraction associated to the ratio $-I_{-45.9}(30)/I_{-46.9}(30) \simeq 0.187$; initially, the continued fraction approaches the ratio $K_{-45.9}(30)/K_{-46.9}(30) \simeq 0.297$. Right: relative deviation between successive approximants of the continued fraction.*

4.8.2 Other examples: Hypergeometric recursions

There are many more not so evident cases for which anomalous behavior occurs, such as the confluent $(++)$ (Gautschi's case) and $(+0)$ recursions and the Gauss hypergeometric recurrences.

Consider, for instance, the confluent $(++)$ recursion corresponding to Gautschi's anomalous convergence. The coefficients of the TTRR are

$$b_n = -\frac{(c+n)(1-c-n+x)}{(a+n)x}, \quad a_n = -\frac{(c+n)(c+n-1)}{(a+n)x}, \tag{4.132}$$

and, as we saw in Example 4.9, a pair of independent solutions is given by $f_n = M(a+n, c+n, x)$ (minimal) and $g_n = (-1)^n \Gamma(c+n) U(a+n, c+n, x)$.

For a and c positive, the b_n coefficient changes sign at $n_0 = [x-c]$, and the coefficient a_n is negative. In addition, the minimal solution has a single pattern of signs around n_0. These are clear signatures of anomalous behavior which indeed takes place particularly for large x [51].

For large x, the confluent recurrence can be related to the Bessel recurrence relation. Let us denote $\lambda = x + 1 - c$; then $\hat{y}_n = y_{n+\lambda}$ satisfies the recurrence relation

$$\hat{y}_{n+1} + \hat{b}_n \hat{y}_n + \hat{a}_n y_{n-1} = 0, \tag{4.133}$$

with

$$\hat{b}_n = \frac{n}{x} f(n,x), \quad \hat{a}_n = -\frac{n+x}{x} f(n,x), \quad f(n,x) = \left(1 + \frac{a-c}{n+x+1}\right)^{-1}, \tag{4.134}$$

and therefore $\hat{b}_n(x) = \frac{n}{x}(1 + \mathcal{O}(x^{-1}))$ and $\hat{a}_n(x) = -1 + \mathcal{O}(x^{-1})$, which, to first order, is essentially the recurrence for modified Bessel functions.

4.8. Anomalous behavior of some recurrences

Then, for $n < n_0$, especially for large x, we should not use backward recursion for the minimal solution f_n; also, forward recursion for some dominant solutions, in particular for g_n, is strongly unstable (but backward recursion is stable when $n < n_0$). In addition, when $n < n_0$ the associated continued fraction for f_n/f_{n-1} will appear to converge initially (for the first $[n_0 - n]$ approximants) to g_n/g_{n-1}. In [51] it is shown that the best accuracy reachable when the continued fraction appears to converge to the "wrong ratio" improves exponentially with x. In this same reference, it is shown that this type of behavior takes place for several other confluent $((+0)$ recurrence) and Gauss hypergeometric recursions, including the recurrence for $(+++)$ Gauss recurrence, which has the $(++)$ recurrence as a particular case, regarded from the point of view of confluence limits between the functions related with these recursions.

4.8.3 A first order inhomogeneous equation

This type of analysis also reveals anomalous behavior for certain first order inhomogeneous equations. In Example 4.19, it was shown that the first order inhomogeneous equation

$$y_n = \frac{1}{n}(-xy_{n-1} + e^{-x}) \tag{4.135}$$

does not have minimal solutions, which suggests that it can be computed by forward recursion. However, the associated continued fraction has the characteristic features of a recurrence relation with possible anomalous behavior:

$$y_{n+1} + \frac{x-n}{n+1}y_n - \frac{x}{n+1}y_{n-1} = 0. \tag{4.136}$$

Indeed, the central coefficient changes sign at $n = [x]$ and the coefficient in front of y_{n-1} is negative for positive x. In fact, it can be expected that the anomalous behavior becomes more important as $\lambda = x$ becomes large. With the replacement $\hat{y}_n = y_{n+\lambda}$, the recurrence is

$$\hat{y}_{n+1} - \frac{n}{\lambda}(1 + \mathcal{O}(\lambda^{-1}))\hat{y}_n - (1 + \mathcal{O}(\lambda^{-1}))\hat{y}_{n-1} = 0, \tag{4.137}$$

which is again essentially the modified Bessel function case. The exponential integral is a positive solution for $n, x > 0$ and is therefore a dominant solution of (4.136). But this second order recurrence admits a minimal solution which has alternating sign for large n. The relation with the Bessel case when x is large shows that one can expect that the minimal solution keeps its pattern of signs around $n = n_0 = [x]$. This implies [51] that for $n < n_0$ the exponential integrals will no longer behave as a truly dominant solution of the second order recurrence (4.136).

This can be explicitly confirmed because the minimal solution of the second order recurrence can be found explicitly. It is

$$f_n = \frac{(-x)^n}{n!}, \tag{4.138}$$

which is the solution of the homogeneous equation for (4.135) (precisely because the minimal solution of (4.136) is the homogeneous equation, the inhomogeneous equation does not possess a minimal solution). Because f_n is always alternating when $x > 0$ and $g_n = E_n(x)$ (the exponential integrals) are positive, then necessarily $|f_n/g_n|$ is increasing when $n < x$ and decreasing when $n > x$.

Although f_n is not a solution of the first order inhomogeneous equation, any linear combination

$$y_n = Cf_n + E_n(x) \tag{4.139}$$

is a solution, and the different relative behavior has implications for the stability of recursion. Clearly, a stable method of computation for $E_n(x)$ is as follows: start at $n = n_0$ and proceed by forward recursion for $n > n_0$ and by backward recursion for $n < n_0$. For an implementation of this method see [76].

4.8.4 A warning

Beware of those recurrences whose coefficients change sign; take extreme care when proceeding past this change of sign.

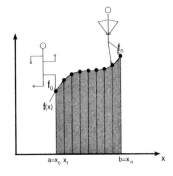

Chapter 5
Quadrature Methods

A journey of a thousand miles begins with one step.
—Lao Tzu, Chinese philosopher, founder of Taoism

One sees great things from the valley, only small things from the peak.
—Gilbert Chesterton, English writer

5.1 Introduction

A great number of special functions can be expressed in terms of integral representations, and usually many representations are available for one function.

In this chapter, we illustrate the use of standard quadrature rules, such as Gauss quadrature and the trapezoidal rule, with the computation of integral representations of well-known special functions and provide hints regarding the selection of quadrature rules depending on the type of integral to be evaluated.

Usually, the computation of a special function by quadrature needs a careful analysis of the properties of the integral and the integrand before applying a quadrature rule. This, in some cases, leads to the necessity of looking for alternatives to certain integral representations. Take, for instance, the case of the real Airy integral

$$\text{Ai}(x) = \frac{1}{\pi} \int_0^\infty \cos\left(\frac{1}{3}t^3 + xt\right) dt. \tag{5.1}$$

This is a convergent integral (as can be checked by integration by parts) for real x. This representation is, however, of no practical use for two main reasons. First, the integrand is infinitely oscillating, which may lead to strong numerical cancellations. Second, the behavior for large x is hidden (see (1.9)), particularly for large $x > 0$.

We will return later to this example and show how alternative expressions can be considered which can be used to accurately compute this function with the two main types of quadratures considered in this chapter: the recursive trapezoidal rule and Gauss quadrature. We discuss the different features of these quadrature rules. For a selection of other quadrature rules we refer to §9.6.

5.2 Newton–Cotes quadrature: The trapezoidal and Simpson's rule

A quadrature rule for evaluating an integral is a formula of the type

$$\int_a^b f(x)\,dx \approx Q(f) = \sum_{i=1}^n w_i f(x_i). \tag{5.2}$$

The numbers w_i are called the *weights of the quadrature rule* and the values x_i are the *nodes of the quadrature rule*.

The basic idea in quadrature consists of approximating the integrand $f(x)$ by a simpler function that can be integrated analytically. A possible choice is the approximation of the function by the polynomial of degree not larger than $n-1$, $P(x)$, which interpolates $f(x)$ at the n nodes x_i (Chapter 3). Integrating, one obtains an interpolatory quadrature formula, namely,

$$\int_a^b f(x)\,dx \approx \int_a^b P(x)\,dx = \int_a^b \left(\sum_{i=1}^n f(x_i) L_i(x)\right) dx = \sum_{i=1}^n w_i f(x_i), \tag{5.3}$$

where

$$w_i = \int_a^b L_i(x)\,dx. \tag{5.4}$$

$L_i(x)$ are the Lagrange fundamental interpolating polynomials (see Chapter 3), that is, the polynomials of degree $n-1$ such that $L_i(x_j) = \delta_{ij}$, $i,j = 1,\dots,n$. We have

$$L_i(x) = \prod_{k=1, k\neq i}^n \frac{x - x_k}{x_i - x_k}. \tag{5.5}$$

Newton–Cotes quadrature rules are those interpolatory formulas based on equally spaced abscissas. The simplest Newton–Cotes quadrature rule is the *trapezoidal rule* which consists of approximating the function $f(x)$ in the interval $[a,b] = [x_1, x_2]$ by the interpolating polynomial of order 1, $P(x)$, such that $P(a) = f(a)$, $P(b) = f(b)$.

This gives the simple rule

$$I(f) = \int_a^b f(x)\,dx \approx Q(f) = \frac{h}{2}(f_1 + f_2), \tag{5.6}$$

where the abbreviated notation $f_i \equiv f(x_i)$ ($x_1 = a$, $x_2 = b$) is used and $h = b - a$. This is the area of the trapezoid in Figure 5.1.

By construction, it is is clear that the trapezoidal rule is exact for all polynomials of degree 1 as maximum. Under this condition, it is said that the trapezoidal rule has *degree of exactness* 1.

Definition 5.1 (degree of exactness[2]). *A quadrature rule* (5.2) *has degree of exactness* m *if it renders exact results when* $f(x)$ *is any polynomial of degree not larger than* m *but it is not exact for all polynomials of degree* $m + 1$.

---

[2]This concept is more usually named *degree of precision*, but the name *degree of exactness* describes more accurately the idea. This is also used, for instance, in [80].

5.2. Newton–Cotes quadrature: The trapezoidal and Simpson's rule

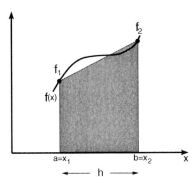

Figure 5.1. *The trapezoidal rule.*

The Lagrange formula for the interpolation error (Chapter 3) can be used to prove that if $f^{(2)}(x)$ is continuous in $[a, b]$, the truncation error for this quadrature rule can be written as

$$R(f) = I(f) - Q(f) = -\frac{h^3}{12} f^{(2)}(c) \qquad (5.7)$$

for a certain $c \in (a, b)$. This also shows that the degree of exactness is one.

Needless to say, the approximations that can be obtained from the simple trapezoidal rule will be generally very poor, unless the integration interval is very small (and hence the step h). There are two obvious directions for possible improvement: one is to divide the integral as a sum over smaller intervals, which is the approach of §5.2.1 and §5.2.2; the second one is to obtain quadrature rules with a higher degree of exactness by interpolating in a higher number of nodes.

If we approximate $f(x)$ by the polynomial of degree not larger than two, interpolating $f(x)$ at the equally spaced abscissas $x_1 = a$, $x_2 = (a+b)/2$, $x_3 = b$, and integrate, we obtain *Simpson's rule*

$$I(f) = \int_a^b f(x)\, dx \approx S(f) = \frac{h}{3}(f_1 + 4f_2 + f_3), \quad h = (b-a)/2, \qquad (5.8)$$

and an elementary analysis shows that, when the fourth derivative is continuous, the error can be written as

$$R(f) = I(f) - S(f) = -\frac{h^5}{90} f^{(4)}(c) \qquad (5.9)$$

for a certain $c \in (a, b)$. This shows that the degree of exactness has increased in two unities with respect to the trapezoidal rule. Also the exponent in h has increased in two unities, which means that the error will be smaller than that for the trapezoidal rule for small enough h.

One can, of course, continue with this process and build quadrature rules with more interpolation nodes. For n equally spaced nodes $x_{i+1} - x_i = h, i = 1, \ldots, n-1$, the degree of exactness is n when n is odd and $n-1$ when it is even. The weights can be written explicitly as [237]

$$w_i = h \frac{(-1)^{n-i}}{(i-1)!(n-i)!} \int_1^n (t-1)(t-2) \cdots (t-i+1)(t-i-1) \cdots (t-n) \, dt. \quad (5.10)$$

Remark 4. The weights of the quadrature rule satisfy $\sum_{i=1}^n w_i = b - a$ for any interpolatory rule because $\sum_{i=1}^n L_i(x) = 1$, which means that any rule is exact if $f(x)$ is a constant function (as it should be!).

For several reasons, however, considering Newton–Cotes rules for a high number of nodes is not very practical. First, if the goal is to obtain the highest possible degree of exactness, then Gauss quadrature is the best option, as we will see in §5.3. Second, higher degree of exactness, as we defined it, does not necessarily mean higher accuracy, and the Runge phenomenon (discussed in Chapter 3) gives a clear warning on the dangers of using equally spaced abscissa interpolation; higher accuracy is easier to obtain by using compound Newton–Cotes rules, that is, by subdividing the interval of integration into smaller subintervals. But even if one is interested in obtaining high-degree Newton–Cotes rules, this can be achieved by combining rules with different values of h, say, h and $h/2$, and using the recurrent (compound) trapezoidal rule (§5.2.2); this leads to the Romberg method (see §9.6.1).

5.2.1 The compound trapezoidal rule

Let us consider an equally spaced partition of the interval $[a, b]$,

$$a = x_0 < x_1 < \cdots < x_n = b, \quad x_k = x_0 + ih, \quad i = 0, \ldots, n, \quad h = (b-a)/n, \quad (5.11)$$

and apply the trapezoidal rule to each of the subintervals $[x_{i-1}, x_i], i = 1, \ldots, n$. Adding the contributions, we have the approximation

$$\int_a^b f(x) \, dx = \sum_{i=0}^{n-1} \int_{x_i}^{x_{i+1}} f(x) \, dx \approx T_n(f) = \frac{h}{2}(f_0 + f_n) + h \sum_{i=1}^{n-1} f_i. \quad (5.12)$$

Remark 5. Observe that now the nodes x_i are enumerated starting from $i = 0$. In other words, n now represents the number of subintervals, while the number of nodes is $n + 1$. We will use this notation for compound rules (also in §5.2.2).

$T_n(f)$ is called the *compound trapezoidal rule* over n subintervals, as shown in Figure 5.2. The truncation error for this quadrature rule is given in the next theorem, which can be easily proved from (5.7).

5.2. Newton–Cotes quadrature: The trapezoidal and Simpson's rule

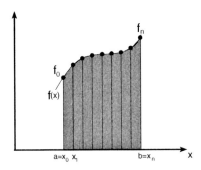

Figure 5.2. *The compound trapezoidal rule.*

Theorem 5.2. *Let $f(x)$ be a function with continuous second derivative on the interval $[a, b]$, and let $a = x_0 < x_1 < \cdots < x_n = b$, $x_i = x_0 + ih$, $i = 0, 1, \ldots, n$, be an equally spaced partition of $[a, b]$. Then*

$$\int_a^b f(x)\,dx = \frac{h}{2}(f_0 + f_1) + h \sum_{i=1}^{n-1} f_i + R_n, \tag{5.13}$$

where $f_i = f(x_i)$ and

$$\exists \tau \in [a, b] : R_n = -\frac{(b-a)h^2}{12} f^{(2)}(\tau) = -\frac{(b-a)^3}{12n^2} f^{(2)}(\tau). \tag{5.14}$$

Proof. Considering (5.6) and (5.7), for each subinterval $[x_{i-1}, x_i]$, $i = 1, \ldots, n$, and adding, we have

$$\int_{x_0}^{x_n} f(x)\,dx = \frac{h}{2}(f_0 + f_1) + h \sum_{i=1}^{n-1} f_i + R_n, \tag{5.15}$$

with

$$R_n = -\sum_{i=1}^{n} \frac{h^3}{12} f^{(2)}(c^{(i)}) = -\frac{(b-a)h^2}{12} \frac{1}{n} \sum_{i=1}^{n} f^{(2)}(c^{(i)}) \tag{5.16}$$

and $c^{(i)} \in [x_{i-1}, x_i]$, $i = 1, \ldots, n$. Observing that

$$\min_{x \in [a,b]} f^{(2)}(x) \le \frac{1}{n} \sum_{i=1}^{n} f^{(2)}(c^{(i)}) \le \max_{x \in [a,b]} f^{(2)}(x), \tag{5.17}$$

and applying the intermediate value theorem, we obtain

$$\exists \tau \in [a, b] : f^{(2)}(\tau) = \frac{1}{n} \sum_{i=1}^{n} f^{(2)}(c^{(i)}). \quad \square \tag{5.18}$$

Table 5.1. *Relative errors using the trapezoidal rule for computing (5.21) for $x = 5$.*

n	R_n
4	$-.12 \cdot 10^{-000}$
8	$-.48 \cdot 10^{-006}$
16	$-.11 \cdot 10^{-021}$
32	$-.13 \cdot 10^{-062}$
64	$-.13 \cdot 10^{-163}$
128	$-.53 \cdot 10^{-404}$

Remark 6. The expression for the truncation error in (5.16) indicates that when doubling the number of subintervals, the error will be approximately reduced by a factor of 4.

A more practical estimate of the error for the compound trapezoidal rule is the following, which is called an asymptotic error estimate:

$$\widehat{R}_n(f) = -\frac{(b-a)^2}{12n^2}(f'(b) - f'(a)) = -\frac{h^2}{12}(f'(b) - f'(a)). \qquad (5.19)$$

This estimate is better as n is larger (h is smaller). This estimate is easy to check from the proof of the previous theorem. Indeed, the term $h \sum_{i=1}^{n} f^{(2)}(c^{(i)})$ in (5.16) approximates the integral $\int_a^b f^{(2)}(x)\,dx = f'(b) - f'(a)$ and tends to the value of the integral when $n \to \infty$ (and, consequently, when $h \to 0$). We can write this estimate as

$$\int_a^b f(x)\,dx = T_n(f) - \frac{h^2}{12}(f'(b) - f'(a)) + r(h), \qquad (5.20)$$

where $r(h)$ goes to zero as $h \to 0$ faster than h^2 (abbreviated, $r(h) = o(h^2)$).

A first and interesting conclusion that can be inferred from (5.20) is that in the case $f'(a) = f'(b)$ one can expect a faster convergence of the trapezoidal rule in comparison to the case $f'(a) \neq f'(b)$. Let us show this with an example.

Example 5.3. Let us consider the evaluation of the Bessel function $J_0(x)$,

$$\pi J_0(x) = \int_0^\pi \cos(x \sin t)\,dt = h + h \sum_{j=1}^{n-1} \cos[x \sin(hj)] + R_n, \qquad (5.21)$$

where $h = \pi/n$.

Table 5.1 shows the errors for the computation of (5.21) for $x = 5$. We observe that the error R_n is much smaller than the upper bound that can be obtained from (5.14). From the estimate of the error (5.19) and because the integrand $f(x)$ satisfies $f'(0) = f'(\pi)$, one can expect that (5.14) is improved and that, because of (5.20), the error decreases faster than quadratically. However, the situation is much better than expected and the following upper bound can be given [222] which describes faithfully the observed error:

$$|R_n| \leq 2e^{x/2} \frac{(x/2)^{2n}}{(2n)!}. \quad \blacksquare \qquad (5.22)$$

5.2. Newton–Cotes quadrature: The trapezoidal and Simpson's rule

This example suggests that the trapezoidal rule can be much better than expected for certain types of integrals. The crucial result to understand this behavior is the Euler–Maclaurin formula, as we will explain next.

Before this and for later use (Romberg quadrature, §9.6.1) we show the parallel of Theorem 5.2 and (5.19) for Simpson's rule (the proof is very similar).

Theorem 5.4. *Let $f(x)$ be a function with continuous fourth derivative in $[a, b]$ and let $a = x_0 < x_1 < \cdots < x_n = b$, $x_i = x_0 + ih$, $i = 0, 1, \ldots, n = 2m$. Then*

$$\int_a^b f(x)\,dx = \tfrac{1}{3}h(f_0 + f_n) + \tfrac{2}{3}h\sum_{i=1}^{m-1} f_{2i} + \tfrac{4}{3}h\sum_{i=1}^{m} f_{2i-1} + E_n^S(f), \qquad (5.23)$$

where $f_i = f(x_i)$ and

$$\exists \tau \in [a, b] : E_n^S(f) = -\frac{(b-a)h^4}{180}f^{(4)}(\tau) = -\frac{(b-a)^5}{180n^4}f^{(4)}(\tau). \qquad (5.24)$$

The error can be estimated for large n as

$$\widehat{E}_n^S(f) = -\frac{(b-a)^4}{180n^4}(f^{(3)}(b) - f^{(3)}(a)) = -\frac{h^4}{180}(f^{(3)}(b) - f^{(3)}(a)), \qquad (5.25)$$

in the sense that (assuming that $f^{(3)}(a) \neq f^{(3)}(b)$)

$$\lim_{n \to \infty} \frac{\widehat{E}_n^S(f)}{E_n^S(f)} = 1. \qquad (5.26)$$

5.2.2 The recurrent trapezoidal rule

The compound trapezoidal rule, although being very simple, has several nice properties. One important feature is that, as we will now explain, the compound trapezoidal rule can be computed in an efficient recursive way. This allows refinement of the computation by adding additional nodes without the need of recomputing previously evaluated function values; in addition, this recursiveness provides an automatic control of the accuracy.

We will see that when halving the step size h the new quadrature rule can be evaluated efficiently.

Let us consider the partition

$$a \leq x_0 < x_1 < \cdots < x_n = b, \quad n = 2m, \quad m \in \mathbb{N}. \qquad (5.27)$$

The trapezoidal rule with step size h ($h = x_{i+1} - x_i$), $T(f, h)$, can be written as

$$\int_a^b f(x)\,dx \approx T(f, h) = \frac{h}{2}\sum_{i=0}^{n-1}(f_i + f_{i+1}). \qquad (5.28)$$

On the other hand,

$$T(f, 2h) = h\sum_{i=0}^{m-1}(f_{2i} + f_{2i+2}). \qquad (5.29)$$

Hence,
$$T(f,h) = \frac{T(f,2h)}{2} + h\sum_{i=1}^{m} f_{2i-1}. \tag{5.30}$$

In other words, when halving h the new trapezoidal rule can be obtained just by adding the values of $f(x)$ in the new nodes, multiplying by the new h, and adding the previous trapezoidal rule divided by 2. This can be summarized in the following algorithm.

ALGORITHM 5.1. Recursive trapezoidal rule.
Input: $\epsilon > 0$, $f(x)$, b, a, $n \in \mathbb{N}$.

Output: $I \approx \int_a^b f(x)\,dx$.

- $h = (b-a)/n;\ I = \frac{h}{2}(f(a) + f(b));$

- IF $n > 1$ THEN $I = I + h\sum_{j=1}^{n-1} f(a+jh);$

- $\Delta = 1 + \epsilon;$

- DO WHILE $\Delta > \epsilon$

    $h = h/2;\ I_0 = I;$

    $I = I_0/2 + h\sum_{j=1}^{n} f(a + (2j-1)h);$

    $n = 2n;$

    $\Delta = |I - I_0|;$

Remark 7. The termination criterion is determined by the difference between two consecutive computations $\Delta = |I - I_0|$. This is a reasonable criterion when a given absolute error tolerance is demanded. When instead we want to demand a given relative accuracy, we should consider $\Delta = |1 - I_0/I|$ (but only when I does not become too small).

The recurrent scheme can also be applied for computing a quadrature rule of higher order (Simpson's rule) from quadrature rules of lower order (trapezoidal). The generalization of this process is behind the Romberg quadrature rule (see §9.6.1).

5.2.3 Euler's summation formula and the trapezoidal rule

The following result can be easily derived from the results of §11.2 and explains why for certain integrals the trapezoidal rule can be expected to be much better than predicted by Theorem 5.2 and Remark 6.

5.2. Newton–Cotes quadrature: The trapezoidal and Simpson's rule

Theorem 5.5 (Euler–Maclaurin). *Let $f(x)$ be a function with $2m+2$ continuous derivatives in $[x_0, x_n]$. Then, given the compound trapezoidal rule for n intervals $T_n(h)$, (5.12), we have*

$$\int_{x_0}^{x_n} f(x)\, dx = T_n(f) + R_n(f), \tag{5.31}$$

where the truncation error admits the expansion

$$\begin{aligned} R_n(f) = &\sum_{l=1}^{m} \frac{B_{2l}}{(2l)!} h^{2l} \left(f^{(2l-1)}(x_0) - f^{(2l-1)}(x_n) \right) \\ &- \frac{B_{2m+2}}{(2m+2)!}(x_n - x_0) h^{2m+2} f^{(2m+2)}(\zeta) \end{aligned} \tag{5.32}$$

for some ζ in $[x_0, x_n]$. B_k are the Bernoulli numbers. The first few numbers with even index are $B_0 = 1$, $B_2 = 1/6$, $B_4 = -1/30$ (see §11.2 for further details).

The Euler–Maclaurin formula provides an expansion of the error term in powers of h. Observe how for the case $m = 1$ this theorem is related to (5.20).

Notice that as more derivatives of f are equal at a and b, the trapezoidal rule progressively improves in accuracy if h is small enough. This explains the exceptionally fast convergence in Example 5.3, where the integrand is analytic and periodic on the real line and the integral is taken over a full period. We can take any m in (5.32) and all terms in the series vanish. Only the last term (depending on some ζ) survives. In fact we have the following result.

Theorem 5.6. *If f is periodic and has a continuous kth derivative, and if the integral is taken over a full period, then*

$$|R_n| = \mathcal{O}\left(n^{-k}\right), \quad n \to \infty. \tag{5.33}$$

Because for the Bessel function in Example 5.3 (5.33) is true for any k, we can expect that the error may decrease exponentially (as we already know). The same happens when we have a C^∞-function with vanishing derivatives of all orders at a and b. For example, the trapezoidal rule for the integral

$$\int_{-1}^{1} e^{-\frac{1}{1-x^2}} dx \tag{5.34}$$

is again exceptionally efficient. The integrand can be continued as a periodic C^∞-function on the real line with interval of periodicity $[-1, 1]$ and Theorem 5.6 also applies in this case.

Even more important (at least regarding the computation of special functions), also for infinite range integrals

$$\int_{-\infty}^{\infty} f(x)\, dx, \tag{5.35}$$

the error of the trapezoidal rule decays very fast under certain analyticity and fast decay conditions for $f(x)$. As we will see in §5.4 and §5.5, one can take advantage of this for computing a good number of contour integrals defining special functions. Also, considering special changes of the variable of integration, one can transform other types of integrals and put them into a form suitable for the trapezoidal rule (see §5.4.2).

These facts, together with the recursivity of the rule, will favor the use of the trapezoidal rule for computing a good number of special functions. In fact, as we will see at the end of this chapter, many special functions have integral representations for which the trapezoidal rule becomes very efficient (if not optimal).

5.3 Gauss quadrature

The key idea behind Gauss quadrature is a clever choice of the interpolation nodes in order to maximize the degree of exactness of the quadrature rule. We will consider integrals of the form

$$\int_a^b f(x)w(x)\,dx, \tag{5.36}$$

where $w(x)$ is a weight function, which is defined as follows.

Definition 5.7. *We say that $w(x)$ is a weight function on $[a, b]$ if it is nonnegative in any open interval included in $[a, b]$ (where $a = -\infty$ and/or $b = \infty$ are accepted) and*

$$\int_a^b |x|^n w(x)\,dx < \infty, \quad n = 0, 1, 2, \ldots. \tag{5.37}$$

For computing integrals such as (5.36) we will approximate $f(x)$ by the polynomial $f_{n-1}(x)$ of degree not larger than $n-1$, which interpolates $f(x)$ at n distinct nodes $x_1, \ldots, x_n$. That is, we write

$$f(x) \approx f_{n-1}(x) = \sum_{i=1}^n f(x_i) L_i(x), \tag{5.38}$$

where $L_i(x)$ is given in (5.5), and approximate the original integral by the integration of the polynomial

$$I(f) = \int_a^b f(x)w(x)\,dx \approx Q(f) = \sum_{i=1}^n w_i f(x_i), \tag{5.39}$$

where

$$w_i = \int_a^b L_i(x)w(x)\,dx. \tag{5.40}$$

Similarly to Definition 5.1, we will say that the quadrature $Q(f)$ in (5.39) has degree of exactness m, if $I(f) = Q(f)$ for f any polynomial of degree not larger than m, but it is not exact for all polynomials of degree $m + 1$.

Independently of how we choose the n distinct nodes, the interpolatory rule with n nodes has degree of exactness at least equal to $n - 1$, because if $f(x)$ is a polynomial of degree not larger than $n - 1$, the polynomial $f_{n-1}(x)$ which interpolates $f(x)$ at the n distinct points is, by Lagrange's formula (Theorem 3.2), $f(x)$ itself. Conversely, if the degree of exactness is $n - 1$ or larger, then the rule is interpolatory.

5.3. Gauss quadrature

But we can do more than twice better than this! Indeed, when approximating an integral by a quadrature rule (5.2) we have $2n$ parameters w_i, x_i. With $2n$ parameters, one can obtain quadrature rules which are exact for polynomials of degree $2n - 1$ (which have $2n$ coefficients). This is the maximal possible degree of exactness. As we will see, this selection of nodes is possible and in some cases relatively simple to implement by using methods of numerical linear algebra.

A first and naive method of trying to build such a quadrature rule would be to compute the integrals for x^k, $k = 0, \ldots, 2n - 1$ (called moments), and equating to the value given by the quadrature rule (which is exact for these monomials)

$$\mu_k = \int_a^b x^k w(x)\,dx = Q(x^k) = \sum_{i=1}^n w_i x_i^k, \quad k = 0, \ldots, 2n - 1, \qquad (5.41)$$

and solve this set of $2n$ nonlinear equations for the $2n$ unknown values x_i and w_i.

Sadly, however, these nonlinear systems are badly conditioned and are numerically unstable [80]. A more elegant and effective formulation is based on the theory of orthogonal polynomials, which leads to the reformulation of the computation of nodes and weights as an eigenvalue problem (Golub–Welsch algorithm).

Next, we provide a short overview of the very basic theory of orthogonal polynomials and Gauss quadrature. We do not consider in detail variants such as Gauss–Lobatto, Gauss–Radau, and Gauss–Kronrod quadrature formulas, although a brief description is given in §9.6.3. We refer to [80] for more details on these and related quadrature formulas.

5.3.1 Basics of the theory of orthogonal polynomials and Gauss quadrature

Given a weight function $w(x)$ on $[a, b]$ (see Definition 5.7), we can define an inner product in the set of polynomials. Given two polynomials p and q, we denote

$$\langle p, q \rangle = \int_a^b p(x)q(x)w(x)\,dx, \qquad (5.42)$$

which is called the *inner product* or *scalar product* of p and q.

This is an operation which is bilinear, symmetric, and positive definite. Then, the set of polynomials of degree equal to or smaller than N, $\mathbb{P}_N$ (with N any natural number), is a Euclidean vector space of dimension $N + 1$ (that is, a linear vector space of dimension $N + 1$ equipped with a scalar product); one possible basis of this space is the set of the monomials $\{x^k\}_{k=0}^N$.

The basic theory of finite Euclidean spaces is taught in any elementary linear algebra course. We will take profit of this theory. As usual, we define the norm of a vector (a polynomial p) by writing

$$\|p\| = +\sqrt{\langle p, p \rangle}, \qquad (5.43)$$

and we will say that two polynomials p and q are orthogonal with respect to the chosen inner product when

$$\langle p, q \rangle = 0. \qquad (5.44)$$

Given a basis of the vector space $\mathbb{P}_N$, $\{p_i\}_{i=0}^N$, it is called orthogonal with respect to the chosen inner product if it satisfies

$$\langle p_i, p_j \rangle = n_i \delta_{ij}, \quad i, j = 0, \ldots, N, \tag{5.45}$$

$n_i \neq 0$, $i = 0, \ldots, N$. When $n_i = 1$, $i = 0, \ldots, N$, the basis is said to be *orthonormal*. Given any basis $\{q_k\}_{k=0}^N$, the *Gram–Schmidt orthogonalization* method provides a simple rule for building an orthogonal basis, as we describe in the following algorithm.

ALGORITHM 5.2. Gram–Schmidt orthogonalization.
Input: a basis $\{q_k\}_{k=0}^N$ of $\mathbb{P}_N$ (for instance, with $q_k(x) = x^k$).

Output: an orthogonal base $\{p_k\}_{k=0}^N$ and an orthonormal base $\{\tilde{p}_k\}_{k=0}^N$ of $\mathbb{P}_n$.

$\tilde{p}_0(x) = q_0(x)/\|q_0\|$

Repeat for $k = 1, \ldots, N$

$$p_k(x) = q_k(x) - \sum_{j=0}^{k-1} \langle q_k, \tilde{p}_j \rangle \tilde{p}_j(x)$$

$\tilde{p}_k(x) = p_k(x)/\|p_k\|$

Then $\{\tilde{p}_k\}_{k=0}^N$ ($\{p_k\}_{k=0}^N$) is an orthonormal (orthogonal) basis when $\{q_k\}_{k=0}^N$ is a basis.

Although there are infinitely many ways to choose an orthogonal basis for the vector space $\mathbb{P}_N$, there is only one orthogonal basis $\{p_k\}_{k=0}^N$ such that p_k are monic polynomials of degree k, $k = 0, \ldots, N$ (monic means that the highest degree coefficient is 1, that is, $p_k(x) = x^k + a_{k-1} x^{k-1} + \cdots$). This unique set of orthogonal polynomials $\{p_k\}_{k=0}^N$ can be built by applying the Gram–Schmidt orthogonalization method starting with the set $\{q_k\}_{k=0}^N$ where $q_k(x) = x^k$. From now on, $\{p_k\}_{k=0}^N$ will denote this set of monic orthogonal polynomials obtained from Algorithm 5.2. Also, following the notation of this algorithm, $\{\tilde{p}_k\}_{k=0}^N$ will denote the set of orthonormal polynomials given by $\tilde{p}_k(x) = p_k(x)/\|p_k\|$, $k = 0, 1, \ldots, N$.

Any polynomial of degree m, $h_m(x)$, can be written uniquely as a linear combination of the set of orthonormal polynomials $\{\tilde{p}_k\}_{k=0}^m$ in the following way:

$$h_m(x) = \sum_{k=0}^m \langle \tilde{p}_k, h_m \rangle \tilde{p}_k(x). \tag{5.46}$$

In addition, using (5.45), we see that $\langle h_m, p_n \rangle = 0$ for h_m any polynomial of degree m with $m < n$; furthermore, p_n is the only monic orthogonal polynomial of degree n which is orthogonal to all polynomials of degree smaller than n.

We can now establish the following fundamental result in the theory of orthogonal polynomials and Gauss quadrature.

5.3. Gauss quadrature

Theorem 5.8. *Given $w(x)$, a weight function on $[a, b]$, there exists a unique family of monic polynomials, $\{p_k\}$, k being the degree of the polynomials, such that*

$$\int_a^b p_n(x) p_m(x) w(x) \, dx \neq 0 \iff n = m, \tag{5.47}$$

and we say that $\{p_k\}_{k \in \mathbb{N}}$ is the family of monic orthogonal polynomials corresponding to the weight function $w(x)$.

Furthermore, p_n is the only nth-degree monic polynomial which is orthogonal to all polynomials of degree smaller than n.

All the zeros of the polynomials of this family are real and lie in (a, b).

Proof. All that remains to be proved is that p_n, $n \geq 1$, has n real zeros in the interval (a, b).

Because $\langle p_n, p_0 \rangle = 0$, we have

$$\int_a^b p_n(x) w(x) \, dx = 0, \tag{5.48}$$

and therefore p_n changes sign in the interval (a, b) at least once. Let us suppose that p_n changes sign k times ($k \leq n$ as corresponds to a polynomial of degree n). That is, we assume that p_n has k zeros $x_1 < x_2 < \cdots < x_k$ of odd multiplicity in (a, b).

Because $p_n(x) r_k(x) \geq 0$ in (a, b), where $r_k(x) = (x-x_1)(x-x_2)\cdots(x-x_k)$, we have

$$\langle p_n, r_k \rangle = \int_a^b p_n(x) r_k(x) w(x) \, dx > 0. \tag{5.49}$$

But because p_n is orthogonal to all polynomials of degree smaller than n, it follows that $k \geq n$, and, hence, $k = n$ (because we also have $k \leq n$). $\square$

Theorem 5.8 is the main tool for proving the following theorem, which plays a central role in building Gauss quadrature rules, although it does not provide a convenient method of computation except for a low number of nodes.

Theorem 5.9. *Let $w(x)$ be a weight function on $[a, b]$ and let p_n be the monic polynomial of degree n such that*

$$\int_a^b x^k p_n(x) w(x) \, dx = 0, \quad k = 0, \ldots, n - 1. \tag{5.50}$$

Let $x_1, \ldots, x_n$ be the zeros of p_n and let w_i be defined by

$$w_i = \int_a^b L_i(x) w(x) \, dx, \quad L_i(x) = \prod_{k=1, k \neq i}^n \frac{x - x_k}{x_i - x_k}, \tag{5.51}$$

where $i = 1, 2, \ldots, n$. Then, the quadrature rule

$$\int_a^b f(x) w(x) \, dx \approx Q_n^G(f) = \sum_{i=1}^n w_i f(x_i) \tag{5.52}$$

is exact for polynomials of degree less than or equal to $2n - 1$.

If the function f has a continuous derivative $f^{(2n)}$ in $[a, b]$, then

$$\exists \lambda \in (a, b) : \int_a^b f(x)w(x)\,dx = Q_n^G(f) + \gamma_n \frac{f^{(2n)}(\lambda)}{(2n)!}, \tag{5.53}$$

where γ_n is given by

$$\gamma_n = \int_a^b p_n(x)^2 w(x)\,dx. \tag{5.54}$$

The numbers w_i are called the *Christoffel numbers* or weights of the Gauss quadrature rule. Several other representations of these numbers are known; see, for example, [219, Thm. 6.5].

Proof. Using Theorem 5.8, we only need to justify that the interpolatory rule based on the n nodes $x_1, \ldots, x_n$ is exact for polynomials of degree less than or equal to $2n - 1$ and verify the expression for the error term.

Let us assume that f is a polynomial of degree not larger than $2n - 1$ (that is, $f \in \mathbb{P}_{2n-1}$); we are proving now that the quadrature rule is exact for f. Dividing $f(x)$ by $p_n(x) = (x - x_1) \cdots (x - x_n)$, we have

$$f(x) = r(x) p_n(x) + g(x), \quad r \in \mathbb{P}_{n-1}, \quad g \in \mathbb{P}_{n-1}. \tag{5.55}$$

Hence,

$$\int_a^b f(x)w(x)\,dx = \int_a^b g(x)w(x)\,dx + \int_a^b r(x) p_n(x) w(x)\,dx. \tag{5.56}$$

But $\langle r, p_n \rangle = 0$ because r has degree smaller than n; therefore the second integral at the right-hand side vanishes. Also, because $g \in \mathbb{P}_{n-1}$, the first integral can be evaluated exactly with the n-point interpolatory formula. Furthermore, because $p_n(x_i) = 0$, $i = 1, \ldots, n$, from (5.55) we have $f(x_i) = g(x_i)$, $i = 1, \ldots, n$. Therefore,

$$\int_a^b f(x)w(x)\,dx = \int_a^b g(x)w(x)\,dx = \sum_{i=1}^n w_i g(x_i) = \sum_{i=1}^n w_i f(x_i), \tag{5.57}$$

and we have proved that the quadrature rule is indeed exact when f is any polynomial of degree not larger than $2n - 1$.

For proving the formula for the error term, we will consider Hermite interpolation. From Theorem 3.3, we see that, given f, a function with continuous second derivative in $[a, b]$, there exists a unique polynomial h_{2n-1} of degree not larger than $2n - 1$, such that

$$f(x_i) = h_{2n-1}(x_i) \quad \text{and} \quad f'(x_i) = h'_{2n-1}(x_i). \tag{5.58}$$

In addition, there exists $\zeta_x \in (a, b)$ (depending on x), such that

$$f(x) = h_{2n-1}(x) + \frac{f^{(2n)}(\zeta_x)}{(2n)!}(x - x_1)^2 \cdots (x - x_n)^2. \tag{5.59}$$

Therefore

$$\int_a^b f(x)w(x)\,dx = \int_a^b h_{2n-1}(x)w(x)\,dx + \int_a^b \frac{f^{(2n)}(\zeta_x)}{(2n)!} p_n(x)^2 w(x)\,dx. \tag{5.60}$$

5.3. Gauss quadrature

Now, because the rule is exact for polynomials of degree not larger than $2n - 1$, and considering (5.58),

$$\int_a^b h_{2n-1}(x)w(x)\,dx = \sum_{i=1}^n w_i h_{2n-1}(x_i) = \sum_{i=1}^n w_i f(x_i) = Q_n^G(f). \quad (5.61)$$

On the other hand, because $p_n(x)^2 w(x)$ is nonnegative, we can apply the generalized mean value theorem for integrals and get

$$\int_a^b \frac{f^{(2n)}(\zeta_x)}{(2n)!} p_n(x)^2 w(x)\,dx = \frac{f^{(2n)}(\lambda)}{(2n)!} \int_a^b p_n(x)^2 w(x)\,dx \quad (5.62)$$

for some $\lambda \in (a, b)$.

Equations (5.60), (5.61), and (5.62) lead to (5.54), and the theorem is proven. $\square$

The first steps in order to apply a Gauss quadrature rule consist in the computation of the corresponding zeros and weights. Although the previous theorem provides a representation of both the nodes and weights, such expressions are of very limited practical use. First, one should compute the orthogonal polynomial $p_n(x)$, then compute its zeros (see Chapter 7), and then compute the integrals defining the weights. In general, this is a hard computation; however, for Chebyshev polynomials, the construction is simple.

Example 5.10 (Gauss–Chebyshev quadrature). As discussed in Chapter 3, the Chebyshev polynomials of the first kind, $T_n(x)$, are orthogonal with respect to the weight function $w(x) = (1-x^2)^{-1/2}$ in the interval $[-1, 1]$. The polynomial of degree at most $n - 1$ which interpolates a function $f(x)$ at the n Chebyshev points (the roots of $T_n(x)$) can be written

$$f_{n-1}(x) = \sum_{i=0}^{n-1}{}' c_i T_i(x). \quad (5.63)$$

The prime means that the first term has to be multiplied by $1/2$, and (see (3.55))

$$c_i = \frac{2}{n}\sum_{k=1}^n f(x_k) T_i(x_k), \quad x_k = \cos((k - 1/2)\pi/n), \quad k = 1, \ldots, n. \quad (5.64)$$

The Gauss–Chebyshev rule $Q^{G-C}(f)$ consists in the approximation

$$I(f) = \int_{-1}^1 f(x)(1-x^2)^{-1/2}dx \approx Q^{G-C}(f) = \int_{-1}^1 f_{n-1}(x)(1-x^2)^{-1/2}dx \quad (5.65)$$

and then, using (5.63) and the orthogonality of Chebyshev polynomials,

$$Q^{G-C}(f) = \frac{1}{2}c_0\pi = \frac{\pi}{n}\sum_{k=1}^n f(x_k). \quad \blacksquare \quad (5.66)$$

For computing the orthogonal polynomials, TTRRs (see also Chapter 4) are a useful tool. Furthermore, as we will see, both the zeros of the orthogonal polynomial and the weights can be computed in an easy way from the coefficients of the recursion. This is the Golub–Welsch algorithm [97], which we discuss later.

Theorem 5.11. *The monic orthonormal polynomials $\{p_k\}$ associated with the weight function $w(x)$ on the interval $[a, b]$ satisfy the recurrence relation*

$$p_1(x) = (x - B_0)p_0(x),$$
$$p_{k+1}(x) = (x - B_k)p_k(x) - A_k p_{k-1}(x), \quad k = 1, 2, \ldots,$$
(5.67)

where

$$A_k = \frac{\|p_k\|^2}{\|p_{k-1}\|^2}, \quad k \geq 1, \quad B_k = \frac{\langle xp_k, p_k \rangle}{\|p_k\|^2}, \quad k \geq 0.$$
(5.68)

Proof. Because $p_{k+1}(x) - xp_k(x)$ is a polynomial of maximal degree k, we can write

$$p_{k+1}(x) - xp_k(x) = \sum_{i=0}^{k} \xi_i p_i(x).$$
(5.69)

Taking the scalar product with p_j, $j \leq k$, and because $\langle p_{k+1}, p_j \rangle = 0$,

$$-\langle xp_k, p_j \rangle = \xi_j \|p_j\|^2.$$
(5.70)

But if $j \leq k - 2$, then $\langle xp_k, p_j \rangle = \langle p_k, xp_j \rangle = 0$, because the degree of xp_j is smaller than k and, hence, it is orthogonal to $p_k(x)$. Therefore

$$p_{k+1}(x) - xp_k(x) = \xi_k p_k(x) + \xi_{k-1} p_{k-1}(x)$$
(5.71)

and

$$B_k = -\xi_k = \frac{\langle xp_k, p_k \rangle}{\|p_k\|^2},$$
(5.72)

while

$$A_k = -\xi_{k-1} = \frac{\langle xp_k, p_{k-1} \rangle}{\|p_{k-1}\|^2} = \frac{\langle p_k, xp_{k-1} \rangle}{\|p_{k-1}\|^2} = \frac{\|p_k\|^2}{\|p_{k-1}\|^2}.$$
(5.73)

The last equality comes from the fact that $xp_{k-1}(x)$ is a monic polynomial of degree k and that, hence, $xp_{k-1}(x) = p_k(x) + \sum_{i=0}^{k-1} \eta_i p_i(x)$.

Finally, it is obvious that $p_1(x) = (x - B_0)p_0(x)$ and that B_0 can be computed as before. □

Remark 8. Of course, not only the monic orthogonal polynomials satisfy a TTRR. The normalization of the polynomials can be chosen differently; writing $p_k = \lambda_k \hat{p}_k$, where $\lambda_k = \|p_k\|$ if $\{\hat{p}_i\}$ is the set of orthonormal polynomials $\{\tilde{p}_i\}$, we can rewrite the recurrence relation as

$$\frac{\lambda_1}{\lambda_0} \hat{p}_1(x) + (B_0 - x)\hat{p}_0(x) = 0,$$
$$\frac{\lambda_{k+1}}{\lambda_k} \hat{p}_{k+1}(x) + (B_k - x)\hat{p}_k(x) + A_k \frac{\lambda_{k-1}}{\lambda_k} \hat{p}_{k-1}(x) = 0, \quad k = 1, 2, \ldots,$$
(5.74)

5.3. Gauss quadrature

and for the case of orthonormal polynomials we have

$$\alpha_1 \tilde{p}_1(x) + \beta_0 \tilde{p}_0(x) = x\tilde{p}_0(x),$$
$$\alpha_{k+1} \tilde{p}_{k+1}(x) + \beta_k \tilde{p}_k(x) + \alpha_k \tilde{p}_{k-1}(x) = x\tilde{p}_k(x), \quad k = 1, 2, \ldots,$$
(5.75)

with $\alpha_k = \sqrt{A_k} = \|p_k\|/\|p_{k-1}\| = F_{k-1}/F_k$, $\beta_k = B_k = \langle x\tilde{p}_k, \tilde{p}_k \rangle$, where the constants F_k are the coefficients of the highest degree of $\tilde{p}_k(x)$ (it is easy to see that $F_k = 1/\|p_k\|$).

For any other normalization λ_k the polynomials, obviously, still satisfy a TTRR.

In the general case, the coefficients of the recurrence relation can be obtained in parallel with the monic orthogonal polynomials by means of (5.67) and (5.68); this is the so-called Stieltjes procedure, which we describe as follows.

ALGORITHM 5.3. Stieltjes procedure.

Input: $a, b, w(x)$.

Output: $B_0, A_i, B_i, p_i(x), i = 1, 2, \ldots, n$.

- $p_{-1}(x) = 0$; $A_0 = 0$;

- $p_0(x) = 1$, $B_0 = \langle x, 1 \rangle / \langle 1, 1 \rangle$;

- DO $i = 0, \ldots, n - 1$:

    $p_{i+1}(x) = (x - B_i) p_i(x) - A_i p_{i-1}(x)$.
    $A_{i+1} = \|p_{i+1}\|^2 / \|p_i\|^2$; $B_{i+1} = \langle xp_{i+1}, p_{i+1} \rangle / \|p_{i+1}\|^2$.

Notice, however, that each evaluation of a new coefficient in Algorithm 5.3 involves the computation of the type of integrals we would precisely like to compute with our Gauss quadrature formula.

There are, however, some very special cases (called classical cases) for which the recursion coefficients can be given in an explicit analytical form. The three main families correspond to Jacobi, Hermite, and generalized Laguerre quadratures. This is summarized in Table 5.2, together with some important particular cases

In all these cases, the orthogonal polynomials can be written in terms of hypergeometric functions (for Jacobi polynomials and special cases we can use the Gauss hypergeometric functions and for the (generalized) Laguerre and Hermite polynomials the confluent hypergeometric functions). The standard definition of these polynomials [219, Chap. 6] is neither in the form of orthonormal polynomials nor monic polynomials. The TTRR can be written as

$$a_0 P_1(x) + b_0 P_0(x) = x P_0(x),$$
$$a_n P_{n+1}(x) + b_n P_n(x) + c_n P_{n-1}(x) = x P_n(x), \quad n \geq 1,$$
(5.76)

where $P_n(x)$ represents either $H_n(x)$, $L_n^{(\alpha)}(x)$, or $P_n^{(\alpha,\beta)}(x)$.

Table 5.2. *Classical orthogonal polynomials and the corresponding weight function $w(x)$.*

Polynomial	Interval	$w(x)$
Jacobi $P_n^{(\alpha,\beta)}(x)$	$[-1, 1]$	$(1-x)^\alpha (1+x)^\beta$ $\alpha, \beta > -1$
Gegenbauer (Jacobi $\alpha = \beta = \lambda - 1/2$) $C_n^\lambda(x)$	$[-1, 1]$	$(1-x^2)^{\lambda - 1/2}$ $\lambda > -1/2$
Legendre (Jacobi $\alpha = \beta = 0$) $P_n(x)$	$[-1, 1]$	1
Chebyshev of the first kind (Jacobi $\alpha = \beta = -1/2$) $T_n(x)$	$[-1, 1]$	$(1-x^2)^{-1/2}$
Generalized Laguerre $L_n^\alpha(x)$	$[0, \infty)$	$x^\alpha e^{-x}$ $\alpha > -1$
Laguerre (generalized Laguerre $\alpha = 0$) $L_n(x)$	$[0, \infty)$	e^{-x}
Hermite $H_n(x)$	$(-\infty, \infty)$	e^{-x^2}

The coefficients are as follows.

1. $P_n^{(\alpha,\beta)}(x)$ (Jacobi).

$$a_0 = \frac{2}{\alpha + \beta + 2}, \quad b_0 = \frac{\beta - \alpha}{\alpha + \beta + 2},$$
$$a_n = \frac{2(n+1)(n+\alpha+\beta+1)}{(L_n+1)(L_n+2)}, \quad b_n = \frac{\beta^2 - \alpha^2}{L_n(L_n+2)}, \quad n \geq 1,$$
$$c_n = \frac{2(n+\alpha)(n+\beta)}{L_n(L_n+1)}, \quad n \geq 1,$$
$$L_n = 2n + \alpha + \beta.$$

(5.77)

2. $L_n^{(\alpha)}(x)$ (generalized Laguerre).

$$a_n = -(n+1), \quad b_n = (2n+\alpha+1), \quad n \geq 0,$$
$$c_n = -(n+\alpha), \quad n \geq 1.$$

(5.78)

5.3. Gauss quadrature

3. $H_n(x)$ (Hermite).

$$a_n = 1/2, \quad b_n = 0, \quad n \geq 0,$$
$$c_n = n, \quad n \geq 1. \tag{5.79}$$

There are other cases for which the coefficients are analytically expressible in terms of simple functions. The first obvious cases are those in which a change of variables takes an integral into one of the classical cases mentioned before (as, for instance, in the case of shifted Legendre polynomials, orthogonal in $[0, 1]$ with respect to $w(x) = 1$). Other known cases are the generalized Hermite polynomials ($w(x) = |x|^{2\mu} e^{-x^2}$, $\mu > -1/2$, in $(-\infty, \infty)$) and the Meixner–Pollaczek polynomials ($w(x) = (2\pi)^{-1} \exp((2\phi - \pi)x)|\Gamma(\lambda + ix)|^2$, $\lambda > 0$, $0 < \phi < \pi$ in $(-\infty, \infty)$). In other cases, methods are available for a direct computation of the coefficients, as in the case of the half-range generalized Hermite polynomials [10] ($w(x) = x^{\gamma} e^{-x^2}$, $\gamma > -1$, in $(0, \infty)$), for which certain nonlinear recurrences can be used to compute the coefficients of the TTRR. For Gauss quadrature on a vertical line in the complex plane (see §11.5.1) much information is also available.

In other cases, one can always consider the Stieltjes procedure (Algorithm 5.3). But because computing the coefficients numerically is a quite nontrivial task, in most occasions the practical use of Gauss quadrature is limited to the cases in which one of the classical weights appears explicitly in the integrand.

The importance of being able to compute the recursion coefficients is based on the fact that this information can be used for computing both the nodes and weights in a single and straightforward setup, as we explain next.

5.3.2 The Golub–Welsch algorithm

The Golub–Welsch algorithm [97] is a method for computing the nodes and weights of Gauss quadrature formulas by solving an eigenvalue problem (proposed in [239]) based on the QR algorithm. Here, we will give details not on the diagonalization problem but only on the statement of the problem, which can be solved by using standard linear algebra packages [132, 150].

The starting point for computing the nodes and weights of a Gaussian n-point rule is the recurrence relation (5.75). We take $x = x_j$, one of the nodes of the rule, that is, one of the zeros of $\tilde{p}_n$, and write (5.75) for $k \leq n-1$, taking into account in the last equation that $\tilde{p}_n(x_j) = 0$. We have

$$\begin{cases} \alpha_1 \tilde{p}_1(x_j) + \beta_0 \tilde{p}_0(x_j) = x_j \tilde{p}_0(x_j), \\ \alpha_2 \tilde{p}_2(x_j) + \beta_1 \tilde{p}_1(x_j) + \alpha_1 \tilde{p}_0(x_j) = x_j \tilde{p}_1(x_j), \\ \vdots \\ \beta_{n-1} \tilde{p}_{n-1}(x_j) + \alpha_{n-1} \tilde{p}_{n-2}(x_j) = x_j \tilde{p}_{n-1}(x_j), \end{cases} \tag{5.80}$$

which can be written in matrix form as

$$\begin{pmatrix} \beta_0 & \alpha_1 & 0 & \cdots & & 0 \\ \alpha_1 & \beta_1 & \alpha_2 & & & \\ 0 & \alpha_2 & \beta_2 & & & \vdots \\ \vdots & & \ddots & & \alpha_{n-1} & \\ 0 & \cdots & & & \alpha_{n-1} & \beta_{n-1} \end{pmatrix} \begin{pmatrix} \tilde{p}_0(x_j) \\ \tilde{p}_1(x_j) \\ \vdots \\ \tilde{p}_{n-2}(x_j) \\ \tilde{p}_{n-1}(x_j) \end{pmatrix} = x_j \begin{pmatrix} \tilde{p}_0(x_j) \\ \tilde{p}_1(x_j) \\ \vdots \\ \tilde{p}_{n-2}(x_j) \\ \tilde{p}_{n-1}(x_j) \end{pmatrix}, \quad (5.81)$$

or

$$\mathbf{J}\hat{\mathbf{p}}(x_j) = x_j \hat{\mathbf{p}}(x_j), \quad (5.82)$$

where $\mathbf{J}$, the *Jacobi matrix*, is the $n \times n$ matrix of the coefficients of the recurrence relation and

$$\hat{\mathbf{p}}(x_j) = (\tilde{p}_0(x_j), \tilde{p}_1(x_j), \ldots, \tilde{p}_{n-1}(x_j))^T. \quad (5.83)$$

It follows that

$$\tilde{p}_n(x_j) = 0 \iff x_j \text{ is an eigenvalue of } \mathbf{J}. \quad (5.84)$$

In other words, the n zeros of $\tilde{p}_n$, $x_1, x_2, \ldots, x_n$ (that is, the nodes of the corresponding Gauss quadrature), can be obtained by calculating the eigenvalues of the matrix $\mathbf{J}$. A similar situation happens with the weights of the Gauss quadrature, which can be computed from the eigenvectors.

Since $\tilde{p}_i \tilde{p}_k$, $i, k = 0, 1, \ldots, n-1$, are polynomials of degree $\leq 2n-2$, the integral

$$\langle \tilde{p}_i, \tilde{p}_k \rangle = \int_a^b \tilde{p}_i(x) \tilde{p}_k(x) w(x) \, dx \quad (5.85)$$

can be exactly computed using the n-nodes Gauss quadrature. Therefore,

$$\delta_{ik} = \langle \tilde{p}_i, \tilde{p}_k \rangle = \sum_{j=1}^{n} w_j \tilde{p}_i(x_j) \tilde{p}_k(x_j). \quad (5.86)$$

This expression can be written in matrix form as

$$\mathbf{P}^T \mathbf{W} \mathbf{P} = \mathbf{I}, \quad (5.87)$$

where

$$\mathbf{W} = \begin{pmatrix} w_1 & 0 & \cdots & 0 \\ 0 & w_2 & & \vdots \\ \vdots & & \ddots & \\ 0 & \cdots & & w_n \end{pmatrix}, \quad \mathbf{P} = \begin{pmatrix} \tilde{p}_0(x_1) & \cdots & \tilde{p}_{n-1}(x_1) \\ \vdots & & \vdots \\ \tilde{p}_0(x_n) & \cdots & \tilde{p}_{n-1}(x_n) \end{pmatrix}. \quad (5.88)$$

From (5.87), it follows that $\mathbf{P}$ is invertible and therefore $\mathbf{W}^{-1} = \mathbf{P}\mathbf{P}^T$. Then, we have

$$\frac{1}{w_j} = \sum_{k=0}^{n-1} \left(\tilde{p}_k(x_j) \right)^2 = \|\hat{\mathbf{p}}(x_j)\|_E^2, \quad (5.89)$$

where $\|.\|_E$ is the usual Euclidean norm.

5.3. Gauss quadrature

On the other hand, given $\hat{\phi}^{(j)} = (\phi_1^{(j)}, \ldots, \phi_n^{(j)})^T$, an eigenvector of the matrix $\mathbf{J}$ associated to the eigenvalue x_j, there exists a constant C such that

$$\hat{\phi}^{(j)} = C\hat{\mathbf{p}}(x_j) = C\left(\tilde{p}_0(x_j), \ldots, \tilde{p}_{n-1}(x_j)\right). \tag{5.90}$$

This is so because the eigenvector associated to each eigenvalue is unique (up to a constant multiplicative factor).

The value of C can be obtained by considering

$$1 = \langle \tilde{p}_0(x), \tilde{p}_0(x) \rangle = \tilde{p}_0^2 \int_a^b w(x)\,dx = \tilde{p}_0^2 \mu_0. \tag{5.91}$$

It follows that, $\tilde{p}_0 = 1/\sqrt{\mu_0}$ and

$$\hat{\phi}^{(j)} = \phi_1^{(j)} \frac{1}{\tilde{p}_0} \begin{pmatrix} \tilde{p}_0(x_j) \\ \tilde{p}_1(x_j) \\ \vdots \\ \tilde{p}_{n-2}(x_j) \\ \tilde{p}_{n-1}(x_j) \end{pmatrix} = \sqrt{\mu_0}\,\phi_1^{(j)}\tilde{\mathbf{p}}(x_j). \tag{5.92}$$

Therefore,

$$w_j = \frac{1}{\|\tilde{\mathbf{p}}(x_j)\|_E^2} = \mu_0 \frac{(\phi_1^{(j)})^2}{\|\hat{\phi}^{(j)}\|^2}. \tag{5.93}$$

In the case of nonorthonormal orthogonal polynomials, the starting point is a TTRR of the form

$$x P_k(x) = a_k P_{k+1}(x) + b_k P_k(x) + c_k P_{k-1}(x), \quad c_0 P_{-1}(x) = 0. \tag{5.94}$$

Proceeding in the same way, one observes that the zeros of P_n are the eigenvalues of the matrix

$$\mathcal{J} = \begin{pmatrix} b_0 & a_0 & 0 & . & . & 0 \\ c_1 & b_1 & a_1 & 0 & . & 0 \\ 0 & c_2 & b_2 & a_2 & . & 0 \\ . & . & . & . & . & 0 \\ . & . & . & . & . & a_{n-2} \\ 0 & 0 & 0 & . & c_{n-1} & b_{n-1} \end{pmatrix}. \tag{5.95}$$

However, for computing the weights as before, it is necessary to build the Jacobi matrix for the orthonormal polynomials. This can be done in a simple way. The orthogonal polynomials P_i can be related with the orthonormal polynomials $\tilde{p}_i$ by $P_i(x) = \lambda_i p_i(x)$. In this way, and substituting into (5.94), we have

$$x p_i(x) = a_i \frac{\lambda_{i+1}}{\lambda_i} p_{i+1}(x) + b_i p_i(x) + c_i \frac{\lambda_{i-1}}{\lambda_i} p_{i-1}(x). \tag{5.96}$$

Then, by considering

$$a_i \frac{\lambda_{i+1}}{\lambda_i} = c_{i+1} \frac{\lambda_i}{\lambda_{i+1}} \quad \Longrightarrow \quad \frac{\lambda_{i+1}}{\lambda_i} = \sqrt{\frac{c_{i+1}}{a_i}}, \tag{5.97}$$

we arrive at a TTRR of the form

$$xp_i(x) = \alpha_{i+1} p_{i+1}(x) + \beta_i p_i(x) + \alpha_i p_{i-1}(x), \tag{5.98}$$

where

$$\alpha_0 p_{-1}(x) = 0; \quad \beta_i = b_i, \quad i \geq 0; \quad \alpha_i = \sqrt{c_i a_{i-1}}, \quad i \geq 1. \tag{5.99}$$

This recurrence relation has the form of the recursion for the orthonormal polynomials and, furthermore, indeed is necessarily the recursion for orthonormal polynomials.

We see that the corresponding Jacobi matrix $\mathbf{J}$ (see (5.81)) is given by

$$\mathbf{J} = \begin{pmatrix} b_0 & \sqrt{a_0 c_1} & & & \\ \sqrt{a_0 c_1} & b_1 & & & \\ & & \ddots & & \\ & & & & \sqrt{a_{n-2} c_{n-1}} \\ & & & \sqrt{a_{n-2} c_{n-1}} & b_{n-1} \end{pmatrix}. \tag{5.100}$$

Of course, this matrix has the same eigenvalues as (5.95) (they are related by an equivalence transformation), but now we can apply our recipe for computing weights (see (5.93)) using (5.100) instead of (5.95).

Example 5.12. Legendre polynomials satisfy a TTRR of the form

$$(i+1)P_{i+1}(x) = (2i+1)xP_i(x) - iP_{i-1}(x) \tag{5.101}$$

or

$$xP_i(x) = \frac{i+1}{2i+1} P_{i+1}(x) + \frac{i}{2i+1} P_{i-1}(x), \quad i = 1, 2, \ldots, \tag{5.102}$$

with $P_0(x) = 1$, $P_1(x) = x$. We see that $b_i = 0$, $i = 0, 1, \ldots$, and

$$\left. \begin{aligned} a_i &= \frac{i+1}{2i+1}, \quad i \geq 0, \\ c_i &= \frac{i}{2i+1}, \quad i \geq 1, \end{aligned} \right\} \implies \alpha_i = \frac{i}{\sqrt{4i^2 - 1}}, \quad i \geq 1. \tag{5.103}$$

For $n = 2$, the $\mathbf{J}$ matrix is given by

$$\mathbf{J} = \begin{pmatrix} 0 & 1/\sqrt{3} \\ 1/\sqrt{3} & 0 \end{pmatrix} \implies x_1 = 1/\sqrt{3}, \; x_2 = -1/\sqrt{3}, \tag{5.104}$$

and, taking into account that $\mu_0 = \int_a^b w(x)\, dx = \int_{-1}^1 dx = 2$, we have

$$\hat{\phi}^{(1)} = \begin{pmatrix} 1 \\ 1 \end{pmatrix} \implies w_1 = \mu_0 \frac{1}{1+1} = 1, \tag{5.105}$$

$$\hat{\phi}^{(2)} = \begin{pmatrix} 1 \\ -1 \end{pmatrix} \implies w_2 = \mu_0 \frac{1}{1+1} = 1. \quad \blacksquare \tag{5.106}$$

5.3. Gauss quadrature

The classical formulas

For computing nodes and weights of Gauss quadrature, the only information needed are the coefficients α_i and β_i of (5.98) and $\mu_0 = \int_a^b w(x)\,dx$, which for classical quadrature formulas can be expressed in analytic form. We give this information for the three classical cases and summarize the Golub–Welsch algorithm.

1. Hermite polynomials: $\mu_0 = \sqrt{\pi}$.

$$\beta_j = 0, \quad j \geq 0; \quad \alpha_j = \sqrt{j/2}, \quad j \geq 1. \tag{5.107}$$

2. Laguerre polynomials: $\mu_0 = \Gamma(\alpha + 1)$.

$$\beta_0 = \alpha + 1, \quad \beta_j = \beta_{j-1} + 2, \quad \alpha_j = \sqrt{j(j+\alpha)}, \quad j \geq 1. \tag{5.108}$$

3. Jacobi polynomials: $\mu_0 = 2^{\alpha+\beta+1}\Gamma(\alpha+1)\Gamma(\beta+1)/\Gamma(\alpha+\beta+1)$.

$$\beta_0 = \frac{\beta - \alpha}{2 + \beta + \alpha}, \quad \beta_j = \frac{\beta^2 - \alpha^2}{(2j+\alpha+\beta)(2j+\alpha+\beta+2)}, \quad j \geq 1;$$

$$\alpha_j = \frac{2}{2j+\alpha+\beta}\sqrt{\frac{j(j+\alpha)(j+\beta)(j+\alpha+\beta)}{(2j+\alpha+\beta+1)(2j+\alpha+\beta-1)}}, \quad j \geq 1. \tag{5.109}$$

The basic steps of the Golub–Welsch algorithm can be summarized as follows.

ALGORITHM 5.4. Golub–Welsch algorithm.
Input: $\mu_0 = \int_a^b w(x)\,dx; \alpha_1, \ldots, \alpha_{n-1}; \beta_0, \ldots, \beta_{n-1}$.
Output: $x_1, \ldots, x_n; w_1, \ldots, w_n$.

- Build $\mathbf{J}$ from $\alpha_1, \ldots, \alpha_{n-1}, \beta_0, \ldots, \beta_{n-1}$, (5.81).

- Compute the eigenvalues $\rho_1, \ldots, \rho_n$ and $\mathbf{v_1}, \ldots, \mathbf{v_n}$, the corresponding eigenvectors of $\mathbf{J}$.

- DO $i = 1, \ldots, n$:

  $x_i = \rho_i$.
  $w_i = \mu_0 \dfrac{(\mathbf{v}_i(1))^2}{\|\mathbf{v}_i\|^2}$.

5.3.3 Example: The Airy function in the complex plane

We consider the integral representation given by Gautschi in [79], that is,

$$\text{Ai}(z) = \frac{z^{-\frac{1}{4}} e^{-\zeta}}{\sqrt{\pi}\Gamma(\frac{5}{6})} \int_0^\infty \left(1 + \frac{t}{2\zeta}\right)^{-\frac{1}{6}} t^{-\frac{1}{6}} e^{-t}\,dt, \tag{5.110}$$

where
$$\zeta = \tfrac{2}{3}z^{\frac{3}{2}}, \quad |\mathrm{ph}\,\zeta| < \pi \implies |\mathrm{ph}\,z| < \tfrac{2}{3}\pi. \tag{5.111}$$

This integral representation is suitable for computational purposes. The term $e^{-\zeta}$, which gives the dominant contribution, is in front of the integral; the integral itself is monotonic if $z > 0$. Additionally, we observe that the generalized Laguerre weight function with parameter $\alpha = -1/6$ appears explicitly in the integrand and that the additional factor $f(t) = (1 + t/\zeta)^{-1/6}$ has small derivatives when $|z|$ is large; therefore, the truncation error is expected to be small. Hence, we expect that Gauss–Laguerre is suitable for not too small $|z|$.

For evaluating the integral in (5.110), Gautschi [79] suggests for $z \geq 2$ a generalized Gauss–Laguerre quadrature rule with Laguerre parameter $\alpha = -\tfrac{1}{6}$, and reports that a 22-point Gauss rule yields 14-digit precision accuracy and an 83-point formula yields 28-digit precision accuracy.

In [90] we investigated this integral for complex values also. Considering complex z is not a difficulty, because we can separate the integrand into real and imaginary parts. However, for complex z a singularity in the integrand takes place when $t = -2\zeta$. We found that a 40-point Gauss rule yields 13- to 14-digit accuracy in the sector

$$|\theta| \leq \tfrac{1}{2}\pi, \quad |z| \geq 1, \quad \theta = \mathrm{ph}\,z, \tag{5.112}$$

but at the boundary of the sector described in (5.111), when $\theta \to \pm\tfrac{2}{3}\pi$, the singularity at $t = -2\zeta$ reaches the positive real axis, and accuracy is therefore lost.

By turning the path of integration we can always avoid this situation. In §5.5, we will give more details on turning the path of integration for contour integrals. In fact, if $|\theta| \in [\tfrac{1}{2}\pi, \tfrac{2}{3}\pi]$, we turn the path, and use the representation

$$\mathrm{Ai}(z) = \left(\frac{e^{i\tau}}{\cos\tau}\right)^{\frac{5}{6}} \frac{z^{-\frac{1}{4}}e^{-\zeta}}{\sqrt{\pi}\,\Gamma(\tfrac{5}{6})} \int_0^\infty e^{-it\tan\tau}\left(1 + \frac{t}{2\zeta'}\right)^{-\frac{1}{6}} t^{-\frac{1}{6}}e^{-t}\,dt, \tag{5.113}$$

where
$$\zeta' = \zeta e^{-i\tau}\cos\tau, \quad \tau = \frac{3}{2}\left(\theta - \frac{1}{2}\pi\right), \tag{5.114}$$

τ being the angle for turning the path of integration.

In this way, the Gauss–Laguerre rule can be used for the complete sector $|\theta| \leq \tfrac{2}{3}\pi$.

5.3.4 Further practical aspects of Gauss quadrature

Gauss quadrature is in some sense the best method, which does not necessarily mean that this is so in practice. Gauss quadrature is the way to make the degree of exactness as large as possible. Sure, for computing an integral

$$\int_a^b f(x)w(x)\,dx, \tag{5.115}$$

in which f can be accurately approximated by a polynomial, Gauss quadrature is a very good method if the nodes and weights for $w(x)$ can be accurately computed.

5.4. The trapezoidal rule on $\mathbb{R}$

This computation, however, is not immediate, although for the classical weights the problem is easier, because the coefficients of the associated recurrence relations are known analytically and then the equivalent matrix-eigenvalue problem (Golub–Welsch) can be formulated exactly. For nonclassical weights, the coefficients of the recursion should be also numerically computed before applying the matrix method (also, methods of modified moments for computing nodes and weights are available [80]).

Even for the classical cases, the matrix method has its limitations. The matrix-eigenvalue method usually works reasonably well when a not very large number of nodes is required. However, for a high number of nodes the condition of the problem may worsen. In particular, it has been described that for Gauss–Legendre quadrature, loss of precision is expected, in particular close to $x = \pm 1$, where the zeros tend to cluster quadratically. For such cases, alternative methods of computation of the zeros and weights could be of interest. For instance, one can consider Newton or fixed point methods (Chapter 7) for computing the nodes, and alternative formulas for the weights based on the Christoffel–Darboux formula. For instance, from [219, eq. (6.9)] we have

$$w_j = -\frac{F_{n+1}}{F_n} \frac{1}{\tilde{p}_{n+1}(x_j)\tilde{p}'_n(x_j)}. \tag{5.116}$$

See, for instance, [204] for a discussion of the Gauss–Legendre case.

Apart from the difficulties in computing nodes and weights, a drawback of the Gauss method is its inherent lack of flexibility. One has to fix the number of nodes n and then compute the nodes and weights in the (preferably founded) hope that with n nodes one can obtain the required accuracy. This is partly corrected by the Gauss–Kronrod algorithm (see §9.6.3).

However, when a Gauss quadrature can be found and accuracy can be checked, Gauss quadrature is always an option which must be considered in the computation of an integral, particularly when the integral is of the form of a classical weight times a function which can be well approximated by a polynomial.

5.4 The trapezoidal rule on $\mathbb{R}$

Until now we have considered integrals over finite intervals except for the Gauss–Hermite and Gauss–Laguerre quadrature formulas. For integrals over $\mathbb{R}$ the trapezoidal rule again may be very efficient and accurate, as is known since the work of Goodwin [98].

The trapezoidal rule on $\mathbb{R}$ is the natural extension of the compound trapezoidal rule. We write

$$\int_{-\infty}^{\infty} f(x)\,dx = h \sum_{j=-\infty}^{\infty} f(hj) + R(h), \tag{5.117}$$

where $h > 0$. We choose $x = 0$ as one of the nodes, but this is not necessary and the nodes could be shifted by an amount $0 < d < h$. For even functions $f(x) = f(-x)$, it will be always interesting to take into account the symmetry in order to save computations. In this case, we write

$$\int_{-\infty}^{\infty} f(x)\,dx = 2h \left(\frac{1}{2}f_0 + \sum_{j=1}^{\infty} f_j\right) + R(h), \tag{5.118}$$

where $f_j = f(jh)$. A similar arrangement can be done for functions symmetric around another x value, say x_0.

Also, if a function does not have any symmetry, we can always symmetrize the computation and write it as the integral of an even function. For any real function $f(x)$ integrable in $(-\infty, \infty)$, we can write

$$\int_{-\infty}^{\infty} f(x)\,dx = \frac{1}{2}\int_{-\infty}^{\infty} (f(x) + f(-x))\,dx = \int_{0}^{\infty} (f(x) + f(-x))\,dx. \tag{5.119}$$

Therefore, when we consider integrals of real functions, we can assume that $f(x)$ is even, without loss of generality.

As we will see next, for this type of integral, the trapezoidal rule is very efficient (in fact, optimal) and the error generally decreases exponentially when the function is analytic and bounded in a strip around the real axis.

Before proving this asymptotic behavior we will present a method for estimating the error of quadrature rules based on contour integrals in the complex plane.

5.4.1 Contour integral formulas for the truncation errors

Let us consider a quadrature formula

$$\int_a^b f(x)w(x)\,dx = \sum_k w_k f(x_k) + R, \tag{5.120}$$

where a and/or b could be infinite and, depending on this, the sum may have a finite or an infinite number of terms. We include a weight function for greater generality, although in this section we are mainly concerned with the trapezoidal rule.

For the moment, we consider a finite interval and discuss later the infinite case. Assume now that $f(z)$ is analytic in a domain D containing the interval $[a, b]$ and let $x \in [a, b]$. Using Cauchy's integral formula we can write

$$R = \frac{1}{2\pi i} \left[\int_a^b w(x) \left(\int_C \frac{f(z)}{z-x} dz \right) dx - \sum_k \int_C \frac{w_k}{z-x_k} f(z)\,dz \right], \tag{5.121}$$

where C is a counterclockwise contour around $[a, b]$ and contained in D; see Figure 5.3. Interchanging the order of integration in the first term, and considering a finite number of nodes $x_k, k = 1, \ldots, n$, we can write

$$R = \frac{1}{2\pi i} \int_C \Phi(z) f(z)\,dz, \tag{5.122}$$

where

$$\Phi(z) = \int_a^b \frac{w(x)}{z-x} dx - \sum_{k=1}^n \frac{w_k}{z-x_k}. \tag{5.123}$$

5.4. The trapezoidal rule on $\mathbb{R}$ 149

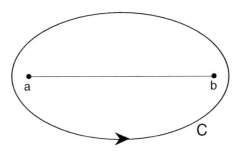

Figure 5.3. *The path of integration in* (5.121) *and* (5.122).

The function $\Phi(z)$ is the characteristic function of the error and can be used for analyzing the accuracy of a quadrature method.

The degree of exactness of a quadrature rule can be interpreted in terms of the asymptotic behavior of this function as $z \to \infty$. Indeed, when $\Phi(z) = \mathcal{O}(z^{-n-2})$, $n \in \mathbb{N}$, the degree of exactness of the quadrature rule is n. To see this, observe that for all polynomials of degree not larger than n the integral in (5.122) vanishes by Cauchy's theorem. Apart from giving an interesting connection between rational approximation and Gauss quadrature, this property suggests that the behavior of $\Phi(z)$ for large z is crucial: the faster it decays, the greater the accuracy.

For integrals in $\mathbb{R}$ we can use these same ideas with a few modifications. Let us consider $w(x) = 1$ in this case, although the generalization to other weights is obvious. We have a quadrature

$$\int_{-\infty}^{\infty} f(x)\, dx = \sum_{k} w_k f(x_k) + R, \qquad (5.124)$$

where the nodes may tend to infinity. We can use the same ideas as before but modifying the contour of integration. We choose a path Γ consisting of two lines parallel to the x-axis, one below the axis and one above and both at a distance a from $\mathbb{R}$. This is the right choice when $f(z)$ is analytic in an open region containing the strip

$$\{x + iy \mid x \in \mathbb{R},\ |y| \leq a\}, \qquad (5.125)$$

and when the contributions over the lines $\pm \xi + iy$, $|y| \leq a$, tend to zero as $\xi \to \infty$ (it is sufficient that $f(x + iy)$ goes to zero as $x \to \infty$ uniformly with respect to y in $|y| \leq a$). Then, as in (5.121), we have

$$R = \frac{1}{2\pi i} \left(\int_{\Gamma} f(z) \left[\int_{-\infty}^{\infty} \frac{w(x)}{z - x}\, dx \right] dz - \sum_{k} \int_{\Gamma} \frac{w_k}{z - x_k} f(z)\, dz \right). \qquad (5.126)$$

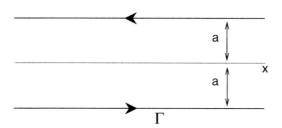

Figure 5.4. *The integration path in* (5.130).

Behavior of the trapezoidal rule in $\mathbb{R}$

Let us consider the particular case of the trapezoidal rule for (5.126). In this case $w(x) = 1$, $w_k = h$, and $x_k = kh$, $k \in \mathbb{Z}$. Hence, the second term becomes (see [2, eq. (4.3.91)])

$$h \sum_{k=-\infty}^{\infty} \int_\Gamma \frac{f(z)}{z - kh} dz = h \int_\Gamma \left\{ \frac{1}{z} + \sum_{k=1}^{\infty} \left(\frac{1}{z - kh} + \frac{1}{z + kh} \right) \right\} f(z) \, dz \\ = \int_\Gamma \cot(\pi z/h) f(z) \, dz. \quad (5.127)$$

The x-integral in (5.126) will be interpreted as follows (a rigorous approach can be based on taking finite integrals as a start). We write

$$\int_{-\infty}^{\infty} \frac{1}{z - x} dx = -\int_{-\infty}^{\infty} \frac{x + z}{x^2 - z^2} dx = -z \int_{-\infty}^{\infty} \frac{1}{x^2 - z^2} dx, \quad (5.128)$$

where $z \notin \mathbb{R}$. Evaluating the third integral by using residue calculus, we find

$$\int_{-\infty}^{\infty} \frac{1}{z - x} dx = -i\pi \mathrm{sign}(\Im z). \quad (5.129)$$

Using (5.127) and (5.129) we obtain

$$R = \frac{1}{2\pi i} \int_\Gamma \Phi(z) f(z) \, dz, \quad \text{where} \quad \Phi(z) = -\pi \left[\cot(\pi z/h) + i\mathrm{sign}(\Im z) \right], \quad (5.130)$$

and Γ is the above-described path, shown in Figure 5.4.

We notice that $|\Phi(z)|$ becomes exponentially small as $|\Im z|$ becomes large and that

$$|\Phi(z)| \sim 2\pi \exp(-2\pi |\Im z|/h), \quad |\Im z| \to \infty, \quad (5.131)$$

which is telling us that the trapezoidal rule (5.117) is an interesting quadrature rule for computing integrals on $\mathbb{R}$. Furthermore, if the closest singularity of $f(z)$ to the real axis is such that it is at a distance a from this axis, we can expect an error which can be estimated to be $\mathcal{O}(\exp(-2\pi a/h))$. We will refine this estimate in the next theorem.

Theorem 5.13. *Let $f(z)$ be an analytic function in an open set containing the strip*

$$\{z = x + iy \mid x \in \mathbb{R}, \ -a \le y \le a\}, \quad (5.132)$$

5.4. The trapezoidal rule on $\mathbb{R}$

where

$$\int_{-\infty}^{\infty} |f(x+iy)|\, dx \tag{5.133}$$

is convergent. Then $R(h)$ of (5.118) satisfies

$$R(h) = \int_{-\infty}^{\infty} \frac{f(x+iy)}{1 - \exp[-2\pi i(x+iy)/h]}\, dx \\ + \int_{-\infty}^{\infty} \frac{f(x-iy)}{1 - \exp[2\pi i(x-iy)/h]}\, dx \tag{5.134}$$

for any y with $0 < y \leq a$. Moreover, if $f(x)$ is real for real x, then

$$|R(h)| \leq \frac{e^{-\pi a/h}}{\sinh(\pi a/h)} \int_{-\infty}^{\infty} |f(x \pm ia)|\, dx. \tag{5.135}$$

Proof. Equation (5.134) is just the explicit representation of (5.130) with the corresponding integration contour (parallel lines at a distance $y \leq a$ from the real axis).

With respect to the bound, both integrals can be bounded in a similar way. For the first one we have

$$\left| \int_{-\infty}^{\infty} \frac{f(x+ia)}{1 - \exp[-2\pi i(x+ia)/h]}\, dx \right| \leq \int_{-\infty}^{\infty} \frac{|f(x+ia)|}{|1 - \exp(-2\pi ix/h)\exp(2\pi a/h)|}\, dx \\ \leq \int_{-\infty}^{\infty} \frac{|f(x+ia)|}{|1 - \exp(2\pi a/h)|}\, dx \\ = \frac{e^{-\pi a/h}}{2\sinh(\pi a/h)} \int_{-\infty}^{\infty} |f(x+ia)|\, dx, \tag{5.136}$$

similarly to the second integral in (5.134), now with $f(x+ia)$ replaced by $f(x-ia)$; but $|f(x+ia)| = |f(x-ia)|$ due to the Schwartz reflection principle, and applying the triangular inequality the result is proven. $\square$

This shows clearly that the trapezoidal rule is very convenient. Furthermore, it can be proven that the trapezoidal rule is optimal compared to any other rule with the same average of nodes per unit length.

From (5.135) it follows that the error in the trapezoidal rule when $h \to 0$ is $\mathcal{O}(e^{-2\pi a/h})$, as we suggested before. Large values of a generally result in small errors, but the quantities $M_a(f)$ may influence the behavior of the error when a becomes large. Often the best bounds for the error are not obtained necessarily for the largest a and there is a value optimizing the error. For example, let us consider

$$f(x) = e^{-wx^2} g(x), \quad w > 0. \tag{5.137}$$

Then we have

$$\int_{-\infty}^{\infty} |f(x \pm ia)|\, dx = e^{wa^2} \int_{-\infty}^{\infty} e^{-wx^2} |g(x \pm ia)|\, dx \\ \leq e^{wa^2} \int_{-\infty}^{\infty} |g(x \pm ia)|\, dx. \tag{5.138}$$

Therefore we obtain in this case

$$|R(h)| \leq \frac{e^{-\pi a/h + wa^2}}{\sinh(\pi a/h)} \int_{-\infty}^{\infty} |g(x \pm ia)|\, dx. \tag{5.139}$$

The function $-2\pi a/h + wa^2$, considered as a function of a, is minimal for

$$a = \frac{\pi}{wh}. \tag{5.140}$$

Assuming that this value of a is possible and neglecting the dependence on a of the integral, we have

$$|R(h)| = \mathcal{O}\left(e^{-\pi^2/wh^2}\right), \quad h \to 0. \tag{5.141}$$

This exponential behavior for the error was first observed by Goodwin [98].

Practical aspects

For computing an integral on $\mathbb{R}$ by means of the trapezoidal rule we have to make two choices: first, the selection of the step h (which can be refined recursively, as discussed in §5.2.2); second, the integral (or the sum) must be truncated to a finite integral or a finite sum.

This can be done in several ways. The first possibility is to replace the integral by a finite integral and neglecting intervals containing $\pm \infty$. For simplicity, let us assume that $f(x)$ is even. Then, for sufficiently large x_0, we consider the following approximations:

$$\int_{-\infty}^{\infty} f(x)\, dx \approx 2 \int_{0}^{x_0} f(x)\, dx \approx 2h \left[\frac{1}{2} f_0 + \sum_{i=1}^{n} f_i\right], \quad h = x_0/n. \tag{5.142}$$

This is a good approach if we can somehow estimate the tail of the integral or find a practical bound. After reducing the interval of integration, the recursive trapezoidal rule can be used for computing the integral on $[0, x_0]$.

The second possibility is to select h and then fix the number of terms in the sum n by using some other criterion. For instance, if the terms rapidly decrease and are monotonic, then we can stop when a term is less than a certain small number. If h is small enough, this will give an estimate of the error. This method, however, breaks in some way the recursivity of Algorithm 5.1. This approach is considered in Example 5.14.

When it is possible, the first option seems preferable, although two problems may arise: if the interval is too small, accuracy is lost; if it is too large, an excessive number of function evaluations is computed and efficiency is lost.

Another way to tackle the problem and to reduce the possible drawbacks in the selection of the interval is to consider adaptivity. One would start with an excessively large interval of the computations. By keeping track of partial integrals over smaller subintervals, one can discard portions of the integral giving negligible contributions. This leads to more involved algorithms and to the necessity of some data storage.

Depending on the application, one method could be better than another one. Selecting a fixed interval is a common practice and, when no reliable estimates are available, it is usually possible to perform other numerical checks to certify the correctness in the selection of the interval. This allows us to apply Algorithm 5.1 as it is. This is the approach considered, for instance, in [86, 89, 94].

5.4. The trapezoidal rule on $\mathbb{R}$

Table 5.3. *Computation of $R(h)$ in (5.144) for $x = 5$.*

h	j_0	$R_0(h)$
1	2	$-0.18\,10^{-001}$
1/2	5	$-0.24\,10^{-006}$
1/4	12	$-0.65\,10^{-015}$
1/8	29	$-0.44\,10^{-032}$
1/16	67	$-0.19\,10^{-066}$
1/32	156	$-0.55\,10^{-136}$
1/64	355	$-0.17\,10^{-272}$

Changing the variable of integration (see §5.4.2) is also useful in reducing the typical intervals to be considered.

Example 5.14 (modified Bessel function $K_0(x)$). Consider the modified Bessel function

$$K_0(x) = \tfrac{1}{2} \int_{-\infty}^{\infty} e^{-x \cosh t}\, dt. \tag{5.143}$$

We have

$$e^x K_0(x) = \tfrac{1}{2} h + h \sum_{j=1}^{\infty} e^{-x(\cosh(hj)-1)} + R_0(h). \tag{5.144}$$

For $x = 5$ and several values of h we obtain the values of Table 5.3 (j_0 denotes the number of terms used in the series in (5.144)). We see that already for $h = \tfrac{1}{4}$ and 12 function evaluations, IEEE double precision can be obtained. ∎

5.4.2 Transforming the variable of integration

Given the optimality of the trapezoidal rule for the computation of certain integrals in $\mathbb{R}$, it is natural to ask whether other types of integrals can be transformed by changing the variable of integration to a form suitable for the trapezoidal rule.

Let us, for instance, consider an integral over a finite interval,

$$I(f) = \int_{-1}^{1} f(x)\, dx. \tag{5.145}$$

If we consider a new variable u related to x by $x = \phi(u)$, ϕ being a function mapping the interval $[-\infty, \infty]$ onto $[-1, 1]$, we can write

$$I(f) = \int_{-\infty}^{\infty} f(\phi(u)) \phi'(u)\, du, \tag{5.146}$$

and apply the trapezoidal rule. We can try to choose ϕ in order to make the trapezoidal rule as successful as possible.

The tanh-rule

For instance, if we take $x = \tanh u$ or

$$u = \frac{1}{2} \log\left(\frac{1+x}{1-x}\right) = \tanh^{-1}(x), \tag{5.147}$$

we obtain

$$I(f) = \int_{-\infty}^{\infty} f(\tanh u) \frac{1}{\cosh^2 u} \, du. \tag{5.148}$$

The integrand is exponentially decaying, which is interesting for applying the trapezoidal rule. We write

$$I(f) = h \sum_{k=-\infty}^{\infty} f(\tanh kh) \frac{1}{\cosh^2 kh} + R^{\tanh}(h), \tag{5.149}$$

which is named the *tanh-rule*.

The change of variable introduces, however, singularities at $u = \pm i(k + 1/2)\pi$, $k \in \mathbb{Z}$ (the zeros of $\cosh u$). The singularities closest to the real axis are at a distance $d = \pi/2$. Therefore, when the function $f(\tanh u)$ does not have singularities in the u-plane at a distance smaller than $\pi/2$, we have, according to Theorem 5.13, the error estimate $R^{\tanh}(h) = \mathcal{O}(\exp(-\pi^2/h))$.

We can give an estimate of the error in terms of the number of function evaluations. For simplicity, we assume that the integral is even (which is no restriction when $f(x)$ is real for real x) and that we approximate by integrating in $[-u_0, u_0]$, $u_0 = Nh$. That is, we take

$$I(f) = \int_{-u_0}^{u_0} f(\tanh u) \frac{1}{\cosh^2 u} \, du + \epsilon_t, \tag{5.150}$$

and consider the trapezoidal rule for this reduced integral. The truncation error ϵ_t will be governed by $1/\cosh^2 u_0 = \mathcal{O}(\exp(-2u_0)) = \mathcal{O}(\exp(-2Nh))$. On the other hand, we have $R(h) = \mathcal{O}(\exp(-\pi^2/h))$. By fixing the number of points N, the error for the truncation of the interval increases as h decreases, while the contrary happens for $R(h)$; the best situation is when both errors are similar. We have $\epsilon_t \sim R(h)$ and then $h \sim \pi/\sqrt{2N}$; therefore

$$R^{\tanh}(h) = \mathcal{O}(\exp(-\pi^2/h)) = \mathcal{O}(\exp(-\pi\sqrt{2N})). \tag{5.151}$$

The two main limiting factors are the truncation of the integration interval and the distance of the singularities introduced to the real axis.

The erf-rule

We have the possibility of using a change of variable for which no additional singularities are introduced, such that the second limitation is reduced. Consider the error function

$$\operatorname{erf} z = \frac{2}{\sqrt{\pi}} \int_0^z e^{-z^2} \, dz, \quad z \in \mathbb{C}, \tag{5.152}$$

5.4. The trapezoidal rule on $\mathbb{R}$

which is an entire function that maps $[-\infty, \infty]$ onto $[-1, 1]$. The change of variable $x = \operatorname{erf} u$ transforms (5.145) into

$$I(f) = \frac{2}{\sqrt{\pi}} \int_{-\infty}^{\infty} g(u) e^{-u^2} du, \quad g(u) = f(\operatorname{erf} u), \tag{5.153}$$

which, as we know, is suitable for the trapezoidal rule under certain analyticity conditions for $g(u)$.

If f is an entire function (analytic at all finite points in $\mathbb{C}$), then $g(u)$ is also entire and a bound for the error of the trapezoidal rule is given by (5.141), which decays very fast. Again, we can study the dependence on the number of function evaluations by setting $R(h) \sim \epsilon_t$. This gives

$$R^{\operatorname{erf}}(h) = \mathcal{O}(\exp(-\pi N)). \tag{5.154}$$

This shows that the application of the *erf-rule*

$$I(f) = \sum_{k=-\infty}^{\infty} f(\operatorname{erf}(kh)) \exp(-k^2 h^2) + R^{\operatorname{erf}}(h), \tag{5.155}$$

with the sum conveniently truncated, is a very efficient method for entire functions. There exist very efficient methods to compute the error function, such as, for instance, Chebyshev rational approximations [42]. This variable transformation was considered in [85, 86] for computing parabolic cylinder functions.

The idea of transforming the variable so that the performance of the trapezoidal rule is optimized was considered by Schwartz in [192] and enhanced by Takahashi and Mori [206, 207].

These ideas can also be applied to cases for which the function $f(x)$ in (5.145) is singular at the extreme points $x = \pm 1$ (of course, provided that the integral converges). Depending on the mapping function $\phi(u)$, these or other singularities will result in several transformed singular points in the transformed complex plane (in the u variable). Apart from the singularities at $\pm \infty$, the multivaluedness of the inverse function ϕ^{-1} will give additional singularities away from the real axis.

For instance, if $f(x) = (1 - x^2)^{-\alpha} g(x)$, $0 < \alpha < 1$, with g an entire function, the transformed integral for the erf-rule reads

$$\int_{-1}^{1} f(x) \, dx = \frac{2}{\sqrt{\pi}} \int_{-\infty}^{\infty} [\operatorname{erfc}(-u)\operatorname{erfc}(u)]^{-\alpha} g(\operatorname{erf} u) e^{-u^2} du, \tag{5.156}$$

where we use the complementary error function $\operatorname{erfc} z = 1 - \operatorname{erf} z = 2 - \operatorname{erfc}(-z)$ to control the relative error in $1 \pm \operatorname{erf} u$ at $u = \mp \infty$. The singularities at $u = \pm \infty$ are of no concern, given the accompanying exponential e^{-u^2}. Additional singularities occur because of the zeros of $\operatorname{erfc} z$. For these zeros, see §7.6.3 and Table 7.3. As discussed in §5.5, the error is essentially determined by the position of these transformed singularities as well as the position of the possible singularities of $\phi'(u)$ (the transformation $x = \phi(u) = \operatorname{erf} u$ does not produce such singularities).

For this type of integral with endpoint singularities, one can prove that the performance of the tanh-rule does not change because the nearest singularity to the real axis is still coming from the function $1/\cosh^2 u$, which is introduced by changing the variable. Contrarily, the erf-rule worsens when endpoint singularities occur; the change of variable does not introduce any singularity, but in the transformed plane the singularities coming from ± 1 do contribute to the error [206]. In such a case, because $R(h) = \mathcal{O}(\exp(-2\pi d/h))$, with d the distance of the singularity to the origin, and by setting $R(h) \sim \epsilon_f = \mathcal{O}(\exp(-kN^2h^2))$, with k a factor depending on how the endpoint singularities of f behave, we have

$$\widetilde{R}^{\text{erf}}(h) = \mathcal{O}(\exp(-cN^{2/3})), \tag{5.157}$$

where c is a constant. Now the behavior of the error, as a function of N, is something between the tanh-rule and the erf-rule for entire functions.

In conclusion, the erf-rule is superior to the tanh-rule and it is particularly efficient for entire functions.

Double exponential formulas

Following this tendency, it appears that the faster ϕ' decays, the better the performance. In fact, it was proved that a doubly exponential decay, such that

$$|f(\phi(u))\phi'(u)| = \mathcal{O}\left(\exp(-\alpha \exp|u|)\right), \quad |u| \to \infty, \tag{5.158}$$

appears to be optimal [92]. For the change of variable

$$x = \tanh\left(\tfrac{1}{2}\pi \sinh u\right), \tag{5.159}$$

where the constant $\tfrac{1}{2}\pi$ is convenient in the analysis, it is seen by performing a similar analysis to before that the error behaves as

$$R^{\text{DE}}(h) = \mathcal{O}\left(\exp\left(-\frac{\pi d N}{2\log(dN)}\right)\right), \tag{5.160}$$

which is better than the erf-rule, (5.157), when endpoint singularities occur.

Changes of variables can also be used for speeding the convergence of intervals in $\mathbb{R}$ or in $(0, \infty)$ when the integrands are slowly decaying. For instance, for an integral

$$I(f) = \int_{-\infty}^{\infty} f(x)\,dx, \tag{5.161}$$

in which $f(x)$ decays algebraically as $|x| \to \infty$, we can apply

$$x = \sinh\left(\frac{\pi}{2}\sinh u\right), \tag{5.162}$$

and the resulting integrand will decay as a double exponential.

Additional changes of variable are as follows [154]:

$$\begin{aligned} I(f) &= \int_0^{\infty} f(x)\,dx \Rightarrow x = \exp\left(\tfrac{1}{2}\pi \sinh u\right), \\ I(f) &= \int_0^{\infty} e^{-x} f(x)\,dx \Rightarrow x = \exp\left(u - \exp(-u)\right). \end{aligned} \tag{5.163}$$

5.5 Contour integrals and the saddle point method

So far, we have considered numerical quadrature for real integrands and real intervals of integration. However, many special functions (with real or complex variables) can be defined by contour integrals in the complex plane with complex integrands. For computing these integrals (in fact, also for computing many real integrals), it is important to take special precautions when rapidly oscillating integrands appear.

In this section, we will discuss how contour integrals can be modified in order to improve the stability of numerical quadrature. These ideas, particularly the saddle point method and the steepest descent contours, are frequently encountered in the asymptotic analysis of special functions, but we will use them as an analytical tool for obtaining stable integral representations (free of oscillations).

Example 5.15 (the cosine transform of the Gaussian). This simple example will illustrate the difficulties encountered when computing rapidly oscillating integrals. Consider the integral

$$I(\lambda) = \int_{-\infty}^{\infty} e^{-t^2} \cos(2\lambda t)\, dt = \sqrt{\pi} e^{-\lambda^2}. \tag{5.164}$$

Taking $\lambda = 10$ we get

$$I(10) \approx 0.6593662990 \cdot 10^{-43}. \tag{5.165}$$

If we try to compute this integral for $\lambda = 10$ by any of the quadrature methods discussed in this chapter, the result will be completely wrong, unless we can use greater than 43-digit precision in the computation. The reason is clear: the maximal value of the integrand is 1 (at $x = 0$), but the value of the integral is much smaller. When computing the integral by, let's say, the trapezoidal rule, not so small positive and negative values of the rapidly oscillating function are canceled in order to reach such a tiny result. Therefore, loss of accuracy is inevitable, and when computing the integral with, for instance, 15-digit precision, a result of the order of 10^{-15} will be obtained instead of (5.165).

This severe loss of accuracy becomes worse as λ increases. In order to avoid it, some preliminary analysis is necessary. The analysis needed for this case is simple (and leads to the exact result).

We write

$$I(\lambda) = \int_{-\infty}^{\infty} e^{-t^2 + 2i\lambda t}\, dt = e^{-\lambda^2} \int_{-\infty}^{\infty} e^{-(t - i\lambda)^2}\, dt. \tag{5.166}$$

Now, by invoking Cauchy's theorem, it is an easy exercise in complex analysis to show that with the substitution $t = s + i\lambda$ the integral can be written as

$$I(\lambda) = e^{-\lambda^2} \int_{-\infty}^{\infty} e^{-s^2}\, ds. \tag{5.167}$$

No oscillations occur now. Moreover, the small factor $e^{-\lambda^2}$ is in front of the integral, and we do not get a tiny result by cancellation but by computing a real exponential. The resulting s-integral can now be efficiently computed with the trapezoidal rule (see §5.4), although this is completely unnecessary in this case because we know that the result is $\sqrt{\pi}$.

The substitution $t = s+i\lambda$ is, in this case, equivalent to shifting the path of integration upward into the complex plane until we reach the point $t = i\lambda$; indeed, the new path of integration is given by $t = s + i\lambda$ with s real. This shifting can be done because of the following two points. First, no singularities of the integrand are crossed when moving the integration path in this way. Second, the contributions at infinity are zero, meaning that the endpoints of the path can be moved.

The point $t = i\lambda$ is a saddle point and the new path of integration is a steepest descent path. We will give detailed information on these concepts in the following section. ∎

Many integral representations for special functions are in terms of complex contour integrals. For evaluating integrals along a complex contour we can try to parametrize directly this contour by means of a real variable that can be used for integration. However, as illustrated in the previous example, sometimes it is very convenient to use methods of complex analysis to prepare the integrals before numerical quadrature is considered.

In particular, it will be interesting to modify the integrals so that imaginary parts disappear from the dominant exponential functions, as we did when transforming (5.166) into (5.167). Steepest descent and saddle point methods will be useful for this.

5.5.1 The saddle point method

The saddle point method is a useful tool from asymptotic analysis for the evaluation of contour integrals in the complex plane of the form

$$F(\lambda) = \int_{\mathcal{L}} e^{-\lambda \psi(z)} \eta(z)\, dz, \quad (5.168)$$

where ψ and η are analytic functions of z in a domain $\mathcal{D}$ in the complex plane and $\mathcal{L}$ is a contour in $\mathcal{D}$ with endpoints at infinity. We will take λ as a positive real parameter; for our numerical discussion, we can skip the parameter λ, but it is useful for describing the behavior of the integral as the oscillations become faster.

The evaluation of (5.168) may be problematic because of the oscillatory behavior of the integrand, due to the term $e^{-i\lambda \Im \psi(z)}$. However, for integrals such as (5.168) with an analytic integrand, there is a lot of freedom in choosing the path of integration, and if we could deform or change the original contour $\mathcal{L}$ into a new contour $\mathcal{C}$ in $\mathcal{D}$ such that

$$\Im \psi(z) = a = \text{constant} \quad \text{for } z = x + iy \in \mathcal{C}, \quad (5.169)$$

then we have

$$F(\lambda) = e^{-i\lambda a} \int_{\mathcal{C}} e^{-\lambda \Re \psi(z)} \eta(z)\, dz, \quad (5.170)$$

and the remaining exponential part $e^{-\lambda \Re \psi(z)}$ will no longer be oscillating. This is what we did in our previous example. This is also the starting point for computing many asymptotic expansions as the parameter λ becomes large. Our main interest now, however, is of a numerical nature.

5.5. Contour integrals and the saddle point method

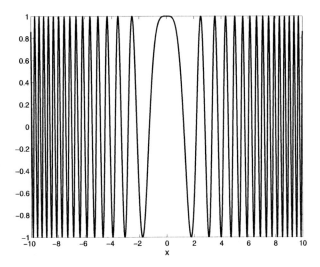

Figure 5.5. *Real part of the integrand in* (5.171) *for* $\lambda = 1$.

When the endpoints of the integration paths $\mathcal{C}$ and $\mathcal{L}$ coincide, then, by Cauchy's theorem, we can deform the contour $\mathcal{L}$ into $\mathcal{C}$, and the right-hand sides of (5.168) and (5.170) are equal, because we have assumed that the integrand is analytic in $\mathcal{D}$. If, as is usually the case, the endpoints do not coincide, we have to verify that the integral over the paths joining specific endpoints of $\mathcal{L}$ and $\mathcal{C}$ cancel. For paths having both endpoints at infinity, which is the case we are considering, it is usually sufficient to find a bound for the amplitude of the integrand as z tends to infinity; when this bound decays fast enough, the contribution caused by changing the endpoints will vanish. For doing this, it will be crucial to understand the behavior of $\Re \psi(z)$ in the complex plane.

We now give another example showing how to avoid oscillations in the integrand by an adequate change in the integration path.

Example 5.16 (an oscillatory Gaussian integral). Consider the evaluation of

$$F(\lambda) = \int_{-\infty}^{\infty} e^{-i\lambda z^2} \, dz, \quad \lambda > 0. \tag{5.171}$$

Figure 5.5 shows a plot of the real part of the integrand (5.171) for $\lambda = 1$ and $z \in [-10, 10]$. The amplitude of the integrand is 1 along the real axis, but the integral is convergent because oscillations tend to cancel, particularly the oscillations away from the origin.

For obtaining an integrand free from oscillations, we can try to turn the path of integration, by setting $z = \rho e^{\pm i\pi/4}$, with real ρ. As we explain next, only one of the two options is possible—the minus sign—and taking this option, we have

$$F(\lambda) = e^{-i\pi/4} \int_{-\infty}^{\infty} e^{-\lambda \rho^2} \, d\rho. \tag{5.172}$$

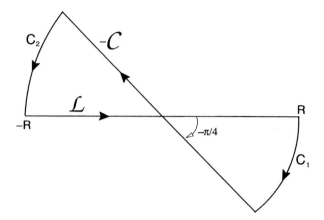

Figure 5.6. *The path of integration in* (5.174).

Indeed, writing $z = x + iy$ the integrand reads

$$f(z) = e^{-i\lambda z^2} = e^{2\lambda xy}e^{-i\lambda(x^2 - y^2)}. \tag{5.173}$$

Therefore, in the sectors where $xy \leq \delta < 0$ the integrand decays exponentially at infinity. Then, we can turn the original path of integration in the clockwise direction to the line $x = -y$ because the contributions at infinity will vanish. Indeed, by Cauchy's theorem the integral over the closed circuit of Figure 5.6 is zero. Then

$$\int_{-R}^{R} f(x)\,dx + \int_{C_1} f(z)\,dz + \int_{R}^{-R} f(e^{-i\pi/4}\rho)e^{-i\pi/4}d\rho + \int_{C_2} f(z)\,dz = 0, \tag{5.174}$$

and taking $R \to +\infty$ the contributions of the arcs C_1 and C_2 tend to zero. To see this, we observe that the contribution from C_1 reads

$$iR \int_{0}^{-\pi/4} e^{-i\lambda R^2 e^{2i\theta} + i\theta}\,d\theta, \tag{5.175}$$

which can be bounded as

$$R \int_{0}^{\pi/4} e^{-\lambda R^2 \sin(2\theta)}\,d\theta \leq R \int_{0}^{\pi/4} e^{-\frac{1}{2}\lambda R^2 \theta}\,d\theta. \tag{5.176}$$

Integration by parts shows that the last integral tends to zero, as $R \to +\infty$. The same holds for the contribution of C_2.

Contrarily, we cannot turn the integration path counterclockwise to reach the diagonal $x = y$ because the integrand is unbounded as z tends to infinity in the sectors where $xy \geq \delta > 0$.

The integral in (5.172) can be evaluated. Taking $\rho = \sqrt{\lambda}t$ we have

$$F(\lambda) = e^{-i\pi/4} \frac{1}{\sqrt{\lambda}} \int_{-\infty}^{\infty} e^{-t^2}\,dt = e^{-i\pi/4} \sqrt{\frac{\pi}{\lambda}}, \tag{5.177}$$

5.5. Contour integrals and the saddle point method

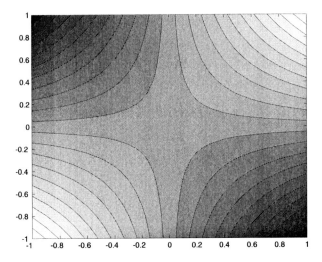

Figure 5.7. *Contour lines of $|f(z)| = |e^{-i\lambda z^2}| = e^{2\lambda xy}$. The dark shaded regions correspond to small values of $|f(z)|$ (valleys) and the light regions to large values (hills).*

which becomes smaller as λ increases. By turning the path of integration, we have managed to obtain the λ-dependence as a factor in front of the integral (which we can compute exactly), instead of coming from cancellations in a rapidly oscillating integrand. ∎

From Example 5.16 we learn that contours should be modified such that they enter the valleys of the integrand and that, as a consequence, the contributions at infinity vanish. Let us have a closer look at the amplitude of the integrand of (5.171). In Figure 5.7, the contour lines of

$$|f(z)| = \left|e^{-\lambda\psi(z)}\right| = \left|e^{-\lambda i z^2}\right| = e^{2\lambda xy} \tag{5.178}$$

are plotted. The dark gray levels represent the valleys (small values of the integrand) of the landscape defined by $|f(z)|$, while the light gray levels represent the hills.

The alternative path of integration that replaces the original one goes from one valley ($x < 0$, $y > 0$) into the other ($x > 0$, $y < 0$), following the most rapid direction of descending values of $|f(z)|$ and passing through the origin $z_0 = (0, 0)$; z_0 is a saddle point of the amplitude $|f(z)|$, which lies between the two valleys and on the path of steepest descent ($x = -y$). There is no other path different from $x = -y$ that goes through the two valleys on which $f(z)$ has constant phase.

We have considered two simple examples with a few specific features, and these are far from being particular properties; the features encountered in the previous examples are quite general and will show up in many other cases. Indeed, when trying to evaluate an integral

$$\int_{\mathcal{L}} e^{\phi(z)} \, dz \tag{5.179}$$

by deforming $\mathcal{L}$ (with both endpoints at infinity) into a path where $\Im\phi(z)$ is constant, first we need to deform it into a path $\mathcal{C}$ that runs down into the valleys of $|e^{\phi(z)}|$, so that the contributions at infinity vanish. As we will see, when $\phi(z)$ is analytic there is not much choice when trying to deform the contour in such a way: the path should be a steepest descent path and it should run through a saddle point of $\phi(z)$, z_0, that is, a point such that $\phi'(z_0) = 0$.

For understanding these facts, we must recall some basic concepts of complex analysis. Let us write,

$$\phi(z) = R(x, y) + iI(x, y), \quad z = x + iy. \tag{5.180}$$

If $\phi(z)$ is differentiable, it is a well-known fact that

$$\phi'(z) = R_x + iI_x = I_y - iR_y, \tag{5.181}$$

where the subindices mean partial differentiations. These relations imply that the Cauchy–Riemann equations must be satisfied:

$$R_x = I_y, \quad R_y = -I_x. \tag{5.182}$$

As for any other differentiable function, the gradients of $R(x, y)$ and $I(x, y)$, defined as

$$\nabla R = (R_x, R_y), \quad \nabla I = (I_x, I_y), \tag{5.183}$$

give the directions of most rapid variation of R and I. By the Cauchy–Riemann equations, we observe that at each point (x, y), the gradients of R and I are orthogonal and that the directions of fastest variation of R and I are orthogonal; this also means that the lines $R = $ const. are orthogonal to the lines $I = $ const. As a consequence, given a point (x_0, y_0), the line passing through this point and satisfying $I(x, y) = I(x_0, y_0)$ is the path of most rapid variation for R which passes through this point.

The functions $R(x, y)$ and $I(x, y)$ are the so-called *conjugate harmonic functions* and, when they have continuous second derivatives, they satisfy the potential equation

$$\Delta R = \Delta I = 0, \quad \Delta = \frac{\partial^2}{\partial x^2} + \frac{\partial^2}{\partial y^2}. \tag{5.184}$$

Being harmonic, the functions $R(x, y)$ and $I(x, y)$ cannot have local extrema and all the critical points must be saddle points.

This discussion can be translated to the integrand in (5.179). Let us denote $f(z) = e^{-\phi(z)}$. The function

$$|f(z)| = e^{-R(x, y)} \tag{5.185}$$

defines a modular landscape with valleys and hills, the curves $R(x, y) = $ const. being the level lines of the landscape (these are the curves shown in Figure 5.7). The lines $I(x, y) = $ const. are orthogonal to these lines and may be called *lines of steepest descent* (or *ascent*); $f(z)$ has a constant phase along these lines.

As a conclusion, if the new path of integration $\mathcal{C}$ for computing (5.179) must be chosen in such a way that the phase of $\phi(z)$ is constant, then it must follow a path orthogonal to the level curves $R(x, y) = $ const. In particular, when we have a path with both endpoints at infinity, the path should go down into the valleys of the amplitude $|f(z)|$ at both extremes. This integration path, joining two valleys, will pass through a saddle point of $|f(z)|$ (and $R(x, y)$), that is, through a point where

$$R_x(x_0, y_0) = R_y(x_0, y_0) = 0, \tag{5.186}$$

and then $\phi'(z_0) = 0$, where $z_0 = x_0 + iy_0$.

5.5. Contour integrals and the saddle point method

Therefore, we will have to look for integration paths satisfying $\Im\phi(z) = \Im\phi(z_0)$.

Example 5.17 (example 5.16 revisited). For evaluating

$$\int_{-\infty}^{\infty} e^{\phi(z)}\, dz = \int_{-\infty}^{\infty} e^{-i\lambda z^2}\, dz \qquad (5.187)$$

without oscillations in the integrand, we obtain the saddle point from $\phi'(z_0) = -i\lambda z_0 = 0$ and then set

$$\Im\phi(z) = \Im\phi(z_0) \Rightarrow x^2 - y^2 = 0. \qquad (5.188)$$

The path $y = x$ runs uphill and should not be considered. We must consider the path of steepest descent $y = -x$, which can be used because, as discussed before, it is possible to deform the original contour into this steepest descent path.

It is interesting to observe that the original path of integration already passed through the saddle point, but not following the path of steepest descent. The direction of crossing of the saddle is essential. ∎

Example 5.18 (example 5.15 revisited). We consider a slightly modified version of this example. Let

$$G(\lambda) = \int_{-\infty}^{\infty} e^{-\lambda(z^2 - 2iz)}\, dz = \frac{1}{\sqrt{\lambda}} I(\sqrt{\lambda}), \qquad (5.189)$$

where $I(\lambda)$ is the integral (5.164).

In this case, we have $\phi(z) = -\lambda(z^2 - 2iz)$. There is a single saddle point $z = i$. The steepest descent or ascent curves are $\Im\phi(z) = $ const., that is, $x(y - 1) = $ const. The steepest descent line $y = 1$, joining the valleys, passes through the saddle point. Figure 5.8 shows the contour lines of $|f(z)| = |e^{-\lambda(z^2 - 2iz)}|$.

The steepest descent path $y = 1$ can be considered because shifting the original path of integration upward is possible since the integrand decays exponentially as $|x| \to +\infty$ with constant y.

Thus we have

$$G(\lambda) = \int_{-\infty}^{\infty} e^{-\lambda((t-i)^2 + 1)}\, dt = \frac{e^{-\lambda}}{\sqrt{\lambda}} \int_{-\infty}^{\infty} e^{-s^2}\, ds = \sqrt{\pi}\frac{e^{-\lambda}}{\sqrt{\lambda}}. \quad \blacksquare \qquad (5.190)$$

Remark 9. The value of the integral for large λ in Example 5.18 is much smaller than that of the integral in Example 5.17. Shifting the path (on which initially no saddle point occurs) up to a path through the saddle point has a larger effect than turning the path when we are already crossing the saddle point.

5.5.2 Other integration contours

In the examples just considered we can change the path of integration into an optimal path (for numerical evaluations) without passing singularities of the integrand. When the function $\eta(z)$ of (5.168) has poles, which are passed during the modification of the contour, residues should be taken into account. Other types of singularities may yield extra integrals

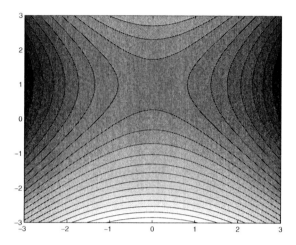

Figure 5.8. *Contour lines of $|f(z)| = |e^{-\lambda(z^2-2iz)}|$.*

around branch cuts. Also, more saddle points may occur, and it is necessary to investigate which one should be used for the saddle point analysis, and perhaps the new contour may pass through more than one saddle point.

Another complication may arise when we cannot deform the contour directly through one of the saddle points. Consider

$$F_\alpha(\lambda) = \int_\alpha^\infty e^{-\lambda(z^2-2iz)} \, dz, \tag{5.191}$$

where $\alpha = \beta + i\gamma$ may be any complex number and the upper endpoint means that the path tends to infinity $+\infty + ic$ for some $c \in \mathbb{R}$; therefore, it runs down the valley at $x > 0$; see Figure 5.8.

For general α, the saddle point contour $y = 1$, satisfying $I(x, y) = 0$, does not pass through the point α. In order to integrate along a curve of constant phase, we can now consider the steepest descent path passing through $\alpha = \beta + i\gamma$; then, we would integrate along a path such that $I(x, y) = I(\beta, \gamma)$.

From Figure 5.8, we observe that when $\beta > 0$ the steepest path $I(x, y) = I(\beta, \gamma)$, indeed, goes to $+\infty$ and the integral can be computed following this path. Contrarily, when $\beta < 0$ the corresponding steepest descent path runs into the other valley ($-\infty$). In this case, we can write

$$F_\alpha(\lambda) = \int_{-\infty}^\infty e^{-\lambda(z^2-2iz)} \, dz - \int_{-\infty}^\alpha e^{-\lambda(z^2-2iz)} \, dz, \tag{5.192}$$

and the first integral is computed as in Example 5.18, while the second is computed on the path $I(x, y) = I(\beta, \gamma)$.

5.5. Contour integrals and the saddle point method

An example of this type of computation, where one of the endpoints is finite, is provided by Scorer functions [87, 89]. The Scorer functions Hi(z) have the integral representation

$$\text{Hi}(z) = \frac{1}{\pi} \int_0^\infty e^{zw - \frac{1}{3}w^3} \, dw \qquad (5.193)$$

(notice that now z no longer denotes the variable of integration). In [87] (see also §12.2), it is shown that the original contour of integration can be continuously deformed into the steepest descent path when z is in the sector $2\pi/3 \leq \text{ph } z \leq \pi$, but that two pieces have to be considered when $0 \leq \text{ph } z \leq 2\pi/3$ (similarly to what happened in the previous example (5.191) for $F_\alpha(\lambda)$, which depends on α). This also means that the integration path becomes nonsmooth when $\text{ph } z = 2\pi/3$. Also, as we will discuss in Example 5.19, nonsmoothness takes place for Airy functions around $\text{ph } z = 2\pi/3$.

When steepest contours become nonsmooth, sometimes it is convenient to replace the steepest path by a smooth modified contour sharing with the steepest path some essential properties. In the case of a saddle point contour, we should demand that it has the same endpoints at infinity (in the same valleys) and that it passes through the saddle point and in the same direction (this is indeed essential, as we learned in Example 5.17). Also, for other steepest contours (like the Scorer case), alternative paths can be found for avoiding nonsmoothness (see [89]).

5.5.3 Integrating along the saddle point contours and examples

We now give some hints on how to integrate over saddle point contours and provide two examples (Airy functions and parabolic cylinder functions).

In order to simplify the notation, we consider, as before, an integral of the form

$$I = \int_\mathcal{C} e^{\phi(w)} \, dw, \qquad (5.194)$$

although a factor $\eta(w)$, as in (5.168), could be considered as well. $\mathcal{C}$ is a steepest descent contour passing through one saddle point w_0, such that $\phi'(w_0) = 0$, where the prime means differentiation with respect to w.

Because the path runs through two valleys and $\Re\phi$ reaches a local maximum along the path when crossing the saddle point w_0 (which is not an extremum of $\Re\phi$ as a function of two variables), we expect that the dominant contribution to the integral will come from the contributions around the saddle point. Additionally, because $\phi'(w_0) = 0$, we expect that

$$I = e^{\phi(w_0)} \int_\mathcal{C} e^{\phi(w) - \phi(w_0)} dw = e^{\phi(w_0)} \int_\mathcal{C} e^{\frac{1}{2}\phi''(w_0)(w-w_0)^2 + \mathcal{O}((w-w_0)^3)} \, dw. \qquad (5.195)$$

Away from the saddle, the integrand decays rapidly. Quite often it decreases exponentially or faster (in our examples, it behaves like a Gaussian).

For computing the integral, the first step is the factorization of the leading factor, coming from the saddle point. We write

$$I = e^{\phi(w_0)} \int_\mathcal{C} e^{\psi(w)} \, dw, \qquad (5.196)$$

where
$$\psi(w) = \phi(w) - \phi(w_0) = \psi_r(w) + i\psi_i(w). \tag{5.197}$$

But on the steepest descent path we have
$$\psi_i(w) = 0, \tag{5.198}$$

and therefore
$$I = e^{\phi(w_0)} \int_C e^{\psi_r(w)} dw, \tag{5.199}$$

and the integrand will decrease as $w \to \infty$ along the path C (and in the fastest possible way).

Because the main contribution to the integrand will come from the saddle point $w = w_0$, it is important to accurately compute the integrand around this point, particularly because cancellations take place in $\psi(w)$ around $w = w_0$, given that $\psi(w_0) = \psi'(w_0) = 0$. For this, it is convenient to use coordinates relative to the saddle point $w_0 = u_0 + iv_0$ by writing $w = u + iv$ and
$$u = u_0 + \sigma, \quad v = v_0 + \tau, \quad \sigma \in \Delta_\sigma, \quad \tau \in \Delta_\tau, \tag{5.200}$$

where Δ_σ and Δ_τ are real bounded or unbounded intervals such that the complete contour C can be described by this parametrization.

After parametrizing the path C, the integral can be split up into real and imaginary parts. The quadrature methods considered in this chapter can be used for computing the integrals. In particular, because the resulting integrals usually have fast exponential decays, an interesting and efficient choice is the compound trapezoidal rule.

Of course, the selection of parametrization will depend on the curve; if, for instance, σ can explicitly be written as a function of τ (as will happen in the next example), then we can write
$$I = e^{\phi(w_0)} \int_{\Delta_\tau} e^{\psi_r(\sigma,\tau)} \left(\frac{d\sigma}{d\tau} + i \right) d\tau, \tag{5.201}$$

where $\psi_r(\sigma, \tau)$ follows from $\psi_r(w)$ after using (5.200), and σ as a function or τ can be obtained by solving (5.198). If (5.198) cannot be solved analytically, the methods considered in Chapter 7 could be considered (see, for instance, [91, 95]). For computing $d\sigma/d\tau$, the implicit function theorem can be invoked by writing
$$\frac{d\sigma}{d\tau} = -\frac{d\psi_i/d\tau}{d\psi_i/d\sigma}. \tag{5.202}$$

When this is used, special precautions should be taken to avoid cancellations near the saddle point because at the saddle point ($\tau = \sigma = 0$) we have $d\psi_i/d\tau = d\psi_i/d\sigma = 0$.

As mentioned before, it may happen that the path becomes nonsmooth in some cases and that the derivative may then become undefined at a certain point. In this case, modified contours could be considered, as we comment on in the next example.

Example 5.19 (again, the complex Airy function). In §5.3.3 we have discussed an efficient quadrature method for complex Airy functions based on Gauss–Laguerre quadrature. Also, saddle point techniques can be applied for obtaining integral representations which are suitable for applying the trapezoidal rule (see §5.4). This gives a very flexible and efficient algorithm with adjustable precision.

5.5. Contour integrals and the saddle point method

We consider

$$\text{Ai}(z) = \frac{1}{2\pi i} \int_{\mathcal{C}} e^{\frac{1}{3}w^3 - zw} \, dw, \tag{5.203}$$

valid for all complex z, but which we consider for ph $z \in [0, \frac{2}{3}\pi]$. $\mathcal{C}$ is a contour starting at $\infty e^{-i\pi/3}$ and terminating at $\infty e^{+i\pi/3}$ (the half-lines $\rho e^{\pm i\pi/3}$, $\rho \geq 0$, are the two central lines of two of the three valleys of the amplitude of the integrand, the third one being the negative axis). Considering only the sector ph $z \in [0, \frac{2}{3}\pi]$ is enough for computing the Airy function in the whole complex plane, because it satisfies the connection formula

$$\text{Ai}(z) + e^{-2\pi i/3}\text{Ai}(ze^{-2\pi i/3}) + e^{2\pi i/3}\text{Ai}(ze^{2\pi i/3}) = 0. \tag{5.204}$$

Let

$$\phi(w) = \tfrac{1}{3}w^3 - zw. \tag{5.205}$$

The saddle points are $w_0 = \sqrt{z}$ and $-w_0$ and follow from solving $\phi'(w) = w^2 - z = 0$. When ph $z \in [0, \frac{2}{3}\pi]$, w_0 will be in the half-plane $\Re z > 0$ and $-w_0$ in $\Re z < 0$; because we are interested in a deformation of a contour in the right half-plane $\Re z > 0$, we will be interested in the saddle point contours passing through the saddle in this half-plane.

The saddle point contour (the path of steepest descent) that runs through the saddle point w_0 is defined by

$$\Im[\phi(w)] = \Im[\phi(w_0)]. \tag{5.206}$$

We write

$$z = x + iy = re^{i\theta}, \quad w = u + iv, \quad w_0 = u_0 + iv_0. \tag{5.207}$$

Then

$$u_0 = \sqrt{r} \cos \tfrac{1}{2}\theta, \quad v_0 = \sqrt{r} \sin \tfrac{1}{2}\theta, \quad x = u_0^2 - v_0^2, \quad y = 2u_0 v_0. \tag{5.208}$$

The path of steepest descent through w_0 is given by the equation

$$u = u_0 + \frac{(v - v_0)(v + 2v_0)}{3\left[u_0 + \sqrt{\tfrac{1}{3}(v^2 + 2v_0 v + 3u_0^2)}\right]}, \quad -\infty < v < \infty. \tag{5.209}$$

Integrating with respect to $\tau = v - v_0$ (and writing $\sigma = u - u_0$) we obtain

$$\text{Ai}(z) = \frac{e^{-\zeta}}{2\pi i} \int_{-\infty}^{\infty} e^{\psi_r(\sigma, \tau)} \left(\frac{d\sigma}{d\tau} + i\right) d\tau, \tag{5.210}$$

where $\zeta = \tfrac{2}{3}z^{\frac{3}{2}}$,

$$\sigma = \frac{\tau(\tau + 3v_0)}{3\left[u_0 + \sqrt{\tfrac{1}{3}(\tau^2 + 4v_0 \tau + 3r)}\right]}, \quad -\infty < \tau < \infty, \tag{5.211}$$

and

$$\psi_r(\sigma, \tau) = \Re[\phi(w) - \phi(w_0)] = u_0(\sigma^2 - \tau^2) - 2v_0 \sigma \tau + \tfrac{1}{3}\sigma^3 - \sigma \tau^2. \tag{5.212}$$

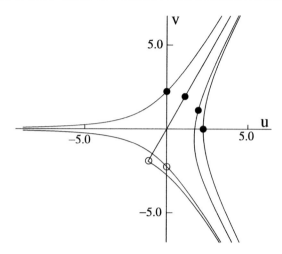

Figure 5.9. *Saddle point contours for* ph $z = 0$, $\frac{1}{3}\pi$, $\frac{2}{3}\pi$, π *and* $r = 5$.

Near the saddle point (where σ and τ are small), σ given in (5.211) should be expanded in powers of τ to avoid cancellations when evaluating (5.212).

Details of the saddle point contours for $r = 5$ and a few θ-values are shown in Figure 5.9. The saddle points are located on the circle with radius $\sqrt{r}$ and four w_0 are indicated by small black dots (two saddles $-w_0$ are indicated by open dots). The saddle point on the positive real axis corresponds to the case ph $z = 0$. It is interesting to notice that when ph $z = 2\pi/3$ the saddle point contour passes through the two saddle points $\pm w_0$, but it is nonsmooth at the saddle $-w_0$. The case $2\pi/3 < $ ph $z \leq \pi$ is not necessary for computing Ai(z) (because of (5.204)), but it is instructive to see how the saddle point contour splits into two contours, each one passing through one saddle. The two saddles on the imaginary axis are for the case ph $z = \pi$ and the corresponding two contours are also shown in Figure 5.9. Contributions from the two saddle points (when ph $z = \pi$) give the oscillatory behavior of Ai(z) when $z < 0$.

The nonsmooth behavior of the path of integration as w_0 crosses the line ph $z = 2\pi/3$ is related to the Stokes phenomenon (see §2.4.3). When w_0 crosses this line, we have $\Im\phi(w_0) = \Im\phi(-w_0)$, and indeed the path of integration runs through both w_0 and $-w_0$. Contributions from $-w_0$ become more and more relevant as ph $z \to \pi$.

What is of more concern, from a numerical point of view, is the nonsmoothness of the path of integration as the Stokes line ph $z = 2\pi/3$ is crossed. This causes difficulties in the evaluation of $d\sigma/d\tau$. A way to solve this problem is to consider modified contours running through the same valleys and passing through the saddle and with the same directions. If an appropriate modified contour is chosen, the oscillations of the integrand will not be completely avoided, but they will be of no concern, and the nonsmoothness of the derivative can be eliminated. ∎

Example 5.20 (parabolic cylinder function). As a second example of the evaluation of contour integrals, we consider the parabolic cylinder function $U(a, x)$ for positive x and a.

5.5. Contour integrals and the saddle point method

The starting point is the integral (see [2, eq. (19.5.4)])

$$U(a, x) = \frac{e^{\frac{1}{4}x^2}}{i\sqrt{2\pi}} \int_C e^{-xs+\frac{1}{2}s^2} s^{-a} \frac{ds}{\sqrt{s}}, \tag{5.213}$$

where C is a vertical line on which $\Re s > 0$. On C we have $-\frac{1}{2}\pi < \text{ph } s < \frac{1}{2}\pi$, and the many-valued function $s^{-a-1/2}$ assumes its principal value. The transformations

$$x = 2t\sqrt{a}, \quad s = \sqrt{a}\, w \tag{5.214}$$

give

$$U(a, x) = \frac{e^{\frac{1}{4}x^2} a^{\frac{1}{4}-\frac{1}{2}a}}{i\sqrt{2\pi}} \int_C e^{a\phi(w)} \frac{dw}{\sqrt{w}}, \tag{5.215}$$

where

$$\phi(w) = \tfrac{1}{2}w^2 - 2tw - \ln w. \tag{5.216}$$

Notice that we have managed to concentrate into a single exponential the oscillatory behavior of the integrand as a becomes large. In this respect, the $\sqrt{w}$ factor is of no concern.

The saddle points follow from solving

$$\phi'(w) = \frac{w^2 - 2tw - 1}{w} = 0, \tag{5.217}$$

giving saddle points at $w = t \pm \sqrt{t^2 + 1}$. Because the original path is in the half-plane ph $z > 0$, we need to consider the saddle on this half-plane and we keep the upper sign. The steepest descent path then follows from solving $\Im[\phi(w)] = \Im[\phi(w_0)]$ with $w_0 = t + \sqrt{t^2 + 1}$.

In the present case $\Im[\phi(w_0)] = 0$, and we obtain for the saddle point contour the equation

$$\tfrac{1}{2}r^2 \sin 2\theta - 2tr \sin\theta - \theta = 0, \quad \text{where} \quad w = re^{i\theta}, \tag{5.218}$$

which can be solved for $r = r(\theta)$,

$$r = \frac{t + \sqrt{t^2 + \theta \cot\theta}}{\cos\theta}, \quad -\tfrac{1}{2}\pi < \theta < \tfrac{1}{2}\pi, \tag{5.219}$$

giving the contour as shown in Figure 5.10. Then (5.215) can be written as

$$U(a, x) = \frac{e^{\frac{1}{4}x^2 + a\phi(w_0)} a^{\frac{1}{4}-\frac{1}{2}a}}{\sqrt{2\pi}} \int_{-\frac{1}{2}\pi}^{\frac{1}{2}\pi} e^{a\psi(\theta)} g(\theta)\, d\theta, \tag{5.220}$$

where

$$\psi(\theta) = \Re[\phi(w) - \phi(w_0)] = \tfrac{1}{2}r^2 \cos 2\theta - 2tr \cos\theta - \ln r - \phi(w_0), \tag{5.221}$$

and

$$g(\theta) = \Im\left[\frac{1}{\sqrt{w}} \frac{dw}{d\theta}\right] = \Im\left[e^{\frac{1}{2}i\theta} \frac{1}{\sqrt{r}} \left(\frac{dr}{d\theta} + ir\right)\right]$$

$$= \frac{(2\cos\theta + 1)r^2 - 2tr + 1}{4\sqrt{r}\cos\tfrac{1}{2}\theta \sqrt{t^2 + \theta \cot\theta}}. \tag{5.222}$$

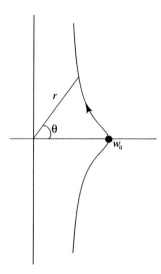

Figure 5.10. *Steepest descent contour for the integral in* (5.215).

If we consider the quantity $\widetilde{\xi}$ defined by

$$\widetilde{\xi} = \tfrac{1}{2}\left[t\sqrt{t^2+1} + \ln(t+\sqrt{t^2+1})\right], \tag{5.223}$$

we have

$$\tfrac{1}{4}x^2 + a\phi(w_0) = a\left[\tfrac{1}{2} - t\sqrt{t^2+1} - \ln(t+\sqrt{t^2+1})\right] = a\left(\tfrac{1}{2} - 2\widetilde{\xi}\right). \tag{5.224}$$

This gives

$$U(a,x) = \frac{a^{\frac{1}{4}} e^{-2a\widetilde{\xi}}}{\sqrt{2\pi}\gamma(a)} \int_{-\frac{1}{2}\pi}^{\frac{1}{2}\pi} e^{a\psi(\theta)} g(\theta)\, d\theta, \tag{5.225}$$

where

$$\gamma(a) = e^{-\frac{1}{2}a} a^{\frac{1}{2}a}. \tag{5.226}$$

In [85, 86], the vertical line passing through the saddle point is considered an integration path. Notice that this path has the same direction as the steepest descent path when crossing the saddle point and runs through the same valleys. The performance of the resulting integral is improved by using the erf-rule (see §5.4.2). ∎

Part II

Further Tools and Methods

Chapter 6

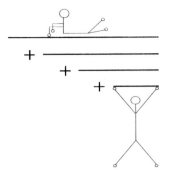

Numerical Aspects of Continued Fractions

*I am Orpheus, and follow Eurydice's steps
through these murky deserts
where no mortal man has ever trod.*
—Third act, *L'Orfeo*, Claudio Monteverdi

6.1 Introduction

For many elementary and special functions, representations as continued fractions exist. In this chapter we give examples for incomplete gamma functions and incomplete beta functions that are useful for numerical computation of these functions. We also give a few special forms and evaluation schemes. For certain continued fractions we give convergence domains in the complex plane and transformations for improving convergence.

Continued fractions play an important role in the theory of recurrence relations for special functions. These aspects were considered in Chapter 4.

6.2 Definitions and notation

Let $\{a_n\}_{n=1}^{\infty}$ and $\{b_n\}_{n=0}^{\infty}$ be two sequences of real or complex numbers. With these numbers we construct a *continued fraction* of the form

$$b_0 + \cfrac{a_1}{b_1 + \cfrac{a_2}{b_2 + \cfrac{a_3}{b_3 + \cfrac{a_4}{b_4 + \cdots}}}} \qquad (6.1)$$

A more convenient notation is

$$b_0 + \frac{a_1}{b_1+} \frac{a_2}{b_2+} \frac{a_3}{b_3+} \frac{a_4}{b_4 + \cdots}. \qquad (6.2)$$

Manipulating continued fractions, and proving properties, is made easy by introducing a set of rational transformations t_p. Let

$$t_0(w) = b_0 + w, \quad t_p(w) = \frac{a_p}{b_p + w}, \quad p = 1, 2, 3, \ldots. \qquad (6.3)$$

Then we have the composed mappings

$$t_0 t_1(w) = b_0 + \frac{a_1}{b_1 + w}, \quad t_0 t_1 t_2(w) = b_0 + \cfrac{a_1}{b_1 + \cfrac{a_2}{b_2 + w}}. \qquad (6.4)$$

By composing more transformations t_p we obtain more initial parts of the continued fraction (6.1). Observe that

$$t_0 t_1 \cdots t_n(0) = t_0 t_1 \cdots t_{n+1}(\infty) \qquad (6.5)$$

and that the *value of the continued fraction* is

$$C = \lim_{n \to \infty} t_0 t_1 \cdots t_n(0) = \lim_{n \to \infty} t_0 t_1 \cdots t_{n+1}(\infty). \qquad (6.6)$$

For the existence of these limits we refer to §6.5.

We call

$$C_n = t_0 t_1 \cdots t_n(0) = b_0 + \frac{a_1}{b_1+} \frac{a_2}{b_2+} \frac{a_3}{b_3 + \cdots} \frac{a_n}{b_n} \qquad (6.7)$$

the nth *approximant* or *convergent* of (6.1). The 0th approximant is $t_0(0) = b_0$.

By mathematical induction it easily follows that

$$t_0 t_1 \cdots t_n(w) = \frac{A_{n-1} w + A_n}{B_{n-1} w + B_n}, \quad n = 0, 1, 2, \ldots, \qquad (6.8)$$

where the quantities A_{n-1}, A_n, B_{n-1}, B_n are independent of w, with $A_{-1} = 1$, $A_0 = b_0$, $B_{-1} = 0$, $B_0 = 1$. The following recursion is available to compute A_n and B_n. For $n = 1, 2, 3, \ldots$ we have

$$A_n = b_n A_{n-1} + a_n A_{n-2}, \quad B_n = b_n B_{n-1} + a_n B_{n-2}. \qquad (6.9)$$

It follows by putting $w = 0$ in (6.8) that

$$C_n = t_0 t_1 \cdots t_n(0) = \frac{A_n}{B_n}, \qquad (6.10)$$

and that the convergents C_n can be computed by the recurrence relations in (6.9). A_n and B_n are called the nth numerator and denominator, respectively.

To show that A_n and B_n indeed satisfy these recursions, mathematical induction can be used again. The relations in (6.9) hold for $n = 1$. If they are true for $n = k \geq 1$, then

$$\begin{aligned} t_0 t_1 \cdots t_{k+1}(w) &= t_0 t_1 \cdots t_k \left(\frac{a_{k+1}}{b_{k+1} + w} \right) \\ &= \frac{A_k w + (b_{k+1} A_k + a_{k+1} A_{k-1})}{B_k w + (b_{k+1} B_k + a_{k+1} B_{k-1})} \\ &= \frac{A_k w + A_{k+1}}{B_k w + B_{k+1}}, \end{aligned} \qquad (6.11)$$

so that the statement is true for $n = k + 1$ and therefore for all n.

The determinant of the transformation $t_0 t_1 \cdots t_n(w)$ in (6.8) is

$$\begin{vmatrix} A_{n-1} & A_n \\ B_{n-1} & B_n \end{vmatrix} = \begin{vmatrix} A_{n-1} & b_n A_{n-1} + a_n A_{n-2} \\ B_{n-1} & b_n B_{n-1} + a_n B_{n-2} \end{vmatrix}$$
$$= -a_n \begin{vmatrix} A_{n-2} & A_{n-1} \\ B_{n-2} & B_{n-1} \end{vmatrix}. \tag{6.12}$$

This gives the determinant formula

$$A_n B_{n-1} - B_n A_{n-1} = (-1)^{n-1} \prod_{k=0}^{n} a_k, \quad n = 0, 1, 2, \ldots, \tag{6.13}$$

where $a_0 = 1$.

We see that the convergence of (6.6) can be described by A_n and B_n. We assume that at most a finite number of denominators B_n vanish. Then the continued fraction in (6.1) or (6.2) is said to converge to $C = \lim_{n \to \infty} C_n$ if this limit exists and is finite.

For these and many other relations between a_n, b_n and A_n, B_n, C_n we refer to [47, 119, 138, 229].

The recurrence relations in (6.9) provide one way to compute approximate values of the continued fraction. In §6.6 we give more details on computations.

6.3 Equivalence transformations and contractions

For numerical purposes it is often convenient to write the continued fraction (6.2) in a different form by means of a so-called equivalence transformation. We write (6.2) as

$$b_0 + \frac{c_1 a_1}{c_1 b_1 +} \frac{c_1 c_2 a_2}{c_2 b_2 +} \frac{c_2 c_3 a_3}{c_3 b_3 +} \frac{c_3 c_4 a_4}{c_4 b_4 + \cdots}, \tag{6.14}$$

where all $c_p \neq 0$. This continued fraction has the same approximants C_n as (6.2), and the nth numerator and denominator of (6.14) are

$$c_1 c_2 \cdots c_n A_n, \quad c_1 c_2 \cdots c_n B_n, \tag{6.15}$$

respectively, where A_n and B_n are the nth numerator and denominator of (6.2). It follows that we can obtain, by suitable choices of c_n, continued fractions of the forms

$$b_0 + \frac{1}{\beta_1 +} \frac{1}{\beta_2 +} \frac{1}{\beta_3 +} \frac{1}{\beta_4 + \cdots} \tag{6.16}$$

and

$$b_0 + \frac{\alpha_1}{1 +} \frac{\alpha_2}{1 +} \frac{\alpha_3}{1 +} \frac{\alpha_4}{1 + \cdots}. \tag{6.17}$$

By the *even part of a continued fraction* we shall mean the continued fraction whose sequence of approximants is the sequence of even approximants of the given continued fraction. Similarly, the *odd part of a continued fraction* is the continued fraction whose sequence of approximants is the sequence of odd approximants of the given continued fraction. The even and odd parts are examples of *contractions* of a given continued fraction. With contractions we can obtain better approximations from the same computing effort.

We show for a simple form how the even and odd parts can be obtained. Consider

$$\frac{1}{1+}\frac{a_2}{1+}\frac{a_3}{1+}\frac{a_4}{1+\cdots}. \tag{6.18}$$

Then the even part of (6.18) is

$$\frac{1}{1+a_2-}\frac{a_2 a_3}{1+a_3+a_4-}\frac{a_4 a_5}{1+a_5+a_6-\cdots}, \tag{6.19}$$

and the odd part is

$$1 - \frac{a_2}{1+a_2+a_3-}\frac{a_3 a_4}{1+a_4+a_5-}\frac{a_5 a_6}{1+a_6+a_7-\cdots}. \tag{6.20}$$

The even and odd parts of (6.2) may be obtained from (6.19) and (6.20) by multiplying them by a_1, adding b_0, and then replacing a_1 by a_1/b_1, and a_n by $a_n/(b_{n-1}b_n)$, $n = 2, 3, 4, \ldots$. These substitutions are related to an equivalence transformation that relates the continued fractions (6.18) and (6.2) to each other.

To prove that (6.19) is the even part of (6.18) consider the linear transformations t_p that generate (6.18). That is,

$$t_1(w) = w, \quad t_p = \frac{1}{1+a_p w}, \quad p = 2, 3, 4, \ldots, \tag{6.21}$$

so that $t_1 t_2 \cdots t_n(1) = A_n/B_n$, the nth approximant of (6.18). Let $s_p(w) = t_p t_{p+1}(w)$, $p = 1, 2, 3, \ldots$. Then

$$s_1(w) = \frac{1}{1+a_2 w}, \quad s_p(w) = 1 - \frac{a_{2p-1}}{1+a_{2p-1}+a_{2p}w}, \quad p = 2, 3, 4, \ldots, \tag{6.22}$$

and $s_1 s_2 \cdots s_p(w) = t_1 t_2 \cdots t_{2p}(1) = A_{2p}/B_{2p}$, the $2p$th approximant of (6.18). Since

$$s_1 s_2 \cdots s_p(1) = \frac{1}{1+a_2-}\frac{a_2 a_3}{1+a_3+a_4-}\cdots\frac{a_{2p-2}a_{2p-1}}{1+a_{2p-1}+a_{2p}}, \tag{6.23}$$

it therefore follows that (6.19) is the even part of (6.18). The proof that (6.20) is the odd part of (6.18) can be made in an analogous way.

Example 6.1 (incomplete gamma function). Consider the continued fraction (see §6.7.1 and [229, p. 353])

$$F(z) = \frac{1}{1+}\frac{az}{1+}\frac{1z}{1+}\frac{(a+1)z}{1+}\frac{2z}{1+}\frac{(a+2)z}{1+}\frac{3z}{1+\cdots}. \tag{6.24}$$

6.3. Equivalence transformations and contractions

The function $F(z)$ is defined by

$$F(z) = \frac{1}{\Gamma(a)} \int_0^\infty \frac{u^{a-1}e^{-u}}{1+zu} du, \quad \Re a > 0, \quad |\text{ph } z| < \pi, \quad (6.25)$$

and is related to the incomplete gamma function (see [219, p. 280])

$$F(z) = z^{-a} e^{1/z} \Gamma(1-a, 1/z), \quad (6.26)$$

where

$$\Gamma(a, z) = \int_z^\infty t^{a-1} e^{-t} dt. \quad (6.27)$$

The continued fraction in (6.24) has the form of (6.18) with

$$a_{2n} = (a+n-1)z, \quad a_{2n+1} = nz, \quad n = 1, 2, 3, \ldots. \quad (6.28)$$

The even part of (6.24) is given by

$$F(z) = \frac{1}{1+az+} \frac{c_2}{d_2+} \frac{c_3}{d_3+\cdots}, \quad (6.29)$$

where, for $n = 2, 3, 4, \ldots$,

$$\begin{aligned} c_n &= -a_{2n-2}a_{2n-1} = -(a+n-2)(n-1)z^2, \\ d_n &= 1 + a_{2n-1} + a_{2n} = 1 + (2n+a-2)z. \end{aligned} \quad (6.30)$$

We can write (6.29) in the form (6.18) with $a_n = c_n/(d_{n-1}d_n)$, and for the incomplete gamma function we obtain, replacing z by $1/z$ and changing $a \to 1-a$,

$$(z+1-a)z^{-a}e^z \Gamma(a, z) = \frac{1}{1+} \frac{\alpha_1}{1+} \frac{\alpha_2}{1+} \frac{\alpha_3}{1+\cdots}, \quad (6.31)$$

where

$$\alpha_n = \frac{n(a-n)}{(z+2n-1-a)(z+2n+1-a)}, \quad n = 1, 2, 3, \ldots. \quad (6.32)$$

This form is used in [78] for computing the function $\Gamma(a, x)$, for $x > 1.5$ and $-\infty < a < \alpha^*(z)$, where $\alpha^*(x) \sim x$ for large x. For positive values of a and z the incomplete gamma function is usually computed together with the other incomplete gamma functions defined by

$$\gamma(a, z) = \int_0^z t^{a-1} e^{-t} dt = \Gamma(a) - \Gamma(a, z), \quad \Re a > 0. \quad (6.33)$$

Roughly speaking, $\Gamma(a, z)$ is smaller than $\gamma(a, z)$ when $z \geq a$, and should be computed first. When $z < a$, $\gamma(a, z)$ should be computed first. To control overflow, the incomplete gamma functions can be defined with normalization factor $\Gamma(a)$. That is,

$$P(a, z) = \frac{\gamma(a, z)}{\Gamma(a)}, \quad Q(a, z) = \frac{\Gamma(a, z)}{\Gamma(a)}, \quad P(a, z) + Q(a, z) = 1. \quad (6.34)$$

These normalized functions are used in mathematical statistics as cumulative distribution functions (for the gamma distribution). See also Chapters 8 and 10.

6.4 Special forms of continued fractions

We consider a few special forms that occur frequently in the theory of special functions.

6.4.1 Stieltjes fractions

A continued fraction of the form

$$C = \frac{a_0}{1-} \frac{a_1 z}{1-} \frac{a_2 z}{1-} \cdots \quad (6.35)$$

is called a *Stieltjes fraction* (*S-fraction*). We say that it *corresponds* to the formal power series

$$f(z) = c_0 + c_1 z + c_2 z^2 + \cdots \quad (6.36)$$

if its nth convergent C_n, when expanded in ascending powers of z, agrees with $f(z)$ up to and including the term with z^{n-1}, $n = 1, 2, 3, \ldots$.

Quotient-difference algorithm

For several special functions the S-fractions are known analytically, but in any case the coefficients can always be calculated from the power series coefficients by means of the *quotient-difference algorithm*, according to the following scheme:

$$\begin{array}{ccccccc}
& & q_1^0 & & & & \\
e_0^1 & & & & e_1^0 & & \\
& & q_1^1 & & & & q_2^0 \\
e_0^2 & & & & e_1^1 & & & & e_2^0 \\
& & q_1^2 & & & & q_2^1 & & \\
e_0^3 & & & & e_1^2 & & & & e_2^1 & & \ddots \\
& & q_1^3 & & & & q_2^2 & & \\
e_0^4 & \vdots & & & e_1^3 & & & & e_2^2 \\
& & \vdots & & & & q_2^3 & \vdots & \ddots \\
\vdots & & & & & & & &
\end{array} \quad (6.37)$$

The first two columns of the scheme are defined by

$$\begin{aligned} e_0^n &= 0, & n &= 1, 2, \ldots, \\ q_1^n &= c_{n+1}/c_n, & n &= 0, 1, \ldots, \end{aligned} \quad (6.38)$$

where $c_n \neq 0$ appear in (6.36). We continue by means of the *rhombus rule*:

$$\begin{aligned} e_j^k &= e_{j-1}^{k+1} + \left(q_j^{k+1} - q_j^k \right), & j &\geq 1, \; k \geq 0, \\ q_j^{k+1} &= q_j^{k+1} \left(e_j^{k+1}/e_j^k \right), & j &\geq 1, \; k \geq 0. \end{aligned} \quad (6.39)$$

The coefficients a_n of the S-fraction (6.35) are then

$$a_0 = c_0, \quad a_1 = q_1^0, \quad a_2 = e_1^0, \quad a_3 = q_2^0, \quad a_4 = e_2^0, \ldots. \tag{6.40}$$

The quotient-difference algorithm is frequently unstable and may require higher precision arithmetic or exact arithmetic with computer algebra. A more stable version of the algorithm is discussed in [202]. For an application to compute Bessel functions and Whittaker functions with this algorithm, see [73]. For the original source, see [191].

6.4.2 Jacobi fractions

A continued fraction of the form

$$C = \frac{\beta_0}{1 - \alpha_0 z -} \frac{\beta_1 z^2}{1 - \alpha_1 z -} \frac{\beta_2 z^2}{1 - \alpha_2 z - \cdots} \tag{6.41}$$

is called a *Jacobi fraction* (*J-fraction*). We say that it is *associated* with the formal power series $f(z)$ in (6.36) if its nth convergent C_n, when expanded in ascending powers of z, agrees with $f(z)$ up to and including the term with z^{2n-1}, $n = 1, 2, 3, \ldots$. The convergent C_n of the Jacobi fraction (6.41) associated with $f(z)$ equals the convergent C_{2n} of the Stieltjes fraction (6.35) corresponding to $f(z)$.

6.4.3 Relation with Padé approximants

The S-fraction and J-fraction are examples of continued fractions related to power series. Now, let us construct polynomials $P_n(z)$ and $Q_m(z)$ of degrees n and m, respectively, associated with the (formal) power series in (6.36), such that the (formal) power series of $f(z)Q_m(z) + P_n(z)$ has an expansion $\sum_{j=0}^{\infty} d_j z^j$, in which $d_j = 0$ for $j = 0, 1, 2, \ldots, n+m$. If we impose a normalization condition and if we require that P_n and Q_m have no common factors, then we can prove that P_n and Q_m are unique.

Frobenius conceived P_n/Q_m as an element of a matrix and Padé developed the theory. We say that P_n/Q_m occupies the position (n, m) of the *Padé table* (see §9.2). It can be proved (see [119, Thm. 5.19]) that the convergents of a continued fraction $f(z)$ occupy the stair step sequence

$$P_0/Q_0, \quad P_1/Q_0, \quad P_1/Q_1, \quad P_2/Q_1, \quad P_2/Q_2, \ldots \tag{6.42}$$

of the Padé table of the power series $\sum_{j=0}^{\infty} c_j z^j$ of $f(z)$.

For more details on Padé tables and approximation problems by means of the table elements, we refer to §9.2.

6.5 Convergence of continued fractions

To consider an easy-to-use result we suppose that in (6.2) all a_k and b_k are positive. Then the convergents C_n show the following pattern:

$$0 < C_0 < C_2 \leq \cdots \leq C_{2n} \leq \cdots \cdots \leq C_{2n+1} \leq \cdots \leq C_3 < C_1. \tag{6.43}$$

In this case we see that convergence means that both $\lim_{n\to\infty} C_{2n}$ and $\lim_{n\to\infty} C_{2n+1}$ exist and are finite and equal. This property then provides excellent numerical checks for terminating the computation: if the limit exists, it is approached from both sides.

Convergence theory of continued fractions is a fascinating topic, and is much more complicated than the theory for series, power series included. We mention four well-known and important theorems. For proofs we refer to the theory given in [119, 138, 229].

Theorem 6.2 (Worpitzky). *Let $|a_n| \leq \frac{1}{4}$ for all n. Then the continued fraction*

$$\frac{a_1}{1+} \frac{a_2}{1+} \frac{a_3}{1+} \frac{a_4}{1+\cdots} \tag{6.44}$$

converges; all approximants C_n satisfy $|C_n| < \frac{1}{2}$ and the value C of the continued fraction satisfies $|C| \leq \frac{1}{2}$.

Theorem 6.3 (Śleżyński–Pringshcim). *Let $|b_n| \geq |a_n| + 1$ for all n. Then the continued fraction*

$$\frac{a_1}{b_1+} \frac{a_2}{b_2+} \frac{a_3}{b_3+} \frac{a_4}{b_4+\cdots} \tag{6.45}$$

converges; all approximants C_n satisfy $|C_n| < 1$ and the value C of the continued fraction satisfies $|C| \leq 1$.

Theorem 6.4 (Van Vleck). *Let $0 < \varepsilon < \frac{1}{2}\pi$ and let b_n satisfy $|\mathrm{ph}\, b_n| \leq \frac{1}{2}\pi - \varepsilon$ for all n. Then all approximants C_n of the continued fraction*

$$\frac{1}{b_1+} \frac{1}{b_2+} \frac{1}{b_3+} \frac{1}{b_4+\cdots} \tag{6.46}$$

are finite and satisfy $|\mathrm{ph}\, C_n| \leq \frac{1}{2}\pi - \varepsilon$. The sequences $\{C_{2n}\}$ and $\{C_{2m+1}\}$ converge to finite values. The continued fraction (6.46) converges if and only if, in addition, $\sum_{n=1}^{\infty} |b_n| = \infty$.

Theorem 6.5 (Van Vleck). *Let $\lim_{n\to\infty} a_n = 0$, $a_n \neq 0$, for all n. Then the S-fraction*

$$\frac{a_0}{1-} \frac{a_1 z}{1-} \frac{a_2 z}{1-\cdots} \tag{6.47}$$

converges to a meromorphic function of z. The convergence is uniform over every closed bounded region containing none of the poles of this function. If $\lim_{n\to\infty} a_n = a \neq 0$, then the S-fraction (6.47) converges, in the domain exterior to the rectilinear cut running from $1/(4a)$ to ∞ in the direction of the vector from 0 to $1/(4a)$, to a function having at most polar singularities in this domain. The convergence is uniform over every closed bounded region exterior to the cut which contains no poles of this function.

6.6 Numerical evaluation of continued fractions

The A_n and B_n introduced in (6.8) can be computed by a forward recurrence algorithm based on the TTRRs in (6.9). The computation of solutions of recurrence relations may be unstable; for details we refer to Chapter 4. Also, the quantities A_n and B_n may grow in magnitude when n is large, and overflow may occur. Because only the ratio $C_n = A_n/B_n$ is needed for approximating the continued fraction, rescaling from time to time of a few successive A_n and B_n can be used to avoid overflow. The same can be done when underflow may occur. On the other hand, we can avoid overflow by rescaling the recurrence relations in (6.9) in such a way that, for example, $A_n/n!$ and $B_n/n!$ are the requested solutions.

The computation of C_n can also be done by using a backward recurrence algorithm. This means the computational scheme

$$u_k = b_k + \frac{a_{k+1}}{u_{k+1}}, \quad k = n-1, n-2, \ldots, 0, \tag{6.48}$$

with the initial value $u_n = b_n$. Then we have $u_0 = C_n$. To obtain a certain accuracy, a priori knowledge is needed of the value of n. This algorithm is in general more stable than the forward algorithm; see [118]. The continued fraction

$$C = \frac{a_0}{1-} \frac{a_1}{1-} \frac{a_2}{1-} \cdots \tag{6.49}$$

can be written in the form

$$C = \sum_{k=0}^{\infty} t_k, \tag{6.50}$$

where (see [229, pp. 17ff.])

$$\begin{aligned} t_0 &= a_0, \quad t_k = \rho_k t_{k-1}, \quad k = 1, 2, 3, \ldots, \\ \rho_0 &= 0, \quad \rho_k = \frac{a_k(1+\rho_{k-1})}{1 - a_k(1+\rho_{k-1})}, \quad k = 1, 2, 3, \ldots. \end{aligned} \tag{6.51}$$

The nth partial sum of the series (the first being t_0) equals the nth convergent of the continued fraction, $n = 1, 2, 3, \ldots$. No scaling problems arise (as in the forward recurrence algorithm) and no a priori information is needed, as in the backward recurrence algorithm. In [78] the forward series algorithm is used for the evaluation of the continued fractions of the incomplete gamma function (see Example 6.1).

For information on algorithms, on convergence in the complex plane, and on ways of accelerating the convergence of continued fractions, see [17, 138]. For the evaluation of special functions by using continued fractions see also [75, §1] and [241, Chap. 4, §5].

6.6.1 Steed's algorithm

A modification of the forward algorithm in which efficiency is obtained in the computation of the convergents $C_n = A_n/B_n$ is Steed's algorithm; see [11], where this algorithm is used for the evaluation of Coulomb wave functions.

ALGORITHM 6.1. Steed's algorithm.

Input: $b_0, b_1, \ldots; a_1, a_2, \ldots$.

Output: $C_n, n = 0, 1, \ldots$.

- $C_0 = b_0$, $D_1 = 1/b_1$, $\nabla C_1 = a_1 D_1$, $C_1 = C_0 + \nabla C_1$

- DO $n = 2, 3, \ldots$

$$D_n = \frac{1}{D_{n-1} a_n + b_n}$$
$$\nabla C_n = (b_n D_n - 1) \nabla C_{n-1}$$
$$C_n = C_{n-1} + \nabla C_n$$

This can be iterated until $\nabla C_n / C_n$ is sufficiently small.

To verify Steed's algorithm, observe that $D_n = B_{n-1}/B_n$, and that, by using the recurrence relation in (6.9),

$$C_n = \frac{A_n}{B_n} = D_n \left(b_n C_{n-1} + a_n \frac{A_{n-2}}{B_{n-1}} \right). \tag{6.52}$$

The next steps are

$$\begin{aligned}
\nabla C_n &= C_n - C_{n-1} \\
&= (b_n D_n - 1) C_{n-1} + a_n D_n \frac{A_{n-2}}{B_{n-1}} \\
&= (b_n D_n - 1) \left(C_{n-1} - \frac{A_{n-2}}{B_{n-2}} \right) \\
&= (b_n D_n - 1)(C_{n-1} - C_{n-2}) \\
&= (b_n D_n - 1) \nabla C_{n-1},
\end{aligned} \tag{6.53}$$

where we have used $b_n D_n - 1 = -a_n B_{n-2}/B_n$.

Apart from efficiency aspects, Steed's algorithm uses the quantities D_n, which are ratios of the B_n quantities, and this may avoid overflow that may occur in the computation of A_n and B_n when the recurrence relations in (6.9) are used straightforwardly.

Example 6.6 (Steed's algorithm for the incomplete gamma function). We apply Steed's algorithm for the evaluation of the continued fraction in (6.31) with coefficients given in (6.32). We have $b_0 = 0$, $b_n = 1$, $n \geq 1$, $a_1 = 1$, and $a_{n+1} = \alpha_n$, $n \geq 1$. The initial conditions for the algorithm are

$$\begin{aligned} C_0 = b_0 = 0, \quad D_1 = 1/b_1 = 1, \quad \nabla C_1 = a_1 D_1 = 1, \\ C_1 = A_1/B_1 = b_0 + a_1/b_1 = 1. \end{aligned} \tag{6.54}$$

We verify the relation

$$zG(a+1, z) - aG(a, z) - 1 = 0, \quad G(a, z) = z^{-a} e^z \Gamma(a, z), \tag{6.55}$$

6.6. Numerical evaluation of continued fractions

Table 6.1. *Verification of the relation in (6.55) for some values of z and a by using Steed's algorithm; n is the number of iterations.*

z	a	n	Error in (6.55)	z	a	n	Error in (6.55)
1.0	−5.5	33	$0.16\,10^{-09}$	50.0	−50.5	7	$0.30\,10^{-12}$
1.0	0.5	38	$0.40\,10^{-09}$	50.0	0.5	5	$0.30\,10^{-14}$
5.0	−5.5	15	$0.21\,10^{-10}$	50.0	45.5	24	$0.95\,10^{-10}$
5.0	0.5	12	$0.16\,10^{-10}$	100.0	−100.5	5	$0.16\,14^{-12}$
5.0	4.5	11	$0.89\,10^{-10}$	100.0	0.5	4	$0.48\,10^{-11}$
10.0	−15.5	11	$0.68\,10^{-11}$	100.0	95.5	31	$0.20\,10^{-14}$
10.0	0.5	8	$0.15\,14^{-11}$	500.0	−500.5	5	$0.27\,10^{-09}$
10.0	9.5	11	$0.20\,10^{-10}$	500.0	0.5	3	$0.40\,10^{-14}$
25.0	−25.5	8	$0.21\,10^{-11}$	500.0	495.5	54	$0.17\,10^{-08}$
25.0	0.5	6	$0.92\,10^{-12}$	1000.0	0.5	3	$0.10\,10^{-15}$
25.0	24.5	18	$0.29\,10^{-10}$	1000.0	999.5	69	$0.21\,10^{-08}$

which follows from the recurrence relation for the incomplete gamma function

$$\Gamma(a+1, z) = a\Gamma(a, z) + z^a e^{-z}. \tag{6.56}$$

In Table 6.1 we give the relative error of relation (6.55) for a few values of z and a. The number n gives the maximal number of iterations, corresponding with the first time that $\delta_n = |\nabla C_n / C_n|$ is smaller than 10^{-10}. We see that this criterion fails when $z = 1$, $z = 500, a = -500.5, a = 495.5$, and $z = 1000, a = 999.5$. Notwithstanding that, the continued fraction expansion is an excellent tool for computing this incomplete gamma function. Observe that we do not use it for $a \geq z$. For that case it is better to compute first $\gamma(a, z)$; then $\Gamma(a, z)$ follows from the relation $\Gamma(a, z) = \Gamma(a) - \gamma(a, z)$.

In Chapter 8, §8.3.3, we give an algorithm that is very efficient for large values of z and a, which can be used for both $\gamma(a, z)$ and $\Gamma(a, z)$. This gives a suitable "crossing domain" where the continued fraction (for $\Gamma(a, z)$) and the convergent power series (for $\gamma(a, z)$) become less efficient because of poor convergence.

6.6.2 The modified Lentz algorithm

From a computational point of view, the evaluation of the continued fraction using Steed's algorithm can be strongly affected by roundoff errors if cancellations in the denominators of the convergents C_n occur. Let us consider, as an example, the computation of the continued fraction for the Bessel functions $J_\nu(x)$ which follows from applying repeatedly the relation (see (4.37))

$$\rho_\nu = \frac{J_\nu(x)}{J_{\nu-1}(x)} = \frac{1}{\frac{2\nu}{x} - \rho_{\nu+1}}. \tag{6.57}$$

As can be immediately seen from (6.57), the denominator of the convergent C_2 will vanish for $x^2 - 4\nu(\nu + 1) = 0$. If we take, for instance, $\nu = 10$, the denominator of C_2

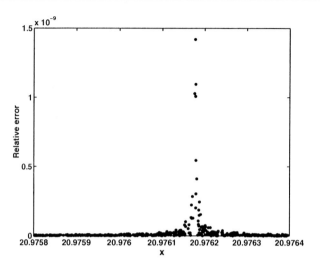

Figure 6.1. *Relative error in the computation of the ratio $J_{10}(x)/J_9(x)$ for $x \in (20.9758, 20.9764)$ when comparing the direct computation of the ratio with the continued fraction representation evaluated using Steed's algorithm.*

vanishes for $x = \sqrt{440} \approx 20.9762$. What is the effect of such cancellation when applying Steed's algorithm? As can be seen in Figure 6.1, a loss of significant digits occurs when approaching $x = 20.9762$. The figure shows the relative error in the computation of the ratio $J_{10}(x)/J_9(x)$ for $x \in I = (20.9758, 20.9764)$ by comparing the direct computation with the continued fraction (computed by using Steed's algorithm). We take $\delta_n = |\nabla C_n/C_n| = 10^{-15}$ and 500 random points inside the interval I.

An alternative to Steed's algorithm is the *modified Lentz algorithm* [223]. In this algorithm, the convergents C_n are computed as $C_n = C_{n-1} E_n D_n$, where E_n and D_n are the ratios A_n/A_{n-1} and B_{n-1}/B_n, respectively. The ratios D_n were already computed in Steed's algorithm using a recurrence relation. On the other hand, the ratios E_n can also be evaluated by applying the following recurrence relation:

$$E_n = b_n + \frac{a_n}{E_{n-1}}, \qquad (6.58)$$

which also follows from (6.9).

When the quantities $E_n = b_n + a_n/E_{n-1}$ or the denominators of the quantities $D_n = 1/(D_{n-1}a_n + b_n)$ are close to zero (or zero), the possible overflows can be avoided by taking

$$E_n = \epsilon, \quad D_{n-1}a_n + b_n = \epsilon, \qquad (6.59)$$

where ϵ is a small parameter such that $1/\epsilon$ does not overflow.

ALGORITHM 6.2. Modified Lentz algorithm.

Input: $b_0, b_1, \ldots; a_1, a_2, \ldots; \epsilon$.

Output: C_n, $n = 0, 1, \ldots$.

- $C_0 = b_0$;
- IF $b_0 = 0$ THEN $C_0 = \epsilon$;
- $E_0 = C_0$; $D_0 = 0$.
- DO $n = 1, 2, \ldots$

    $D_n = b_n + a_n D_{n-1}$;
    IF $D_n = 0$ THEN $D_n = \epsilon$;
    $E_n = b_n + a_n / E_{n-1}$;
    IF $E_n = 0$ THEN $E_n = \epsilon$;
    $D_n = 1/D_n$;
    $H_n = E_n D_n$;
    $C_n = C_{n-1} H_n$;

This can be iterated until $\delta_n = |H_n - 1|$ is less than the prescribed accuracy for the computation of the continued fraction.

Notice that the conditions $D_n = 0$, $E_n = 0$ are never met exactly and that in a real implementation one should consider more realistic checks; the conditions should be replaced by $|D_n| <$ *tiny* (and similarly for E_n), with *tiny* a small number close to underflow, but not too close. If the coefficients a_n may become large, one should also check that $b_n + a_n D_n$ and $b_n + a_n/E_{n-1}$ do not overflow; in practice, if they do not take huge values, it is enough to consider ϵ such that $1/\epsilon$ is quite large, but not too close to overflow.

In comparison to Steed's algorithm, the contamination with roundoff errors in the computation of the continued fraction, using the modified Lentz algorithm in those cases when the denominators of the convergents C_n vanish, will be much less significant. As an illustration, we repeat now the calculation of the continued fraction for the ratio $J_{10}(x)/J_9(x)$ for $x \in (20.9758, 20.9764)$, but now using the modified Lentz algorithm. In Figure 6.2 we show the relative error (much more uniform in comparison to Figure 6.1) in the computation of the ratio when comparing the direct computation with the continued fraction.

6.7 Special functions and continued fractions

In the literature (see in particular [47, 119, 123, 138, 229]), many examples are given of continued fractions for elementary and special functions. In this section we consider two important cases: the incomplete gamma function and the Gauss hypergeometric function. The latter case gives a continued fraction for the incomplete beta function. Additional examples of the use of continued fractions as computational tools for special functions can be found in Chapters 4 and 12 (Algorithms 12.3–12.8 and 12.10).

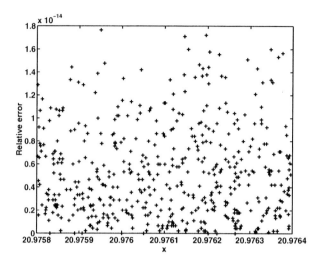

Figure 6.2. *Relative error in the computation of the ratio $J_{10}(x)/J_9(x)$ for $x \in (20.9758, 20.9764)$ when comparing the direct computation of the ratio with the continued fraction representation evaluated using the modified Lentz algorithm.*

6.7.1 Incomplete gamma function

We derive the fraction (6.24) for the incomplete gamma function (see also (6.26)) in the form

$$z^{-a}e^z\Gamma(a,z) = \frac{1}{z+}\frac{1-a}{1+}\frac{1}{z+}\frac{2-a}{1+}\frac{2}{z+}\frac{3-a}{1+}\cdots. \qquad (6.60)$$

This example is due to Legendre. The starting point is a different integral representation,

$$\Gamma(a,z) = \frac{e^{-z}}{\Gamma(1-a)}\int_0^\infty \frac{e^{-zt}t^{-a}}{t+1}\,dt, \quad \Re a < 0, \quad \Re z > 0. \qquad (6.61)$$

Now let

$$U^{\nu,\rho} = \int_0^\infty e^{-zt}t^\nu(1+t)^\rho\,dt, \quad \Re\nu > -1, \quad \Re z > 0. \qquad (6.62)$$

Then according to (6.61) we have

$$U^{-a,-1} = \Gamma(1-a)e^z\Gamma(a,z). \qquad (6.63)$$

Integrating by parts in (6.62), writing $t^\nu\,dt = \frac{1}{\nu+1}dt^{\nu+1}$, we obtain

$$zU^{\nu+1,\rho} = (\nu+1)U^{\nu,\rho} + \rho U^{\nu+1,\rho-1}, \qquad (6.64)$$

with

$$\frac{U^{\nu+1,\rho}}{U^{\nu,\rho}} = \frac{\nu+1}{z - \rho\dfrac{U^{\nu+1,\rho-1}}{U^{\nu+1,\rho}}} \qquad (6.65)$$

6.7. Special functions and continued fractions

as a variant. On the other hand, we can write in (6.62)

$$t^{\nu+1}(1+t)^\rho = t^{\nu+1}(1+t)^{\rho-1} + t^{\nu+2}(1+t)^{\rho-1}, \qquad (6.66)$$

so that (6.65) can be written as

$$\frac{U^{\nu+1,\rho}}{U^{\nu,\rho}} = \frac{\nu+1}{z - \dfrac{\rho}{1 + \dfrac{U^{\nu+2,\rho-1}}{U^{\nu+1,\rho-1}}}}. \qquad (6.67)$$

Applying this formula repeatedly we obtain a formal expansion in the form of a continued fraction. From (6.63) it follows that

$$\frac{U^{1-a,-1}}{U^{-a,-1}} = \frac{\Gamma(2-a)\Gamma(a-1,z)}{\Gamma(1-a)\Gamma(a,z)} = \frac{(1-a)\Gamma(a-1,z)}{\Gamma(a,z)}. \qquad (6.68)$$

From this and the relation $\Gamma(a+1,z) = a\Gamma(a,z) + z^a e^{-z}$, we get

$$\frac{U^{1-a,-1}}{U^{-a,-1}} = -1 + \frac{e^{-z} z^{a-1}}{\Gamma(a,z)}. \qquad (6.69)$$

Thus, with $\nu = -a$, $\rho = -1$, (6.67) finally gives the result in (6.60). Observe that this method works because we have a simple recurrence relation for the incomplete gamma function. The continued fraction (6.60) converges for all complex a and all $z \neq 0$ in the cut plane $|\mathrm{ph}\, z| < \pi$. The same holds for the transformed fraction (6.31).

6.7.2 Gauss hypergeometric functions

The contiguous relation

$$_2F_1\begin{pmatrix} a, b \\ c \end{pmatrix}; z\end{pmatrix} = {}_2F_1\begin{pmatrix} a, b+1 \\ c+1 \end{pmatrix}; z\end{pmatrix} - \frac{a(c-b)}{c(c+1)} z\, {}_2F_1\begin{pmatrix} a+1, b+1 \\ c+2 \end{pmatrix}; z\end{pmatrix} \qquad (6.70)$$

can be verified by comparing coefficients of corresponding powers of z in the left- and right-hand sides. This relation may be written in the form

$$\frac{{}_2F_1\begin{pmatrix} a, b+1 \\ c+1 \end{pmatrix}; z\end{pmatrix}}{{}_2F_1\begin{pmatrix} a, b \\ c \end{pmatrix}; z\end{pmatrix}} = \frac{1}{1 - \dfrac{a(c-b)}{c(c+1)} z\, \dfrac{{}_2F_1\begin{pmatrix} a+1, b+1 \\ c+2 \end{pmatrix}; z\end{pmatrix}}{{}_2F_1\begin{pmatrix} a, b+1 \\ c+1 \end{pmatrix}; z\end{pmatrix}}}. \qquad (6.71)$$

We interchange a and b in (6.71), and afterwards replace b by $b+1$ and c by $c+1$. This gives

$$\frac{{}_2F_1\left(\begin{array}{c}a+1,\,b+1\\c+2\end{array};z\right)}{{}_2F_1\left(\begin{array}{c}a,\,b+1\\c+1\end{array};z\right)}=\frac{1}{1-\dfrac{(b+1)(c-a+1)}{(c+1)(c+2)}z\dfrac{{}_2F_1\left(\begin{array}{c}a+1,\,b+2\\c+3\end{array};z\right)}{{}_2F_1\left(\begin{array}{c}a+1,\,b+1\\c+2\end{array};z\right)}}. \tag{6.72}$$

The quotient on the left-hand side of (6.72) is the same as the quotient of hypergeometric functions appearing in the denominator of the right-hand side of (6.71). Also, if a, b, c are replaced by $a+1, b+1, c+2$, respectively, in (6.71), the quotient on the left-hand side becomes equal to the quotient of hypergeometric functions appearing in the denominator of the right-hand side of (6.72). On applying first one identity and then the other, we obtain by successive substitution the continued fraction of Gauss,

$$\frac{{}_2F_1\left(\begin{array}{c}a,\,b+1\\c+1\end{array};z\right)}{{}_2F_1\left(\begin{array}{c}a,\,b\\c\end{array};z\right)}=\frac{1}{1-}\frac{a_1z}{1-}\frac{a_2z}{1-}\frac{a_3z}{1-\cdots}, \tag{6.73}$$

where, for $n = 0, 1, 2, \ldots$,

$$a_{2n+1}=\frac{(a+n)(c-b+n)}{(c+2n)(c+2n+1)},\quad a_{2n+2}=\frac{(b+n+1)(c-a+n+1)}{(c+2n+1)(c+2n+2)}. \tag{6.74}$$

When $a_n = 0$ for some n, then the continued fraction of Gauss terminates, and the quotient on the left-hand side of (6.73) is a rational function of z, which is equal to the terminating continued fraction.

Because $\lim_{n\to\infty} a_n = \frac{1}{4}$, it follows from Theorem 6.5 that the continued fraction (6.73) converges throughout the z-plane exterior to the cut along the real axis from 1 to $+\infty$, except possibly at certain isolated points, it is equal to the left-hand side of (6.73) in the neighborhood of the origin, and it furnishes the analytic continuation of this function into the interior of the cut plane.

The incomplete beta function

If we put $b = 0$ in (6.73), we obtain a continued fraction of $F(a, 1; c; z)$, because $F(a, 0; c; z) = 1$. This gives a continued fraction for the *incomplete beta function* defined by

$$B_x(p,q)=\int_0^x t^{p-1}(1-t)^{q-1}\,dt,\quad \Re p>0,\quad \Re q>0, \tag{6.75}$$

6.7. Special functions and continued fractions

and usually $0 \le x \le 1$; when $x < 1$ the condition on q can be omitted. The beta integral is obtained when we take $x = 1$, that is,

$$B(p, q) = \int_0^1 t^{p-1}(1-t)^{q-1}\, dt = \frac{\Gamma(p)\Gamma(q)}{\Gamma(p+q)}, \quad \Re p > 0, \quad \Re q > 0. \tag{6.76}$$

In terms of the hypergeometric functions we have

$$\begin{aligned}
B_{x(p,q)} &= \frac{x^p}{p}\, {}_2F_1\!\left(\begin{matrix} p,\ 1-q \\ p+1 \end{matrix}; x\right) \\
&= \frac{x^p(1-q)^{q-1}}{p}\, {}_2F_1\!\left(\begin{matrix} 1,\ 1-q \\ p+1 \end{matrix}; \frac{x}{x-1}\right) \\
&= \frac{x^p(1-x)^q}{p}\, {}_2F_1\!\left(\begin{matrix} p+q,\ 1 \\ p+1 \end{matrix}; x\right).
\end{aligned} \tag{6.77}$$

These representations give the analytic continuation for complex x.

By using the third representation and (6.73), we obtain

$$B_x(p, q) = \frac{x^p(1-x)^q}{p}\left(\frac{1}{1+}\,\frac{d_1}{1+}\,\frac{d_2}{1+}\,\frac{d_3}{1+}\cdots\right), \tag{6.78}$$

where, for $n = 0, 1, 2, \ldots$,

$$d_{2n+1} = -\frac{(p+n)(p+q+n)}{(p+2n)(p+2n+1)}x, \quad d_{2n+2} = \frac{(n+1)(q-n-1)}{(p+2n+1)(p+2n+2)}x. \tag{6.79}$$

When $p > 1$ and $q > 1$, the maximum of the integrand in (6.75) occurs at $x_0 = (p-1)/(p+q-2)$, and the best numerical results are obtained when $x \le x_0$. When $x_0 < x \le 1$, we use the reflection relation with the beta integral (see (6.76))

$$B_x(p, q) = B(p, q) - B_{1-x}(q, p). \tag{6.80}$$

From a numerical point of view the continued fraction (6.78) has an interesting property of the convergents: C_{4n} and C_{4n+1} are less than this value of the continued fraction and C_{4n+2}, C_{4n+3} are greater than this value. This gives excellent control of the convergence of an algorithm that uses (6.78).

Chapter 7

Computation of the Zeros of Special Functions

To those who ask what the infinitely small quantity in mathematics is, we answer that it is actually zero.
—L. Euler

I don't agree with mathematics; the sum total of zeros is a frightening figure.
—Stanislaw J. Lec, 1962

7.1 Introduction

The zeros of special functions appear in a great number of applications in mathematics, physics, and engineering, from the computation of Gauss quadrature rules [97] in the case of orthogonal polynomials to many applications in which boundary value problems for second order ordinary differential equations arise. In some sense, the computation of the zeros of special functions has nothing special: to compute the roots of an equation $y(x) = 0$, with $y(x)$ special or not, we can apply well-known methods (bisection, secant, Newton–Raphson, or whatever) once we know how to compute the function $y(x)$ (and in the case of Newton–Raphson also its derivative) accurately enough.

However, as is generally the case with nonlinear equations, some information on the location of the zeros is desirable, particularly when applying rapidly convergent (but often unpredictable) methods like Newton–Raphson or higher order methods. This is generally true for nonlinear equations, even for "elementary" equations, as in the following example, which we use as a "Leitmotif" in §7.2 and §7.3 for illustrating some approximation methods.

Example 7.1. Compute the first 6 zeros of

$$\kappa x \sin(x) - \cos x = 0 \tag{7.1}$$

for $\kappa = 0.01, 0.1, 1, 10$. ∎

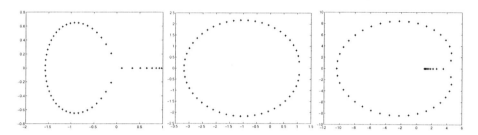

Figure 7.1. *Zeros of the Jacobi polynomials $P_{50}^{(2,-42.5)}(x)$ (left), $P_{50}^{(2,-52)}(x)$ (center), and $P_{50}^{(2,-63.5)}(x)$ (right) in the complex plane.*

The zeros of special functions usually appear nicely arranged, forming clear patterns from which a priori information can be found. For instance, the zeros of Jacobi polynomials $P_n^{(\alpha,\beta)}(x)$ are all real and in the interval $(-1, 1)$ for $\alpha > -1$ and $\beta > -1$, and they satisfy other regularity properties (such as, for instance, interlacing with the zeros of the derivative and with the zeros of contiguous orders). As α and/or β become smaller than -1, some or all of the n zeros escape from the real axis, forming a regular pattern in the complex plane (see Figure 7.1).

These regular patterns formed by the zeros is a common feature of "classical" special functions and beyond [34]. The regularity in the distribution of zeros helps in the design of specific algorithms with good convergence properties. In addition, for many special functions, accurate a priori approximations are available. This a priori information, when wisely applied, will save computation time and avoid divergent (or even chaotic) algorithms. In a fortunate situation, as in the case of Bessel functions, asymptotic approximations provide accurate enough starting values for higher order Newton–Raphson methods; see §10.6 and [210].

In §7.3 we discuss the basics of asymptotic approximations for the zeros of some Bessel functions and other special functions. In addition, we will find cases for which it is not necessary to compute values of these functions themselves in order to obtain their zeros. This is the case for the classical orthogonal polynomials, the zeros of which are the exact eigenvalues of real tridiagonal symmetric matrices with very simple entries. Also, there are other functions (again Bessel functions $J_\nu(x)$ are among them) for which the problem of computing zeros is not exactly an eigenvalue problem for a (finite) matrix, but it can be approximated by it. This is discussed in §7.4.

The discussion on fixed point methods is concluded in §7.5 with a brief description of certain methods introduced in [84, 193]. These are (like Newton–Raphson) second order methods that converge with certainty (as is not usually the case for Newton–Raphson), producing all the zeros in any real interval. Also, the only information required consists of some elementary coefficients (the matrix entries for the eigenvalue method) in those cases for which the eigenvalue method applies. These methods are applicable to a wider range of functions than the eigenvalue methods, but their convergence has been proved only for real zeros.

Finally, in §7.6 we give further asymptotic examples for the zeros of Airy functions, Scorer functions, and error functions.

7.2 Some classical methods

We consider the computation of real zeros of real functions, although the methods in this section can be generalized to complex zeros (see §7.2.3).

7.2.1 The bisection method

The most elementary method for computing real zeros of functions is the bisection method, which is a direct consequence of *Bolzano's theorem*: if a function $f(x)$ which is continuous in $[a, b] = [a_0, b_0]$ is such that $f(a)f(b) < 0$, then there is at least one zero of $f(x)$ in (a, b), that is, a value $x_0 \in (a, b)$ such that $f(x_0) = 0$. Bisection provides a method to compute with certainty one of such zeros. It consists in halving the interval and comparing the sign of the function at $c = (a+b)/2$ with the sign at a and b; when $f(a)f(c) < 0$, the halving process continues with the interval $[a_1, b_1] = [a, c]$ (where there is a zero), otherwise it continues with $[a_1, b_1] = [b, c]$ (unless the rare event $f(c) = 0$ takes place, which solves the problem exactly). The process continues until a small enough interval is obtained $[a_n, b_n]$ with opposite signs of $f(x)$ at the endpoints of the interval. A zero lies inside this interval which can be approximated by $c_n = (a_n + b_n)/2$ with absolute error smaller than $(b_n - a_n)/2 = (b-a)/2^{n+1}$.

The good news regarding this simple method is that it converges with certainty to a zero when an interval where $f(x)$ changes sign is located, but also that, in some sense, it is optimal [198]. If, for instance, all the information available for the function is that it is continuous and changes sign in $[a, b]$ (and function values can be computed) bisection is the optimal choice, at least in a minimax sense. The bad news is that the convergence is rather slow because the error bound is reduced by only a factor 2 in each iteration. The speed of convergence is independent of the function; it depends only on the location of the root.

7.2.2 The fixed point method and the Newton–Raphson method

If possible, it is preferable to use more specialized methods making use of additional information on the functions in such a way that greater rates of convergence can be obtained. Because we are dealing with special functions, that is, functions with special properties, there is hope that this is achievable.

The Newton–Raphson method is a particular case of a wider family of methods known as fixed point methods. The basic idea behind these methods consists in finding an iteration function $T(x)$ such that the zero α (or zeros) of $f(x)$ to be computed satisfies $T(\alpha) = \alpha$ (that is to say, α is a fixed point of $T(x)$) and generating the successive approximations to α with the iterative process $x_{i+1} = T(x_i)$, starting from some x_0. Under certain conditions, convergence to a fixed point (or a zero of $f(x)$) will be guaranteed.

Theorem 7.2 (fixed point theorem). *Let $T(x)$ be a continuous and differentiable function in $I = [a, b]$ such that $T([a, b]) \subset [a, b]$ and such that $|T'(x)| \leq M < 1$ for all $x \in I$. Then, there exists only one $\alpha \in I$ such that $T(\alpha) = \alpha$ and for any $x_0 \in [a, b]$ the sequence*

$x_{n+1} = T(x_n)$, $n = 0, 1, \ldots$, converges to α (which is the only fixed point of $T(x)$ in $[a, b]$). The error after the nth iteration can be bounded in the following way:

$$|x_n - \alpha| \leq \frac{M^n}{1 - M}|x_1 - x_0|. \tag{7.2}$$

A detailed proof of the existence and uniqueness of the fixed point and the convergence of the sequences x_n can be found elsewhere; it is a straightforward application of the mean value theorem. We have

$$|x_{n+1} - x_n| = |T(x_n) - T(x_{n-1})| = |T'(\zeta)||x_n - x_{n-1}| \leq M|x_n - x_{n-1}|, \tag{7.3}$$

where ζ is a value between x_n and x_{n-1}. Then $|x_{n+1} - x_n| \leq M|x_n - x_{n-1}| \leq \cdots \leq M^n|x_1 - x_0|$, which can be used to prove that x_n (for any $x_0 \in I$) is a Cauchy sequence and therefore convergent; the proof that the fixed point is unique follows similar lines.

Regarding the derivation of the error bound we have

$$|x_n - \alpha| = |T(x_{n-1}) - T(\alpha)| = |T'(\zeta)||x_{n-1} - \alpha| \tag{7.4}$$

for some ζ between α and x_{n-1}. Therefore,

$$|x_n - \alpha| \leq M|x_{n-1} - \alpha| \leq \cdots \leq M^n|x_0 - \alpha| \tag{7.5}$$

and because $|x_0 - \alpha| \leq |x_1 - x_0| + |x_1 - \alpha| \leq |x_1 - x_0| + M|x_0 - \alpha|$, then $|x_0 - \alpha| \leq |x_1 - x_0|/(1 - M)$, and we get the error bound.

This theorem, with some modifications, also holds for complex variables.

The main question to be answered for considering the practical application of fixed point methods is, can we gain speed of convergence with respect to bisection? The answer is, yes, at least locally, but be prepared to lose the nice global convergence of bisection. Let us quantify the speed of convergence in the following way.

Definition 7.3. *Let $\{x_n\}_{n=0}^{\infty}$ be a sequence which converges to α and such that $x_n \neq \alpha$ for high enough n. We will say that the sequence converges with order $p \geq 1$ when there exists a constant $C > 0$ such that*

$$\lim_{n \to +\infty} \frac{|\epsilon_{n+1}|}{|\epsilon_n|^p} = C, \tag{7.6}$$

where $\epsilon_n = x_n - \alpha$, $n = 0, 1, \ldots$; C is called the asymptotic error constant.

Another possible definition is: a method is of order at least p if there exists $C > 0$ ($C < 1$ if $p = 1$) such that $|\epsilon_{k+1}| \leq C|\epsilon_k|^p$ for large enough values of k.

The higher the order, the faster the convergence (when the method converges). With respect to bisection, the best we can expect is linear convergence and never more than that [122] (although we could be infinitely lucky to strike the zero in a finite number of steps). This is independent of $f(x)$; it depends only on the location of the zero and on the initial interval.

7.2. Some classical methods

For a fixed point method, higher orders of convergence can be obtained by considering iterations $T(x)$ with a number of vanishing derivatives at α. Indeed, considering Taylor series, we have

$$x_{n+1} = T(\epsilon_n + \alpha) = T(\alpha) + T'(\alpha)\epsilon_n + \frac{1}{2}T''(\alpha)\epsilon_n^2 + \cdots, \qquad (7.7)$$

and if the first nonvanishing derivative of $T(x)$ is the pth derivative, then

$$\epsilon_{n+1} = \frac{T^{(p)}(\alpha)}{p!}\epsilon_n^p + \mathcal{O}(\epsilon_n^{p+1}) \quad \text{as } \epsilon_n \to 0, \qquad (7.8)$$

and then the method would be of order p with asymptotic error constant

$$C = T^{(p)}(\alpha)/p!. \qquad (7.9)$$

It is easy to check that if a fixed point method is at least of order 1 for the convergence to a certain fixed point α, then, when $T(x)$ is continuously differentiable around this point, there is an interval containing α for which the method converges with certainty for any initial value x_0 in this interval. The problem is that, for some iterations, this interval could be quite tiny and the iterative process can become uncontrollable when the initial value is not wisely chosen.

A particular example of a fixed point method is the *Newton–Raphson method*, which consists in considering the fixed point iteration

$$T(x) = x - \frac{f(x)}{f'(x)} \qquad (7.10)$$

in an interval where $f'(x) \neq 0$.

Because if $f(\alpha) = 0$ and $f'(\alpha) \neq 0$, then $T'(\alpha) = 0$ and $T''(\alpha) = f''(\alpha)/f'(\alpha)$, and we see that the Newton–Raphson method has at least order of convergence 2, and it has exactly order of convergence 2 when $f''(\alpha) \neq 0$, with the asymptotic error constant

$$C = \frac{T''(\alpha)}{2} = \frac{f''(\alpha)}{2f'(\alpha)}. \qquad (7.11)$$

This constitutes an improvement in the speed of convergence, when the method does converge, with respect to the safe but slow bisection method. However, initial guesses should be good enough in order to guarantee convergence. Particularly, when a value x_n is reached for which the derivative x_n becomes very small, the method can become unpredictable. For an illustration, let us consider Example 7.1 for $\kappa = 1$.

Example 7.4 (Example 7.1, first attempt: Newton–Raphson). Let us compute the first positive root of $f(x) = x \sin x - \cos x$.

The first positive root is the first intersection point of the straight line $y = x$ and the graph of $y = \cotan x$. It is evident that the first root is inside the interval $(0, \pi/2)$. Considering, for instance, $x_0 = 1$, we obtain 16-digit precision to the first root, $\alpha = 0.8603335890193798$, with 6 Newton–Raphson iterations. Had we tried initial values incautiously, we would have observed quite erratic behavior. For instance, for $x_0 = 2.1$ we have convergence to the first negative root, but for $x_0 = 2.2$ we have convergence to the 7th negative root and for $x_0 = 2.3$ to the 20th positive root. The reason for this erratic behavior is that $f(x) = x\sin x - \cos x$ has a maximum at $x \simeq 2.29$. Also, when considering small starting values, the method would be unpredictable because the function $f(x)$ has a minimum at $x = 0$. ∎

Fixed point methods can also be built for approximating the solutions of the equation of the previous example.

Example 7.5 (Example 7.1, second attempt: fixed point method). Let us consider the computation of the zeros of $\kappa x \sin x - \cos x = 0$ by a fixed point method.

We can write the equation as $\tan x = 1/(\kappa x)$ and inverting we have

$$x = \arctan\left(\frac{1}{\kappa x}\right) + m\pi = T_m(x), \quad m = 0, 1, 2, \ldots, \tag{7.12}$$

where, as usual, arctan is valued in the interval $(-\pi/2, \pi/2]$.

Let us denote $I_m = [(m - 1/2)\pi, (m + 1/2)\pi], m \in \mathbb{N}$. Graphical arguments (plotting κx and $1/\tan x$) show that when $\kappa < 0$ the first positive zero is in I_1, while for $\kappa > 0$ it is in $I_0 = [-\pi/2, \pi/2]$; we can also consider, alternatively,

$$T_0(x) = \frac{\pi}{2} - \arctan(\kappa x). \tag{7.13}$$

Differentiating $T_m(x)$ of (7.12) gives

$$T'_m(x) = -\frac{\kappa}{1 + \kappa^2 x^2}. \tag{7.14}$$

The fixed point method $T_m(x)$, $m = 0, 1, 2, \ldots$, satisfies $T_m(I_m) \subset I_m$. Also, if $|\kappa| < 1$, then $|T'_m(x)| \leq \kappa < 1$ and the fixed point method for T_m converges to the only root of $\kappa x \sin x - \cos x$ in I_m. If $\kappa < -1$, it is easy to see that convergence is also certain.

However, when $\kappa > 1$, it is not possible to decide the convergence to the first positive zero, which lies in I_0.

Additionally, we can check that for large m, and using the bound from Theorem 7.2, the error in the approximation after the nth iteration is $\mathcal{O}(m^{-2n})$. Indeed, large zeros are close to the values $m\pi$ and for such large zeros $T'_m(x) \sim 1/(\kappa x^2) = \mathcal{O}(m^{-2})$, which means that there exists $M = Cm^{-2}$ for a given $C > 0$ such that $|T'_m(x)| \leq M < 1$. Applying the error bound of Theorem 7.2, we get the desired result. ∎

When higher derivatives of the function are available, which is the case for functions which are solutions of ordinary differential equations, higher order Newton-like methods can be built. The order can be high but the unpredictability tends to be high too; however, they are useful in some applications, as illustrated in Chapter 10, §10.6.

7.2.3 Complex zeros

For complex variables, the fixed point theorem can still be used. Also, generalized bisection is possible when a method to count zeros inside a square in the complex plane is available. One would start with a rectangle containing one or several zeros and subdivide the rectangle into smaller ones, counting again the zeros and proceeding iteratively until the zeros are bracketed inside small enough rectangles. For polynomials, one can consider the principle of argument together with Sturm sequences [240]; for complex zeros of Bessel functions, Cauchy's theorem has been used for computing the number of zeros inside each square [126]; see also [127].

Chaotic behavior in the complex plane

We have seen that fixed point methods on the real line, such as the Newton–Raphson method, may diverge or may converge to a zero that is not wanted. This type of behavior also occurs when using fixed point methods in the complex plane, whether or not complex zeros are sought.

For example, computing the zeros of the cubic polynomial $z^3 - 1$, with zeros 1 and $e^{\pm 2\pi i/3}$, by using Newton–Raphson with the iteration

$$z_{n+1} = z_n - \frac{z_n^3 - 1}{3z_n^2}, \quad n = 0, 1, 2, \ldots, \tag{7.15}$$

and with starting value z_0, is a very unpredictable process.

When we start with an arbitrary starting point $z_0 \in \mathbb{C}$, convergence may be unpredictable, in particular when we take z_0 near the negative axis or near the half-lines with phase $\pm \frac{1}{3}\pi$. In fact, in the neighborhoods of these lines, exotic sets appear. For some points in these sets the fixed point method (7.15) converges to 1; for a different point slightly removed from one of these points the iteration may converge to a different root.

The boundary of the set of points z_0 giving a sequence of iterates $\{z_n\}$ that converges to a particular zero has a very complicated structure. It is called a *Julia set*. This Julia set is a *fractal*, which has a self-similar structure that repeats on all scales of magnification. Figure 7.2 shows the points for which convergence to the root $z = 1$ takes place. See [54] for further information.

7.3 Local strategies: Asymptotic and other approximations

Because the bisection method may be safe but certainly slow and higher order fixed point methods may be fast but uncertain, a good strategy consists in using fast methods with initial iteration values accurate enough to guarantee convergence. These initial estimates could be obtained, for instance, from bisection; however, in many cases, a priori approximations could make the job more efficient.

Example 7.6 (Example 7.1, third visit: an expansion for the large zeros). Taking again Example 7.5, we can use fixed point iteration to obtain approximate expressions for large

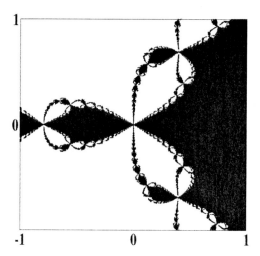

Figure 7.2. *Part of the Julia set of the iteration* (7.15). *In the black domain the starting points z_0 give convergence to the zero $z = 1$.*

zeros (as $m \to +\infty$). Because the arctan function is bounded, we know that $\alpha \sim m\pi$ for fixed and large m. Taking as a first approximation $x_0^{(m)} = m\pi$, $m \in \mathbb{N}$, we proved that convergence of the fixed point method is guaranteed (except, possibly, for the smallest zero) and that, furthermore, the error in the nth iteration is $x_n^{(m)} - \alpha^{(m)} = \mathcal{O}(m^{-2n})$.

If, for instance, we iterate once, then

$$\alpha^{(m)} = \arctan\left(\frac{1}{\kappa m \pi}\right) + m\pi + \mathcal{O}(m^{-2}) = m\pi + \frac{1}{\pi \kappa m} + \mathcal{O}(m^{-2}), \quad (7.16)$$

where we have expanded the arctan function for large m. We can also consider a greater number of iterations. For instance, iterating twice and then expanding (it is advisable to use a symbolic manipulator for this), we have

$$\alpha^{(m)}(\kappa) \sim m\pi + \mu^{-1} - \left(\kappa + \tfrac{1}{3}\right)\mu^{-3} + \mathcal{O}(\mu^{-5}), \quad m \to \infty, \quad (7.17)$$

where $\mu = \pi \kappa m$.

This approximation is accurate for large zeros but is also useful for the first zeros. When $\kappa > 0$ the first zero cannot be computed with this approximation, but if κ is not too small, it gives good starting approximations to the second zero. For instance, for $\kappa = 1$ the relative accuracy is $2\,10^{-5}$ for the second positive zero. For κ negative and not too small, we also have reasonable approximations for the second zero. For instance, for $\kappa = -0.5$ the relative accuracy is 0.001, and it improves as $|\kappa|$ increases. For the first zero, the approximation does not work except when κ becomes sufficiently negative; in this case, $m = 1$ may provide good approximations to the first root. ∎

7.3. Local strategies: Asymptotic and other approximations

This example illustrates the use of an iterative scheme to generate an asymptotic approximation for the large zeros, explained from a fixed point method perspective. From a computational point of view, it is not necessary to keep the full expression for each iteration and then expand. Instead, we can expand in each step and keep only the relevant terms for the next step, as follows.

Example 7.7 (Example 7.1, part 4: large zeros by resubstitution). We start from $\alpha \sim m\pi$. We have

$$\arctan\left(\frac{1}{z}\right) = \frac{1}{z} - \frac{1}{3z^3} + \frac{1}{5z^5} - \cdots, \qquad (7.18)$$

and when we keep only the first term and substitute the first approximation $m\pi$, we obtain

$$\alpha_1 = m\pi + \frac{1}{\kappa m \pi}. \qquad (7.19)$$

Resubstituting again, we obtain

$$\alpha_2 = m\pi + \frac{1}{\kappa \alpha_1} - \frac{1}{3(\kappa \alpha_1)^3} \qquad (7.20)$$

and expanding we get (7.17). ∎

7.3.1 Asymptotic approximations for large zeros

The previous example of approximation of the large (and not so large) zeros of $\kappa x \sin(x) - \cos(x) = 0$ is a particular case of a more general situation for which asymptotic approximations for the roots can be obtained, as we discuss in more detail in §7.6. Indeed, our problem consisted in finding an approximation for the root of an equation

$$f(z) = w, \qquad (7.21)$$

where $f(z)$ has the expansion

$$f(z) \sim z + \sum_{i=0}^{n} \frac{f_i}{z^i} + \mathcal{O}(z^{-n-1}), \quad z \to \infty, \qquad (7.22)$$

and w is a large parameter. In this case, under certain conditions (see [66]) we can guarantee that a solution z of the equation $f(z) = w$ exists, such that

$$z \sim w - F_0 - \frac{F_1}{w} - \frac{F_2}{w^2} - \frac{F_3}{w^3} - \cdots, \quad w \to \infty, \qquad (7.23)$$

where the first coefficients are

$$F_0 = f_0, \quad F_1 = f_1, \quad F_2 = f_0 f_1 + f_2, \quad F_3 = f_0^2 f_1 + f_1^2 + 2 f_0 f_2 + f_3. \qquad (7.24)$$

One can understand the validity of the asymptotic series mentioned above as a consequence of the fixed point theorem. We will consider only the particular case of real zeros. We write $f(x) = w$ with w a large real parameter and $f(x) = x + g(x)$, with

$$g(x) \sim f_0 + \frac{f_1}{x} + \frac{f_2}{x^2} + \cdots. \qquad (7.25)$$

If $f(x)$ is continuous and strictly increasing and $f(a) < w$, it is clear that the equation $f(x) = w$ has one and only one solution x with $x > a$. Furthermore, because $f(x) \sim x$ for large x, this solution satisfies $x \sim w + \mathcal{O}(1)$. Then, if w is large enough, this solution can be computed from the fixed point theorem. Indeed, we write

$$x = T(x), \quad T(x) = w - g(x), \tag{7.26}$$

and assume that the series for $g(x)$ in (7.25) is differentiable. Then $|T'(x)| = |g'(x)| \sim \mathcal{O}(x^{-2})$. Therefore, if we start the iteration of the fixed point method with $x_0 = w$, we will have that $|T'(x_n)| = \mathcal{O}(w^{-2})$ and for large w the iteration will converge to the solution $x = \lim_{n \to \infty} T^{(n)}(w) = T^{(\infty)}(w)$. The fixed point theorem can then be used for estimating the error after the nth iteration, which is

$$|T^{(n)}(w) - T^{(\infty)}(w)| \sim f_0 w^{-2n} \quad \text{as } w \to \infty. \tag{7.27}$$

Also, we have

$$|T^{(n+1)}(w) - T^{(n)}(w)| \sim w^{-2}|T^{(n)}(w) - T^{(n-1)}(w)|. \tag{7.28}$$

This can be used for a direct numerical computation of the root (as in Example 7.5), but also for computing asymptotic (or convergent) series of the zeros (as in Examples 7.6 and 7.7). Indeed, because the error is $\mathcal{O}(w^{-2n})$ after the nth iteration, we can expand to obtain a series in inverse powers of w correct up to the term w^{2n-1}. For instance, the second iteration gives

$$\begin{aligned} x_2 &= T(T(w)) = T(w - g(w)) = w - g(w - g(w)) \\ &= w - g(w)\left[1 - g'(w) + \frac{1}{2}g''(w)g(w)\right] + \mathcal{O}(w^{-4}) \\ &= w - F_0 - \frac{F_1}{w} - \frac{F_2}{w^2} - \frac{F_3}{w^3} + \mathcal{O}(w^{-4}), \end{aligned} \tag{7.29}$$

where the coefficients are as in (7.24). This is a method which is very simple to implement in a symbolic computation package: given the expansion, we only need to iterate and then expand. For each iteration, we get two new correct terms. For computing these expansions, it is not necessary to work with the full expression for each iterate, as shown in Example 7.7.

An alternative way consists in computing the coefficients one by one by using the next result, which we prove in §7.6.1, following along the lines of [66].

Theorem 7.8. *The quantity jF_j is the coefficient of z^{-1} in the asymptotic expansion of $f(z)^j$.*

Example 7.9 (McMahon expansions for the zeros of Bessel functions). The method used in (7.29), preferably with re-expansion in each iteration (as in Example 7.7), can

7.3. Local strategies: Asymptotic and other approximations

be used for computing the large zeros of some special functions. For example, the Bessel functions $J_\nu(z)$ have, as $z \to \infty$, the asymptotic representation [2, eq. (9.2.1)]

$$J_\nu(z) = \sqrt{\frac{2}{\pi z}} \{P(\nu, z) \cos \xi - Q(\nu, z) \sin \xi\}, \quad |\arg z| < \pi, \quad (7.30)$$

where $\xi = z - \nu\pi/2 - \pi/4$ and P and Q have known asymptotic expansions as $z \to \infty$, with $Q/P \sim \frac{\mu-1}{8z}(1 + \mathcal{O}(z^{-1}))$, $\mu = 4\nu^2$. We can compute asymptotic approximations to the large roots by setting $\tan(\xi) = P/Q$. Inverting this equation we obtain

$$z = \beta - \arctan(Q/P), \quad \beta = (s + \nu/2 - 1/4)\pi, \quad s = 1, 2, \ldots. \quad (7.31)$$

The first approximation to the zeros of $J_\nu(x)$ is $j_{\nu,s} \sim \beta$ and the second is

$$j_{\nu,s} \sim \beta - \frac{\mu - 1}{8\beta}, \quad (7.32)$$

and so on.

This process can be repeated as many times as needed. With the use of a symbolic manipulation package (like Maple), one can obtain almost as many terms as needed (limited by the capability of the computer). We can write the expansion as

$$j_{\nu,s} \sim \beta - \frac{\mu - 1}{2} \sum_{i=0}^{\infty} \frac{P_i(\mu)}{(4\beta)^{2i+1}}, \quad (7.33)$$

where $P_i(\mu)$ are polynomials of μ of degree i, which can be written in the form

$$P_i(\mu) = c_i \sum_{k=0}^{i} (-1)^{i-k} a_i^{(k)} \mu^k, \quad c_i \in \mathbb{Q}^+, \quad a_i^{(k)} \in \mathbb{N}. \quad (7.34)$$

Some of these coefficients are shown in the table below.

i	c_i	$a_i^{(0)}$	$a_i^{(1)}$	$a_i^{(2)}$	$a_i^{(3)}$	$a_i^{(4)}$	$a_i^{(5)}$
0	1	1	—	—	—	—	—
1	1/3	31	7	—	—	—	—
2	2/15	3779	982	83	—	—	—
3	1/105	6277237	1585743	153855	6949	—	—
4	2/305	2092163573	512062548	48010494	2479316	70197	—
5	2/3465	8249725736393	1982611456181	179289628602	8903961290	287149133	5592657

The same expansion applies for other solutions of the Bessel differential equation with the appropriate change in the parameter β. For instance, for the zeros of $Y_\nu(x)$ we should consider $\beta = (s + \nu/2 - 3/4)\pi$. ∎

The use of fixed point iterations (or related resubstitutions) is not limited to the case described in (7.21)–(7.24), as the next example illustrates.

Example 7.10. Compute an asymptotic approximation for the solution x of $x - \log(x) = u$ for large u.

We write $x = T(x)$ with $T(x) = u - \log(x)$. We start from $x_0 = u$ and because $|T'(x)| = x^{-1} \sim u^{-1}$, for each iteration we have an extra order in u. Iterating three times we obtain

$$x \sim u - \log(u - \log(u - \log u))$$
$$\sim u - \log(u) \left\{ -1 + \frac{1}{u} + \frac{\log(u) - 2}{2u^2} \left[1 + \mathcal{O}\left(\frac{\log u}{u}\right) \right] \right\} \quad (7.35)$$

as $u \to \infty$. ∎

7.3.2 Other approximations

Asymptotic approximations for large zeros are in many circumstances useful tools for locating almost all the zeros of many functions. For instance, we have given approximations for the equation $\kappa x \sin x - \cos x = 0$ (Example 7.1) which can be used as first approximations to the roots except for the smallest root. This difficulty in computing small zeros comes as no surprise given that the expansion (7.17) was intended for computing large zeros.

Similarly, in the literature it has been observed for Bessel functions that special strategies are convenient for computing the smallest zeros, particularly the first zero. In fact, for the first zero of the Bessel functions $J_\nu(x)$, we cannot use asymptotic expansions for large zeros as $\nu \to -1^+$; in this case we have $j_{\nu,1} \to 0^+$. Series expansions in powers of $\nu + 1$ have been obtained in [181]. Another possibility when accurate approximations are needed for the first few zeros $j_{\nu,s}$ is to compute highly accurate values by some other method with guaranteed convergence and then build a Chebyshev series expansion from these values; this approach has been considered in [180].

For Example 7.1 (which for $\kappa = -1$ gives the zeros of the Bessel function $Y_{3/2}(x)$) the first zero cannot be computed from the large zeros expansion when κ is positive. Also, when κ is small, the first zeros cannot be computed from (7.17), even when the fixed point iteration will be convergent. The problem can also be solved satisfactorily for small κ by rewriting conveniently the fixed point iteration.

Example 7.11 (Example 7.1 for the 6th time: small κ). We compute an approximation for small κ, valid for the smallest zeros of $\kappa x \sin x - \cos x$.

We write

$$x = \arctan\left(\frac{1}{\kappa x}\right) + m\pi = \left(m + \frac{1}{2}\right)\pi - \arctan(\kappa x), \quad m = 0, 1, 2, \ldots, \quad (7.36)$$

and consider the convergent fixed point iteration for $x_{n+1} = T(x_n)$, with

$$T(x) = \Lambda \pi - \arctan(\kappa x), \quad \Lambda = \left(m + \frac{1}{2}\right). \quad (7.37)$$

7.3. Local strategies: Asymptotic and other approximations

For small κ and moderate x, $|T'(x)| \sim \kappa$ and, starting from $x_0 = \Lambda\pi$, we get an extra $\mathcal{O}(\kappa)$ for each iteration. Iterating three times, we get

$$\alpha^{(m)}(\kappa) \sim \Lambda\pi \left[1 - \kappa + \kappa^2 + \left(-1 + \frac{\Lambda^2\pi^2}{3}\right)\kappa^3 + \mathcal{O}(\kappa^4) \right], \quad \kappa \to 0. \tag{7.38}$$

This expansion is useful for small values of κ and for not very large zeros, for which our previous approach does not work. ∎

Example 7.12 (Example 7.1, part 7: smallest zero for $\kappa > 0$). The expansions for large m, (7.17), and for small κ, (7.38), provide accurate enough values for starting Newton–Raphson, except when κ is large. This is a special case because the first zero goes to 0 as $\kappa \to \infty$ and none of the previous expansions works. A crude but sufficient approximation for large κ consists in expanding and solving

$$\kappa x \sin x - \cos x \simeq -1 + (\kappa + 1/2)x^2 - (\kappa/6 + 1/24)x^4 = 0. \tag{7.39}$$

We expand the smallest positive solution for x in inverse powers of κ, getting

$$\alpha^{(0)} \sim \frac{1}{\sqrt{\kappa}}\left(1 - \frac{1}{6\kappa}\right). \tag{7.40}$$

Because $\alpha^{(0)} \sim 1/\sqrt{\kappa}$, the fixed point method does not work as before as $\kappa \to +\infty$ because $|T'(x)| = \kappa/(1 + \kappa^2 x^2) \sim \kappa/(1 + \kappa) \sim 1$. However, we can still get an expansion from the fixed point method $T(x) = \arctan(1/(\kappa x))$ (the fact that $|T'| \sim 1$ but with values smaller than 1 is a good indication). Choosing

$$x_0 = \frac{1}{\sqrt{\kappa}}\left(1 + \frac{\alpha}{\kappa}\right), \tag{7.41}$$

where α will be determined, we see that the expansions for the first two iterations are

$$x_1 = \frac{1}{\sqrt{\kappa}}\left(1 - \frac{\alpha + 1/3}{\kappa} + \mathcal{O}(\kappa^{-2})\right), \quad x_2 = \frac{1}{\sqrt{\kappa}}\left(1 + \frac{\alpha}{\kappa} + \mathcal{O}(\kappa^{-2})\right), \tag{7.42}$$

and this oscillating behavior is maintained in successive iterations, in the sense that the first two terms are equal for $x_1, x_3, \ldots$, and the same happens for $x_2, x_4, \ldots$. Then choosing $\alpha = -1/6$ the second term becomes stabilized and the first two terms are the same for all $x_k, k \in \mathbb{N}$. This process can be continued. In this way, starting now from

$$x_0 = \frac{1}{\sqrt{\kappa}}\left(1 - \frac{1}{6\kappa} + \frac{\beta}{\kappa^2}\right), \tag{7.43}$$

we get that the third term is stable for $\beta = 11/360$. Continuing the process, we get

$$\alpha^{(0)} = \frac{1}{\sqrt{\kappa}}\left(1 - \frac{1}{6}\kappa^{-1} + \frac{11}{360}\kappa^{-2} - \frac{17}{5040}\kappa^{-3} \right. \\ \left. - \frac{281}{604800}\kappa^{-4} + \frac{44029}{119750400}\kappa^{-5} + \cdots\right). \tag{7.44}$$

∎

With the previous expansion for the first zero, we can now give a satisfactory solution to Example 7.1. These examples show how different strategies have to be considered depending on the zeros.

Example 7.13 (Example 7.1: 8th and final part). With the three approximations considered, sufficiently accurate expansions are available, which guarantees convergence of, for instance, the Newton–Raphson method. The following table provides a solution to the example problem described at the beginning of the chapter. The three approximations are labeled as follows: (1) first three terms of the asymptotic expansion for large zeros, (2) first three terms of the expansion for the first zeros when κ is small, and (3) first six terms of the expansion for the smallest zero for large κ. ■

κ	1	2	3	4	5	6
0.01	$2\,10^{-7}$ (2)	$6\,10^{-6}$ (2)	$2\,10^{-6}$ (2)	$4\,10^{-5}$ (2)	$6\,10^{-5}$ (2)	$1\,10^{-4}$ (2)
0.1	$4\,10^{-4}$ (2)	$4\,10^{-3}$ (2)	$1\,10^{-2}$ (2)	$5\,10^{-3}$ (1)	$2\,10^{-3}$ (1)	$6\,10^{-4}$ (1)
1	$1\,10^{-4}$ (3)	$2\,10^{-5}$ (1)	$5\,10^{-5}$ (1)	$5\,10^{-6}$ (1)	$9\,10^{-7}$ (1)	$2\,10^{-7}$ (1)
10	$1\,10^{-10}$ (3)	$2\,10^{-6}$ (1)	$3\,10^{-8}$ (1)	$3\,10^{-9}$ (1)	$5\,10^{-10}$ (1)	$1\,10^{-10}$ (1)

The situation for "not so elementary" functions can be very similar to that of Example 7.13. The most well-known case is probably that of the zeros of Bessel functions. Most available algorithms are based on several expansions depending on which zeros are computed (large or not so large) and depending on the order ν of the Bessel function.

For moderate orders ν, McMahon expansions (Example 7.9) give accurate enough values for initiating a Newton–Raphson iteration. Even for the first zeros, these expansions are able to produce acceptable values [210]. However, the use of specific approximations for the first zeros can help in improving the performance, particularly for $J_\nu(x)$ as $\nu \to -1^+$ [180, 181].

In addition, for large ν McMahon expansions fail. This can be expected given that the polynomials $P_i(\mu)$ become very large as $\mu = 4\nu^2$ increases. In this case, one should consider expansions for large z which are uniformly valid with respect to ν, also for large ν. Expansions for the zeros based on uniform Airy-type asymptotic expansions are advisable [40, 41, 168, 182, 210].

We see that for computing zeros of special functions, particularly when they depend on a parameter, usually a combination of different approximations and methods are necessary for obtaining accurate enough approximations of the zeros for the whole possible range of parameters. We observed this situation for our "Leitmotif" example, but this is also the case for other special functions (like Bessel functions).

There are, however, other possible strategies which work globally for certain (and important) cases and which may be efficient enough. In the next two sections, we describe two different global strategies; one of them is based on linear algebra and the other one on (again) fixed point iterations.

7.4 Global strategies I: Matrix methods

Many families of special functions, depending on at least one parameter, satisfy TTRRs (see Chapter 4) of the form

$$A_n(x)y_{n+1}(x) + B_n(x)y_n(x) + C_n(x)y_{n-1}(x) = 0, \quad n = 0, 1, \ldots. \tag{7.45}$$

In some cases these recurrence relations can be used as the starting point for the computation of the zeros of the solutions. As an important class for applications we consider recurrence relations of the form

$$a_n y_{n+1}(x) + b_n y_n(x) + c_n y_{n-1}(x) = g(x) y_n(x), \quad n = 0, 1, \ldots, \tag{7.46}$$

where the coefficients a_n, b_n, c_n do not depend on x and the function g does not depend on n. When $g(x) = x$ we are in the class of orthogonal polynomials. Considering the first N relations of (7.46) we can write these relations in the form

$$\mathbf{J}_N \mathbf{Y}_N(x) + a_{N-1} y_N(x) \mathbf{e}_N + c_0 y_{-1}(x) \mathbf{e}_1 = g(x) \mathbf{Y}_N(x), \tag{7.47}$$

where $\mathbf{e}_1 = (1, 0, \ldots, 0)^T$ and $\mathbf{e}_N = (0, \ldots, 0, 1)^T$ are N-vectors,

$$\mathbf{Y}_N(x) = \begin{pmatrix} y_0(x) \\ \vdots \\ y_{N-1}(x) \end{pmatrix}, \tag{7.48}$$

and $\mathbf{J}_N$ is the tridiagonal matrix (the *Jacobi matrix*),

$$\mathbf{J}_N = \begin{pmatrix} b_0 & a_0 & 0 & . & . & 0 \\ c_1 & b_1 & a_1 & 0 & . & 0 \\ 0 & c_2 & b_2 & a_2 & . & 0 \\ . & . & . & . & . & 0 \\ . & . & . & . & . & a_{N-2} \\ 0 & 0 & 0 & . & c_{N-1} & b_{N-1} \end{pmatrix}. \tag{7.49}$$

We notice that (7.47) reminds us of the problem of finding those vectors $\mathbf{Y}$ and those values λ for which the equation

$$\mathbf{J}_N \mathbf{Y}_N = \lambda \mathbf{Y}_N \tag{7.50}$$

is satisfied. Indeed, given a value $x = x_0$, (7.47) is the equation for an eigenvalue problem with eigenvalue $g(x_0)$ if both $a_{N-1} y_N(x_0) = 0$ and $c_0 y_{-1}(x_0) = 0$ for this value of x. This can be accomplished when x_0 is a zero of $y_N(x)$ (or of $y_{-1}(x)$) and, for any reason, $c_0 y_{-1}(x) = 0$ (or $a_{N-1} y_N(x) = 0$).

7.4.1 The eigenvalue problem for orthogonal polynomials

For the particular case of a family of orthogonal polynomials $P_n(x) = y_n(x)$, it is well known that they satisfy a TTRR of the form

$$xy_n(x) = a_n y_{n+1}(x) + b_n y_n(x) + c_n y_{n-1}(x), \quad n = 0, 1, \ldots, \quad (7.51)$$

with all the coefficients a_n and c_n real, of the same sign, and always different from zero ($a_n = 1$ for monic polynomials), real b_n, and $y_{-1}(x) = 0$. As a consequence of this we can prove the following result, which is essentially the eigenvalue method for computing the zeros of P_N.

Theorem 7.14. *Let $\{P_k\}_{k\in\mathbb{N}}$ be a sequence of polynomials generated from the recurrence relation (7.51), with $P_{-1} = 0$. The roots of the equation $P_N(x) = 0$ are the eigenvalues of the matrix*

$$\mathbf{S}_N = \begin{pmatrix} b_0 & \alpha_1 & 0 & . & . & 0 \\ \alpha_1 & b_1 & \alpha_2 & 0 & . & 0 \\ 0 & \alpha_2 & b_2 & \alpha_3 & . & 0 \\ . & . & . & . & . & 0 \\ . & . & . & . & . & \alpha_{N-1} \\ 0 & 0 & 0 & . & \alpha_{N-1} & b_{N-1} \end{pmatrix}, \quad (7.52)$$

where $\alpha_k = \sqrt{c_k a_{k-1}}$, $k = 1, 2, \ldots, N-1$.

Proof. This theorem follows from (7.47), denoting $y_k = P_k$. Because $P_{-1} = 0$ we see that

$$P_N(x_0) = 0 \iff \mathbf{J}_N \mathbf{Y}_N(x_0) = x_0 \mathbf{Y}_N(x_0). \quad (7.53)$$

This means that the N eigenvalues of the $N \times N$ Jacobi matrix in (7.49) are the N zeros of the orthogonal polynomial P_N of degree N. The eigenvalue problem can be written in symmetric form by applying a similarity transformation. To see this, we define a new family of functions $\tilde{y}_n(x)$ given by $y_n(x) = \lambda_n \tilde{y}_n(x)$, with $\lambda_n > 0$; this family satisfies a TTRR

$$x\tilde{y}_n(x) = \tilde{a}_n \tilde{y}_{n+1}(x) + b_n \tilde{y}_n(x) + \tilde{c}_n \tilde{y}_{n-1}(x), \quad (7.54)$$

with

$$\tilde{a}_n = \mu_n a_n, \quad \tilde{c}_n = c_n / \mu_{n-1}, \quad \mu_k = \lambda_{k+1}/\lambda_k. \quad (7.55)$$

The Jacobi matrix for the new set of functions $\{\tilde{y}_k(x)\}$ thus becomes symmetric by taking $\tilde{a}_k = \tilde{c}_{k+1}, k = 0, 1, \ldots, N - 1$; this can be accomplished by taking

$$\mu_k = \sqrt{\frac{c_{k+1}}{a_k}} \quad (7.56)$$

and then

$$\tilde{y}_n(x_0) = \tilde{P}_N(x_0) = 0 \iff \mathbf{S}_N \tilde{\mathbf{Y}}_N(x_0) = x_0 \tilde{\mathbf{Y}}_N(x_0). \quad (7.57)$$

7.4. Global strategies I: Matrix methods

Of course, the eigenvalues of the new (symmetric and tridiagonal) Jacobi matrix $\mathbf{S}_N$ are identical to those of the original matrix $\mathbf{J}_N$ (the zeros of $y_n(x)$ and $\tilde{y}_n(x)$ are the same). □

The symmetric Jacobi matrices for the classical orthogonal polynomials are given explicitly in Chapter 5, §5.3.2.

The fact that the matrices $\mathbf{J}_N$ in (7.49) and $\mathbf{S}_N$ are similar can also be seen as a consequence of the following theorem.

Theorem 7.15. *Given two $N \times N$ square tridiagonal matrices $\mathbf{A}$ and $\mathbf{B}$, if they have the same diagonal elements ($a_{ii} = b_{ii}$), if all the off-diagonal elements are different from zero, and if they satisfy $a_{ij}a_{ji} = b_{ij}b_{ji}$, then $\mathbf{A}$ and $\mathbf{B}$ are similar.*

Notice that Theorem 7.14 (and also Theorem 7.15) shows that all the eigenvalues, and hence the zeros of $P_N(x)$, are real when all the a_k and c_k have the same sign; this is so because the matrix $\mathbf{S}_N$ is real and symmetric in that case. The diagonalization of real symmetric tridiagonal matrices is one of the simplest and best understood problems of numerical linear algebra. Depending on how many zeros one needs, different strategies can be convenient. If one does not need to compute all the zeros and/or a priori approximations are to be used, Sturm bisection may be a simple and effective procedure. Otherwise, when all the zeros are needed, a diagonalization of the Jacobi matrix through the QR algorithm seems to be preferable.

We will not enter into a detailed description of the numerical methods for diagonalizing tridiagonal matrices. Instead, we consider this as a "black box" procedure to be filled with our favorite diagonalization routines. Two related and reliable possibilities are MATLAB built-in commands and LAPACK [132] (Linear Algebra Package) in Fortran. This method is simple to implement, particularly if we are not diagonalizing the matrix by our own means, and it is efficient. In addition, one can also obtain the complex zeros, providing our routine for diagonalization does the work accurately enough, which is the usual situation with MATLAB or LAPACK. There are, however, some situations where numerical instabilities occur, as happens in the case of densely packed eigenvalues and very small eigenvalues compared to the matrix norm. Also, some Jacobi matrices may be ill conditioned, as is the case of computing the complex zeros of some Bessel polynomials [172].

7.4.2 The eigenvalue problem for minimal solutions of TTRRs

In some circumstances, as first shown in [99], one can compute zeros of functions with infinitely many zeros, like Bessel functions, by diagonalizing a finite matrix which approximates a compact operator (an "infinite matrix" with "small enough" elements). The main idea goes as follows.

We consider a family of functions satisfying a TTRR, which we can write in matrix form (see (7.47)). Differently from orthogonal polynomials, $c_0 y_{-1}(x)$ will not vanish identically; instead, we consider that $y_{-1}(x)$ is the function whose zeros are sought. Then, we have for all $N > 0$

$$y_{-1}(x_0) = 0 \Rightarrow \mathbf{J}_N \mathbf{Y}(x_0) = g(x_0)\mathbf{Y}(x_0) + a_{N-1} y_N(x_0)\mathbf{e}_N. \qquad (7.58)$$

We can always write the equation in symmetric tridiagonal form assuming that none of the tridiagonal elements vanishes, similarly to how it was done in the case of orthogonal polynomials. By assuming that, for large enough N, $a_{N-1} y_N(x_0)$ is sufficiently small (in a sense to be made explicit later) and that it can be dropped, we could write

$$y_{-1}(x) = 0 \Rightarrow \mathbf{J}_N \mathbf{Y}(x_0) \approx g(x_0) \mathbf{Y}(x_0), \qquad (7.59)$$

and assuming that $g(x)$ is an invertible function we can obtain the value of the zero x_0 from $g(x_0)$, which should be an approximate eigenvalue of $\mathbf{J}_N$.

Example 7.16 (the zeros of the Bessel function $J_\nu(x)$). As an example, let us consider the case of Bessel functions $J_\nu(x)$, which are the minimal solution of the TTRR

$$\frac{1}{\nu}(J_{\nu+1}(x) + J_{\nu-1}(x)) = \frac{2}{x} J_\nu(x). \qquad (7.60)$$

Let us take $y_{-1}(x) = J_\nu(x)$ and therefore $y_n = J_{\nu+n+1}(x)$. Then if $J_\nu(x_0) = 0$, we see that (7.59) is satisfied with $\mathbf{J}_N$ a real $N \times N$ tridiagonal matrix which can be easily read from the TTRR. We can write the equivalent tridiagonal system by performing the same operation as done in Theorem 7.14; this is equivalent to writing the TTRR for $y_m(x) = \sqrt{m} J_m(x)$,

$$\alpha_{m+1} y_{m+1} + \alpha_m y_{m-1} = \frac{2}{x} y_m, \qquad (7.61)$$

where

$$\alpha_k = \frac{1}{\sqrt{k(k-1)}}. \qquad (7.62)$$

Then $J_\nu(x_0) = 0$ implies that

$$\mathbf{S}_N \mathbf{Y}_N(x_0) = \lambda \mathbf{Y}_N(x_0) + \alpha_{\nu+N+1} y_{\nu+N+1}(x_0) \mathbf{e}_N, \qquad (7.63)$$

where

$$\lambda = g(x_0), \quad g(x) = 2/x, \qquad (7.64)$$

and

$$\mathbf{S}_N = \begin{pmatrix} 0 & \alpha_{\nu+2} & 0 & . & . & . & 0 \\ \alpha_{\nu+2} & 0 & \alpha_{\nu+3} & 0 & . & . & 0 \\ 0 & \alpha_{\nu+3} & 0 & \alpha_{\nu+4} & . & . & 0 \\ . & . & . & . & . & . & 0 \\ . & . & . & . & . & . & \alpha_{\nu+N} \\ 0 & 0 & 0 & . & . & \alpha_{\nu+N} & 0 \end{pmatrix}, \qquad (7.65)$$

7.4. Global strategies I: Matrix methods

$$\mathbf{Y}_N(x_0) = \begin{pmatrix} \sqrt{\nu+1}J_{\nu+1}(x_0) \\ \sqrt{\nu+2}J_{\nu+2}(x_0) \\ \cdot \\ \cdot \\ \cdot \\ \sqrt{\nu+N}J_{\nu+N}(x_0) \end{pmatrix}, \quad (7.66)$$

and

$$\alpha_{\nu+N+1}y_{\nu+N+1}(x_0) = \frac{1}{\sqrt{\nu+N}} J_{\nu+N+1}(x_0). \quad (7.67)$$

From [2, eq. (9.3.1)] we obtain

$$J_\nu(x) \sim \frac{1}{\sqrt{2\pi\nu}} \left(\frac{ez}{2\nu}\right)^\nu, \quad \nu \to +\infty, \quad (7.68)$$

and it follows that $\alpha_n J_n(x) \to 0$. With this, the limit $N \to \infty$ can be considered to yield an eigenvalue problem for an infinite matrix $\mathbf{S}_\infty$ (known as a *compact operator in a Hilbert space*),

$$\mathbf{S}_\infty \mathbf{Y}_\infty(x_0) = \lambda \mathbf{Y}_\infty(x_0), \quad \lambda = g(x_0). \quad (7.69)$$

This makes sense because, considering (7.68), we see that

$$||\mathbf{Y}_\infty(x_0)|| = \left(\sum_{n=1}^\infty (\sqrt{\nu+n}J_{\nu+n}(x_0))^2\right)^{1/2} < \infty, \quad (7.70)$$

and it is also easy to see that

$$||\mathbf{S}_\infty|| = \left(\sum_{i=1}^\infty \sum_{j=1}^\infty (\mathbf{S}_\infty)_{ij}\right)^{1/2} < \infty \quad \text{for } \nu > -1. \quad (7.71)$$

This can be used to bound the eigenvalues of $\mathbf{S}_\infty$ because $|\lambda \mathbf{Y}_\infty| = ||\mathbf{S}_\infty \mathbf{Y}_\infty|| \leq ||\mathbf{S}_\infty||||\mathbf{Y}_\infty||$ and then $|\lambda| \leq ||\mathbf{S}_\infty||$. Equation (7.69) implies that, by considering matrices $\mathbf{S}_N$, we expect that for large enough N the eigenvalues can be used to compute zeros of the Bessel function $J_\nu(x)$. The largest eigenvalue, $\lambda_N^{(1)}$, will give the approximation $j_{\nu,1} = g^{-1}(\lambda_N^{(1)})$ for the smallest positive zero of $J_\nu(x)$. ∎

General formulation for compact infinite matrices

A more rigorous formulation of the method outlined earlier for Bessel functions requires that we admit compact matrices of infinite size to come into play. We assume now that taking $N \to \infty$ makes sense and $\mathbf{J}_N \to \mathbf{J}$ with $\mathbf{J}$ being an infinite matrix (in fact, an operator in a Hilbert space), and we let $\mathbf{Y}_N \to \mathbf{Y}$. For this to make sense the y_N must tend to zero rapidly enough as $N \to \infty$. In this case, (7.58) yields

$$\mathbf{J}\mathbf{Y}(\mathbf{x_0}) = g(x_0)\mathbf{Y}(\mathbf{x_0}). \quad (7.72)$$

Under adequate circumstances (Theorem 7.17), (7.72) represents an eigenvalue problem for an infinite matrix (operator) with eigenvalues $g(x_0)$, x_0 being the zero of $y_{-1}(x)$. Therefore, with the additional assumption that $g(x)$ is an invertible function, this can be used to compute the zeros of $y_{-1}(x)$. Of course, because diagonalizing an infinite matrix is, except in a few cases, out of human reach, an approximation method is necessary for computing approximate eigenvalues of the infinite matrix $\mathbf{J}$. This is accomplished by diagonalizing "large enough" matrices to approximate the problem with enough accuracy. We have the following theorem (adapted from [108, 153]; see also [99]).

Theorem 7.17. *Let $\{y_k(x)\}$, $k = -1, 0, 1, \ldots$, be a solution of the three-term recurrence relation*

$$\alpha_{n+1} y_{n+1}(x) + \beta_n y_n(x) + \alpha_n y_{n-1}(x) = g(x) y_n(x), \quad n = 0, 1, \ldots, \quad (7.73)$$

with $\alpha_i \neq 0$, $\sum_i^\infty \alpha_i^2 + \sum_i^\infty \beta_i^2 < \infty$, and $\sum_k y_k(x)^2 < \infty$. Let $\mathbf{J}$ be the infinite matrix

$$\mathbf{J} = \begin{pmatrix} \beta_0 & \alpha_1 & & & \mathbf{0} \\ \alpha_1 & \beta_1 & \alpha_2 & & \\ & \alpha_2 & \beta_2 & \ddots & \\ \mathbf{0} & & & \ddots & \ddots \end{pmatrix}, \quad (7.74)$$

and let $\{\mathbf{A}_n\}$ denote the sequence of matrices $\mathbf{A}_n$ formed by the first n rows and columns of $\mathbf{J}$. Then, if x_0 is a zero of $y_{-1}(x) = 0$ and $\mathbf{Y} = (y_0(x_0), y_1(x_0), \ldots)^T \neq \mathbf{0}$, we have the following:

1. *$x_0 = g^{-1}(\lambda)$, where λ is the eigenvalue of $\mathbf{J}$ corresponding to the eigenvector $\mathbf{Y}$, that is,*

$$\mathbf{J}\mathbf{Y} = \lambda \mathbf{Y}. \quad (7.75)$$

2. *There exists a sequence of eigenvalues $\lambda^{(n)}$ of the successive matrices $\mathbf{A}_n$ which converges to λ.*

3. *If y_n/y_{n-1} is bounded for all sufficiently large n, we have the following error estimate:*

$$\lambda - \lambda_n = \alpha_n \frac{y_n y_{n-1}}{\mathbf{Y}^T \mathbf{Y}} (1 + o(1)) \quad \text{as } n \to \infty. \quad (7.76)$$

Theorem 7.17 has been applied to Bessel functions $J_\nu(x)$ and to regular Coulomb wave functions $F_L(\eta, \rho)$ (see [107]). These are both minimal solutions of their corresponding TTRRs. This, as could be expected, is not a coincidence and Theorem 7.17 is valid only for minimal solutions because of the following theorem.

Theorem 7.18. *Let $\{y_n^{(1)}\}$, $n \in \mathbb{N}$, be a solution of the TTRR*

$$\alpha_{n+1} y_{n+1} + \beta_n y_n + \alpha_n y_{n-1} = 0, \quad \alpha_i \neq 0, \quad \sum_i \alpha_i^2 < \infty, \quad (7.77)$$

7.4. Global strategies I: Matrix methods

which is an element of the Hilbert space ℓ^2 (i.e., it is such that $\sum_k (y_k^{(1)})^2 < \infty$). Then any other solution of (7.77), linearly independent of $\{y_n^{(1)}\}$, is unbounded for large n. Therefore, $\{y_n^{(1)}\}$ is a minimal solution of the recurrence relation (7.77).

Theorem 7.17 can be considered for real or complex zeros. In the case when all α_i, β_i are real, **J** is symmetric and real and therefore all the eigenvalues will be real. This is the case with the aforementioned examples. Next, we consider the case of conical functions as an illustration.

Example 7.19 (zeros of conical functions). The conical function $P_{-1/2+i\tau}^m(x)$ is the minimal solution of a TTRR (see (4.72)), which can be written as

$$-\frac{1}{2(m+1)} P_\beta^{m+2}(x) - \left(\frac{\tau^2 + (1/2+m)^2}{2(m+1)} \right) P_\beta^m(x) = \frac{x}{\sqrt{x^2-1}} P_\beta^{m+1}(x), \quad (7.78)$$

where $\beta = -1/2 + i\tau$. Using this recurrence relation, the coefficients of the symmetric matrix of the equivalent tridiagonal system in (7.59) are given by

$$\alpha_m = \frac{1}{2} \sqrt{\frac{\tau^2 + (m+1/2)^2}{m(m+1)}}. \quad (7.79)$$

In other words, $P_\beta^m(x_0) = 0$ implies that

$$\mathbf{S}_N \mathbf{Y}_N(x_0) \sim \lambda \mathbf{Y}_N(x_0) \quad \text{for large enough } N, \quad (7.80)$$

where

$$\lambda = g(x_0), \quad g(x) = \frac{x}{\sqrt{x^2-1}}, \quad (7.81)$$

and

$$\mathbf{S}_N = \begin{pmatrix} 0 & \alpha_{m+1} & 0 & . & . & . & 0 \\ \alpha_{m+1} & 0 & \alpha_{m+2} & 0 & . & . & 0 \\ 0 & \alpha_{m+2} & 0 & \alpha_{m+3} & . & . & 0 \\ . & . & . & . & . & . & 0 \\ . & . & . & . & . & . & \alpha_{m+N} \\ 0 & 0 & 0 & . & . & \alpha_{m+N} & 0 \end{pmatrix}, \quad (7.82)$$

$$\mathbf{Y}_N(x_0) = \begin{pmatrix} P_\beta^{m+1}(x_0) \\ P_\beta^{m+2}(x_0) \\ . \\ . \\ . \\ P_\beta^{m+N}(x_0) \end{pmatrix}. \quad (7.83)$$

Table 7.1. *First three zeros found of $P^{10}_{-1/2+i\tau}(x)$ using the matrix method for $\tau = 2, 10$.*

τ	First 3 zeros of $P^{10}_{-1/2+i\tau}(x)$	Zeros of $P^{10}_{-1/2+i\tau}(x)$ (matrix method)
2	$x_1 = 22.48899483$	$x_1^M = 22.48899483$
	$x_2 = 109.1581737$	$x_2^M = 109.6594128$
	$x_3 = 525.3027307$	$x_3^M = 319.1320827$
10	$x_1 = 1.942777791$	$x_1^M = 1.942777791$
	$x_2 = 2.684075396$	$x_2^M = 2.684075396$
	$x_3 = 3.681526452$	$x_3^M = 3.681526452$

The matrix problem is ill conditioned for small values of τ. In Table 7.1 we show for a fixed m ($m = 10$) and two values of τ ($\tau = 2, 10$) the first three zeros found with the matrix method using a matrix size of $N = 500$. ∎

As happened for the zeros of Bessel functions (Example 7.16), also in Example 7.19 the eigenvalues appear as symmetrical pairs of positive and negative values. This is not the best situation for computational purposes, because spurious negative eigenvalues are computed. In the next section, we explain how to repair this situation for the case of Bessel functions.

Example 7.20 (the zeros of $J_\nu(x)$ revisited). Let us return to the example of the zeros of the Bessel functions $J_\nu(x)$. For computational purposes, as discussed in [9], instead of starting with the TTRR (7.60) it is better to iterate this relation once and obtain

$$\frac{1}{x^2} J_\nu(x) = \frac{1}{2\nu} \left[\frac{1}{2(\nu+1)} J_{\nu+2}(x) \right. \\ \left. + \left(\frac{1}{2(\nu+1)} + \frac{1}{2(\nu-1)} \right) J_\nu(x) + \frac{1}{2(\nu-1)} J_{\nu-2}(x) \right]. \tag{7.84}$$

This is a TTRR for $J_\nu(x)$ with spacing 2 in the order. In this way, the functions $y_n(x)$, which are defined by

$$y_n(x) = \sqrt{2n+2+\nu} \, J_{2n+2+\nu}(x), \tag{7.85}$$

satisfy the TTRR given in Theorem 7.17 with $g(x) = 1/x^2$,

$$\alpha_n = \frac{1}{4(2n+3+\nu)\sqrt{(2n+2+\nu)(2n+4+\nu)}} \tag{7.86}$$

and

$$\beta_n = \frac{1}{2} \left(\frac{1}{(2n+3+\nu)(2n+1+\nu)} \right). \tag{7.87}$$

7.5. Global strategies II: Global fixed point methods

With the restriction $\nu > -2$, the corresponding tridiagonal matrix is real. Then, the eigenvalues are real and its eigenvectors are orthogonal. In the limit $N \to \infty$ with x fixed, $\alpha_N y_{N+1} \to 0$ and the eigenvalues of the matrix converge to $1/j_{\nu,i}^2$, $j_{\nu,i}$ being the ith root of $J_\nu(x)$. Then, in this case, the eigenvalues are all positive. This is the reason why the TTRR of (7.84) is preferable to the first TTRR (7.60): by using the original one, the eigenvalues appear as symmetrical positive and negative roots. Therefore, for the same number of zeros, we will need a matrix twice as large as the one required for the second recurrence. ∎

7.5 Global strategies II: Global fixed point methods

Matrix eigenvalue methods have been successful in providing a direct method of computation of zeros for some special functions without the need for computing function values. Only the coefficients of the TTRR were needed. The fixed point methods to be described next share the same property and can also be applied to other solutions of the differential equation. The starting point is, in some sense, something in the middle between a TTRR and the second order ordinary differential equation: a first order difference-differential system.

7.5.1 Zeros of Bessel functions

We will take the example of Riccati–Bessel functions $c_\nu(x)$ related to Bessel functions

$$\mathcal{C}_\nu(x) = \cos\alpha J_\nu(x) - \sin\alpha Y_\nu(x), \tag{7.88}$$

by taking

$$c_\nu(x) = \sqrt{x}\mathcal{C}_\nu(x). \tag{7.89}$$

The functions $c_\nu(x)$ are solutions of the differential equation

$$y_\nu''(x) + A_\nu(x)y_\nu(x) = 0, \quad A_\nu(x) = 1 - \frac{\nu^2 - 1/4}{x^2}, \tag{7.90}$$

and have an infinite number of real positive zeros. Riccati–Bessel functions satisfy the first order difference-differential system,

$$\begin{aligned} c_\nu'(x) &= -\eta_\nu(x)c_\nu(x) + c_{\nu-1}(x), \\ c_{\nu-1}'(x) &= \eta_\nu(x)c_{\nu-1}(x) - c_\nu(x), \end{aligned} \tag{7.91}$$

where

$$\eta_\nu(x) = \frac{\nu - 1/2}{x}. \tag{7.92}$$

We will use this system to build a fixed point method which is convergent regardless of the initial guesses. The method will be, as the Newton–Raphson method is, quadratically convergent, however, with the essential advantage of converging with certainty. Later we will show that by applying some simple transformations, we can write similar systems for a broad set of special functions including, but not limited to, all the special functions discussed in the present chapter. We define the ratio $H_\nu(x) = c_\nu(x)/c_{\nu-1}(x)$, which satisfies

$$H_\nu'(x) = 1 + H_\nu(x)^2 - 2\eta_\nu(x)H_\nu(x). \tag{7.93}$$

The function $H_\nu(x)$ has the same zeros as $c_\nu(x)$, because the zeros of $c_\nu(x)$ and $c_{\nu-1}(x)$ cannot coincide (in fact, (7.91) can be used to prove that they are interlaced). With these few ingredients, one can prove that the fixed point iteration $x_{n+1} = T(x_n)$, with

$$T(x) = x - \arctan(H_\nu(x)), \tag{7.94}$$

converges, as described in the following theorem.

Theorem 7.21. *Given any x_0 between two consecutive zeros of $c_{\nu-1}(x)$, the sequence $x_{n+1} = T(x_n)$, $n = 0, 1, \ldots$, with $T(x)$ as in (7.94) as iterating function, converges to the zero of $c_\nu(x)$, c_ν, between two such zeros, that is,*

$$\lim_{n \to \infty} x_n = c_\nu. \tag{7.95}$$

The convergence is quadratic with asymptotic error constant $\eta_\nu(c_\nu)$.

Proof. For proving convergence, we will consider the case $\nu > 1/2$, for which $\eta_\nu(x) > 0$; the case $\nu < 1/2$ is analogous and the case $\nu = 1/2$ is trivial ($H_{1/2}(x) = \tan(x + \phi)$).

Due to the interlacing of the zeros of $c_\nu(x)$ and $c_{\nu-1}(x)$, $H_\nu(x)$ is a function with zeros and singularities interlaced. On account of (7.93), $H'(c_\nu) = 1$; therefore the zeros are simple and $H_\nu(x)$ is positive to the right of the zero and negative to the left. In addition we have

$$T'(x) = 2\eta_\nu(x) \frac{H}{1 + H^2}, \tag{7.96}$$

and $T'(x)$ has the same sign as $H_\nu(x)$.

Let us denote by $c_{\nu-1}^{\text{left}}$ and $c_{\nu-1}^{\text{right}}$ the zeros of $c_{\nu-1}(x)$ (singularities of $H_\nu(x)$) close to c_ν and such that $c_{\nu-1}^{\text{left}} < c_\nu < c_{\nu-1}^{\text{right}}$ (see Figure 7.3).

Consider first x_0 in $(c_\nu, c_{\nu-1}^{\text{right}})$. Because $H_\nu(x_0) > 0$ then

$$x_0 - x_1 = \arctan(H_\nu(x_0)) > 0; \tag{7.97}$$

also, by the mean value theorem $x_1 - \alpha = T(x_0) - T(\alpha) = T'(\zeta)(x_0 - \alpha)$, where ζ is in (α, x_0), and then, because $T'(\zeta) > 0$, $x_1 - \alpha > 0$. Therefore $\alpha < x_1 < x_0$, and by induction $\alpha < x_n < \cdots < x_1 < x_0$ for any n. In other words, the sequence converges monotonically when $x_0 \in (c_\nu, c_{\nu-1}^{\text{right}})$.

We now consider that x_0 lies in the interval $(c_{\nu-1}^{\text{left}}, c_\nu)$. Let us prove that x_1 is in $(c_\nu, c_{\nu-1}^{\text{right}})$, which shows that the sequence $\{x_1, x_2, \ldots\}$ converges monotonically to c_ν. We prove this fact in three steps.

1. $x_1 > c_\nu$ because $T'(x) < 0$ between x_0 and c_ν and on account of the mean value theorem $x_1 > \alpha$.

2. $x_1 < c_\nu + \pi/2$ because $\arctan(H_\nu(x_0)) > -\pi/2$.

3. $c_{\nu-1}^{\text{right}} - c_\nu > \pi/2$ because $H_\nu(x) > 0$ in $(c_\nu, c_{\nu-1}^{\text{right}})$ and then

$$H'_\nu(x) = 1 + H_\nu(x)^2 - 2\eta_\nu H_\nu < 1 + H_\nu^2. \tag{7.98}$$

Because $H(c_\nu) = 0$ and $H'(c_\nu) = 1$, and considering the previous inequality, it is clear that the graph of $H_\nu(x)$ lies below the graph of $y(x) = \tan(x - c_\nu)$ at the right of c_ν (see Figure 7.3). Therefore, $c_{\nu-1}^{\text{right}} > \pi/2 + c_\nu$.

7.5. Global strategies II: Global fixed point methods   215

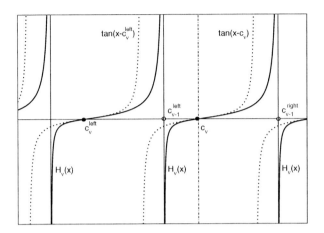

Figure 7.3. *The function H_ν, solution of the nonlinear ordinary differential equation $H'_\nu(x) = 1 + H_\nu(x)^2 - 2\eta_\nu(x)H_\nu(x)$, $\eta > 0$, is compared against the principal branches of certain* tan *functions.*

All that remains to be proved is the order of convergence. From (7.96) we see that $T'(c_\nu) = 0$, and taking another derivative, we obtain $T''(c_\nu) = 2\eta_\nu(c_\nu)$. Therefore, the asymptotic error constant (see (7.9)) is $\eta_\nu(c_\nu)$. □

Because the zeros of $c_\nu(x)$ and $c_{\nu-1}(x)$ are interlaced, the previous theorem guarantees convergence to a zero of $c_\nu(x)$, provided the starting value x_0 is larger than the smallest positive zero of $c_{\nu-1}(x)$. If x_0 is smaller than such zero, we could find convergence to 0 or that the method gives negative values, and it is convenient, but not necessary, to consider some rough estimation of the first zero to avoid this.

The previous theorem provides also a clue regarding how to compute all the zeros in a certain interval. We give a result for $\nu > 1/2$ (and thus $\eta > 0$), but a similar result can be given for $\nu < 1/2$.

Theorem 7.22 (backward sweep). *Let $\eta_\nu(x) > 0$ ($\nu > 1/2$). If c_ν is a zero of $c_\nu(x)$ but not the smallest one, then the sequence $x_{n+1} = T(x_n)$ with $x_0 = c_\nu - \pi/2$ converges to the largest zero of $c_\nu(x)$ smaller than c_ν, c_ν^{left}.*

Proof. Because $H'_\nu = 1 + H_\nu^2 - 2\eta_\nu H_\nu$, $\eta_\nu > 0$, comparing the graph of H_ν with that of the function $\tan(x - c_\nu)$ (Figure 7.3), we see that $0 < c_\nu - c_{\nu-1}^{left} < \pi/2$. And similarly to the proof of Theorem 7.21 (see also Figure 7.3) $c_{\nu-1}^{left} - c_\nu^{left} > \pi/2$. Therefore $x_0 = c_\nu - \pi/2 \in (c_\nu^{left}, c_{\nu-1}^{left})$.

But, according to Theorem 7.21, convergence to c_ν^{left} is certain (and monotonic). □

This gives a simple recipe to compute all the zeros of $c_\nu(x)$ inside an interval $[a, b]$, $a > 0$, based on the ratio $H_\nu(x) = c_\nu(x)/c_{\nu-1}(x)$ because, once a zero is computed, the next one is obtained by subtracting $\pi/2$ and applying the fixed point method again.

ALGORITHM 7.1. Zeros of cylinder functions of order $v > 1/2$.
Input: $[a, b]$, ϵ.

Output: Zeros of $\mathcal{C}_v(x)$ in $[a, b]$: $c_n < c_{n-1} < \cdots < c_1$.

- $i = 1; F = 1; \Delta = \pi/2$
- IF $H_v(b) < 0$ THEN $x = b - \pi/2$ ELSE $x = b$
- DO WHILE $x \geq a$:

  $E = 1 + \epsilon$

  DO WHILE $(F = 1)$ and $(E > \epsilon)$

      $x_p = x; x = x - \arctan(H_v(x)); E = |1 - x/x_p|;$

      IF $x < a$ THEN $F = 0$

  IF $(F = 1)$ THEN $c_i = x; i = i + 1; x = x - \Delta$.

This algorithm implements Theorems 7.21 and 7.22. It is important to take into account that $H_v(x)$ is increasing together with the monotonic convergence from starting values such that $H_v(x_0) > 0$. This allows us to compute all the zeros by monotonically convergent sequences.

Remark 10. The distance between consecutive zeros of $\mathcal{C}_v(x)$ is larger than π when $v > 1/2$. This is a consequence of the fact that the functions are solutions of (7.90), which is an equation of the form $y''(x) + A_v(x)y(x) = 0$ with $0 < A_v(x) < 1$. The solutions of the ordinary differential equation have, by Sturm's comparison theorem, zeros separated by distances larger than π (see [219, 230]).

Because of this, the step $\Delta = \pi/2$ can be replaced by the better step $\Delta = \pi$, which also guarantees monotonic convergence.

Let us consider a numerical illustration of the method.

Example 7.23. Computation of the zeros of $\mathcal{C}_v(x) = \cos(\pi/3)J_v(x) - \sin(\pi/3)Y_v$, $v = 7$, in the interval $[10, 27]$. We take the ratio $H_v(x) = \mathcal{C}_v(x)/\mathcal{C}_{v-1}(x)$. The following computations are carried out when applying the previous algorithm for computing the zeros with 10^{-15} relative accuracy.

1. Because $H_7(27) = -0.11$, the initial choice $x = 27$ would give convergence to a zero of $\mathcal{C}_v(x)$ greater than $b = 27$. Then, we take $x = b - \pi/2$, which, by the comparison arguments considered in Theorem 7.21, is larger than the largest zero in $[a, b]$ and $H_v(x) > 0$, which guarantees monotonic convergence; indeed $H_{27}(x) = 3.32$.

2. Iterating, we observe that the fifth and the sixth iterations of $x_{n+1} = T(x_n)$ differ by a relative error less than 10^{-15}. Therefore, six iterations provide the largest zero, $c_1 = 23.84210124377931$.

3. Take $x = c_1 - \pi$. $H_7(x) = 20.7 > 0$ and we have monotonic convergence. After the fifth iteration we have an accurate enough value $c_2 = 20.5319724058051$.

4. Proceeding as above, we obtain the zeros $c_3 = 17.1487388622114$ and $c_4 = 13.6210725334372$.

5. The next step is $x = c_4 - \pi$, where $H_\nu(x) = 0.57 > 0$, and the method will converge monotonically if there exists a zero smaller than x. But the first iteration gives $x = 9.95 < 10$. Then, the flag F is set to the value 0 and the program terminates, giving the four zeros inside the interval $[10, 27]$ (in decreasing order). ∎

When $\nu < 1/2$, the function $\eta_\nu(x)$ is negative. In this case, a similar algorithm can be considered, but the zeros would be generated in increasing order. In addition, because when $|\nu| < 1/2$ the functions are solutions of $y''(x) + A_\nu(x)y(x) = 0$ with $A_\nu(x) > 1$, the distance between zeros of $C_\nu(x)$ is smaller than π for these ν-values, and we cannot consider the improvement of Remark 10. Therefore, we have the following algorithm.

ALGORITHM 7.2. Zeros of cylinder functions of order $\nu < 1/2$.
Input: $[a, b]$, ϵ.

Output: Zeros of $C_\nu(x)$ in $[a, b]$: $c_1 < c_2 < \cdots < c_n$.

- $i = 1$; $F = 1$; $\Delta = \pi/2$

- IF $H_\nu(a) > 0$ THEN $x = a + \pi/2$ ELSE $x = a$

- DO WHILE $x \leq b$:

    $E = 1 + \epsilon$
    DO WHILE $(F = 1)$ and $(E > \epsilon)$
    $\quad x_p = x$; $x = x - \arctan(H_\nu(z))$; $E = |1 - x/x_p|$;
    $\quad$ IF $x > b$ THEN $F = 0$
    IF $(F = 1)$ THEN $c_i = z$; $i = i + 1$; $x = x + \Delta$;

Improving the step Δ

The fixed point method converges quite rapidly because it is quadratically convergent, as is the Newton–Raphson method, with the crucial advantage that the fixed point method converges globally and with certainty. But it is important for a fast convergence to have initial approximations not too far from the values of the zeros.

Recalling the defining differential equation (7.90), $y''_\nu(x) + A_\nu(x)y_\nu(x) = 0$, we observe that $A'_\nu(x) > 0$ when $\nu > 1/2$. Using Sturm's comparison theorem, this means that, because $A_\nu(x)$ increases, the solutions oscillate more rapidly as x increases. In other words, the distance between consecutive zeros tends to decrease as x increases. Therefore, after computing the first two zeros (in decreasing order) in Algorithm 7.1 ($c_1 > c_2$), the initial estimate $x = c_2 - \Delta > c_3$, $\Delta = c_1 - c_2 > \pi$ can be used for computing c_3; the same can be considered for the successive zeros. This is an improvement which leads to a significant reduction in the number of required steps.

On the other hand, because for $v < 1/2$ the function $A_v(x)$ is decreasing, the distance between consecutive zeros increases as larger zeros are considered. Therefore, after the first two zeros are obtained ($c_1 < c_2$), we can give a starting value for the next zero $x = c_2 + \Delta < c_3$, with $\Delta = c_2 - c_1 > \pi/2$, and the same can be considered for the successive values.

Remark 11. For computing the zeros of $c_v(x)$ one can also consider the system

$$c_v'(x) = -\eta_v(x)c_v(x) - c_{v+1}(x),$$
$$c_{v+1}'(x) = \eta_v(x)c_{v+1}(x) + c_v(x),$$
(7.99)

and the related function $H_v(x) = -c_v(x)/c_{v-1}(x)$, which satisfies (7.93) with $\eta_v(x) = -(v+1/2)/x$. However, for the case $v > 1/2$, because $\eta_v(x) < 0$ the computation of zeros should be performed in increasing order, and the step cannot be improved.

Computation of the ratios $H_v(x)$

The fixed point method for computing zeros of cylinder functions can be applied for computing the zeros of general Bessel functions

$$\mathcal{C}_v(x) = \cos \alpha J_v(x) - \sin \alpha Y_v(x)$$
(7.100)

or the Riccati–Bessel functions $c_v(x) = \sqrt{x}\mathcal{C}_v(x)$.

All that is needed is a method for computing the ratios $H_v(x) = \mathcal{C}_v(x)/\mathcal{C}_{v-1}(x)$, that is, the ratio of two consecutive values of a solution of the TTRR (7.60). As discussed in Chapter 4, depending on the type of solution (minimal or not), different strategies need to be considered. For the particular case of the minimal solution, $J_v(x)$, there is a continued fraction representation for this ratio in terms of the recurrence coefficients, as is the case with any minimal solution of a TTRR (Theorem 4.7). Namely, we have

$$\frac{J_v(x)}{J_{v-1}(x)} = \cfrac{1}{\frac{2v}{x} - \cfrac{1}{\frac{2(v+1)}{x} - \cdots}},$$
(7.101)

which can be computed with the techniques explained in Chapter 6, particularly with the modified Lentz algorithm (Algorithm 6.2).

For computing zeros of the minimal solutions, the fixed point method can be based solely on the coefficients of the TTRR, same as for the matrix methods. The advantage of the present method is that for other nonminimal solutions the method can still be used (provided a numerical method for computing the ratio $H_v(x)$ is available).

7.5.2 The general case

The methods described for Bessel functions can be applied to more general functions. For Bessel functions, as we already saw, there exist asymptotic approximations which can be used to provide accurate enough approximations for starting high order iterative methods (for instance, see [210]). When accurate asymptotic approximations are available, it is always a wise idea to use them. However, for more general cases, asymptotic approximations which guarantee convergence for all parameter values could be hard to find. Fortunately, the fixed point methods are general enough to solve the problem for a large set of special functions. On the other hand, even when asymptotic approximations are available, fixed point methods (and also matrix methods) are attractive because of their simplicity.

For more general cases, as we discuss next, Algorithms 7.1 and 7.2 can still be used with some modifications. Indeed, the fixed point methods can be used to compute the zeros of functions y and w, which are solutions of the first order system

$$\begin{aligned} y' &= \alpha y + \delta w, \\ w' &= \beta w + \gamma y, \end{aligned} \quad (7.102)$$

with continuous coefficients in an interval I and such that $\delta\gamma < 0$. The continuity of the solutions and the coefficients is enough to guarantee that the zeros of y and w are interlaced [193]. On the other hand, the condition $\delta\gamma < 0$ is an oscillatory condition, which necessarily has to be met when y or w have at least two zeros in the interval I [81, 193].

With these restrictions, we can define new functions and a new variable such that the first order system (7.102) is similar to the Bessel function case. First, we consider a change of the dependent functions

$$y(x) = \lambda_y(x)\bar{y}(x), \quad w(x) = \lambda_w(x)\bar{w}(x), \quad (7.103)$$

with $\lambda_y(x) \neq 0$, $\lambda_w(x) \neq 0$ for all $x \in I$, in such a way that $\bar{y}$ and $\bar{w}$ satisfy

$$\begin{aligned} \bar{y}' &= \bar{\alpha}\,\bar{y} + \bar{\delta}\,\bar{w}, \\ \bar{w}' &= \bar{\beta}\,\bar{w} + \bar{\gamma}\,\bar{y}, \end{aligned} \quad (7.104)$$

with $\bar{\delta} > 0$ and $\bar{\delta} = -\bar{\gamma}$. This is accomplished by choosing

$$\lambda_y = \text{sign}(\delta)\lambda_w\sqrt{-\delta/\gamma}. \quad (7.105)$$

The new functions $\bar{y}$ and $\bar{w}$ have obviously the same zeros as y and w.

When we now use a change of variable

$$z(x) = \int \bar{\delta}(x)\,dx, \quad (7.106)$$

the system reads

$$\begin{pmatrix} \dot{\bar{y}} \\ \dot{\bar{w}} \end{pmatrix} = \begin{pmatrix} \bar{a} & 1 \\ -1 & \bar{b} \end{pmatrix} \begin{pmatrix} \bar{y} \\ \bar{w} \end{pmatrix}, \quad (7.107)$$

where $\bar{a} = \bar{\alpha}/\bar{\delta}$, $\bar{b} = \bar{\beta}/\bar{\delta}$, and the dots mean derivative with respect to z. This new system resembles the Bessel case in the sense that the off-diagonal coefficients are ± 1. Now the ratio $H(z) = \bar{y}/\bar{w}$ satisfies the first order nonlinear ordinary differential equation

$$\dot{H} = 1 + H^2 - 2\eta H, \quad \eta = (\bar{b} - \bar{a})/2, \tag{7.108}$$

similarly to the case for Bessel functions. Because in proving the results for the fixed point method for Bessel functions we used only an equation like (7.108) together with the interlacing properties of the zeros, the algorithms for computing zeros of Bessel functions can be applied for computing the zeros of solutions y (or of w) of any system of the form (7.102) satisfying the following for x in the interval I:

1. The coefficients α, β, γ, δ and the solutions y and w are differentiable.
2. $\delta\gamma < 0$.

With this, we can summarize the transformations as follows.

Theorem 7.24. *Let $y(x)$, $w(x)$ be solutions of the system (7.102) with differentiable coefficients in the interval I and $\delta\gamma < 0$, and let*

$$H(x) = \operatorname{sign}(\delta)|\gamma/\delta|^{1/2} y(x)/w(x). \tag{7.109}$$

With the change of variable $z(x) = \int \sqrt{|\delta\gamma|}\, dx$, $H(x(z))$ satisfies

$$\dot{H} = 1 + H^2 - 2\eta H, \tag{7.110}$$

with η given by

$$\eta = \frac{\beta - \alpha}{2}\dot{x} + \frac{1}{4}\frac{d}{dz}\log\left|\frac{\delta}{\gamma}\right| = \left(\frac{\beta - \alpha}{2} + \frac{1}{4}\frac{d}{dx}\log\left|\frac{\delta}{\gamma}\right|\right)\dot{x}, \tag{7.111}$$

or, in terms of the original coefficients and variable,

$$\eta(x) = \frac{1}{2\sqrt{-\delta\gamma}}\left[\beta - \alpha + \frac{1}{2}\left(\frac{\delta'}{\delta} - \frac{\gamma'}{\gamma}\right)\right]. \tag{7.112}$$

Remark 12. The similarity with Riccati–Bessel functions becomes even more evident when we consider an additional change of the dependent variables, taking $\bar{y}(z) = v(z)\tilde{y}(z)$ and $\bar{w}(z) = v(z)\tilde{w}(z)$, with $v(z) = \exp\left(\int^z \frac{1}{2}(\bar{a}(\zeta) + \bar{b}(\zeta))\, d\zeta\right)$. Then, we have

$$\begin{aligned}\dot{\tilde{y}} &= -\eta\tilde{y} + \tilde{w},\\ \dot{\tilde{w}} &= \eta\tilde{w} - \tilde{y}.\end{aligned} \tag{7.113}$$

In the particular case of Riccati–Bessel functions we have $\eta = (\nu - 1/2)/x$ and no change of the independent variable is necessary. When we consider this additional change, $\tilde{y}$ and $\tilde{w}$ satisfy second order ordinary differential equations in normal form (as Riccati–Bessel functions do). We have

$$\ddot{\tilde{y}} + \tilde{A}(z)\tilde{y} = 0, \quad \tilde{A}(z) = 1 + \dot{\eta} - \eta^2, \tag{7.114}$$

and similarly for w with $\tilde{A}(z) = 1 - \dot{\eta} - \eta^2$.

7.5. Global strategies II: Global fixed point methods

The transformations considered for relating the general case to the Riccati–Bessel case are Liouville transformations (see also §2.2.4).

The parameter η describes the oscillatory nature of the solutions in a way similar to the conditions on the sign of $\tilde{A}(z(x))$ and $\delta\gamma$.

Theorem 7.25. *If $|\eta| > 1$ in an interval I, then $H(z(x))$ has at most one zero and one singularity in this interval.*

Also, we have the following result [84].

Theorem 7.26. *If $|\eta(x)| < 1 - \epsilon$, $\epsilon > 0$, for all x greater than x_0, then $H(z)$ has infinitely many interlaced zeros and singularities for $z > z(x_0)$. Therefore, $y(x)$ and $w(x)$ have infinitely many zeros.*

As happened for Bessel functions, this method can be applied to first order systems relating certain function $y_n(x)$ (n a parameter) and its derivative with $y_{n-1}(x)$ (or y_{n+1}) and its derivative.

Summarizing the transformations, we have the following method for computing zeros of a solution y of (7.102).

ALGORITHM 7.3. Zeros of special functions from a fixed point method.

1. Consider $z(x) = \int \sqrt{|\delta\gamma|}dx$, $H(x) = \text{sign}(\delta)\sqrt{-\frac{\gamma}{\delta}}\frac{y(x)}{w(x)}$, and $\eta(x)$ given by (7.112).
2. Divide the interval I into subintervals where $\eta(x)$ does not change sign.
3. Apply Algorithm 7.1 for those subintervals where $\eta(x) > 0$ and Algorithm 7.2 for those subintervals where $\eta(x) < 0$, but replacing the variable x by $z(x)$ and $H_\nu(x)$ by $H(z) = H(x(z))$. The zeros are computed in the z variable.
4. Invert the change of variable $z(x)$ to obtain the zeros in the original variable.

It can be proved (by considering Remark 12) that, under quite broad circumstances, $\eta(x)$ will change sign at most once [193, 84], and then only one subdivision of the interval is necessary. On the other hand, it is also possible to use the improved iteration step in most occasions.

Let us illustrate the method with two examples.

Example 7.27 (Laguerre polynomials $L_n^{(\alpha)}(x)$). The differential equation for the Laguerre polynomials $L_n^{(\alpha)}(x)$ is [219, p. 150]

$$xy'' + (\alpha + 1 - x)y' + ny = 0. \tag{7.115}$$

Taking $y(x) = L_n^{(\alpha)}(x)$ and $w(x) = L_{n-1}^{(\alpha)}(x)$, we can verify that these functions satisfy (7.91) with (we denote α of (7.102) by $\hat{\alpha}$)

$$\begin{aligned}\hat{\alpha} &= 1 - (n+1+\alpha)/x, & \beta &= (n+1)/x, \\ \delta &= (n+1)/x, & \gamma &= -(n+1+\alpha)/x.\end{aligned} \tag{7.116}$$

A possible change of variable is $z(x) = \int \sqrt{-\delta\gamma} dx = \sqrt{(n+1+\alpha)(n+1)} \log(x)$. We have

$$\eta(x) = \frac{1}{2\sqrt{(n+1)(n+1+\alpha)}} (2(n+1) + \alpha - x), \quad (7.117)$$

which changes sign at $x = 2(n+1) + \alpha = x_c$, and $\eta(x)$ is negative for large values of x. Therefore, Algorithm 7.2 can be used to compute zeros in intervals $[z(x_c), z(b)]$ (zeros are computed in increasing order) and Algorithm 7.1 should be considered in $[z(a), z(x_c)]$. Because the transformed ordinary differential equation (Remark 12) in the variable z reads

$$\ddot{\tilde{y}} + \tilde{A}(z)\tilde{y} = 0, \quad \tilde{A}(z) = 1 + \dot{\eta} - \eta^2, \quad (7.118)$$

we have $\dot{\tilde{A}}(z) = \frac{d}{dx}(\frac{d\eta}{dx}\frac{dx}{dz} - \eta^2)\frac{dx}{dz}$ and from here it is easy to check that $\tilde{A}(z)$ is decreasing for $z > z(x_e)$, $x_e = x_c - 1$, and increasing for $z < z(x_e)$.

Therefore when considering Algorithm 7.1 or 7.2 we have

$$H(z) = \sqrt{1 + \frac{\alpha}{n+1}} \frac{L_n^{(\alpha)}(x(z))}{L_{n+1}^{(\alpha)}(x(z))} = \sqrt{1 + \frac{\alpha}{n+1}} \frac{L_n^{(\alpha)}(\exp(z/\lambda))}{L_{n+1}^{(\alpha)}(\exp(z/\lambda))},$$
$$z(x) = \lambda \log(x), \quad (7.119)$$
$$\lambda = \sqrt{(n+1)(n+\alpha+1)}.$$

In $[z(x_c), z(b)]$ the improved iteration step can be used (after the first two zeros have been computed) because $\tilde{A}(z)$ is decreasing (and then the distance between consecutive zeros increases). The same is true for the computation of the zeros in $[z(a), z(x_c)]$ after the first two zeros of $H(z)$, smaller than $z(x_c)$, are computed (in decreasing order) when the second zero is smaller than $z(x_e)$ (which is a condition easily verified to be true, particularly for large n).

For a numerical illustration, let us consider the computation of the zeros of $L_{50}^{-0.6}(x)$ with a relative precision of 10^{-15}. We have $x_c = 101.4$ and we divide the search for zeros into two intervals: $[z(a), z_c]$, $z_c \simeq 234.18$, and $[z_c, z(b)]$. For $z > z(x_c)$ we compute zeros in increasing order using Algorithm 7.2 (through Algorithm 7.3), with the improved step Δ after the second zero has been computed. In the z variable the first steps would be as follows.

1. $H(z_c) \simeq -0.79 < 0$, and then we have monotonic convergence to a zero of $H(z)$ larger than z_c. After iterating four times $z_{n+1} = T(z_n) = z_n - \arctan(H(z_n))$, we see that the third and fourth iterations agree within the relative precision: $z^{(1)} = 234.85448310115433$.

2. Take $z_0 = z^{(1)} + \pi/2$. After five iterations we get $z^{(2)} = 238.0325962975257$.

3. The improved step $\Delta = z^{(2)} - z^{(1)}$ can be considered. We take $z_0 = z^{(2)} + \Delta$ and after five iterations we get $z^{(3)} = 241.2306728015528$.

4. We can proceed in the same way until $z(b)$ is reached.

By inversion of the change of variables we get the three smallest zeros larger than x_c: 109.39937721151, 116.52254033290, 124.206020228638 (15 digits).

7.5. Global strategies II: Global fixed point methods

The zeros smaller than x_c can be obtained using Algorithm 7.1 in $[z(a), z_c]$. Because, as explained above, $H(z_c) < 0$, one should start with $z_0 = z_c - \pi/2$, which guarantees convergence to the largest zero smaller than z_c. Then, the method proceeds similarly as above, with zeros generated in decreasing order.

This fixed point method is not the only possibility for computing zeros of $y(x) = L_n^{(\alpha)}(x)$. We could, for instance, also consider the system relating $y(x) = L_n^{(\alpha)}(x)$ with $w(x) = L_{n+1}^{(\alpha-1)}(x)$. It turns out that for small x, this is an interesting alternative [81]. ∎

Example 7.28 (zeros of conical functions). Conical functions are special cases of Legendre functions, namely, of degrees $-1/2 + i\tau$ with real τ ($P_{-1/2+i\tau}^n(x)$ or $Q_{-1/2+i\tau}^n(x)$). They satisfy the ordinary differential equation

$$y''(x) + B(x)y'(x) + A_n(x)y(x) = 0, \tag{7.120}$$

with coefficients

$$B(x) = 2x/(x^2 - 1), \quad A(x) = [1/4 + \tau^2 - n^2/(x^2 - 1)]/(x^2 - 1). \tag{7.121}$$

We consider $n > 0$ (the differential equation is invariant under the replacement $n \to -n$) and restrict our study to $x > 1$ (where the functions oscillate).

Taking $y(x) = P_{-1/2+i\tau}^n(x)$ and $w(x) = P_{-1/2+i\tau}^{n-1}(x)$, we have a differential system (7.102) with

$$\alpha = \frac{-nx}{x^2 - 1}, \quad \beta = \frac{(n-1)x}{x^2 - 1}, \quad \delta = -\frac{\lambda_n^2}{\sqrt{x^2 - 1}}, \quad \gamma = \frac{1}{\sqrt{x^2 - 1}}, \tag{7.122}$$

where $\lambda_n = \sqrt{(n - 1/2)^2 + \tau^2}$.

The function $\eta(x)$ is given by $\eta(x) = (n - 1/2)x/(\lambda_n \sqrt{x^2 - 1})$. The associated change of variables is

$$z(x) = \lambda_n \cosh^{-1} x, \tag{7.123}$$

and the coefficient $\tilde{A}(z)$ of the equation in normal form reads

$$\tilde{A}(z) = 1 + \dot{\eta}_{-1} - \eta_{-1}^2 = \frac{1}{\lambda_n^2}\left[\tau^2 - \frac{n^2 - 1/4}{\sinh^2(z/\lambda_n)}\right], \tag{7.124}$$

hence $\dot{\tilde{A}}(z)\eta_{-1}(z) > 0$ for $n > 0$, and therefore the improved iteration step can always be used. The computation of zeros will be performed in the backward direction for $n > 1/2$ and in the forward direction for $n < 1/2$ because $\text{sign}(\eta) = \text{sign}(n - 1/2)$.

Notice that from the expression for η (Theorem 7.26) as well as from that of $\tilde{A}(z)$, one deduces the following theorem.

Theorem 7.29. *Conical functions have infinitely many zeros for $x > 1$ and $\tau \neq 0$.*

Conical functions illustrate how the change of variables associated with the fixed point method tends to uniformize the distribution of zeros. In fact, we see that $\tilde{A}(z) \to (\tau/\lambda_n)^2$

as $z \to +\infty$, which means that the difference between zeros in the z variable for large z tends to

$$z_y^{(j+1)} - z_y^{(j)} \to \pi\sqrt{1 + \left(\frac{n-1/2}{\tau}\right)^2}, \quad j \to +\infty. \tag{7.125}$$

In fact, for the particular case $n = 1/2$ we observe that the zeros are equally spaced in the z variable.

When applying the method for computing the zeros of $P^n_{-1/2+i\tau}(x)$ one obtains the numerical results of Table 7.1 (the correct results on the left) using an IEEE double precision Fortran routine. ∎

7.6 Asymptotic methods: Further examples

Many special functions have an infinite number of real and/or complex zeros that can be described by certain patterns and rules, which can be obtained from asymptotic expansions of these functions. Usually, the zero distribution is not as simple as in the case of the sine and cosine functions, where pure periodicity gives a perfect description of the behavior of the zeros, but for, say, Bessel functions, the zero distribution is well understood and can be described analytically.

An important tool for describing the zero distribution, and for obtaining estimates of the zeros, is the reversion of asymptotic expansions for the functions. We explain this method for the Airy functions, and we give further examples for Scorer functions, which are related to Airy functions, and for the error functions. We include some information with references to the literature for other special functions.

In this section we focus on the methods for obtaining asymptotic estimates. Methods for obtaining bounds for the zeros are available for many cases, but we refer to the literature for this information. For example, in [105] general theorems are given on error bounds for the asymptotic approximation of both real and complex zeros of certain special functions.

7.6.1 Airy functions

The Airy functions $Ai(z)$ and $Bi(z)$ are solutions of the differential equation $w'' = zw$. Many properties of these functions are derived in [168, Chaps. 2 and 11]; see also [2, Chap. 10]. In §3.9, we have given the asymptotic expansions for the Airy functions.

From the representations and expansions in (3.9) and (3.136), valid at $+\infty$, we conclude that no sets of zeros can be expected for the function $Ai(z)$ inside the sector $|\text{ph } z| < \pi$ and for $Bi(z)$ inside the sector $|\text{ph } z| < \frac{1}{3}\pi$. We do expect zeros that follow from the representations and expansions in (3.133) and (3.134).

In fact, it is known that $Ai(z)$, $Ai'(z)$, $Bi(z)$, and $Bi'(z)$ each have an infinite number of zeros on the negative real axis, all of which are simple. In ascending order of absolute values they are denoted by a_s, a'_s, b_s, and b'_s, respectively, with $s = 1, 2, 3, \ldots$. In addition, $Bi(z)$ and $Bi'(z)$ have an infinite number of zeros in the neighborhood of the half-lines ph $z = \pm\frac{1}{3}\pi$ (the boundaries of the sector where the expansions at $+\infty$ are valid). It is possible to find asymptotic expansions of all these zeros for large values of s, which already give information on the location of the early zeros (small values of s).

7.6. Asymptotic methods: Further examples

To find these expansions we need to revert the asymptotic expansions for the Airy functions and those of their derivatives that are valid on the negative real axis. In §3.9, we have given these expansions, and we see that at a zero of $\mathrm{Ai}(-x)$ we have from (3.133) that

$$\tan\left(\xi + \frac{1}{4}\pi\right) = \frac{g(x)}{f(x)} \sim \frac{5}{72}\xi^{-1} - \frac{39655}{1119744}\xi^{-3} + \cdots. \quad (7.126)$$

Hence

$$\xi + \frac{1}{4}\pi - s\pi = \arctan\frac{g(x)}{f(x)} \sim \frac{5}{72}\xi^{-1} - \frac{1105}{31104}\xi^{-3} + \cdots, \quad (7.127)$$

where s now denotes an arbitrary integer. By reverting the expansion in (7.127) (for details we refer to the subsection below), we derive an expansion for large values of s

$$\xi \sim \left(s - \frac{1}{4}\right)\pi + \frac{5}{72}\left(s - \frac{1}{4}\right)^{-1} - \frac{1255}{31104}\left(s - \frac{1}{4}\right)^{-2} + \cdots. \quad (7.128)$$

Since $x = (\frac{3}{2}\xi)^{2/3}$ we conclude that for large s the equation $\mathrm{Ai}(-x) = 0$ is satisfied by $x = T(t)$, where $t = \frac{3}{8}\pi(4s - 1)$ and

$$T(t) \sim t^{2/3}\left(1 + \frac{5}{48}t^{-2} - \frac{5}{36}t^{-4} + \cdots\right), \quad t \to \infty. \quad (7.129)$$

In Table 7.2 we give values of the first five zeros of $\mathrm{Ai}(x)$ obtained from the asymptotic expansion in (7.129) by using all the terms shown. The values $a_s^{(0)}$ are obtained from (7.129), and the values $a_s^{(1)}$ are obtained by using one Newton–Raphson step. We also show values of the Airy function at the approximated values $a_s^{(0)}$ and the relative errors in values $a_s^{(0)}$. The computations are done in Maple with default 10-digit arithmetic. We observe that, although the asymptotic analysis is for large values of s, for $s = 1$ we obtain three significant digits by using (7.129). We have assumed that (7.128) and (7.129) give the approximations of the zeros, starting with $s = 1$. In (7.127), however, s is an arbitrary integer, and it is not at all obvious that $s = 1$ is the correct starting value. A further analysis, based on the phase

Table 7.2. *Zeros of* $\mathrm{Ai}(x)$ *computed from* (7.129).

s	$a_s^{(0)}$	$\mathrm{Ai}(a_s^{(0)})$	$a_s^{(1)}$	Rel. error
1	−2.337534466	0.0004017548	−2.338107411	$0.25\ 10^{-3}$
2	−4.087939499	−0.0000079870	−4.087949444	$0.24\ 10^{-5}$
3	−5.520558833	0.0000008610	−5.520559828	$0.18\ 10^{-6}$
4	−6.786707898	−0.0000001749	−6.786708090	$0.28\ 10^{-7}$
5	−7.944133531	0.0000000532	−7.944133587	$0.70\ 10^{-8}$

principle for counting the zeros and poles of analytic functions, can be used for this case to prove that in a certain large interval $[-s_0, 0]$ exactly s zeros occur. For details, see [66]. In this reference further information can be found on the zeros of $\text{Ai}'(z)$, $\text{Bi}(z)$, and $\text{Bi}'(z)$. Also information is given on how to obtain expansions for the associated values $\text{Ai}'(a_s)$, $\text{Ai}(a'_s)$, $\text{Bi}'(a_s)$, and $\text{Bi}(a'_s)$, and similarly for the complex zeros of $\text{Bi}(z)$ and $\text{Bi}'(z)$.

Reverting asymptotic series

We recall a few steps from §7.3.1 on how to revert the asymptotic expansion in (7.127) and show how to find coefficients in the reverted expansion. Assume we wish to solve an equation $w = f(z)$ for large values of w, where f has an asymptotic expansion of the form

$$f(z) \sim z + f_0 + f_1 z^{-1} + f_2 z^{-2} + f_3 z^{-3} + \cdots \quad (7.130)$$

as $z \to \infty$ in a certain sector in the complex plane. Then the equation $w = f(z)$ has a solution $z = F(w)$, and we can find an asymptotic expansion of w of the form

$$z \sim w - F_0 - F_1 w^{-1} - F_2 w^{-2} - F_3 w^{-3} + \cdots, \quad (7.131)$$

where the coefficients F_j are constants. The first few are

$$F_0 = f_0, \quad F_1 = f_1, \quad F_2 = f_0 f_1 + f_2, \quad F_3 = f_0^2 f_1 + f_1^2 + 2 f_0 f_2 + f_3. \quad (7.132)$$

For details that (7.131) is a valid asymptotic expansion, when the function f satisfies certain conditions, we refer to [66]. There are many ways for obtaining the coefficients F_j in the new expansion (7.131). One straightforward method is based on substituting (7.131) with the unknown coefficients F_j into (7.130), with the left-hand side replaced with w, expanding formally all powers z^{-k} in power series of w^{-1}, and equating equal powers of w^{-1} in the expansion. In practice we may take advantage of computer algebra packages, such as Maple, to perform these calculations. Also, the evaluation of coefficients when going from the expansion in (7.126) to that of (7.127) can easily be done with the help of computer algebra.

There is another way of computing the coefficients F_j. Let P_N denote the polynomial

$$P_N(\zeta) = 1 + f_0 \zeta + f_1 \zeta^2 + f_2 \zeta^3 + \cdots + f_N \zeta^{N+1}, \quad (7.133)$$

where f_j appear in (7.130) and with N a positive integer. If $N \geq j > 0$, then jF_j of (7.131) is the coefficient of ζ^{j+1} of the jth power of the polynomial $P_N(\zeta)$. To verify this we first transform the point at infinity to the origin by means of the substitutions $z = 1/\zeta$, $w = 1/\omega$. Then the equation $w = f(z)$, with f having expansion (7.130), becomes

$$\frac{1}{\omega} \sim \frac{1}{\zeta} + f_0 + f_1 \zeta + f_2 \zeta^2 + \cdots. \quad (7.134)$$

This can be written in a formal way as an identity,

$$\omega = \frac{\zeta}{P(\zeta)}, \quad P(\zeta) = \lim_{N \to \infty} P_N(\zeta), \quad (7.135)$$

7.6. Asymptotic methods: Further examples

where $P_N(\zeta)$ is introduced in (7.133). Next we assume that $P(\zeta)$ is an analytic function in a neighborhood of the origin of which the first terms of the power series are given in (7.133), that in (7.134) the asymptotic sign is replaced by an equal sign, and that (7.131) can be written in the form

$$\frac{1}{\zeta} - \frac{1}{\omega} = -\sum_{j=0}^{\infty} F_j \omega^j, \qquad (7.136)$$

where the series also represents an analytic function in the neighborhood of the origin. When we differentiate this relation with respect to ω and multiply by ω, we obtain

$$-\frac{\omega}{\zeta^2}\frac{d\zeta}{d\omega} + \frac{1}{\omega} = -\sum_{j=1}^{\infty} jF_j \omega^j. \qquad (7.137)$$

By using Cauchy's integral and (7.135) we can write, for $j > 0$,

$$jF_j = \frac{1}{2\pi i} \int \left(\frac{\omega}{\zeta^2}\frac{d\zeta}{d\omega} - \frac{1}{\omega} \right) \frac{d\omega}{\omega^{j+1}} = \frac{1}{2\pi i} \int \frac{d\zeta}{\zeta^2 \omega^j} = \frac{1}{2\pi i} \int \frac{P^j(\zeta)}{\zeta^{j+2}} d\zeta, \qquad (7.138)$$

where the integral is taken around a small circle around the origin. We have used (7.135) and the fact that the term $-1/\omega$ in the first integral does not give a contribution. The third integral says that jF_j is the coefficient of ζ^{j+1} in the expansion of $P^j(\zeta)$. The same result holds when $P^j(\zeta)$ is replaced by $P_N^j(\zeta)$, $N \geq j$, where $P_N(z)$ is introduced in (7.133). Convergent infinite series and analytic functions are not really needed for this result, but when using these concepts the proof is rather elegant.

7.6.2 Scorer functions

Scorer functions are related to Airy functions. They are particular solutions of the inhomogeneous Airy differential equation $w'' - zw = \pm 1/\pi$. We have for real values of z

$$w'' - zw = -1/\pi, \quad \text{with solution} \quad \text{Gi}(z) = \frac{1}{\pi}\int_0^\infty \sin\left(zt + \frac{1}{3}t^3\right) dt, \qquad (7.139)$$

and for general complex z

$$w'' - zw = 1/\pi, \quad \text{with solution} \quad \text{Hi}(z) = \frac{1}{\pi}\int_0^\infty e^{zt - \frac{1}{3}t^3} dt. \qquad (7.140)$$

Analytic continuation for Gi(z) follows from the relation

$$\text{Gi}(z) + \text{Hi}(z) = \text{Bi}(z). \qquad (7.141)$$

Detailed information on these functions can be found in [2, Chap. 10] and, in particular in connection with the zeros, in [93]. We use the asymptotic expansions

$$\text{Gi}(z) \sim \frac{1}{\pi z}\left[1 + \frac{1}{z^3}\sum_{s=0}^{\infty} \frac{(3s+2)!}{s!(3z^3)^s} \right], \quad z \to \infty, \quad |\text{ph } z| \leq \frac{1}{3}\pi - \delta, \qquad (7.142)$$

$$\mathrm{Hi}(z) \sim -\frac{1}{\pi z}\left[1+\frac{1}{z^3}\sum_{s=0}^{\infty}\frac{(3s+2)!}{s!(3z^3)^s}\right], \quad z\to\infty, \quad |\mathrm{ph}(-z)|\le\frac{2}{3}\pi-\delta, \quad (7.143)$$

δ being an arbitrary small positive constant. Inside the indicated sectors no sets of zeros can occur, except for possibly isolated zeros with small modulus. When we write (cf. (7.141)) $\mathrm{Gi}(-x) = \mathrm{Bi}(-x) - \mathrm{Hi}(-x)$, and use the asymptotic expansion for Bi and Hi valid on the negative real axis, we obtain compound expansions valid on the negative axis, where indeed zeros of this function occur. Gi also has complex zeros (see [93]), but we concentrate here on those on the negative real axis. We write

$$\mathrm{Hi}(-x) = \frac{1}{\pi x}Ha(x), \quad Ha(x) \sim 1 - \sum_{s=0}^{\infty}\frac{h_s}{x^{3(s+1)}}, \quad h_s = (-1)^s\frac{(3s+2)!}{s!\,3^s}, \quad (7.144)$$

and for the asymptotic expansion of $\mathrm{Bi}(-x)$ we have

$$\mathrm{Bi}(-x) = \pi^{-1/2}x^{-1/4}\left\{\cos\left(\xi+\tfrac{1}{4}\pi\right)f(x) + \sin\left(\xi+\tfrac{1}{4}\pi\right)g(x)\right\}, \quad \xi=\tfrac{2}{3}x^{3/2}. \quad (7.145)$$

We refer the reader to (3.133) and (3.134) for more details. We explain the method by taking $Ha(x) = 1$, $f(x) = 1$, and $g(x) = 0$. This gives for the equation $\mathrm{Gi}(-x) = \mathrm{Bi}(-x) - \mathrm{Hi}(-x) = 0$ a first equation

$$\cos\left(\xi+\frac{1}{4}\pi\right) = \frac{1}{\sqrt{\pi}\,x^{3/4}}. \quad (7.146)$$

Using $x^{3/4} = \sqrt{3\xi/2}$, we obtain

$$\cos\left(\xi+\frac{1}{4}\pi\right) = \sqrt{\frac{2}{3\pi\xi}}. \quad (7.147)$$

For large values of ξ solutions occur when the cosine function is small. We put

$$\xi = \xi_s + \varepsilon, \quad \xi_s = \left(s-\tfrac{3}{4}\right)\pi, \quad s=1,2,3,\ldots. \quad (7.148)$$

The equation for ε reads

$$\sin\varepsilon = \frac{c_s}{\sqrt{\xi_s+\varepsilon}} = \frac{c_s t}{\sqrt{1+\varepsilon t^2}}, \quad t=1/\sqrt{\xi_s}, \quad c_s=(-1)^s\sqrt{2/(3\pi)}. \quad (7.149)$$

For small values of t this equation can be solved by substituting a power series $\varepsilon = \varepsilon_1 t + \varepsilon_2 t^2 + \cdots$, and the coefficients can be obtained by standard methods. For example, $\varepsilon_1 = c_s$. By using the asymptotic expansions for $Ha(x)$, $f(x)$, and $g(x)$ a few extra technicalities are introduced. With the help of a computer algebra package the general coefficients ε_s are easy to calculate. Finally we find for $x = (3\xi/2)^{2/3}$, and for g_s, the negative zeros of $\mathrm{Gi}(x)$, the expansion

$$g_s \sim -\left(\tfrac{3}{2}\xi_s\right)^{\frac{2}{3}}\left[1+\varepsilon_1 t^3 + \varepsilon_2 t^4 + \cdots\right]^{2/3}, \quad s=1,2,3,\ldots, \quad (7.150)$$

7.6. Asymptotic methods: Further examples

or

$$g_s \sim -[(3\pi(4s-3)/8]^{\frac{2}{3}}\left[1+\frac{\gamma_3}{t^3}+\frac{\gamma_4}{t^4}+\cdots\right], \quad t=\sqrt{(s-3/4)\pi}, \tag{7.151}$$

where

$$\gamma_3 = \frac{2c_s}{3}, \quad \gamma_4 = \frac{5}{108}, \quad \gamma_5 = \frac{c_s^3}{9}, \quad \gamma_6 = -\frac{4c_s^2}{9},$$
$$\gamma_7 = \frac{c_s(81c_s^4-1060)}{1620}, \quad \gamma_8 = -\frac{189c_s^4+20}{729}. \tag{7.152}$$

The expansion in (7.151) reduces to the expansion of the zeros b_s of $\mathrm{Bi}(z)$ if we take $c_s = 0$. For $\mathrm{Gi}'(-x)$ a similar method can be used. $\mathrm{Gi}(z)$ and $\mathrm{Gi}'(z)$ have an infinite number of complex zeros just below the half-line $\mathrm{ph}\, z = \frac{1}{3}\pi$ and at the conjugate values. $\mathrm{Hi}(z)$ and $\mathrm{Hi}'(z)$ have an infinite number of complex zeros just above the half-line $\mathrm{ph}\, z = \frac{1}{3}\pi$ and at the conjugate values. Approximations for all these zeros can be found in [93].

7.6.3 The error functions

The error functions are important in mathematical statistics and probability theory in connection with the normal distribution [2, Chap. 26], in physical problems, and they play an important role as main approximants in uniform asymptotic analysis; see [243, Chap. 7]. The zero distribution of the error functions will give information on the zero distribution of the function that is approximated. We use the following definitions:

$$\mathrm{erf}\, z = \frac{2}{\sqrt{\pi}}\int_0^z e^{-t^2}\,dt, \quad \mathrm{erfc}\, z = \frac{2}{\sqrt{\pi}}\int_z^\infty e^{-t^2}\,dt, \tag{7.153}$$

where z is any complex number. We have the symmetry relations

$$\mathrm{erf}\, z + \mathrm{erfc}\, z = 1, \quad \mathrm{erf}(-z) = -\mathrm{erf}\, z, \quad \mathrm{erfc}(-z) = 2 - \mathrm{erfc}\, z. \tag{7.154}$$

The *Fadeeva* or *plasma-dispersion function* [2, Chap. 7]

$$w(z) = \frac{1}{\pi i}\int_{-\infty}^\infty \frac{e^{-t^2}}{t-z}\,dt, \quad \Im z > 0, \tag{7.155}$$

is also related to the error functions. We have the relation

$$w(z) = e^{-z^2}\mathrm{erfc}(-iz), \tag{7.156}$$

which defines $w(z)$ for all complex z. The asymptotic expansion of $\mathrm{erfc}\, z$ is given by

$$\mathrm{erfc}\, z \sim \frac{e^{-z^2}}{z\sqrt{\pi}}\sum_{k=0}^\infty \frac{(-1)^k(2k)!}{k!(2z)^{2k}} = \frac{e^{-z^2}}{z\sqrt{\pi}}\left(1-\frac{1}{2}z^{-2}+\frac{3}{4}z^{-4}-\frac{15}{8}z^{-6}+\cdots\right), \tag{7.157}$$

as $z\to\infty$, $|\mathrm{ph}\, z|\le \frac{3}{4}\pi - \delta$, where δ is a small positive number.

The complementary error function

First we show that in certain parts of the complex plane the complementary error function cannot have zeros. We write $z = x + iy$.

Lemma 7.30. erfc z has no zeros in the sector $\frac{3}{4}\pi \leq \mathrm{ph}\, z \leq \frac{5}{4}\pi$.

Proof. From (7.153) we obtain

$$\mathrm{erfc}\, z = \frac{2e^{-z^2}}{\sqrt{\pi}} \int_0^\infty e^{-t^2-2zt}\, dt. \tag{7.158}$$

Let z be in the sector $-\frac{1}{4}\pi \leq \mathrm{ph}\, z \leq \frac{1}{4}\pi$. We have $|e^{-z^2}| = e^{-(x^2-y^2)} \leq 1$ and $|e^{-2zt}| \leq 1$. This gives $|\mathrm{erfc}\, z| \leq 1$. Hence,

$$|\mathrm{erfc}(-z)| = |2 - \mathrm{erfc}(z)| \geq 2 - 1 = 1, \tag{7.159}$$

and the lemma follows. □

Lemma 7.31. erfc z has no zeros when $x \geq 0$.

Proof. If $x = 0$, we have

$$\mathrm{erfc}(iy) = 1 - \mathrm{erf}(iy) = 1 - \frac{2i}{\sqrt{\pi}} \int_0^y e^{t^2}\, dt, \tag{7.160}$$

which cannot vanish for real values of y. Next, if $x > 0$, we have from (7.155) and (7.156)

$$e^{z^2} \mathrm{erfc}\, z = \frac{1}{\pi} \int_{-\infty}^\infty \frac{e^{-t^2}}{t - iz}\, dt = u(x, y) - iv(x, y), \tag{7.161}$$

where, as easily follows from $1/(t-iz) = (t+y+ix)/[(t+y)^2 + x^2]$ and taking $s = t + y$ as the new variable of integration,

$$\begin{aligned} u(x, y) &= \frac{2xe^{-y^2}}{\pi} \int_0^\infty e^{-s^2} \cosh(2ys) \frac{ds}{s^2 + x^2}, \\ v(x, y) &= \frac{2e^{-y^2}}{\pi} \int_0^\infty e^{-s^2} \sinh(2ys) \frac{s\, ds}{s^2 + x^2}. \end{aligned} \tag{7.162}$$

It follows that $u(x, y) > 0$ and that $v(x, y) > 0$ when $x > 0$ and $y > 0$. This proves the lemma. □

With these two lemmas, and considering the sector of validity of the asymptotic expansion (7.157), we expect that the large zeros must have phases that tend to $\pm\frac{3}{4}\pi$. In fact we have the following. The complementary error function erfc z has an infinite number of simple zeros in z_n in the second quadrant just above the diagonal $x + y = 0$ with conjugate values $\bar{z}_n$ in the third quadrant. Next we describe how to obtain the asymptotic expansion for the zeros z_n of erfc z. First we give the result. Let

$$\lambda = \sqrt{(n - 1/8)\pi}, \quad \mu = -\ln(2\lambda\sqrt{2\pi}). \tag{7.163}$$

7.6. Asymptotic methods: Further examples

Then

$$z_n \sim e^{\frac{3}{4}\pi i}\sqrt{2}\lambda \sum_{k=0}^{\infty} \frac{c_k}{\lambda^{2k}}, \quad n \to \infty, \tag{7.164}$$

where the first few coefficients are given by

$$\begin{aligned}
c_0 &= 1, \quad c_1 = \tfrac{1}{4}i\mu, \quad c_2 = \tfrac{1}{32}(\mu^2 + 2\mu + 2), \\
c_3 &= -\tfrac{1}{128}i(\mu^3 + 4\mu^2 + 8\mu + 7), \\
c_4 &= -\tfrac{1}{6144}(15\mu^4 + 92\mu^3 + 288\mu^2 + 516\mu + 440), \\
c_5 &= \tfrac{1}{24576}i(21\mu^5 + 176\mu^4 + 764\mu^3 + 2094\mu^2 + 3596\mu + 3062).
\end{aligned} \tag{7.165}$$

To obtain this result we try to solve the equation erfc $z = 2$ for $x > 0$, which gives the zeros of erfc$(-z)$. We use the asymptotic expansion given in (7.157). For a first orientation, consider the equation

$$ze^{z^2} = \frac{1}{2\sqrt{\pi}}. \tag{7.166}$$

We write $z^2 = 2\pi i(n - 1/8) + \zeta$, and obtain for ζ the equation

$$e^{-\zeta} = 2\lambda\sqrt{2\pi}\sqrt{1 - i\zeta/(2\lambda^2)}, \tag{7.167}$$

where λ is given in (7.163). Assuming that $\zeta/\lambda^2 = o(1)$ for large n, we take the approximation $\zeta = \mu$, where μ is given in (7.163). This gives for z the first approximation

$$z = e^{\frac{1}{4}\pi i}\sqrt{2}\lambda\sqrt{1 - i\zeta/(2\lambda^2)} \sim e^{\frac{1}{4}\pi i}\sqrt{2}\lambda\left(1 - \tfrac{1}{4}i\mu\lambda^{-2}\right), \quad n \to \infty. \tag{7.168}$$

This is the beginning of an asymptotic expansion

$$z \sim e^{\frac{1}{4}\pi i}\sqrt{2}\lambda \sum_{k=0}^{\infty} \frac{d_k}{\lambda^{2k}}, \quad n \to \infty, \tag{7.169}$$

of which $d_1 = -\tfrac{1}{4}i\mu$. For solving the equation erfc $z = 2$, using the asymptotic expansion in (7.157), we obtain the asymptotic relation

$$2\sqrt{\pi}ze^{z^2} \sim \sum_{k=0}^{\infty} \frac{(-1)^k(2k)!}{k!(2z)^{2k}}, \tag{7.170}$$

in which we substitute expansion (7.169). We expand both sides in powers of λ^{-2} and equate equal powers to obtain the coefficients d_k. For a suitable start of this procedure, it is convenient to write

$$2\sqrt{\pi}ze^{z^2} = Se^{2i\lambda^2(S^2-1-2d_1/\lambda^2)}, \tag{7.171}$$

Table 7.3. *Zeros of* erfc z *computed from* (7.164).

n	$z_n^{(0)}$	$z_n^{(1)}$	Rel. error
1	$-1.354703419 + 1.991467105\,i$	$-1.354810113 + 1.991466820\,i$	$0.44\,10^{-4}$
2	$-2.177043673 + 2.691148423\,i$	$-2.177044906 + 2.691149024\,i$	$0.40\,10^{-6}$
3	$-2.784387721 + 3.235330579\,i$	$-2.784387613 + 3.235330868\,i$	$0.72\,10^{-7}$
4	$-3.287410885 + 3.697309569\,i$	$-3.287410789 + 3.697309702\,i$	$0.33\,10^{-7}$
5	$-3.725948777 + 4.106107217\,i$	$-3.725948719 + 4.106107285\,i$	$0.16\,10^{-7}$

Table 7.4. *Zeros of* erf z *computed from* (7.169).

n	$z_n^{(0)}$	$z_n^{(1)}$	Rel. error
1	$1.450350585 + 1.880986597\,i$	$1.450616020 + 1.880942905\,i$	$0.11\,10^{-3}$
2	$2.244653236 + 2.616577820\,i$	$2.244659274 + 2.616575141\,i$	$0.19\,10^{-5}$
3	$2.839740398 + 3.175628464\,i$	$2.839741047 + 3.175628100\,i$	$0.17\,10^{-6}$
4	$3.335460607 + 3.646174455\,i$	$3.335460735 + 3.646174376\,i$	$0.30\,10^{-7}$
5	$3.769005535 + 4.060697257\,i$	$3.769005567 + 4.060697234\,i$	$0.71\,10^{-8}$

where S denotes the series in (7.169). The first few coefficients are

$$d_0 = 1, \quad d_1 = -\tfrac{1}{4}i\mu, \quad d_2 = \tfrac{1}{32}(\mu^2 + 2\mu + 2),$$
$$d_3 = \tfrac{1}{128}i(\mu^3 + 4\mu^2 + 8\mu + 7),$$
$$d_4 = -\tfrac{1}{6144}(15\mu^4 + 92\mu^3 + 288\mu^2 + 516\mu + 440),$$
$$d_5 = -\tfrac{1}{24576}i(21\mu^5 + 176\mu^4 + 764\mu^3 + 2094\mu^2 + 3596\mu + 3062).$$
(7.172)

The expansion in (7.169) is for the solution of erfc $z = 2$, that is, for the zeros of erfc$(-z)$. For the zeros of erfc z we multiply the expansion in (7.169) by -1, and change afterwards i into $-i$, also in the coefficients d_k. This gives the expansion in (7.164) for the zeros in the second quadrant. In Table 7.3 we give values of the first five zeros of erfc z obtained from the asymptotic expansion shown in (7.164) by using the terms up to and including c_4. The values $z_n^{(0)}$ are obtained from (7.164), the values $z_n^{(1)}$ are obtained by using one Newton–Raphson step. The computations are done in Maple with default 10-digit arithmetic. We observe that, although the asymptotic analysis is for large values of n, for $n = 1$ we obtain four significant digits by using (7.164) with the terms up to and including c_4.

7.6. Asymptotic methods: Further examples

The error function

The error function erf z has a simple zero at $z = 0$ and an infinite number of simple zeros z_n in the first quadrant just above the diagonal $x - y = 0$; $\bar{z}_n$, $-z_n$, and $-\bar{z}_n$ are also zeros. To obtain asymptotic estimates of the zeros we try to solve the equation erfc $z = 1$. The procedure is the same as that for solving the equation erfc $z = 2$, and we obtain an expansion as given in (7.169), with the same expressions of the coefficients d_k, of which the first few are given in (7.172). However, in this case, μ of (7.163) has to be replaced by $\mu = -\ln(\lambda\sqrt{2\pi})$. In Table 7.4 we give values of the first five zeros of erf z obtained from the asymptotic expansion (7.169) with $\mu = -\ln(\lambda\sqrt{2\pi})$, by using the terms up to and including d_4. The values $z_n^{(1)}$ are obtained by using one Newton–Raphson step. The computations are done in Maple with default 10-digit arithmetic.

The analysis of this section is partly based on [69], where also the first 100 zeros of erf z and $w(z)$ (with 11 digits) are given. Table 2 of that paper does not give the zeros of erfc z, as indicated, but rather those of $w(z)$; to obtain the zeros of erfc z the x- and y-values should be interchanged.

7.6.4 The parabolic cylinder function

The method for the error functions can also be used for the parabolic cylinder function $U(a, z)$, of which the complementary error function is a special case. We have

$$U\left(\frac{1}{2}, z\right) = \sqrt{\pi/2}\, e^{\frac{1}{4}z^2} \operatorname{erfc}(z/\sqrt{2}). \quad (7.173)$$

When $a > -\frac{1}{2}$ there is an infinite number of complex zeros z_n of $U(a, z)$ that approach the ray ph $z = \frac{3}{4}\pi$ as $n \to \infty$, and there is a conjugate set. See [188] for this as well as many other results for asymptotic estimates of the zeros of special functions.

To obtain asymptotic expansions of the complex zeros of $U(a, z)$, the compound asymptotic expansions of $U(a, z)$ valid near these rays have to be inverted. For these compound expansions see [63, Vol. 2, p. 123], where the expansions are given for $D_\nu(z) = U(-\frac{1}{2} - \nu, z)$; an overview of these asymptotic expansions is given in [221].

7.6.5 Bessel functions

For these functions many results are available. The asymptotic expansions of the large real zeros of $J_\nu(x)$ and $Y_\nu(x)$ and those of their derivatives are called McMahon expansions. See Example 7.9 and [2, p. 371]. In [74] bounds for the error term in McMahon's asymptotic approximation of the positive zeros $j_{\nu,k}$ of the Bessel function $J_\nu(x)$ are derived. In [61] upper and lower bounds for the zeros of the *cylinder functions* $C_\nu(x) = J_\nu(x)\cos(\alpha) + Y_\nu(x)\sin(\alpha)$ are given.

The McMahon expansions are valid for fixed order ν. For large values of ν, it is better to use approximations that are obtained from Airy-type asymptotic expansions of the Bessel functions. For expansions of the real zeros valid for large values of ν, see again [2, p. 371]. In [210] these expansions are used in an algorithm. For numerical investigations on the complex ν-zeros of Hankel functions by using Airy-type expansions, see [41].

7.6.6 Orthogonal polynomials

The standard method for computing the zeros of orthogonal polynomials, which is based on the recurrence relation for the polynomials, is described in §7.4. When the degree or other parameters are large, a priori estimates obtained from asymptotic expansions may be useful when starting a numerical algorithm.

Asymptotic expansions holding properly inside the interval of orthogonality usually contain trigonometric functions as main approximants, and reversion of the expansion may be rather easy. Expansions for intervals including the finite endpoints of the interval of orthogonality or so-called turning points are usually given as uniform expansions containing Bessel functions, Airy functions, parabolic cylinder functions, and so on. In these cases the reversion is more complicated. For example, see [136], where information on the zeros is obtained for large-degree Meixner–Pollaczek polynomials.

Example 7.32 (Laguerre polynomials $L_n^{(\alpha)}(x)$ for large α). We recall that the well-known limit relation between the Laguerre and Hermite polynomials is [4, p. 285]

$$\lim_{\alpha \to \infty} \left[\alpha^{-n/2} L_n^{(\alpha)} \left(\alpha + x\sqrt{2\alpha} \right) \right] = \frac{(-1)^n}{n!} 2^{-n/2} H_n(x). \tag{7.174}$$

This gives the opportunity to approximate the zeros of the Laguerre polynomial with large α in terms of the zeros of the Hermite polynomial. However, for $n = 10$ and $\alpha = 100$ these approximations are rather poor, with the best result for the sixth zero 104.8493571 of the Laguerre polynomial which corresponds to the first positive zero 0.3429013272 of the Hermite polynomial, giving the approximation 112.1126502, with only one correct leading digit.

A much better result follows from a uniform asymptotic approximation of the Laguerre polynomial in terms of the Hermite polynomial, and the errors in the approximations are quite the same for all zeros. The description of the uniform approximation requires extra technical details, but this extra work is quite rewarding. We introduce the following quantities:

$$\kappa = n + \frac{1}{2}(\alpha + 1), \quad \tau = \frac{\alpha}{2n + \alpha + 1}, \quad \rho = \sqrt{2(1-\tau)} \tag{7.175}$$

and

$$x_1 = \tfrac{1}{2} - \tfrac{1}{2}\sqrt{1-\tau^2}, \quad x_2 = \tfrac{1}{2} + \tfrac{1}{2}\sqrt{1-\tau^2}, \quad R = \sqrt{(x_2-x)(x-x_1)}. \tag{7.176}$$

Then the approximation reads

$$L_n^{(\alpha)}(4\kappa x) = A_n^\alpha(x) \left[H_n(\eta\sqrt{\kappa}) + o(1) \right], \quad \kappa \to \infty. \tag{7.177}$$

Details on $A_n^\alpha(x)$ and the term $o(1)$ are given in [215]. The quantity η is implicitly defined by the equation

$$\frac{1}{2}\eta\sqrt{\rho^2 - \eta^2} + \frac{1}{2}\rho^2 \arcsin\frac{\eta}{\rho} = 2R - \tau \arctan\frac{2x - \tau^2}{2\tau R} - \arctan\frac{1-2x}{2R}, \tag{7.178}$$

7.6. Asymptotic methods: Further examples

Table 7.5. *Number of correct decimal digits in the approximations of zeros of $L_{(10)}^{(\alpha)}(x)$.*

m	\multicolumn{8}{c}{α}							
	0	1	5	10	25	50	75	100
1	1.7	2.3	3.2	3.7	4.4	5.0	5.3	5.6
2	2.4	2.7	3.4	3.8	4.5	5.0	5.4	5.6
3	2.8	3.0	3.5	3.9	4.5	5.1	5.4	5.6
4	3.0	3.2	3.6	4.0	4.6	5.1	5.4	5.7
5	3.2	3.4	3.8	4.1	4.6	5.1	5.5	5.7
6	3.4	3.5	3.9	4.2	4.7	5.2	5.5	5.7
7	3.5	3.6	4.0	4.2	4.7	5.2	5.5	5.8
8	3.7	3.8	4.1	4.3	4.8	5.3	5.6	5.8
9	3.8	3.9	4.1	4.4	4.9	5.3	5.6	5.8
10	3.9	4.0	4.2	4.5	4.9	5.4	5.6	5.8

where $x_1 \leq x \leq x_2$ and $-\rho \leq \eta \leq \rho$. In [215] similar relations are given outside these intervals, but for the zeros we need the ones we have indicated here, because the Laguerre polynomial in (7.177) has its zeros in $[x_1, x_2]$. The relation between η and x is one-to-one, with $\eta(x_1) = -\rho$ and $\eta(x_2) = \rho$.

Now, let $\ell_{n,m}^{(\alpha)}$, $h_{n,m}$ denote the mth zeros of $L_n^{(\alpha)}(z)$, $H_n(z)$, $m = 1, 2, \ldots, n$, respectively. For given α and n, we can compute

$$\eta_{n,m} = \frac{h_{n,m}}{\sqrt{\kappa}}, \quad m = 1, 2, \ldots, n. \tag{7.179}$$

Upon inverting (7.178) we can obtain $x_{n,m}$, giving the estimate

$$\ell_{n,m}^{(\alpha)} \sim 4\kappa x_{n,m}, \quad m = 1, 2, \ldots, n. \tag{7.180}$$

From properties of the Hermite polynomials [219, p. 168] it follows that all zeros $h_{n,m}$ are located in the interval $[-\sqrt{2n+1}, \sqrt{2n+1}]$. It follows that the numbers $\eta_{n,m}$ of (7.179) belong to the interval $[-\rho, \rho]$, when n is large, $\alpha \geq 0$. The estimate (7.180) is valid, for $\kappa \to \infty$, uniformly with respect to $m \in \{1, 2, \ldots, n\}$ when $\alpha \geq \alpha_2 n$, where α_2 is a fixed positive number.

In Table 7.5 we show for $n = 10$ the number of correct decimal digits in the approximation (7.180). That is, we show

$$\log_{10} \left| \frac{\ell_{10,m}^{(\alpha)} - \tilde{\ell}_{10,m}^{(\alpha)}}{\ell_{10,m}^{(\alpha)}} \right|, \quad m = 1, 2, \ldots, 10, \tag{7.181}$$

where $\tilde{\ell}_{10,m}^{(\alpha)}$ is the approximation obtained from (7.180). It follows that the large zeros are better approximated than the small zeros. Furthermore, large values of α give better approximations, and the approximations are uniform with respect to m. ∎

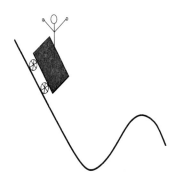

Chapter 8

Uniform Asymptotic Expansions

If I wanted you to understand, I would have explained it better.
—Johan Cruyff, Dutch Soccer Master

Writing efficient algorithms for special functions may become problematic when several large parameters are involved. In particular, problems arise when functions suddenly change their behavior, say from monotonic to oscillatory behavior. For many special functions of mathematical physics, powerful uniform asymptotic expansions are available which describe precisely how the functions behave, which are valid for large domains of the parameters, and which provide tools for designing high-performance computational algorithms.

In this chapter we discuss how to use uniform asymptotic expansions. We explain why these expansions are useful, and why they are usually difficult to handle in numerical algorithms. First we consider uniform expansions of the incomplete gamma functions. These functions are important in probability theory, but also in physical problems. Another important class concerns the functions having a turning point in their defining differential equation, in which case Airy-type expansions arise. As important examples we discuss Airy-type expansions for Bessel functions.

8.1 Asymptotic expansions for the incomplete gamma functions

We recall the definitions of the incomplete gamma functions,

$$\gamma(a, z) = \int_0^z t^{a-1} e^{-t}\, dt, \quad \Gamma(a, z) = \int_z^\infty t^{a-1} e^{-t}\, dt, \tag{8.1}$$

where for the first integral we need $\Re a > 0$ and for both integrals we assume that $|\operatorname{ph} z| < \pi$.

Integrating by parts in the second integral gives

$$\Gamma(a, z) = -\int_z^\infty t^{a-1}\, d\, e^{-t} = z^{a-1} e^{-z} + (a-1) \int_z^\infty t^{a-2} e^{-t}\, dt. \tag{8.2}$$

Repeating this we find for $n = 1, 2, 3, \ldots$,

$$\Gamma(a, z) = z^{a-1} e^{-z} \left[1 + \frac{a-1}{z} + \frac{(a-1)(a-2)}{z^2} + \cdots \right. \\ \left. + \frac{(a-1)(a-2) \cdots (a-n+1)}{z^{n-1}} \right] + R_n(a, z), \quad (8.3)$$

where

$$R_n(a, z) = (a-1)(a-2) \cdots (a-n) \int_z^\infty t^{a-n-1} e^{-t} \, dt. \quad (8.4)$$

For positive a and z we can easily find a bound for the remainder. If $a > n+1$, the integrand has a maximum at $t_0 = a - n - 1$. If $a \leq n + 1$, the integrand is decreasing on $t > 0$. Anyhow, if $z > a - n$, we can integrate in the integral in (8.4) with respect to the variable $p = t + (n - a) \ln t$, which gives

$$R_n(a, z) = (a-1)(a-2) \cdots (a-n) \int_{p_0}^\infty e^{-p} \frac{dp}{t + n - a}, \quad (8.5)$$

where $p_0 = z + (n - a) \ln z$. Because $t \geq z$ in (8.4) we have $t + n - a \geq z + n - a$, and we obtain

$$R_n(a, z) \leq \frac{(a-1)(a-2) \cdots (a-n)}{z^n} \frac{z}{(z+n-a)} z^{a-1} e^{-z}, \quad z > a - n. \quad (8.6)$$

This shows the asymptotic character of the expansion (8.3) when $n > a$ as $z \to \infty$. However, the condition $z > a - n$ is not enough to make it a useful expansion. When a is also large, say $a \sim z$, then the ratios of successive terms in the expansion (8.3) are of order $\mathcal{O}(1)$, and, hence, the terms are not even becoming small. We say that the expansion in (8.3) does not hold uniformly with respect to $a > 0$. It does hold, however, uniformly for a in compact intervals.

For the function $\gamma(a, z)$ we can also obtain an asymptotic representation. Integration by parts now starts with

$$\gamma(a, z) = \frac{1}{a} \int_0^z e^{-t} \, dt^a = \frac{1}{a} z^a e^{-z} + \frac{1}{a} \int_0^z t^a e^{-t} \, dt. \quad (8.7)$$

This is the beginning of the convergent expansion

$$\gamma(a, z) = \frac{1}{a} z^a e^{-z} \sum_{n=0}^\infty \frac{z^n}{(a+1)(a+2) \cdots (a+n)}. \quad (8.8)$$

This expansion has an asymptotic character when a is large, and again we see that the asymptotic property does not hold uniformly with respect to $z > 0$ (although the expansion is convergent for all finite z). Both expansions in (8.3) and (8.8) have their limitations with respect to which domains we can use them in for numerical computations. But they share one nice property: the coefficients can be computed very easily.

8.2 Uniform asymptotic expansions

The asymptotic expansions of the incomplete gamma functions $\Gamma(a, z)$ and $\gamma(a, z)$ given in the previous section become useless when both parameters a and z are of the same size. The representation for $\Gamma(a, z)$ in (8.3) is valid for any a and z (with the usual condition $|\mathrm{ph}\, z| < \pi$), but we can use it as an asymptotic representation only when $|z| \gg |a|$. As mentioned after (8.6), the ratio of successive terms in the representation (8.3) are of order $\mathcal{O}(1)$ when $z \sim a$.

Nonuniform convergence is well known in convergent power series, for example, in the expansion in (8.8), but also in the expansions of the Gauss hypergeometric function, which is defined by the power series in (2.73), which converges for $|z| < 1$ and, in fact, for any a, b, or c (with the usual exception $c \neq 0, -1, -2, \ldots$). But when one or more of these parameters become large, convergence is slowed down, and efficient computation with this series is not possible. Moreover, stability problems may occur. A similar problem occurs for the J_ν-Bessel function, which has an expansion that follows from (2.70): large z-values have a bad influence on the convergence, whereas large ν-values are in favor of convergence.

For many special functions alternatives can be found when two or several parameters become large in convergent or asymptotic expansions, and usually we need expansions of a completely different type when we request that the expansions are valid for several large parameters. Usually, the asymptotics concentrate on one large parameter, say z, and hopefully the expansion remains valid when another parameter, say a, becomes large as well. We say that the expansion holds for large z, uniformly with respect to a in a domain that may be unbounded, or may be more restricted in size, say of order $\mathcal{O}(z)$.

One essential feature of uniform expansions is the role of certain special functions in the expansions. In the standard, nonuniform, expansions usually only elementary functions occur, such as the exponential and trigonometric functions. In uniform expansions we usually need higher transcendental functions, such as Airy functions, error functions, Fresnel integrals, and so on. The proper choice of these special functions is not always clear without a further study of asymptotic analysis.

For functions defined by differential equations the presence of a *turning point* is an important feature. A turning point at $z = 0$ is visible in the equation $w''(z) - \lambda z p(z) w(z) = 0$, with $p(0) \neq 0$, p analytic at $z = 0$, and λ a large parameter. In this case Airy functions can be used, which are solutions of the equation $w''(z) - zw(z) = 0$, the simplest equation showing a turning point. The Airy equation for the turning point problem is a so-called *comparison equation*. A turning point separates an interval in which the solutions are of exponential type from one in which they oscillate. Airy functions are used in the asymptotic analysis of Bessel functions, because a particular form of the Bessel differential equation reads

$$w''(z) - \nu^2 z p(z) w(z) = 0, \quad p(z) = \frac{1 - e^{2z}}{z}, \tag{8.9}$$

which has a turning point at $z = 0$. The Bessel function $J_\nu(\nu e^z)$ is a solution of this equation. The qualitative behavior of this Bessel function near the turning point is similar to that of the Airy function $\mathrm{Ai}(z)$ at $z = 0$: there is a change from oscillatory behavior to exponentially decreasing behavior when z crosses the turning point, and this effect becomes more noticeable when ν is large. A proper description of the asymptotic behavior of $J_\nu(\nu e^z)$ for large values of ν with z in an interval containing the turning point $z = 0$ is not possible when only elementary functions are used. The use of the Airy function is essential then. This leads to uniform Airy-type expansions of the Bessel functions, as explained in §8.4.

In general, uniform expansions (with special functions as main approximants) are more powerful than the standard expansions in which elementary functions are used. The latter can usually be derived from the uniform expansions by assuming certain growth conditions on the extra parameters and by re-expanding the uniform expansion. On the other hand (i.e., on the negative side), the uniform expansions are more difficult to evaluate numerically. As we will see in later sections, the coefficients in the expansions are quite complicated compared with those in the standard expansions; we need the evaluation of extra special functions, such as error functions and Airy functions, and usually quite complicated mappings are needed to bring an integral or differential equation into a form that can be viewed as a standard form to start the uniform asymptotics. For differential equations a mapping is used that is inspired by a Liouville transformation, as discussed in §2.2.4.

For Bessel functions Airy-type expansions can also be obtained by using integral representations (see again §8.4), and usually for many other special functions the starting point can be a differential equation or an integral. For the incomplete gamma functions discussed in the previous section both approaches have been investigated, and in the next section we give a few (formal) steps on how to obtain a uniform expansion in which the complementary error function plays the role of the main approximant.

8.3 Uniform asymptotic expansions for the incomplete gamma functions

We will now show that uniform asymptotic expansions for the incomplete gamma functions are important additions to the several methods we have mentioned already for computing these functions. In §6.7.1 we have derived a continued fraction for $\Gamma(a, z)$, and in Example 6.6 we have concluded that this continued fraction is an excellent tool for computing this incomplete gamma function. When the parameters a and z become large the computations become less efficient, in particular when $a > z$. In §8.1 we have concluded that the standard asymptotic expansion of $\Gamma(a, z)$ becomes useless when a becomes as large as the large parameter z.

We consider a uniform expansion that can be used for both $\Gamma(a, z)$ and $\gamma(a, z)$, and for all large values of a and z, also for complex values, but we continue the discussion for real positive parameters.

The incomplete gamma functions are related to cumulative distribution functions of probability theory, with the gamma distribution as underlying distribution. In that area the incomplete gamma functions also appear in the form of the chi-square probability functions with definitions (see [117, Chaps. 17–18])

$$\begin{aligned} P\left(\chi^2 | \nu\right) &= \frac{1}{2^{\frac{1}{2}\nu}\Gamma(\frac{1}{2}\nu)} \int_0^{\chi^2} t^{\frac{1}{2}\nu-1} e^{-\frac{1}{2}t} \, dt, \\ Q\left(\chi^2 | \nu\right) &= \frac{1}{2^{\frac{1}{2}\nu}\Gamma(\frac{1}{2}\nu)} \int_{\chi^2}^{\infty} t^{\frac{1}{2}\nu-1} e^{-\frac{1}{2}t} \, dt. \end{aligned} \tag{8.10}$$

8.3. Uniform asymptotic expansions for the incomplete gamma functions

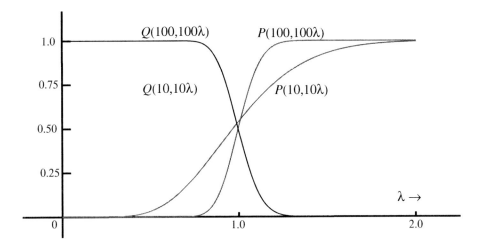

Figure 8.1. *The functions $P(a, \lambda a)$ and $Q(a, \lambda a)$ for $\lambda \in [0, 2]$ and $a = 10$ and $a = 100$.*

The relation with the incomplete gamma functions is

$$P(\chi^2|\nu) = P(a, x), \quad Q(\chi^2|\nu) = Q(a, x), \quad \nu = 2a, \quad \chi^2 = 2x, \tag{8.11}$$

where we have used the ratios

$$P(a, z) = \frac{\gamma(a, z)}{\Gamma(a)}, \quad Q(a, z) = \frac{\Gamma(a, z)}{\Gamma(a)}. \tag{8.12}$$

For several reasons it is convenient to work with these normalized functions, for example, because of their role in probability theory, and because no overflow happens for large values of a. We have

$$P(a, z) + Q(a, z) = 1. \tag{8.13}$$

In Figure 8.1 we show the graphs of these functions, where we have used a parameter λ to scale the z variable. In fact we give the graphs of the functions $P(a, \lambda a)$ and $Q(a, \lambda a)$ for $\lambda \in [0, 2]$ and $a = 10$ and $a = 100$. As a increases the graphs become steeper when λ passes the value $\lambda = 1$. This can be understood when we consider the graphs of the integrands in (8.1). The function $t^a e^{-t}$ has a maximal value at $t = a$, and when a is large and z crosses the maximal value the integral for $\gamma(a, z)$ becomes large, whereas that for $\Gamma(a, z)$ becomes small. This can be seen in Figure 8.1, where the value $\lambda = 1$ is special because the scaled functions $P(a, \lambda a)$ and $Q(a, \lambda a)$ change their behavior at $\lambda = 1$.

As is well known in the theory of cumulative distribution functions, many of these functions approach the normal or Gaussian probability functions when certain parameters become large. In probability theory the normal distribution functions are defined by

$$P(z) = \frac{1}{\sqrt{2\pi}} \int_{-\infty}^{z} e^{-\frac{1}{2}t^2} dt, \quad Q(z) = \frac{1}{\sqrt{2\pi}} \int_{z}^{\infty} e^{-\frac{1}{2}t^2} dt, \tag{8.14}$$

with the property $P(z) + Q(z) = 1$.

In our analysis we prefer the notation in terms of the complementary error function, which is defined by

$$\operatorname{erfc} z = \frac{2}{\sqrt{\pi}} \int_z^\infty e^{-t^2}\, dt, \tag{8.15}$$

with the symmetry relation $\operatorname{erfc} z + \operatorname{erfc}(-z) = 2$. We have the relation with the normal distribution function

$$P(z) = \tfrac{1}{2}\operatorname{erfc}(-z/\sqrt{2}), \quad Q(z) = \tfrac{1}{2}\operatorname{erfc}(z/\sqrt{2}). \tag{8.16}$$

8.3.1 The uniform expansion

In [219, pp. 283–286] we have derived the uniform expansion by using saddle point methods for integrals (see also §5.5). In that analysis the complementary error function appeared, because a singularity (a pole) of the integrand approaches the saddle point when $a \sim z$. We summarize the results by giving the following representations:

$$\begin{aligned} Q(a,z) &= \tfrac{1}{2}\operatorname{erfc}(\eta\sqrt{a/2}) + R_a(\eta), \\ P(a,z) &= \tfrac{1}{2}\operatorname{erfc}(-\eta\sqrt{a/2}) - R_a(\eta), \end{aligned} \tag{8.17}$$

where

$$\tfrac{1}{2}\eta^2 = \lambda - 1 - \ln\lambda, \quad \lambda = \frac{z}{a}, \tag{8.18}$$

and

$$R_a(\eta) = \frac{e^{-\tfrac{1}{2}a\eta^2}}{\sqrt{2\pi a}} S_a(\eta), \quad S_a(\eta) \sim \sum_{n=0}^\infty \frac{C_n(\eta)}{a^n}, \tag{8.19}$$

as $a \to \infty$.

The relation between η and λ in (8.18) becomes clear when we expand

$$\lambda - 1 - \ln\lambda = \tfrac{1}{2}(\lambda-1)^2 - \tfrac{1}{3}(\lambda-1)^3 + \tfrac{1}{4}(\lambda-1)^4 + \cdots, \tag{8.20}$$

and in fact the relation in (8.18) can also be written as

$$\eta = (\lambda - 1)\sqrt{\frac{2(\lambda - 1 - \ln\lambda)}{(\lambda - 1)^2}}, \tag{8.21}$$

where the sign of the square root is positive for $\lambda > 0$. For complex values we use analytic continuation. An expansion for small values of $|\lambda - 1|$ reads

$$\eta = (\lambda - 1) - \tfrac{1}{3}(\lambda - 1)^2 + \tfrac{7}{36}(\lambda - 1)^3 + \cdots, \tag{8.22}$$

and, upon inverting this expansion,

$$\lambda = 1 + \eta + \tfrac{1}{3}\eta^2 + \tfrac{1}{36}\eta^3 + \cdots. \tag{8.23}$$

8.3. Uniform asymptotic expansions for the incomplete gamma functions

Note that the symmetry relation $P(a, z) + Q(a, z) = 1$ is preserved in the representations in (8.17) because $\operatorname{erfc} z + \operatorname{erfc}(-z) = 2$.

We give a few steps on determining the coefficients $C_n(\eta)$ in (8.19). Differentiating the relations in (8.17) of $Q(a, z)$ with respect to η gives, on the one hand,

$$\frac{dQ(a,z)}{d\eta} = \frac{dQ(a,z)}{dz}\frac{dz}{d\eta} = -\frac{1}{\Gamma(a)}z^{a-1}e^{-z}\frac{dz}{d\eta}, \tag{8.24}$$

and on the other hand, by using (8.17)–(8.19),

$$\frac{dQ(a,z)}{d\eta} = \left[-\sqrt{\frac{a}{2\pi}}\frac{\lambda-1}{\lambda} - \eta\sqrt{\frac{a}{2\pi}}S_a(\eta) + \frac{1}{\sqrt{2\pi a}}\frac{dS_a(\eta)}{d\eta}\right]e^{-\frac{1}{2}a\eta^2}, \tag{8.25}$$

where we have used

$$\frac{dz}{d\eta} = a\frac{d\lambda}{d\eta} = a\frac{\lambda\eta}{\lambda-1}. \tag{8.26}$$

After a few manipulations we obtain

$$\frac{dS_a(\eta)}{d\eta} - a\eta S_a(\eta) = a\left[1 - \frac{\eta}{(\lambda-1)\Gamma^*(a)}\right], \tag{8.27}$$

where $\Gamma^*(a)$ is defined by

$$\Gamma^*(a) = \sqrt{\frac{a}{2\pi}}e^a a^{-a}\Gamma(a), \quad a > 0, \tag{8.28}$$

and has the expansion

$$\frac{1}{\Gamma^*(a)} \sim \sum_{n=0}^{\infty}\gamma_n a^{-n}, \quad a \to \infty. \tag{8.29}$$

The first few γ_n are

$$\gamma_0 = 1, \quad \gamma_1 = -\frac{1}{12}, \quad \gamma_2 = \frac{1}{288}, \quad \gamma_3 = \frac{139}{51840}. \tag{8.30}$$

The numbers γ_n also appear in the well-known asymptotic expansion of the Euler gamma function. That is,

$$\Gamma^*(a) \sim \sum_{n=0}^{\infty}(-1)^n\gamma_n a^{-n}, \quad a \to \infty. \tag{8.31}$$

We substitute the expansion of $S_a(\eta)$, given in (8.19), and (8.29) into (8.27). Comparing equal powers of a, we obtain

$$C_0(\eta) = \frac{1}{\lambda-1} - \frac{1}{\eta} \tag{8.32}$$

and the recurrence relation

$$\eta C_n(\eta) = \frac{d}{d\eta}C_{n-1}(\eta) + \frac{\eta}{\lambda-1}\gamma_n, \quad n \geq 1. \tag{8.33}$$

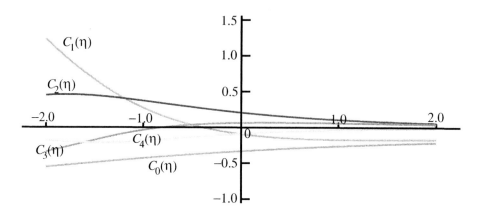

Figure 8.2. *Graphs of the first five coefficients $C_n(\eta)$. Because of scaling we have drawn graphs of $\rho C_n(\eta)$, where $\rho = 1, 50, 50, 100, 100$ for $n = 0, 1, 2, 3, 4$, respectively.*

For $C_1(\eta)$ we have

$$C_1(\eta) = \frac{1}{\eta^3} - \frac{1}{(\lambda-1)^3} - \frac{1}{(\lambda-1)^2} - \frac{1}{12(\lambda-1)}. \tag{8.34}$$

The first two coefficients (and all higher coefficients) have a removable singularity at $\eta = 0$, that is, at $\lambda = 1$ or $z = a$. All $C_n(\eta)$ are analytic at the origin $\eta = 0$.

The expansion in (8.19) has no restrictions on the parameter λ. It holds uniformly with respect to $\lambda \geq 0$ (and for complex values of a and λ). So it is more powerful than, for example, the asymptotic expansion (8.3). However, the computation of the coefficients $C_n(\eta)$ is not as easy as that of the coefficients in (8.3). In particular near the transition point, that is, when $z \sim a$, the removable singularities in the representations of C_0 and C_1 as shown in (8.32) and (8.34) are difficult to handle in numerical computations. All higher coefficients show this type of cancellation, and the removable singularities in C_n are poles of order $2n + 1$.

In Figure 8.2 we show the graphs of $C_n(\eta)$ for $n = 0, 1, 2, 3, 4$, properly scaled in order to get them visible in one figure.

8.3.2 Expansions for the coefficients

In the present discussion we concentrate on the numerical aspects of the expansion of $S_a(\eta)$ in (8.19). We already observed that the coefficients $C_k(\eta)$ in (8.19) are difficult to evaluate near the transition point, and we will give two methods for evaluating the coefficients and the expansion.

The singularities of the mapping $\lambda \to \eta$ (see the relation in (8.18)) follow from the zeros or poles of $d\eta/d\lambda = (\lambda - 1)/(\lambda\eta)$; $\lambda = 0$ is mapped to infinity. A second singular candidate is $\lambda = 1$ with corresponding point $\eta = 0$; but this is a regular point. However, for this mapping we should take into account the values of the logarithm outside the standard

8.3. Uniform asymptotic expansions for the incomplete gamma functions

sector $-\pi \leq \text{ph}\,\lambda \leq \pi$. In fact, when $\lambda = \exp(2\pi i n)$, with $n = \pm 1, \pm 2, \ldots$, the quantity $d\eta/d\lambda$ vanishes. Corresponding η-values satisfying $\frac{1}{2}\eta_n^2 = -2\pi i n$ are singular points of the mapping, and are singular points of the function $S_a(\eta)$ and of the coefficients $C_k(\eta)$.

For $n = \pm 1$ we obtain the η-values $2\sqrt{\pi}\exp(\pm\frac{3}{4}\pi i)$, and it follows that the coefficients $C_n(\eta)$ are analytic inside the disk $|\eta| < 2\sqrt{\pi} = 3.54\ldots$. So, we can expand all coefficients in power series for $|\eta| < 2\sqrt{\pi}$. For numerical applications, these expansions can be used for complex η, say, for $|\eta| \leq 1$. More efficiently, when the variables a and z are real and positive, we can expand the coefficients in terms of Chebyshev polynomials in intervals of the real η-axis.

We give the first terms in the Maclaurin expansions of the first coefficients,

$$\begin{aligned}
C_0(\eta) &= -\tfrac{1}{3} + \tfrac{1}{12}\eta - \tfrac{2}{135}\eta^2 + \tfrac{1}{864}\eta^3 + \tfrac{1}{2835}\eta^4 - \tfrac{139}{777600}\eta^5 + \cdots, \\
C_1(\eta) &= -\tfrac{1}{540} - \tfrac{1}{288}\eta + \tfrac{1}{378}\eta^2 - \tfrac{77}{77760}\eta^3 + \tfrac{1}{4860}\eta^4 - \tfrac{1}{2488320}\eta^5 + \cdots, \\
C_2(\eta) &= \tfrac{25}{6048} - \tfrac{139}{51840}\eta + \tfrac{1}{1296}\eta^2 + \tfrac{1}{497664}\eta^3 - \tfrac{6199}{57736800}\eta^4 + \tfrac{5531}{104509440}\eta^5 + \cdots, \quad (8.35)\\
C_3(\eta) &= \tfrac{101}{155520} + \tfrac{571}{2488320}\eta - \tfrac{54179}{115473600}\eta^2 + \tfrac{41969}{156764160}\eta^3 - \tfrac{20639}{272937600}\eta^4 + \cdots, \\
C_4(\eta) &= -\tfrac{3184811}{3695155200} + \tfrac{163879}{209018880}\eta - \tfrac{8707}{29113344}\eta^2 - \tfrac{47207}{32248627200}\eta^3 + \cdots.
\end{aligned}$$

In the following sections we discuss alternative uniform expansions in which no coefficients occur that are difficult to compute. But first we give another numerical scheme for the expansion discussed above.

8.3.3 Numerical algorithm for small values of η

Instead of expanding each coefficient $C_n(\eta)$ in powers of η, which needs the storage of many coefficients, we expand the function $S_a(\eta)$ of (8.19) in powers of η. The coefficients are functions of a, and we write

$$S_a(\eta) = \sum_{n=0}^{\infty} \alpha_n \eta^n, \qquad (8.36)$$

where the series again converges for $|\eta| < 2\sqrt{\pi}$.

To compute the coefficients α_n, we use the differential equation for $S_a(\eta)$ given in (8.27). Substituting the expansion (8.36) into (8.27), using the coefficients d_n in the expansion

$$\frac{\eta}{\lambda - 1} = \sum_{n=0}^{\infty} d_n \eta^n, \qquad (8.37)$$

we obtain for α_n the recurrence relation

$$\alpha_n = \frac{1}{a}(n+2)\alpha_{n+2} + \frac{d_{n+1}}{\Gamma^*(a)}, \quad n = 0, 1, 2, \ldots. \qquad (8.38)$$

The coefficients d_n follow from the coefficients of $C_0(\eta)$, of which the first few are given in (8.35), because $\eta/(\lambda - 1) = 1 + \eta C_0(\eta)$. We have

$$d_0 = 1, \quad d_1 = -\tfrac{1}{3}, \quad d_2 = -\tfrac{1}{12}, \quad d_3 = -\tfrac{2}{135}, \quad d_4 = \tfrac{1}{864}, \quad d_5 = \tfrac{1}{2835}. \qquad (8.39)$$

With these values we can compute the first terms with odd index,

$$\alpha_1 = \frac{a}{\Gamma^*(a)}\left[\Gamma^*(a) - 1\right],$$

$$\alpha_3 = \frac{a^2}{1\cdot 3\,\Gamma^*(a)}\left[\Gamma^*(a) - 1 - \frac{1}{12a}\right], \quad (8.40)$$

$$\alpha_5 = \frac{a^2}{1\cdot 3\cdot 5\,\Gamma^*(a)}\left[\Gamma^*(a) - 1 - \frac{1}{12a} - \frac{1}{288a^2}\right].$$

We observe (see (8.30) and (8.31)) that the computation of these α_n requires not only the value of $\Gamma^*(a)$, but also that of $\Gamma^*(a)$ with the first terms of the asymptotic expansion subtracted. The higher odd coefficients show the same pattern; more and more terms of the asymptotic expansions have to be subtracted. In fact we recur remainders of the asymptotic expansion of the gamma function. In particular when a is large, this is a very unstable process. The same problems arise with the even coefficients. Note that $\alpha_0 = S_a(0)$, a quantity that can be computed from an asymptotic expansion, and the higher even terms follow from the recursion in (8.38), with more and more terms subtracted in this expansion of $S_a(0)$.

When we use (8.38) in the backward direction the recursion becomes stable. In addition, we don't need the computation of $S_a(0)$ and $\Gamma^*(a)$, because these values follow from the backward recursion process. We only need enough coefficients d_n of (8.37) for this recursion. We give a few details.

We remove $\Gamma^*(a)$ from the recursion in (8.38) by writing

$$\alpha_n = \frac{\beta_n}{\Gamma^*(a)}, \quad n = 0, 1, 2, \ldots, \quad (8.41)$$

which gives for β_n the recursion

$$\beta_n = \frac{1}{a}(n+2)\beta_{n+2} + d_{n+1}, \quad n = 0, 1, 2, \ldots. \quad (8.42)$$

We choose a positive integer N, put $\beta_{N+2} = \beta_{N+1} = 0$, and compute the sequence

$$\beta_N, \beta_{N-1}, \ldots, \beta_1, \beta_0 \quad (8.43)$$

from the recurrence relation (8.42). Because

$$\Gamma^*(a) = 1 + \frac{1}{a}\beta_1, \quad (8.44)$$

we have

$$S_a(\eta) \approx \frac{a}{a+\beta_1}\sum_{n=0}^{N}\beta_n \eta^n \quad (8.45)$$

as an approximation for $S_a(\eta)$.

We verify this algorithm by taking several values of a and $N = 25$ and $N = 35$. In Table 8.1 we give the relative errors of the approximations of $\Gamma(a)$, which are computed by using β_1 in (8.44), and by computing $\Gamma(a)$ from the relation in (8.28). For example, with $N = 25$ and $a = 5$ we obtain

$$\Gamma(a) = 23.999999999892\ldots, \quad \text{with relative error } 0.45\times 10^{-11}. \quad (8.46)$$

8.3. Uniform asymptotic expansions for the incomplete gamma functions

Table 8.1. *Relative errors δ in the computation of $\Gamma(a)$ by using the backward recursion scheme (8.42) for several values of a and $N = 25$ and $N = 35$.*

	a	δ	a	δ	a	δ
$N = 25$	2	$0.39\ 10^{-06}$	8	$0.11\ 10^{-13}$	14	$0.79\ 10^{-17}$
	3	$0.27\ 10^{-08}$	9	$0.24\ 10^{-14}$	15	$0.32\ 10^{-17}$
	4	$0.75\ 10^{-10}$	10	$0.61\ 10^{-15}$	16	$0.14\ 10^{-17}$
	5	$0.45\ 10^{-11}$	11	$0.18\ 10^{-15}$	17	$0.64\ 10^{-18}$
	6	$0.44\ 10^{-12}$	12	$0.58\ 10^{-16}$	18	$0.30\ 10^{-18}$
	7	$0.61\ 10^{-13}$	13	$0.21\ 10^{-16}$	19	$0.15\ 10^{-18}$
$N = 35$	2	$0.80\ 10^{-06}$	8	$0.76\ 10^{-17}$	14	$0.19\ 10^{-21}$
	3	$0.57\ 10^{-09}$	9	$0.82\ 10^{-18}$	15	$0.50\ 10^{-22}$
	4	$0.30\ 10^{-11}$	10	$0.11\ 10^{-18}$	16	$0.15\ 10^{-22}$
	5	$0.50\ 10^{-13}$	11	$0.18\ 10^{-19}$	17	$0.46\ 10^{-23}$
	6	$0.17\ 10^{-14}$	12	$0.35\ 10^{-20}$	18	$0.15\ 10^{-23}$
	7	$0.93\ 10^{-16}$	13	$0.77\ 10^{-21}$	19	$0.54\ 10^{-24}$

We observe that for the larger values of a the scheme gives better approximations, as is the case for the larger value of N.

We have used the approximation in (8.45) for computing the incomplete gamma functions in IEEE double precision for $a \geq 12$ and $|\eta| \leq 1$. We need the storage of 25 coefficients d_n, and in the series in (8.45) we need 25 terms or less.

The value $\eta = -1$ corresponds to $\lambda = 0.30\ldots$, and the value $\eta = 1$ to $\lambda = 2.35\ldots$. In Figure 8.3 we show the area in the (x, a) quarter-plane where we can apply the algorithm to obtain IEEE double precision. The domain is bounded by the lines $a = 12$, $x = 2.35a$, and $x = 0.30a$.

8.3.4 A simpler uniform expansion

The expansion considered in §8.3.1 can be modified to obtain an expansion with coefficients that can be evaluated much easier than the coefficients $C_n(\eta)$. The modified expansion is again valid for large values of a; it is again valid in the transition area $a \sim z$. The restriction on $\lambda = z/a$, however, is $\lambda - 1 = o(a^{-1/3})$ as $a \to \infty$.

We start with the integral (see (8.1))

$$\Gamma(a+1, z) = \int_z^\infty t^a e^{-t}\, dt, \qquad (8.47)$$

and we consider positive parameters a and z. We substitute $t = a(1+s)$. This gives

$$\Gamma(a+1, z) = a^{a+1} e^{-a} \int_\mu^\infty e^{-a[s-\ln(1+s)]}\, ds, \quad \mu = \lambda - 1 = \frac{z-a}{a}. \qquad (8.48)$$

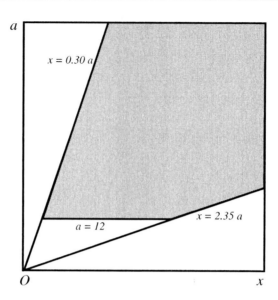

Figure 8.3. *The domain of application (gray) where we can apply the backward recursion scheme (8.42) to obtain IEEE double precision values of $S_a(\eta)$. The gray domain is bounded by the lines $a = 12$, $x = 2.35a$, and $x = 0.30a$.*

The exponential function has a maximum at the origin $s = 0$. We write

$$\Gamma(a+1, z) = a^{a+1} e^{-a} \int_\mu^\infty e^{-\frac{1}{2}as^2 - af(s)} \, ds, \qquad (8.49)$$

where

$$f(s) = s - \tfrac{1}{2}s^2 - \ln(1+s) = \mathcal{O}(s^3), \quad s \to 0. \qquad (8.50)$$

We expand

$$e^{-af(s)} = \sum_{n=0}^\infty D_n(a) s^n, \quad |s| < 1, \qquad (8.51)$$

and upon substituting this expansion in (8.49), we obtain the expansion

$$\Gamma(a+1, z) \sim a^{a+1} e^{-a} \sum_{n=0}^\infty D_n(a) \Phi_n(a, z), \qquad (8.52)$$

where

$$\Phi_n(a, z) = \int_\mu^\infty s^n e^{-\frac{1}{2}as^2} \, ds, \quad n = 0, 1, 2, \ldots. \qquad (8.53)$$

The first two Φ_n are

$$\Phi_0(a, z) = \sqrt{\frac{\pi}{2a}} \operatorname{erfc}(\mu \sqrt{a/2}), \quad \Phi_1(a, z) = \frac{1}{a} e^{-\frac{1}{2} a \mu^2}. \qquad (8.54)$$

By integrating by parts in (8.53) it easily follows that

$$a\Phi_{n+1}(a,z) = n\Phi_{n-1}(a,z) + \mu^n e^{-\frac{1}{2}a\mu^2}, \quad n = 1, 2, 3, \ldots. \tag{8.55}$$

Also for the coefficients we can derive a recurrence relation. We have $D_0(a) = 1$, $D_1(a) = 0$, and by differentiating (8.51), we obtain

$$(n+1)D_{n+1}(a) = aD_{n-2}(a) - nD_n(a), \quad n = 1, 2, 3, \ldots. \tag{8.56}$$

It can be shown that (8.52) is an asymptotic expansion as $a \to \infty$ if $\mu = o(a^{-1/3})$. The uniform expansion in §8.3.1 holds uniformly with respect to $\mu \geq -1$ and is, hence, more powerful. The expansion in (8.52) is of interest because the coefficients $D_n(a)$ can be computed very easily by using the recursion in (8.56). Also the recursion for the Φ_n is quite simple.

The first terms of the expansion in (8.52) are given by Tricomi [225]. The complete expansion, with proofs for complex values of a and z, is derived in [68]. In this reference several other expansions of the incomplete gamma function are considered.

We observe that a similar expansion can be derived for the other incomplete gamma function $\gamma(a+1, z)$. In that case the functions Φ_n should be replaced by the functions Ψ_n defined by

$$\Psi_n(a,z) = \int_{-1}^{\mu} s^n e^{-\frac{1}{2}as^2} \, ds, \quad n = 0, 1, 2, \ldots. \tag{8.57}$$

The function Ψ_0 can be expressed in terms of two error functions, and the other ones follow from integrating by parts.

From a numerical example we conclude that the asymptotic convergence of the expansion (8.52) is quite slow. When we take $z = 100$ and $a = 101$, and we sum the series with 12 terms, we obtain four significant digits. This result does not improve when we take more terms.

8.4 Airy-type expansions for Bessel functions

We recall (see Chapter 1) that Airy functions are solutions of the differential equation

$$w'' - zw = 0, \tag{8.58}$$

and that two linearly independent solutions that are real for real values of z are denoted by $\text{Ai}(z)$ and $\text{Bi}(z)$. Equation (8.58) is the simplest second order linear differential equation that has a simple turning point (at $z = 0$). More general turning point equations have the standard form

$$\frac{d^2 W}{d\zeta^2} = \left[u^2 \zeta + \psi(\zeta)\right] W, \tag{8.59}$$

and the problem is to find an asymptotic approximation of $W(\zeta)$ for large values of u that holds uniformly in a neighborhood of $\zeta = 0$. A first approximation is obtained by neglecting $\psi(\zeta)$, which gives the solutions

$$\text{Ai}\left(u^{2/3}\zeta\right), \quad \text{Bi}\left(u^{2/3}\zeta\right). \tag{8.60}$$

For a detailed discussion of this kind of problem we refer to [168, Chap. 11]. Many solutions of physical problems and many special functions can be transformed into the standard form (8.59). Examples are Bessel functions, Whittaker functions, the classical orthogonal polynomials (in particular Hermite and Laguerre polynomials), and parabolic cylinder functions. The existing uniform expansions for all these functions are powerful in an analytic sense. In several cases rigorous and realistic bounds are given for the remainders of the expansions; cf. [168].

In all known cases the coefficients are difficult to compute in the neighborhood of the turning point, and we saw a similar difficulty in the uniform expansion of the incomplete gamma functions in §8.3. Usually, the turning point is of special interest in the algorithms, since many other methods fail in the turning point area when the parameters are large. In [3] uniform Airy-type expansions are used for the evaluation of Bessel functions. In [151] the implementation of several kinds of asymptotic expansions of the Bessel functions is discussed, however, without referring to Airy-type expansions; for the turning point region that reference proposed numerical quadrature for the Sommerfeld integral of the Hankel functions, after selecting contours of steepest descents.

In this section we discuss two methods for computing the asymptotic series. One method is based on expanding the coefficients in the series into Maclaurin series. We show how to obtain the coefficients of the Maclaurin series for the coefficients of the asymptotic series. In the second method we consider the computation of auxiliary functions that can be computed more efficiently than the coefficients in the first method, and we don't need the tabulation of many coefficients. In some sense this method is similar to the one described for the computation of incomplete gamma functions in §8.3.3.

The methods described in this section are quite general, but we only treat the case of Bessel functions by using the differential equation of the Bessel functions, which has a turning point character when the order and argument of the Bessel functions are equal.

8.4.1 The Airy-type asymptotic expansions

The ordinary Bessel functions $J_\nu(z)$ and $Y_\nu(z)$, and all other Bessel functions, can be expanded in terms of Airy functions. For example, the Bessel function $J_\nu(x)$ has a turning point at $x = \nu$, and for large ν the function is oscillatory for $x > \nu$ and monotonic for $x < \nu$. See Figure 8.4, and compare the graph of $J_\nu(x)$ with the graph of the Airy function $\text{Ai}(x)$ in Figure 1.1.

We give the transformations of the Bessel differential equation

$$z^2 f'' + z f' + \left(z^2 - \nu^2\right) f = 0 \tag{8.61}$$

into the form (8.59). First we change the variable z into νz and apply a Liouville transformation to remove the first derivative term, as explained in §2.2.4. We obtain the equation

$$F'' + \left(\nu^2 \frac{z^2 - 1}{z^2} + \frac{1}{4z^2}\right) F = 0, \tag{8.62}$$

8.4. Airy-type expansions for Bessel functions

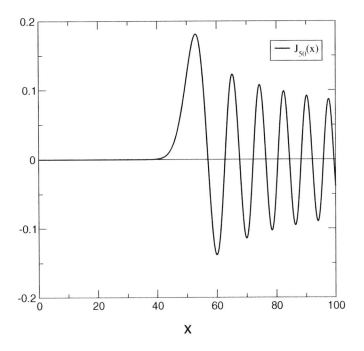

Figure 8.4. *Graph of $J_\nu(x)$ for large ν. Compare with the graph of the Airy function in Figure 1.1.*

with solutions $\sqrt{z}J_\nu(\nu z)$ and $\sqrt{z}Y_\nu(\nu z)$. The turning point character at $z=1$ of this equation is visible now, and we transform this point to the origin by using the Liouville transformation (again see §2.2.4)

$$\zeta\left(\frac{d\zeta}{dz}\right)^2 = \frac{1-z^2}{z^2}, \quad W = \sqrt{\zeta'}w, \tag{8.63}$$

This transformation gives (8.59), with $\psi(\zeta)$ given by

$$\psi(\zeta) = \frac{5}{16\zeta^2} + \frac{\zeta z^2(z^2+4)}{4(z^2-1)^3} \tag{8.64}$$

and solutions

$$\sqrt{z}\sqrt{\zeta'}J_\nu(\nu z), \quad \sqrt{z}\sqrt{\zeta'}Y_\nu(\nu z). \tag{8.65}$$

The comparison equation in the form of (8.59) with $\psi = 0$ has as solutions the Airy functions given in (8.60), and considering further properties of the Bessel functions, the following representations are chosen:

$$J_\nu(\nu z) = \frac{\phi(\zeta)}{\nu^{1/3}}\left[\operatorname{Ai}(\nu^{2/3}\zeta)\, A_\nu(\zeta) + \nu^{-4/3}\operatorname{Ai}'(\nu^{2/3}\zeta)\, B_\nu(\zeta)\right],$$

$$Y_\nu(\nu z) = -\frac{\phi(\zeta)}{\nu^{1/3}}\left[\operatorname{Bi}(\nu^{2/3}\zeta)\, A_\nu(\zeta) + \nu^{-4/3}\operatorname{Bi}'(\nu^{2/3}\zeta)\, B_\nu(\zeta)\right], \tag{8.66}$$

where
$$\phi(\zeta) = \sqrt{2z\zeta'} = \left(\frac{4\zeta}{1-z^2}\right)^{\frac{1}{4}}, \quad \phi(0) = 2^{\frac{1}{3}}. \tag{8.67}$$

The new variable ζ introduced in (8.63) can be written as
$$\frac{2}{3}\zeta^{3/2} = \ln\frac{1+\sqrt{1-z^2}}{z} - \sqrt{1-z^2}, \quad 0 \le z \le 1,$$
$$\frac{2}{3}(-\zeta)^{3/2} = \sqrt{z^2-1} - \arccos\frac{1}{z}, \quad z \ge 1. \tag{8.68}$$

Next we introduce asymptotic expansions for the functions $A_\nu(\zeta)$ and $B_\nu(\zeta)$ of (8.66). It appears, after substituting the Bessel functions in (8.66) into (8.59), that we have the formal expansions
$$A_\nu(\zeta) \sim \sum_{s=0}^{\infty} \frac{a_s(\zeta)}{\nu^{2s}}, \quad B_\nu(\zeta) \sim \sum_{s=0}^{\infty} \frac{b_s(\zeta)}{\nu^{2s}}, \tag{8.69}$$
where the coefficients $a_s(\zeta)$ and $b_s(\zeta)$ satisfy certain recurrence relations. Further details on the coefficients are given later.

Asymptotic representations for the Hankel functions follow from the relations
$$H_\nu^{(1)}(z) = J_\nu(z) + iY_\nu(z), \quad H_\nu^{(2)}(z) = J_\nu(z) - iY_\nu(z), \tag{8.70}$$
and
$$\text{Ai}(z) + i\text{Bi}(z) = 2e^{\pi i/3}\text{Ai}\left(ze^{-2\pi i/3}\right),$$
$$\text{Ai}(z) - i\text{Bi}(z) = 2e^{-\pi i/3}\text{Ai}\left(ze^{2\pi i/3}\right). \tag{8.71}$$

This gives representations for the Hankel functions with the same structure as for the ordinary Bessel functions, with the same functions $A_\nu(\zeta)$ and $B_\nu(\zeta)$.

For the derivatives we have
$$J_\nu'(\nu z) = -\widehat{\phi}(\zeta)\left[\nu^{-4/3}\text{Ai}(\nu^{2/3}\zeta)C_\nu(\zeta) + \nu^{-2/3}\text{Ai}'(\nu^{2/3}\zeta)D_\nu(\zeta)\right],$$
$$Y_\nu'(\nu z) = \widehat{\phi}(\zeta)\left[\nu^{-4/3}\text{Bi}(\nu^{2/3}\zeta)C_\nu(\zeta) + \nu^{-2/3}\text{Bi}'(\nu^{2/3}\zeta)D_\nu(\zeta)\right], \tag{8.72}$$
where
$$\widehat{\phi}(\zeta) = -\frac{d\zeta}{dz}\phi(\zeta) = \frac{2}{z\phi(\zeta)}, \quad \chi(\zeta) = \frac{\phi'(\zeta)}{\phi(\zeta)} = \frac{4 - z^2[\phi(\zeta)]^6}{16\zeta}, \tag{8.73}$$
and
$$C_\nu(\zeta) = \chi(\zeta)A_\nu(\zeta) + A_\nu'(\zeta) + \zeta B_\nu(\zeta),$$
$$D_\nu(\zeta) = A_\nu(\zeta) + \nu^{-2}\chi(\zeta)B_\nu(\zeta) + \nu^{-2}B_\nu'(\zeta). \tag{8.74}$$

Primes denote differentiation with respect to ζ.

The functions $C_\nu(\zeta), D_\nu(z)$ have the expansions
$$C_\nu(\zeta) \sim \sum_{s=0}^{\infty} \frac{c_s(\zeta)}{\nu^{2s}}, \quad D_\nu(\zeta) \sim \sum_{s=0}^{\infty} \frac{d_s(\zeta)}{\nu^{2s}}, \tag{8.75}$$

8.4. Airy-type expansions for Bessel functions

where

$$c_s(\zeta) = \chi(\zeta) a_s(\zeta) + a'_s(\zeta) + \zeta b_s(\zeta),$$
$$d_s(\zeta) = a_s(\zeta) + \chi(\zeta) b_{s-1}(\zeta) + b'_{s-1}(\zeta),$$
(8.76)

The Airy-type asymptotic expansions of this section hold as $\nu \to \infty$, uniformly with respect to $z \in [0, \infty)$. The expansions are valid for complex values of ν and z, but here we concentrate on real values of the parameters. In [168, Chap. 11] these expansions are given with remainders and bounds for the remainders; for the J_ν-function see [168, p. 423].

8.4.2 Representations of $a_s(\zeta), b_s(\zeta), c_s(\zeta), d_s(\zeta)$

The first coefficients a_s, b_s of (8.69) are

$$a_0(\zeta) = 1, \quad b_0(\zeta) = -\frac{5}{48\zeta^2} + \frac{\phi^2(\zeta)}{48\zeta}\left[\frac{5}{1-z^2} - 3\right].$$
(8.77)

Higher coefficients follow from the representations

$$a_s(\zeta) = \sum_{k=0}^{2s} \mu_k \, \zeta^{-3k/2} \, u_{2s-k}(t),$$
$$b_s(\zeta) = -\zeta^{-1/2} \sum_{k=0}^{2s+1} \lambda_k \, \zeta^{-3k/2} \, u_{2s+1-k}(t),$$
(8.78)

where $t = 1/\sqrt{1-z^2}$, $\lambda_0 = \mu_0 = 1$, and

$$\lambda_k = \frac{(2k+1)(2k+3)\cdots(6k-1)}{k!\,(144)^k}, \quad \mu_k = -\frac{6k+1}{6k-1}\lambda_k, \quad k = 1, 2, 3, \ldots.$$
(8.79)

The quantities u_k are given by

$$u_{k+1}(t) = \tfrac{1}{2} t^2 (1-t^2) u'_k(t) + \tfrac{1}{8}\int_0^t (1 - 5\tau^2) u_k(\tau)\, d\tau, \quad k = 0, 1, 2, \ldots,$$
(8.80)

with $u_0(t) = 1$.

The first coefficients c_s, d_s of (8.75) are

$$c_0(\zeta) = \frac{7}{48\zeta} + \frac{\phi^2(\zeta)}{48}\left[9 - \frac{7}{1-z^2}\right], \quad d_0(\zeta) = 1,$$
(8.81)

where $\phi(\zeta)$ is given by (8.67). Higher coefficients follow from the representations

$$c_s(\zeta) = -\zeta^{1/2} \sum_{k=0}^{2s+1} \mu_k \, \zeta^{-3k/2} \, v_{2s+1-k}(t),$$
$$d_s(\zeta) = \sum_{k=0}^{2s} \lambda_k \, \zeta^{-3k/2} \, v_{2s-k}(t),$$
(8.82)

where t, λ_k, and μ_k are as in (8.78)–(8.79), and the quantities v_k can be expressed in terms of the u_k of (8.80),

$$v_k(t) = u_k(t) + t(t^2 - 1)\left[\tfrac{1}{2}u_{k-1}(t) + t\,u'_{k-1}(t)\right], \quad k = 1, 2, \ldots, \tag{8.83}$$

with $v_0(t) = 1$. Explicit representations of $a'_s(\zeta)$, $b'_s(\zeta)$ can be obtained by differentiating the relations in (8.78), but they also follow from the representations for a_s, b_s, c_s, d_s and from (8.76),

$$\begin{aligned} a'_s(\zeta) &= c_s(\zeta) - \chi(\zeta)\,a_s(\zeta) - \zeta\,b_s(\zeta), \\ b'_s(\zeta) &= d_{s+1}(\zeta) - a_{s+1}(\zeta) - \chi(\zeta)\,b_s(\zeta). \end{aligned} \tag{8.84}$$

A recursive scheme for evaluating a_s, b_s is given by

$$\begin{aligned} a''_s(\zeta) + 2\zeta b'_s(\zeta) + b_s(\zeta) - \psi(\zeta)\,a_s(\zeta) &= 0, \\ 2a'_{s+1}(\zeta) + b''_s(\zeta) - \psi(\zeta)\,b_s(\zeta) &= 0, \end{aligned} \tag{8.85}$$

where $a_0(\zeta) = 1$ and $\psi(\zeta)$ is given by (8.64).

The coefficients a_s, b_s, c_s, d_s in (8.69) and (8.75) are complicated expressions. Explicit representations are given in (8.78) and (8.82) in terms of the coefficients u_k of Debye-type asymptotic expansions [2, p. 366]. However, these expressions are difficult to compute near the turning point $z = 1$, or equivalently, near $\zeta = 0$. In [3] all needed coefficients a_s, b_s are expanded in power series of the variable $w^2 = 1 - z^2$.

8.4.3 Properties of the functions A_ν, B_ν, C_ν, D_ν

Using the Wronskian for the Airy functions, namely,

$$\text{Ai}(z)\text{Bi}'(z) - \text{Ai}'(z)\text{Bi}(z) = \frac{1}{\pi}, \tag{8.86}$$

we can invert the relations in (8.66) and obtain

$$\begin{aligned} A_\nu(\zeta) &= \frac{\pi\nu^{1/3}}{\phi(\zeta)}\left[J_\nu(\nu z)\,\text{Bi}'(\nu^{2/3}\zeta) + Y_\nu(\nu z)\,\text{Ai}'(\nu^{2/3}\zeta)\right], \\ B_\nu(\zeta) &= -\frac{\pi\nu^{5/3}}{\phi(\zeta)}\left[J_\nu(\nu z)\,\text{Bi}(\nu^{2/3}\zeta) + Y_\nu(\nu z)\,\text{Ai}(\nu^{2/3}\zeta)\right]. \end{aligned} \tag{8.87}$$

The functions $A_\nu(\zeta)$ and $B_\nu(\zeta)$ are the "slowly varying" parts in the representations in (8.66).

By using (8.59), with $\psi(\zeta)$ as given in (8.64), we can derive the following system of differential equations for the functions $A_\nu(\zeta)$, $B_\nu(\zeta)$:

$$\begin{aligned} A'' + 2\zeta B' + B - \psi(\zeta)A &= 0, \\ B'' + 2\nu^2 A' - \psi(\zeta)B &= 0, \end{aligned} \tag{8.88}$$

8.4. Airy-type expansions for Bessel functions

where primes denote differentiation with respect to ζ. To verify this we write (8.59) in the operator form $\mathcal{L}_\zeta W(\zeta) = 0$. Applying $\mathcal{L}_\zeta$ to

$$W(\zeta) = \text{Ai}(v^{2/3}\zeta)\, A_\nu(\zeta) + v^{-4/3}\text{Ai}'(v^{2/3}\zeta)\, B_\nu(\zeta), \qquad (8.89)$$

we find

$$\begin{aligned}\mathcal{L}_\zeta W(\zeta) = \text{Ai}(v^{2/3}\zeta)[A''_\nu(\zeta) + 2\zeta B'_\nu(\zeta) + B_\nu(\zeta) - \psi(\zeta)A_\nu(\zeta)] \\ + v^{-4/3}\text{Ai}'(v^{2/3}\zeta)[B''_\nu(\zeta) + 2v^2 A'_\nu(\zeta) - \psi(\zeta)B_\nu(\zeta)],\end{aligned} \qquad (8.90)$$

where we have used the differential equation of the Airy functions; cf. (8.58). Because $\mathcal{L}_\zeta W(\zeta) = 0$, the quantities within square brackets in (8.90) must vanish.

A Wronskian for the system (8.88) follows by eliminating the terms with $\psi(\zeta)$. This gives

$$A''B - B''A + B^2 + 2\zeta B'B - 2v^2 A'A = 0, \qquad (8.91)$$

which can be integrated,

$$v^2 A_\nu^2(\zeta) + A_\nu(\zeta)\, B'_\nu(\zeta) - A'_\nu(\zeta)\, B_\nu(\zeta) - \zeta B_\nu^2(\zeta) = v^2. \qquad (8.92)$$

The constant on the right-hand side follows by taking $\zeta = 0$ and from information given later in this section.

By using the Wronskian for the Bessel functions,

$$J_\nu(z)\, Y'_\nu(z) - J'_\nu(z)\, Y_\nu(z) = \frac{2}{\pi z}, \qquad (8.93)$$

it follows that $A_\nu, B_\nu, C_\nu, D_\nu$ are related in the following way:

$$A_\nu(\zeta)\, D_\nu(\zeta) - v^{-2} B_\nu(\zeta)\, C_\nu(\zeta) = 1. \qquad (8.94)$$

By substituting $C_\nu(\zeta), D_\nu(\zeta)$ of (8.74) into (8.94), we again obtain (8.92).

The system in (8.88) is equivalent to a (4×4)-system of first order equations, admitting four independent solutions. The solution $\{A, A', B, B'\}$ that we need satisfies initial conditions at, say, $\zeta = 0$. Exact initial values of A, A', B, B' at $\zeta = 0$ can be obtained from (8.87). They involve values of the Airy functions (and the derivatives thereof) at the origin, and $J_\nu(v), J'_\nu(v), Y_\nu(v), Y'_\nu(v)$. In a numerical scheme for solving the system (8.88), these initial values are needed, up to a certain accuracy.

Expansions of $A_\nu, B_\nu, C_\nu, D_\nu$ at the turning point

It is convenient to collect some information from the literature on the initial values at the turning point $\zeta = 0, z = 1$ of system (8.88), because these values give insight into recurrence relations discussed later. From [2, p. 368] we obtain

$$\begin{aligned}J_\nu(v) &= v^{-1/3} 2^{1/3} \text{Ai}(0)\, S(v) + v^{-5/3} 2^{2/3} \text{Ai}'(0)\, T(v), \\ Y_\nu(v) &= -v^{-1/3} 2^{1/3} \text{Bi}(0)\, S(v) - v^{-5/3} 2^{2/3} \text{Bi}'(0)\, T(v), \\ J'_\nu(v) &= -v^{-2/3} 2^{2/3} \text{Ai}'(0)\, U(v) - v^{-4/3} 2^{1/3} \text{Ai}(0)\, V(v), \\ Y'_\nu(v) &= v^{-2/3} 2^{2/3} \text{Bi}'(0)\, U(v) + v^{-4/3} 2^{1/3} \text{Bi}(0)\, V(v),\end{aligned} \qquad (8.95)$$

in which S, T, U, V denote functions having the following asymptotic expansions:

$$S(\nu) \sim \sum_{n=0}^{\infty} \frac{\alpha_n}{\nu^{2n}}, \quad \alpha_0 = 1, \quad \alpha_1 = -\frac{1}{225},$$
$$T(\nu) \sim \sum_{n=0}^{\infty} \frac{\beta_n}{\nu^{2n}}, \quad \beta_0 = \frac{1}{70}, \quad \beta_1 = -\frac{1213}{1023750},$$
$$U(\nu) \sim \sum_{n=0}^{\infty} \frac{\gamma_n}{\nu^{2n}}, \quad \gamma_0 = 1, \quad \gamma_1 = \frac{23}{3150},$$
$$V(\nu) \sim \sum_{n=0}^{\infty} \frac{\delta_n}{\nu^{2n}}, \quad \delta_0 = \frac{1}{5}, \quad \delta_1 = -\frac{947}{346500}.$$
(8.96)

From the Wronskian in (8.93) it follows that

$$T(\nu) V(\nu) = \nu^2 [S(\nu) U(\nu) - 1]. \tag{8.97}$$

The function $\phi(\zeta)$ defined in (8.67) has the expansion $\phi(\zeta) = 2^{1/3} + \frac{1}{5}\zeta + \mathcal{O}(\zeta^2)$. It follows from (8.66), (8.72), and (8.93) that

$$A_\nu(0) = S(\nu), \qquad A'_\nu(0) = 2^{-1/3}\left[V(\nu) - \tfrac{1}{5}S(\nu)\right],$$
$$B_\nu(0) = 2^{1/3} T(\nu), \quad B'_\nu(0) = -\tfrac{1}{5}T(\nu) + \nu^2 [U(\nu) - S(\nu)].$$
(8.98)

It is easily verified that $A'_\nu(0) = \mathcal{O}(\nu^{-2})$, $B'_\nu(0) = \frac{2}{225} + \mathcal{O}(\nu^{-2})$, as $\nu \to \infty$. Observe that leading terms in $V(\nu) - \tfrac{1}{5}S(\nu)$ and $U(\nu) - S(\nu)$ are canceled.

8.4.4 Expansions for $a_s(\zeta), b_s(\zeta), c_s(\zeta), d_s(\zeta)$

We describe an algorithm for computing the Bessel functions $J_\nu(\nu z), Y_\nu(\nu z)$ and their derivatives in the neighborhood of the turning point $z = 1$ for large values of the parameter ν. We concentrate on the evaluation of the functions $A_\nu, B_\nu, C_\nu, D_\nu$ introduced in §8.4.1 for ζ near the origin, say for $|\zeta| \leq 1$. For real values of ζ this gives an interval in the z-domain around $z = 1$, that is, $[0.39, 1.98]$. A straightforward method is based on using Maclaurin series expansions of the quantities involved in powers of ζ.

The singular points of the functions $z(\zeta), \psi(\zeta), \phi(\zeta), \widehat{\phi}(\zeta), \chi(\zeta)$ and those of coefficients of the asymptotic expansions occur at

$$\zeta^{\pm} = \left(\tfrac{3}{2}\pi\right)^{2/3} e^{\pm i\pi/3}. \tag{8.99}$$

These points correspond to the $z = e^{\mp i\pi}$. It follows that the radius of convergence of the Maclaurin series of these quantities equals $2.81\ldots$. In this section we give the expansions and mention the values of the early coefficients.

It is convenient to start with an expansion of z in powers of ζ. We obtain from (8.63)

$$\zeta z^2 = (1 - z^2)\left(\frac{dz}{d\zeta}\right)^2, \tag{8.100}$$

8.4. Airy-type expansions for Bessel functions

and substitute $z = 1 + z_1 \zeta + \cdots$. This gives $z_1^3 = -1/2$. Using the relations in (8.68) we obtain the correct branch: $z_1 = -2^{-1/3}$. We write

$$\zeta = 2^{1/3} \eta, \tag{8.101}$$

and obtain in a straightforward way the following expansions:

$$z(\zeta) = \sum_{n=0}^{\infty} z_n \eta^n = \left[1 - \eta + \tfrac{3}{10}\eta^2 + \tfrac{1}{350}\eta^3 - \tfrac{479}{63000}\eta^4 + \cdots\right],$$

$$\psi(\zeta) = 2^{1/3} \sum_{n=0}^{\infty} \psi_n \eta^n = 2^{1/3}\left[\tfrac{1}{70} + \tfrac{2}{75}\eta + \tfrac{138}{13475}\eta^2 - \tfrac{296}{73125}\eta^3 + \cdots\right],$$

$$\phi(\zeta) = 2^{1/3} \sum_{n=0}^{\infty} \phi_n \eta^n = 2^{1/3}\left[1 + \tfrac{1}{5}\eta + \tfrac{9}{350}\eta^2 - \tfrac{89}{15750}\eta^3 - \tfrac{4547}{1155000}\eta^4 + \cdots\right], \tag{8.102}$$

$$\widehat{\phi}(\zeta) = 2^{2/3} \sum_{n=0}^{\infty} \widehat{\phi}_n \eta^n = 2^{2/3}\left[1 + \tfrac{4}{5}\eta + \tfrac{18}{35}\eta^2 + \tfrac{88}{315}\eta^3 + \tfrac{79586}{606375}\eta^4 + \cdots\right],$$

$$\chi(\zeta) = 2^{-1/3} \sum_{n=0}^{\infty} \chi_n \eta^n = 2^{-1/3}\left[\tfrac{1}{5} + \tfrac{2}{175}\eta - \tfrac{64}{2625}\eta^2 - \tfrac{30424}{3031875}\eta^3 + \cdots\right].$$

Next we consider the coefficients a_s, b_s that are used in (8.69). We expand

$$a_s(\zeta) = \sum_{t=0}^{\infty} a_s^t \eta^t, \quad b_s(\zeta) = 2^{1/3} \sum_{t=0}^{\infty} b_s^t \eta^t, \tag{8.103}$$

where η is given in (8.101). The coefficients a_s^t, b_s^t are rational numbers. We know that $a_0(\zeta) = 1$. Substituting the expansions in (8.85), we can obtain recursion relations for the coefficients a_s^t, b_s^t. It follows that

$$\begin{aligned} 2(2t+1) b_s^t &= 2\sum_{r=0}^{t} \psi_r a_s^{t-r} - (t+1)(t+2) a_s^{t+2}, \\ 2(t+1) a_{s+1}^{t+1} &= 2\sum_{r=0}^{t} \psi_r b_s^{t-r} - (t+1)(t+2) b_s^{t+2}, \end{aligned} \tag{8.104}$$

where the first coefficients ψ_r are given in the second line of (8.102). These relations are used for fixed $s \geq 0$, while $t = 0, 1, 2, \ldots$. When $s = 0$ the first relation gives $b_0^t = \psi_t/(2t+1)$, $t = 0, 1, 2, \ldots$. We observe that the second relation does not give a value for a_1^0. The same problem happens for all a_s for all $s \geq 1$.

To find a_s^0 we can use (8.92). By substituting the expansions of (8.69), it follows that for $s = 0, 1, 2, \ldots$

$$\begin{aligned} &\sum_{r=0}^{s+1} a_r(\zeta) a_{s+1-r}(\zeta) \\ &+ \sum_{r=0}^{s} \left[a_r(\zeta) b'_{s-r}(\zeta) - a'_r(\zeta) b_{s-r}(\zeta) - \zeta b_r(\zeta) b_{s-r}(\zeta)\right] = 0. \end{aligned} \tag{8.105}$$

Putting $\zeta = 0$ yields

$$2a_{s+1}^0 = -\sum_{r=1}^{s} a_r^0 a_{s+1-r}^0 - \sum_{r=0}^{s} \left[a_r^0 b_{s-r}^1 - a_r^1 b_{s-r}^0 \right], \quad s = 0, 1, 2, \ldots. \quad (8.106)$$

In this way we obtain expansions of which we give the first few terms,

$$a_0(\zeta) = 1,$$
$$a_1(\zeta) = -\frac{1}{225} - \frac{71}{38500}\eta + \frac{82}{73125}\eta^2 + \frac{5246}{3898125}\eta^3 + \frac{185728}{478603125}\eta^4 + \cdots,$$
$$a_2(\zeta) = \frac{151439}{218295000} + \frac{68401}{147262500}\eta - \frac{1796498167}{4193689500000}\eta^2 - \frac{583721053}{830718281250}\eta^3 + \cdots,$$
$$a_3(\zeta) = -\frac{887278009}{2504935125000} - \frac{3032321618951}{9708942993750000}\eta + \cdots,$$
$$b_0(\zeta) = 2^{1/3}\left[\frac{1}{70} + \frac{2}{225}\eta + \frac{138}{67375}\eta^2 - \frac{296}{511875}\eta^3 - \frac{38464}{63669375}\eta^4 + \cdots\right], \quad (8.108)$$
$$b_1(\zeta) = 2^{1/3}\left[-\frac{1213}{1023750} - \frac{3757}{2695000}\eta - \frac{3225661}{6700443750}\eta^2 + \frac{90454643}{336992906250}\eta^3 + \cdots\right],$$
$$b_2(\zeta) = 2^{1/3}\left[\frac{16542537833}{37743205500000} + \frac{115773498223}{162820783125000}\eta + \frac{548511920915149}{1721719224225000000}\eta^2 + \cdots\right],$$
$$b_3(\zeta) = 2^{1/3}\left[-\frac{9597171184603}{25476663712500000} - \frac{430990563936859253}{568167343994250000000}\eta + \cdots\right].$$

(8.107)

Expansions for the coefficients c_s, d_s are not really needed, because these quantities follow from the relations in (8.76) if expansions for the functions on the right-hand sides of (8.76) are available.

Verifying the Maclaurin expansions by numerical experiments

We have used the expansions in (8.103) for $|\zeta| \leq 1$, for obtaining values of $a_s(\zeta)/\nu^{2s}$ and $b_s(\zeta)/\nu^{2s+4/3}$, for $s = 0, 1, 2, \ldots, 5$, with absolute accuracy of 10^{-20} if $\nu \geq 100$. We have used the series in (8.103) with terms up to $t = 45 - 6s$. The evaluated series in (8.103) and those of the derivatives have been used to check the Wronskian in (8.92) for a set of values of ζ on the unit circle. In all cases the Wronskian was verified within a precision of $0.20 \, 10^{-19}$. The same errors have been obtained by calculating the quantities in (8.92) by using the explicit representations in (8.78), and those of the derivatives by using (8.82).

8.4.5 Evaluation of the functions $A_\nu(\zeta)$, $B_\nu(\zeta)$ by iteration

We now concentrate on solving the system of differential equations in (8.88) by using analytical techniques. Instead of expanding the coefficients a_s, b_s of the asymptotic series, we expand the functions $A_\nu(\zeta), B_\nu(\zeta)$ in Maclaurin series. As remarked earlier, the singular points of these functions occur at $\zeta^\pm = (3\pi/2)^{2/3} e^{\pm i\pi/3}$, and the radius of convergence of the series of $A_\nu(\zeta)$ and $B_\nu(\zeta)$ in powers of ζ equals $2.81\ldots$.

8.4. Airy-type expansions for Bessel functions

We expand

$$A_\nu(\zeta) = \sum_{n=0}^{\infty} f_n(\nu)\zeta^n, \quad B_\nu(\zeta) = \sum_{n=0}^{\infty} g_n(\nu)\zeta^n, \quad \psi(\zeta) = \sum_{n=0}^{\infty} h_n\zeta^n. \tag{8.109}$$

The coefficients $f_0, f_1, \ldots, g_0, g_1, \ldots$ are to be determined, with the first elements given in (8.98), while the coefficients h_n are known. The first few h_n follow from (8.101) and the second line in (8.102),

$$h_0 = \tfrac{1}{70} 2^{1/3}, \quad h_1 = \tfrac{2}{75}, \quad h_2 = \tfrac{69}{13475} 2^{2/3}, \quad h_3 = \tfrac{148}{73125} 2^{1/3}. \tag{8.110}$$

Upon substituting the expansions into (8.88), we obtain for $n = 0, 1, 2, \ldots$ the recurrence relations

$$\begin{aligned} (n+2)(n+1) f_{n+2} + (2n+1) g_n &= \rho_n, \quad \rho_n = \sum_{k=0}^{n} h_k f_{n-k}, \\ (n+2)(n+1) g_{n+2} + 2\nu^2(n+1) f_{n+1} &= \sigma_n, \quad \sigma_n = \sum_{k=0}^{n} h_k g_{n-k}. \end{aligned} \tag{8.111}$$

We have already observed that cancellation occurs in the representations in (8.98). All evaluations based on the above recursions for computing higher coefficients f_n, g_n from lower coefficients suffer from cancellations. That is, the recursion relations cannot be used in the *forward direction*, in particular when ν is large. We have observed the same difficulties in §8.3.3

To show what happens in the present case, we give a few details on the first recursion. Take $n = 0$; then we obtain, using $f_0 = A_\nu(0)$, $g_0 = B_\nu(0)$, and (8.98),

$$\begin{aligned} 2f_2 &= f_0 h_0 - g_0 = 2^{1/3} \left[\tfrac{1}{70} S(\nu) - T(\nu) \right], \\ 2g_2 &= g_0 h_0 - 2\nu^2 f_1 = 2^{2/3} \left[\tfrac{1}{70} T(\nu) - \nu^2 \left\{ V(\nu) - \tfrac{1}{5} S(\nu) \right\} \right]. \end{aligned} \tag{8.112}$$

We see from (8.96) that $f_2 = \mathcal{O}(\nu^{-2})$, $g_2 = \mathcal{O}(1)$, as $\nu \to \infty$, whereas the quantities used to compute f_2 are of order $\mathcal{O}(1)$. Also, the term with ν^2 in g_2 is of lower order in the final result. Further use of the recursion makes things worse. In fact, in further steps more and more early terms in the asymptotic expansions of combinations of $S(\nu)$, $T(\nu)$, $U(\nu)$, and $V(\nu)$ are subtracted.

This unstable pulling down of asymptotic series suggests that we use the recursion in (8.111) in the backward direction. When we try to do so, for instance, with false starting values f_N, g_N for some large integer N, a complication arises because of the terms ρ_n, σ_n on the right-hand sides of (8.111). All terms ρ_n, σ_n contain f_k, g_k for $k = 0, 1, 2, \ldots, n$. Hence, recursion in the backward direction is not possible at all. A way out is to consider ρ_n, σ_n as known quantities, and to treat (8.111) as inhomogeneous difference equations.

Solving (8.88) by iteration

A first step in this approach will be to solve the system (8.88) by iteration. That is, we choose an appropriate pair of functions F_0, G_0, and define two sequences of functions $\{F_m\}$, $\{G_m\}$ by writing, for $m = 1, 2, 3, \ldots$,

$$F_m'' + 2\zeta G_m' + G_m = \psi(\zeta) F_{m-1}, \quad G_m'' + 2\nu^2 F_m' = \psi(\zeta) G_{m-1}. \tag{8.113}$$

To study this iterative process we need to know the solutions of the homogeneous equations, that is, of the system

$$F_m'' + 2\zeta G_m' + G_m = 0, \quad G_m'' + 2\nu^2 F_m' = 0. \tag{8.114}$$

One solution is $F = 1, G = 0$. Other solutions of (8.114) follow by eliminating F'' in the first equation by differentiating the second one. The result is

$$G''' - 4\nu^2 \zeta G' - 2\nu^2 G = 0, \tag{8.115}$$

with solutions products of Airy functions

$$\mathrm{Ai}^2(t), \quad \mathrm{Ai}(t)\,\mathrm{Bi}(t), \quad \mathrm{Bi}^2(t), \quad t = \nu^{2/3}\zeta. \tag{8.116}$$

This can easily be verified by using the differential equation of the Airy functions in (8.58). The F-solutions of the homogeneous equations (8.114) follow from integrating the second part in (8.114). Knowing these four linearly independent solutions, we can construct solutions F, G of the inhomogeneous equations corresponding to (8.114), that is, the system (8.88), by using the variation of constants formula, and eventually by constructing Volterra integral equations defining the solutions A, B of (8.88). For details we refer to [220].

We rewrite (8.111) in backward form:

$$\begin{aligned}
f_n &= \frac{1}{2\nu^2}\left[\frac{1}{n}\sigma_{n-1} - (n+1)g_{n+1}\right], \\
g_{n-1} &= dsp\frac{1}{2n-1}\left[\rho_{n-1} - n(n+1)f_{n+1}\right],
\end{aligned} \tag{8.117}$$

where $n \geq 1$. The coefficients are assumed to belong to the functions $F_m(\zeta)$, $G_m(\zeta)$ of the iteration process described by (8.113), while the coefficients ρ_{n-1} and σ_{n-1} are assumed to be known and contain Maclaurin coefficients of $F_{m-1}(\zeta)$, $G_{m-1}(\zeta)$, and $\psi(\zeta)$. Observe that (8.117) does not define f_0. After having computed $f_1, f_2, \ldots$ and $g_0, g_1, g_2, \ldots$ by the backward recursion process, we compute f_0 from the Wronskian (8.92),

$$f_0 = \frac{-g_1 + \sqrt{g_1^2 + 4\nu^2(\nu^2 + f_1 g_0)}}{2\nu^2}, \tag{8.118}$$

where the $+$ sign of the square root is taken because of the known behavior of $F_\nu(0)$ when ν is large; see (8.96).

We give a few steps in the iteration and backward recursion process. Let us start the iterations (8.113) with constant (F_0, G_0) (constant with respect to ζ and ν). The obvious constant choice of (F_0, G_0) is $(1, h_0)$; see (8.96). We use the four coefficients of $\psi(\zeta)$ shown in (8.110) for constructing the ρ and σ coefficients on the right-hand sides of (8.117). We have

$$\rho_n = h_n, \quad \sigma_n = h_0 h_n, \quad n = 0, 1, 2, 3, \quad \rho_n = \sigma_n = 0, \quad n \geq 4. \tag{8.119}$$

8.4. Airy-type expansions for Bessel functions

Then the first iteration gives

$$\begin{aligned}
f_4 &= \tfrac{1}{8} h_0 h_3 \nu^{-2}, & g_3 &= \tfrac{1}{7} h_3, \\
f_3 &= \tfrac{1}{6} h_0 h_2 \nu^{-2}, & g_2 &= \tfrac{1}{5} h_2 - \tfrac{3}{10} h_0 h_3 \nu^{-2}, \\
f_2 &= \tfrac{1}{2}\left(\tfrac{1}{2} h_0 h_1 - \tfrac{3}{7} h_3\right)\nu^{-2}, & g_1 &= \tfrac{1}{3} h_1 - \tfrac{1}{3} h_0 h_2 \nu^{-2}, \\
f_1 &= \tfrac{1}{2}\left(h_0^2 - \tfrac{2}{5} h_2 + \tfrac{3}{5} h_0 h_3 \nu^{-2}\right)\nu^{-2}, & g_0 &= h_0 - \left(\tfrac{1}{2} h_0 h_1 - \tfrac{3}{7} h_3\right)\nu^{-2},
\end{aligned} \qquad (8.120)$$

while f_0 is computed by using (8.118). Expanding the result for f_0, we find

$$f_0 = 1 - \frac{1}{225}\nu^{-2} + \mathcal{O}\left(\nu^{-4}\right), \qquad (8.121)$$

which agrees with the first two terms of the asymptotic expansion of $T(\nu)$ given in (8.96). Also, the first terms of the asymptotic expansions of g_0, f_1, g_1 agree with the first terms of the expansions following from (8.96) and (8.98).

When more coefficients h_k and more iterations are used, the further iterates F_m, G_m have Maclaurin coefficients f_n, g_n, of which the asymptotic expansions with respect to ν are converging to the actual asymptotic expansions of f_n, g_n. In particular, the asymptotic expansions of f_0, g_0 coincide more and more with those following from (8.98). Of course, it is not our goal to obtain the asymptotic expansions of the coefficients f_n, g_n, but this illustrates the analytical nature of the algorithm.

The numerical problem in using the recursions in (8.111) in the forward direction is the influence of dominant solutions of the homogeneous equations of (8.111) (that is, the equations obtained by taking $\rho_n = \sigma_n = 0$). The dominant solutions are the Maclaurin coefficients of the functions given in (8.116), as functions of ζ. The coefficients grow as ν becomes large. The minimal solution is given by $f_0 = 1$ and $f_{n+1} = g_n = 0, n \geq 0$.

From the above observations we infer that the solutions of the inhomogeneous equations (8.111) cannot contain dominant solutions of the homogeneous equations. This explains the unstable character of the forward recursions based on (8.111) and the stable character of the recursion based on the backward form in (8.117). More details on these phenomena can be found in Chapter 4.

Verifying the iterative scheme by numerical experiments

For numerical applications information is needed about the growth of the coefficients f_n, g_n. Since the Maclaurin series in (8.109) have a radius of convergence equal to 2.81..., for all values of ν, the size of the coefficients f_n, g_n is comparable with that of h_n. The number of coefficients f_n, g_n needed in (8.109) also depends on the size of $|\zeta|$. When $|\zeta| = 1$ we need about 45 terms in the Maclaurin series in (8.109) in order to obtain an accuracy of about 20 decimal digits. The ζ-interval $[-1, 1]$ corresponds to the z-interval $[0.39, 1.98]$. When z is outside this interval many other efficient algorithms are available for the computation of $J_\nu(\nu z)$, $Y_\nu(\nu z)$.

Table 8.2. *Relative errors during five iterations (i) of f_0, g_0, f_5, g_5 compared with more accurate values f_0^a, etc. The final column shows the relative error in the Wronskian (8.92) at $\zeta = 1$.*

i	$\|f_0 - f_0^a\|$	$\|g_0 - g_0^a\|$	$\|f_5 - f_5^a\|$	$\|g_5 - g_5^a\|$	Wronskian
			$\nu = 5$		
1	$6.11\,10^{-09}$	$1.76\,10^{-06}$	$1.12\,10^{-03}$	$6.14\,10^{-04}$	$4.36\,10^{-08}$
2	$4.54\,10^{-12}$	$1.03\,10^{-08}$	$6.14\,10^{-06}$	$8.33\,10^{-07}$	$2.05\,10^{-10}$
3	$2.56\,10^{-15}$	$1.60\,10^{-11}$	$1.52\,10^{-08}$	$8.48\,10^{-10}$	$3.29\,10^{-13}$
4	$1.21\,10^{-17}$	$1.83\,10^{-14}$	$6.40\,10^{-12}$	$5.25\,10^{-13}$	$6.72\,10^{-16}$
5	$0.00\,10^{-00}$	$4.19\,10^{-17}$	$4.64\,10^{-14}$	$4.04\,10^{-16}$	$2.23\,10^{-18}$
			$\nu = 10$		
1	$4.24\,10^{-10}$	$1.14\,10^{-07}$	$2.90\,10^{-04}$	$1.63\,10^{-04}$	$2.76\,10^{-09}$
2	$8.50\,10^{-14}$	$8.17\,10^{-10}$	$1.84\,10^{-06}$	$5.76\,10^{-08}$	$1.64\,10^{-11}$
3	$1.45\,10^{-17}$	$3.20\,10^{-13}$	$1.10\,10^{-09}$	$9.06\,10^{-11}$	$6.64\,10^{-15}$
4	$1.08\,10^{-19}$	$9.89\,10^{-17}$	$1.74\,10^{-12}$	$1.25\,10^{-15}$	$8.07\,10^{-18}$
5	$0.00\,10^{-00}$	$1.88\,10^{-19}$	$1.35\,10^{-15}$	$6.74\,10^{-17}$	$3.44\,10^{-19}$
			$\nu = 25$		
1	$1.12\,10^{-11}$	$2.94\,10^{-09}$	$4.70\,10^{-05}$	$2.66\,10^{-05}$	$7.10\,10^{-11}$
2	$3.66\,10^{-16}$	$2.22\,10^{-11}$	$3.09\,10^{-07}$	$1.52\,10^{-09}$	$4.47\,10^{-13}$
3	$0.00\,10^{-00}$	$1.40\,10^{-15}$	$2.95\,10^{-11}$	$2.66\,10^{-12}$	$2.91\,10^{-17}$
4	$0.00\,10^{-00}$	$0.00\,10^{-00}$	$5.63\,10^{-14}$	$7.25\,10^{-18}$	$1.75\,10^{-19}$
5	$0.00\,10^{-00}$	$0.00\,10^{-00}$	$6.45\,10^{-18}$	$3.40\,10^{-19}$	$1.76\,10^{-19}$

We have computed successive iterates of Maclaurin coefficients f_n, g_n defined in (8.109) for different values of ν. To give the algorithm some relevant starting values we have used approximations for f_0, g_0 based on (8.98), with a few terms of $S(\nu)$, $T(\nu)$ of (8.96). Furthermore, we have taken $g_n = h_n/(2n + 1)$, $n \geq 1$, a choice based on taking $A = 1$ in the first line of (8.88), and integrating the resulting relation $(\sqrt{\zeta} B)' = \psi/(2\sqrt{\zeta})$.

During each iteration we start the backward recursions with $f_n = g_{n-1} = 0$, $n \geq 46$, and we compute $f_{45}, g_{44}, f_{44}, g_{43}, \ldots$ by using (8.117). We use h_k, $k = 0, 1, \ldots, 45$, and we recompute the coefficients ρ_k, σ_k, $k = 0, 1, 2, \ldots, 45$, using (8.111) with values f_k, g_k obtained in the previous iteration. In Table 8.2 we show the relative errors in the values f_0, g_0, f_5, g_5, when compared with more accurate values f_0^a, etc. Computations are done with 20 digits. The accurate values are obtained by applying the backward recursion by using 10 iterations. We also give the relative error in the Wronskian (8.92) at $\zeta = 1$ during each iteration.

From Table 8.2 we conclude that for $\nu = 5$ we can already obtain an accuracy of 10^{-10} in the Wronskian after two iterations; further iterations improve the results. For larger values of ν the algorithm is very efficient.

The methods of this chapter can be used for Airy-type asymptotic expansions for other special functions. We mention as interesting cases parabolic cylinder functions, Coulomb wave functions, and other members of the class of Whittaker functions. To stay in the class of Bessel functions, we mention the modified Bessel function of the third kind $K_{i\nu}(z)$ of imaginary order, which plays an important role in the diffraction theory of pulses and in the study of certain hydrodynamical studies. Moreover, this function is the kernel of the Lebedev transform. The same functions A_ν, B_ν, C_ν, D_ν can be used for this case.

8.5 Airy-type asymptotic expansions obtained from integrals

In this section we develop an algorithm for evaluating the coefficients in Airy-type asymptotic expansions that are obtained from integrals with a large parameter. In the case of the Bessel functions considered in §8.4 we could use elements of the theory of differential equations for obtaining properties of the coefficients, or for expanding the slowly varying functions in the asymptotic representations. In the present case the coefficients of the asymptotic series follow from recursive schemes obtained by integration by parts, and we have Cauchy-type integral representations of the coefficients. As an application we consider the Airy-type expansion of the Weber parabolic cylinder function.

As we have seen in the case of the incomplete gamma functions and the Bessel functions earlier in this chapter, the coefficients of uniform asymptotic expansions are difficult to compute, particularly in the neighborhood of the transition or turning points. When the expansions are obtained from differential equations, the coefficients can be obtained (in principle) from recurrence relations for the coefficients. This method has been used in many publications, for example, in [57, 166], and for expansions involving Bessel functions or parabolic cylinder functions similar results are available.

However, having such a recurrence relation for the coefficients does not always give the possibility for obtaining analytic expressions of a number of coefficients, because the recursion involves integrals of previous coefficients together with a function that is not easy to handle. Sometimes the coefficients can be explicitly expressed in terms of coefficients of simpler expansions, because different types of expansions may be valid in overlapping domains. For the Bessel functions, see the relations in (8.78) and (8.82).

For special functions usually the same type of uniform expansion can be obtained by using integral representations of the functions. Sometimes, in a particular problem, the integral is the only tool available for constructing uniform expansions. By using transformations of variables in the integrals, these representations can be transformed into standard forms for which an integration-by-parts procedure can be used to obtain expansions in terms of, for example, Airy functions.

Although it is usually not possible to derive recurrence relations for the coefficients obtained in this way, in all cases for special functions known so far, it is possible to construct a number of coefficients, and only because of the complexity of the problem, which implies limitations with respect to available computer memory when doing symbolic computations, is there an upper bound for this number.

In this section we explain how to obtain coefficients of an Airy-type expansion starting with an integral representation. First we describe the procedure for a general case. For an application we obtain the coefficients of the expansion for the parabolic cylinder functions. Straightforward computations are often complicated by the appearance of algebraic roots in the output or intermediate expressions. In the example of parabolic cylinder functions we avoid computations with algebraic roots by using auxiliary variables.

8.5.1 Airy-type asymptotic expansions

We consider integrals of the form

$$F_\eta(z) = \frac{1}{2\pi i} \int_C e^{z(\frac{1}{3}t^3 - \eta t)} f(t)\, dt, \tag{8.122}$$

where the contour starts at infinity with ph $t = -\pi/3$ and returns to infinity with ph $t = \pi/3$. We assume that the function $f(t)$ is analytic in the neighborhood of the contour; z and η are complex parameters, z is large.

In the case $f(t) = 1$ we obtain the Airy function (see (1.16) and (5.203))

$$\frac{1}{2\pi i} \int_C e^{z(\frac{1}{3}t^3 - \eta t)}\, dt = z^{-\frac{1}{3}} \operatorname{Ai}\left(\eta z^{\frac{2}{3}}\right). \tag{8.123}$$

For more general functions f, the asymptotic expansion of $F_\eta(z)$ can be given in terms of this Airy function. The asymptotic feature of this type of integral is that the phase function $\phi(t) = \frac{1}{3}t^3 - \eta t$ has two saddle points at $\pm\sqrt{\eta}$ that coalesce when $\eta \to 0$, and it is not possible to describe the asymptotic behavior of $F_\eta(z)$ in terms of simple functions when η is small. When the parameter η is positive and bounded away from 0, one can perform a saddle point analysis on (8.122) and use a conformal mapping $\phi(t) - \phi(\sqrt{\eta}) = \frac{1}{2}u^2$ with the condition $u(\sqrt{\eta}) = 0$. We obtain

$$F_\eta(z) = \frac{e^{z\phi(\sqrt{\eta})}}{2\pi i} \int_{-i\infty}^{i\infty} e^{\frac{1}{2}zu^2} g(u)\, du, \tag{8.124}$$

where $g(u) = f(t)\, dt/du$, with $dt/du = u/(t^2 - \eta)$, which is regular at the positive saddle point, but not at the negative saddle point. It follows that when η becomes small, a singularity due to dt/du in the u-plane approaches the origin, and an expansion of dt/du at $u = 0$ will have coefficients that become infinite as $\eta \to 0$. Hence, by using the standard saddle point method we obtain an expansion that is not uniformly valid as $\eta \to 0$.

A modification of the saddle point method is possible by taking into account both saddle points. We give an integration-by-parts procedure that is a variant of Bleistein's method introduced in [18] (for a different class of integrals) and that gives the requested uniform expansion.

8.5. Airy-type asymptotic expansions obtained from integrals

We assume that f is an analytic function in a certain domain G and write

$$f(t) = \alpha_0 + \beta_0 t + (t^2 - \eta)g(t), \qquad (8.125)$$

where

$$\alpha_0 = \frac{1}{2}\left[f(\sqrt{\eta}) + f(-\sqrt{\eta})\right], \quad \beta_0 = \frac{1}{2\sqrt{\eta}}\left[f(\sqrt{\eta}) - f(-\sqrt{\eta})\right]. \qquad (8.126)$$

Clearly $\alpha_0 \to f(0)$, $\beta_0 \to f'(0)$ as $\eta \to 0$. We have the Cauchy integral representation

$$f(t) = \frac{1}{2\pi i}\oint_{\{t\}} \frac{f(s)}{s-t}\,ds, \qquad (8.127)$$

where the contour encircles the point t in the counterclockwise direction. Similarly,

$$\begin{aligned}\alpha_0 &= \frac{1}{2}\left[\frac{1}{2\pi i}\oint_{\{\sqrt{\eta}\}} \frac{f(s)}{s-\sqrt{\eta}}\,ds + \frac{1}{2\pi i}\oint_{\{-\sqrt{\eta}\}} \frac{f(s)}{s+\sqrt{\eta}}\,ds\right] \\ &= \frac{1}{2\pi i}\oint_{\{\pm\sqrt{\eta}\}} \frac{sf(s)}{s^2-\eta}\,ds\end{aligned} \qquad (8.128)$$

and

$$\beta_0 = \frac{1}{2\pi i}\oint_{\{\pm\sqrt{\eta}\}} \frac{f(s)}{s^2-\eta}\,ds, \quad g(t) = \frac{1}{2\pi i}\oint_{\{t,\pm\sqrt{\eta}\}} \frac{f(s)}{(s-t)(s^2-\eta)}\,ds. \qquad (8.129)$$

Upon substituting (8.125) into (8.122), we obtain

$$\begin{aligned}F_\eta(z) = z^{-\frac{1}{3}}\operatorname{Ai}\left(\eta z^{\frac{2}{3}}\right)\alpha_0 - z^{-\frac{2}{3}}\operatorname{Ai}'\left(\eta z^{\frac{2}{3}}\right)\beta_0 \\ + \frac{1}{2\pi i}\int_C e^{z(\frac{1}{3}t^3 - \eta t)}(t^2-\eta)g(t)\,dt.\end{aligned} \qquad (8.130)$$

An integration by parts gives

$$F_\eta(z) = z^{-\frac{1}{3}}\operatorname{Ai}\left(\eta z^{\frac{2}{3}}\right)\alpha_0 - z^{-\frac{2}{3}}\operatorname{Ai}'\left(\eta z^{\frac{2}{3}}\right)\beta_0 - \frac{z^{-1}}{2\pi i}\int_C e^{z(\frac{1}{3}t^3 - \eta t)}f_1(t)\,dt, \qquad (8.131)$$

where $f_1(t) = g'(t)$. Repeating this procedure we obtain the compound expansion

$$F_\eta(z) \sim z^{-\frac{1}{3}}\operatorname{Ai}\left(\eta z^{\frac{2}{3}}\right)\sum_{n=0}^{\infty}(-1)^n \frac{\alpha_n}{z^n} - z^{-\frac{2}{3}}\operatorname{Ai}'\left(\eta z^{\frac{2}{3}}\right)\sum_{n=0}^{\infty}(-1)^n \frac{\beta_n}{z^n}, \qquad (8.132)$$

where the coefficients α_n, β_n are defined as in (8.126) with the function f replaced with f_n, which in turn is defined by the scheme

$$f_{n+1}(t) = g'_n(t), \quad f_n(t) = \alpha_n + \beta_n t + (t^2 - \eta)g_n(t), \qquad (8.133)$$

with $n = 0, 1, 2, \ldots$ and $f_0(t) = f(t)$. The expansion in (8.132) is valid for large values of z and holds uniformly with respect to η in a neighborhood of the origin. A more precise formulation can be given, but more information can be found in the literature; see [168, p. 243].

The functions $f_n(t)$ defined in (8.133) can be represented in the form of Cauchy-type integrals. We have the following theorem.

Theorem 8.1. *Let the rational functions $R_n(s, t, \eta)$ be defined by*

$$R_0(s, t, \eta) = \frac{1}{s-t}, \quad R_{n+1}(s, t, \eta) = \frac{-1}{s^2 - \eta} \frac{d}{ds} R_n(s, t, \eta), \tag{8.134}$$

where $n = 0, 1, 2, \ldots$ and $s, t, \eta \in \mathbb{C}, s \neq t, s^2 \neq \eta$. Let $f_n(t)$ be defined by the recursive scheme (8.133), where f_0 is a given analytic function in a domain G, and let t and $\pm\sqrt{\eta}$ be interior points of G. Then we have

$$f_n(t) = \frac{1}{2\pi i} \oint_{\mathcal{D}} R_n(s, t, \eta) f_0(s) \, ds, \tag{8.135}$$

where $\mathcal{D}$ is a simple closed contour in G that encircles the points t and $\pm\sqrt{\eta}$.

Proof. The proof starts with

$$f_n(t) = \frac{1}{2\pi i} \oint_{\mathcal{D}} R_0(s, t, \eta) f_n(s) \, ds, \tag{8.136}$$

which is just Cauchy's integral representation. We take $n \geq 1$ and use $f_n(s) = g'_{n-1}(s)$ (see (8.133)). This gives

$$f_n(t) = \frac{1}{2\pi i} \oint_{\mathcal{D}} R_0(s, t, \eta) \, dg_{n-1}(s). \tag{8.137}$$

We integrate by parts and in the new integral we use the relation for g_{n-1} that follows from (8.133). We obtain

$$f_n(t) = -\frac{1}{2\pi i} \oint_{\mathcal{D}} \frac{f_{n-1}(s) - \alpha_{n-1} - \beta_{n-1}s}{s^2 - \eta} \frac{d}{ds} R_0(s, t, \eta) \, ds. \tag{8.138}$$

It is easily verified that the contributions from the terms with α_{n-1} and β_{n-1} vanish. This gives

$$f_n(t) = \frac{1}{2\pi i} \oint_{\mathcal{D}} R_1(s, t, \eta) f_{n-1}(s) \, ds. \tag{8.139}$$

Continuing this procedure, we obtain (8.135). □

For the coefficients α_n, β_n we have a similar representation,

$$\alpha_n = \frac{1}{2\pi i} \oint_{\mathcal{D}} A_n(s, \eta) f_0(s) \, ds, \quad \beta_n = \frac{1}{2\pi i} \oint_{\mathcal{D}} B_n(s, \eta) f_0(s) \, ds, \tag{8.140}$$

where $\mathcal{D}$ is a simple closed contour in G that encircles the points $\pm\sqrt{\eta}$ and where $A_n(s, t)$ and $B_n(s, t)$ follow the same recursion (8.134) as the rational functions $R_n(s, t, \eta)$, with initial values

$$A_0(s, \eta) = \frac{s}{s^2 - \eta}, \quad B_0(s, \eta) = \frac{1}{s^2 - \eta}. \tag{8.141}$$

8.5. Airy-type asymptotic expansions obtained from integrals

We see that the coefficients α_n, β_n that play a role in the expansion (8.132) are well defined from an analytical point of view. However, from a computational point of view it may be quite difficult to evaluate the coefficients.

Example 8.2. For a simple rational function like $f_0(t) = 1/(t+1)$ the computations are rather straightforward, and we can even use residue calculus to evaluate the integrals in (8.140):

$$\alpha_n = -A_n(-1, \eta), \quad \beta_n = -B_n(-1, \eta). \tag{8.142}$$

The first few values are in this case

$$\begin{aligned}
\alpha_0 &= -\frac{1}{\eta - 1}, & \beta_0 &= \frac{1}{\eta - 1}, \\
\alpha_1 &= \frac{\eta + 1}{(\eta - 1)^3}, & \beta_1 &= -\frac{2}{(\eta - 1)^3}, \\
\alpha_2 &= -4\frac{2\eta + 1}{(\eta - 1)^5}, & \beta_2 &= 2\frac{\eta + 5}{(\eta - 1)^5}, \\
\alpha_3 &= 4\frac{2\eta^2 + 21\eta + 7}{(\eta - 1)^7}, & \beta_3 &= -40\frac{\eta + 2}{(\eta - 1)^7}, \\
\alpha_4 &= -280\frac{\eta^2 + 4\eta + 1}{(\eta - 1)^9}, & \beta_4 &= 40\frac{\eta^2 + 19\eta + 22}{(\eta - 1)^9}, \\
\alpha_5 &= 280\frac{\eta^3 + 29\eta^2 + 65\eta + 13}{(\eta - 1)^{11}}, & \beta_5 &= -1120\frac{2\eta^2 + 14\eta + 11}{(\eta - 1)^{11}}.
\end{aligned} \tag{8.143}$$

In the present case, with $f(t) = 1/(t+1)$, the value $\eta = 1$ gives singularities for the coefficients. This is because we "interpolate" at $\pm\sqrt{\eta}$, and we cannot interpolate at a singular point of f. ∎

For a more complicated or general function $f_0(t)$, even computer algebra manipulations give complicated expressions which are very difficult to evaluate. In the next section we develop an algorithm for computing the coefficients α_n, β_n when the values of the derivatives of $f_0(t)$ at $t = \pm\sqrt{\eta}$ are available.

8.5.2 How to compute the coefficients α_n, β_n

We explain how the coefficients α_n, β_n of (8.132) can be computed. To avoid the square roots in the formulas we replace η with b^2, and we write (8.133) in the form

$$f_0(t) = f(t), \quad f_{n+1}(t) = g'_n(t), \quad f_n(t) = \alpha_n + \beta_n t + (t^2 - b^2)g_n(t) \tag{8.144}$$

for $n = 0, 1, 2, \ldots$. We assume that the function f is analytic in a domain G, that the series expansions used in this section are convergent in G, and that the points $\pm b$ are inside G. Furthermore, we assume that the coefficients $p_k^{(1)}$, $p_k^{(2)}$ of the expansions

$$f(t) = \sum_{k=0}^{\infty} p_k^{(1)}(t-b)^k, \quad f(-t) = \sum_{k=0}^{\infty} p_k^{(2)}(t-b)^k \tag{8.145}$$

are available.

Theorem 8.3 (algorithm). *Let the coefficients f_k^e, f_k^o be defined by*

$$f_k^e = \frac{1}{2}\left[p_k^{(1)} + p_k^{(2)}\right], \quad f_k^o = \frac{1}{2}\left[p_k^{(1)} - p_k^{(2)}\right], \quad k = 0, 1, 2 \ldots, \tag{8.146}$$

and the coefficients $f_k^{o,e}$ by the recursion

$$b f_k^{o,e} = f_k^o - f_{k-1}^{o,e}, \quad k \geq 0, \tag{8.147}$$

with $f_{-1}^{o,e} = 0$. Next, define the coefficients γ_k, δ_k by

$$\gamma_0 = f_0^e, \quad \delta_0 = f_0^{o,e}, \tag{8.148}$$

and for $k \geq 1$,

$$\begin{aligned}\gamma_k &= \sum_{j=1}^{k} \frac{(-1)^{k-j} j (2k - j - 1)!}{(2b)^{2k-j} k! (k-j)!} f_j^e, \\ \delta_k &= \sum_{j=1}^{k} \frac{(-1)^{k-j} j (2k - j - 1)!}{(2b)^{2k-j} k! (k-j)!} f_j^{o,e}. \end{aligned} \tag{8.149}$$

Finally, for $n \geq 0$ let the coefficients $\gamma_k^{(n)}, \delta_k^{(n)}$ be defined by the recursion

$$\begin{aligned}\gamma_k^{(n+1)} &= (2k+1)\delta_{k+1}^{(n)} + 2b^2(k+1)\delta_{k+2}^{(n)}, \\ \delta_k^{(n+1)} &= 2(k+1)\gamma_{k+2}^{(n)}, \quad k = 0, 1, 2, \ldots,\end{aligned} \tag{8.150}$$

with $\gamma_k^{(0)} = \gamma_k, \delta_k^{(0)} = \delta_k$. Then the coefficients α_n, β_n of expansion (8.132) are given by

$$\alpha_n = \gamma_0^{(n)}, \quad \beta_n = \delta_0^{(n)}, \quad n \geq 0. \tag{8.151}$$

Proof. The coefficients f_k^e, f_k^o occur in the expansions

$$f_e(t) = \sum_{k=0}^{\infty} f_k^e (t-b)^k, \quad f_o(t) = \sum_{k=0}^{\infty} f_k^o (t-b)^k, \tag{8.152}$$

where $f_e(t), f_o(t)$ are the even and odd parts of f,

$$f_e(t) = \frac{1}{2}[f(t) + f(-t)], \quad f_o = \frac{1}{2}[f(t) - f(-t)], \tag{8.153}$$

and the coefficients $f_k^{o,e}$ occur in the expansion

$$\frac{1}{t} f_o(t) = \sum_{k=0}^{\infty} f_k^{o,e} (t-b)^k. \tag{8.154}$$

The coefficients γ_k, δ_k occur in the expansion

$$f(t) = \sum_{k=0}^{\infty} \gamma_k (t^2 - b^2)^k + t \sum_{k=0}^{\infty} \delta_k (t^2 - b^2)^k. \tag{8.155}$$

8.5. Airy-type asymptotic expansions obtained from integrals

Observe that

$$f_e(t) = \sum_{k=0}^{\infty} \gamma_k (t^2 - b^2)^k, \quad f_o(t) = t \sum_{k=0}^{\infty} \delta_k (t^2 - b^2)^k, \quad (8.156)$$

and we will verify the first relation of (8.149). We write

$$\gamma_k = \frac{1}{2\pi i} \oint f_e\left(\sqrt{z + b^2}\right) \frac{dz}{z^{k+1}}, \quad (8.157)$$

where the contour is a small circle around the origin. Also,

$$\gamma_k = \frac{1}{2\pi i} \oint f_e(t) \frac{2t\, dt}{(t+b)^{k+1}(t-b)^{k+1}}, \quad (8.158)$$

where the contour is a small circle around $t = b$.

Substitute here the expansion $f_e(t) = \sum_{j=0}^{\infty} f_j^e (t - b)^j$. Then,

$$\gamma_k = \sum_{j=0}^{k} f_j^e \frac{1}{2\pi i} \oint \frac{2t\, dt}{(t+b)^{k+1}(t-b)^{k+1-j}}. \quad (8.159)$$

Expand

$$\frac{2t}{(t+b)^{k+1}} = \sum_{m=0}^{\infty} q_m (t - b)^m. \quad (8.160)$$

We use (see (2.67))

$$(1 - z)^{-a} = \sum_{n=0}^{\infty} \frac{(a)_m}{m!} z^m = \sum_{n=0}^{\infty} \binom{-a}{m} (-z)^m \quad (8.161)$$

and find

$$q_m = (-1)^m \frac{(k-m)(k+m-1)!}{(2b)^{k+m}\, m!\, k!}. \quad (8.162)$$

When we use (8.160) in (8.159), we only need q_m with $m = k - j$. This gives the first result of (8.149). The proof for δ_k is the same, because $(1/t) f_o(t)$ is again even.

The coefficients $\gamma_k^{(n)}, \delta_k^{(n)}$ are used in

$$f_n(t) = \sum_{k=0}^{\infty} \gamma_k^{(n)} (t^2 - b^2)^k + t \sum_{k=0}^{\infty} \delta_k^{(n)} (t^2 - b^2)^k, \quad (8.163)$$

and the recursions in (8.150) are easily verified, as is the final relation

$$\alpha_n = \gamma_0^{(n)}, \quad \beta_n = \gamma_0^{(n)}, \quad n \geq 0. \qquad \square \quad (8.164)$$

The first few values of the coefficients γ_k, δ_k of the expansion in (8.155) are

$$\begin{aligned}
\gamma_0 &= f_0^e, & \delta_0 &= \frac{1}{b} f_0^o, \\
\gamma_1 &= \frac{1}{2b} f_1^e, & \delta_1 &= \frac{1}{2b^3}(b f_1^o - f_0^o), \\
\gamma_2 &= \frac{1}{8b^3}(2b f_2^e - f_1^e), & \delta_2 &= \frac{1}{8b^5}(2b^2 f_2^o - 3b f_1^o + 3 f_0^o),
\end{aligned} \quad (8.165)$$

and we observe, as in (8.149), negative powers of b. From a computational point of view, this may cause numerical instabilities, because the coefficients are analytic functions of b at $b = 0$.

Example 8.4. When we again take $f(t) = 1/(t+1)$, we obtain

$$\gamma_k = \frac{1}{(1-b^2)^{k+1}}, \quad \delta_k = -\frac{1}{(1-b^2)^{k+1}}, \quad k = 0, 1, 2, \ldots, \qquad (8.166)$$

which follows from

$$\frac{1}{t+1} = \frac{1-t}{(1-b^2) - (t^2-b^2)} = \sum_{k=0}^{\infty} \frac{(t^2-b^2)^k}{(1-b^2)^{k+1}} - t \sum_{k=0}^{\infty} \frac{(t^2-b^2)^k}{(1-b^2)^{k+1}}. \quad \blacksquare \qquad (8.167)$$

From the representations in, for example, (8.149), we conclude that if we apply the algorithm for computing the coefficients α_n, β_n of expansion (8.132), starting with numerical values of the coefficients $p_k^{(1)}, p_k^{(2)}$ of (8.145), we may encounter numerical instabilities when b is small. For this reason, it is important to use exact values of $p_k^{(1)}, p_k^{(2)}$, and computer algebra is of great help here. In the next section we consider a nontrivial case in which obtaining the exact values of the coefficients $p_k^{(1)}, p_k^{(2)}$ of (8.145) also needs special care.

Remark. In order to compute the coefficients α_n, β_n for $n = 0, 1, \ldots, N$ from the relation

$$\alpha_n = \gamma_0^{(n)}, \quad \beta_n = \delta_0^{(n)}, \quad n \geq 0, \qquad (8.168)$$

and the recursion in (8.150), we need the starting values for this recursion γ_k, δ_k for $k = 0, 1, \ldots, 2N$. Hence, as follows from (8.149), we also need $p_k^{(1)}, p_k^{(2)}, k = 0, 1, \ldots, 2N$, in the expansions in (8.145).

8.5.3 Application to parabolic cylinder functions

Weber parabolic cylinder functions are solutions of the differential equation

$$\frac{d^2 y}{dx^2} - \left(\frac{1}{4}x^2 + a\right) y = 0. \qquad (8.169)$$

Airy-type expansions for the solutions of this equation can be found in [166], and are obtained by using the differential equation. In this section we show how to obtain an integral representation like (8.122), and how to apply the algorithm of the previous section for deriving an Airy-type asymptotic expansion.

A standard solution of (8.169) is the integral

$$U(a, x) = \frac{e^{\frac{1}{4}x^2}}{i\sqrt{2\pi}} \int_C e^{-xs + \frac{1}{2}s^2} s^{-a-\frac{1}{2}} ds, \qquad (8.170)$$

where the contour $\mathcal{C}$ is a vertical line in the complex plane with $\Re s > 0$; see (5.213).

We consider large negative values of a, and use Olver's notation,

$$a = -\frac{1}{2}\mu^2, \quad x = \mu t\sqrt{2}. \qquad (8.171)$$

8.5. Airy-type asymptotic expansions obtained from integrals

Changing the variable of integration by writing $s \to \mu s/\sqrt{2}$, we obtain

$$U\left(-\tfrac{1}{2}\mu^2, \mu t\sqrt{2}\right) = \frac{e^{\frac{1}{2}\mu^2 t^2}}{i\sqrt{2\pi}} \left(\frac{\mu}{\sqrt{2}}\right)^{\frac{1}{2}\mu^2 + \frac{1}{2}} \int_{\mathcal{C}} e^{z\phi(s)} s^{-1/2}\, ds, \qquad (8.172)$$

where

$$\phi(s) = \tfrac{1}{2}s^2 - 2st + \ln s, \quad z = \tfrac{1}{2}\mu^2. \qquad (8.173)$$

The saddle points are obtained from the equation $\phi'(s) = 0$, that is, from the equation $s^2 - 2st + 1 = 0$, which gives two solutions:

$$s_\pm = t \pm \sqrt{t^2 - 1}. \qquad (8.174)$$

These points coalesce when $t \to \pm 1$. Observe that in the new variables the differential equation (8.169) transforms into

$$\frac{d^2 y}{dt^2} - \mu^4 (t^2 - 1) y = 0, \qquad (8.175)$$

which has turning points at $t = \pm 1$.

A transformation into the standard form (8.122) can be obtained by writing

$$\phi(s) = \tfrac{1}{3} w^3 - \eta w + A, \qquad (8.176)$$

where η and A have to be determined and do not depend on w. A transformation into the cubic polynomial is first considered in [33]. For further details on the theory of this method we refer to [163, 168, 221, 243].

The parameters η and A are obtained by assuming that the saddle points $s_\pm$ in the s variable should correspond with the saddle points $w_\pm = \pm\sqrt{\eta}$ in the w variable. We write

$$t = \cosh\theta, \quad \text{which gives} \quad s_\pm = e^{\pm\theta}, \qquad (8.177)$$

assuming for the time being that $\theta \geq 0$. We obtain the equations

$$\begin{aligned}\tfrac{1}{2} e^{+2\theta} - 2e^{+\theta} \cosh\theta + \theta &= -\tfrac{2}{3}\eta^{3/2} + A, \\ \tfrac{1}{2} e^{-2\theta} - 2e^{-\theta} \cosh\theta - \theta &= +\tfrac{2}{3}\eta^{3/2} + A,\end{aligned} \qquad (8.178)$$

from which we derive

$$\tfrac{4}{3}\eta^{3/2} = \sinh 2\theta - 2\theta, \quad A = -\tfrac{1}{2} - \cosh^2\theta = -\tfrac{1}{2} - t^2. \qquad (8.179)$$

By using these values of η and A, the w-solution of the equation in (8.176) is uniquely defined. Namely, we use that branch (of the three solutions) that is real for all positive values of s, and $s > 0$ corresponds with $w \in \mathbb{R}$.

After these preparations we obtain the standard form (cf. (8.122))

$$U\left(-\tfrac{1}{2}\mu^2, \mu t\sqrt{2}\right) e^{-zA} = \sqrt{2\pi}\, e^{\frac{1}{2}\mu^2 t^2} \left(\frac{\mu}{\sqrt{2}}\right)^{\frac{1}{2}\mu^2 + \frac{1}{2}} F_\eta(z), \qquad (8.180)$$

where
$$F_\eta(z) = \frac{1}{2\pi i} \int_C e^{z(\frac{1}{3}w^3 - \eta w)} f(w)\, dw, \quad f(w) = \frac{1}{\sqrt{s}} \frac{ds}{dw}. \tag{8.181}$$

Taking into account the mapping in (8.176), we have
$$\frac{ds}{dw} = s \frac{w^2 - b^2}{s^2 - 2ts + 1}, \quad f(w) = \sqrt{s}\, \frac{w^2 - b^2}{s^2 - 2ts + 1}, \quad \eta = b^2. \tag{8.182}$$

As explained in the previous section, for the computation of the coefficients α_n, β_n, we need the coefficients $p_k^{(1)}$, $p_k^{(2)}$ of the expansions (cf. (8.145))
$$f(w) = \sum_{k=0}^\infty p_k^{(1)} (w-b)^k, \quad f(-w) = \sum_{k=0}^\infty p_k^{(2)} (w-b)^k. \tag{8.183}$$

It turns out that $p_0^{(1)} = p_0^{(2)}$. Indeed, consider the expansions
$$s = s_+ + \sum_{k=1}^\infty s_k^+ (w-b)^k, \quad s = s_- + \sum_{k=1}^\infty s_k^- (w+b)^k. \tag{8.184}$$

Using the expression of ds/dw in (8.182) and l'Hôpital's rule, we obtain
$$s_1^+ = s_+ \frac{2b}{2(s_+ - t) s_1^+}, \quad \text{so that} \quad s_1^+ = \frac{\sqrt{bs_+}}{(t^2-1)^{\frac{1}{4}}} = \sqrt{\frac{bs_+}{\sinh\theta}}. \tag{8.185}$$

The square root has the plus sign because ds/dw is positive if $w \in \mathbb{R}$, as follows from the first relation in (8.182) and the properties of the mapping. From the expression (8.183) for $f(w)$ we obtain
$$p_0^{(1)} = f(b) = \frac{s_1^+}{\sqrt{s_+}} = \sqrt{\frac{b}{\sinh\theta}}. \tag{8.186}$$

Analogously,
$$s_1^- = \sqrt{\frac{bs_-}{\sinh\theta}} \quad \text{and} \quad p_0^{(2)} = f(-b) = \sqrt{\frac{b}{\sinh\theta}} = p_0^{(1)}. \tag{8.187}$$

In order to avoid expressions with algebraic roots in the computations, it is convenient to consider expansions like (8.183) for the function $\tilde{f}(w) = f(w)/f(b)$. We denote the corresponding coefficients by $\tilde{p}_k^{(1)}$ and $\tilde{p}_k^{(2)}$. In addition, to avoid algebraic roots in the expansions of (8.184), we replace t by a new variable
$$u = \sqrt{2b}\left(\frac{t-1}{t+1}\right)^{\frac{1}{4}}, \quad \text{so that} \quad t = \frac{4b^2 + u^4}{4b^2 - u^4}. \tag{8.188}$$

Then
$$s_+ = \frac{2b + u^2}{2b - u^2}, \quad s_1^+ = \frac{2b + u^2}{2u}. \tag{8.189}$$

8.5. Airy-type asymptotic expansions obtained from integrals

Other coefficients s_k^+ can be obtained by deriving a recurrence relation for them from the differential equation in (8.182). These are rational functions in u and b. The coefficients s_k^- can be obtained from the corresponding s_k^+ by changing the signs of both u and b. In particular,

$$s_- = \frac{2b - u^2}{2b + u^2}, \quad s_1^- = \frac{2b - u^2}{2u}. \tag{8.190}$$

Further, the coefficients $\tilde{p}_k^{(1)}$ and $\tilde{p}_k^{(2)}$ can be computed using

$$f(w) = \frac{1}{\sqrt{s}}\frac{ds}{dw} = 2\frac{d\sqrt{s}}{dw}. \tag{8.191}$$

Recall that $\sqrt{s}$ satisfies the differential equation $2s\, dS/dw = S\, ds/dw$. It is convenient to compute the power series (in $w-b$) solution S_+ of this equation with $S_+(b) = 4u/(2b-u^2)$. Then $\tilde{f}(w) = dS_+/dw$ and the coefficients $\tilde{p}_k^{(1)}$ are obtained easily. The coefficients $\tilde{p}_k^{(2)}$ can be obtained by changing the sign of both b and u in the expression for $(-1)^k \tilde{p}_k^{(1)}$.

Application of the algorithm of the previous section gives the coefficients α_j, β_j for the expansion of $\tilde{f}(w)$, and these coefficients are rational functions in b and u. We write them in a more compact form as rational functions in $\eta = b^2$ and

$$\xi = \frac{u^4 + 4b^2}{4u^2} = \frac{bt}{\sqrt{t^2 - 1}}. \tag{8.192}$$

The first few coefficients in the expansion (8.132) are

$$\alpha_0 = 1, \quad \beta_0 = 0, \quad \alpha_1 = \frac{1}{48}, \quad \beta_1 = \frac{5\xi^3 - 6\eta\xi - 5}{48\,\eta^2},$$

$$\alpha_2 = \frac{385\xi^6 - 924\eta\xi^4 + 684\eta^2\xi^2 - 143\eta^3 + 70\xi^3 - 84\eta\xi - 455}{4608\,\eta^3},$$

$$\beta_2 = \frac{\beta_1}{48}, \quad \alpha_3 = \frac{\alpha_2}{48} - \frac{2021}{34560}\alpha_1,$$

$$\beta_3 = \frac{425425\xi^9 - 1531530\eta\xi^8 + 2040012\eta^2\xi^5 - 28875\xi^6 - 1189005\eta^3\xi^3}{3317760\,\eta^5}$$

$$+ \frac{69300\eta\xi^4 + 259110\eta^4\xi - 51300\eta^2\xi^2 + 28875\xi^3}{3317760\,\eta^5}$$

$$+ \frac{10725\eta^3 - 34650\eta\xi - 425425}{3317760\,\eta^5},$$

$$\beta_4 = \frac{\beta_3}{48} - \frac{2021}{34560}\beta_2.$$

The linear relations between the coefficients follow from expansion (8.11) in [166], where both power series factors of Ai and Ai′ contain only even powers of our z (in Olver's notation, $z = \frac{1}{2}\mu^2$), but the whole expansion is multiplied by a function $g(z)$ with known asymptotics. Olver also notes that the coefficients in the Airy-type asymptotic expansion of $U(a, x)$ can be linearly determined from the asymptotic expansion (of the same function) in terms of elementary functions; see formulas (8.12), (8.13) in [166].

The coefficients α_n, β_n are analytic functions at $\eta = 0$, and we can expand them in Maclaurin series. The first few coefficients are expanded as follows:

$$\beta_1 = -\frac{9}{560} + \frac{7}{1800}\eta - \frac{1359}{1078000}\eta^2 + \frac{7}{16250}\eta^3 - \frac{152723}{1018710000}\eta^4 + \cdots, \qquad (8.193)$$

$$\alpha_2 = -\frac{199}{115200} + \frac{6849}{4928000}\eta - \frac{737}{1040000}\eta^2 + \frac{46711}{142560000}\eta^3 + \cdots. \qquad (8.194)$$

The radius of convergence equals $(3\pi/2)^{2/3} = 2.81\ldots$. This number follows from the singularity of the first relation given in (8.179), with θ defined in (8.177). The function $\eta(t)$ is singular at $t = -1$.

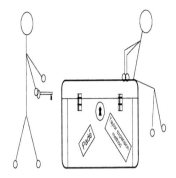

Chapter 9
Other Methods

One day Alice came to a fork in the road and saw a Cheshire cat in a tree. Which road do I take? she asked. Where do you want to go? was his response. I don't know, Alice answered. Then, said the cat, it doesn't matter.
—Lewis Carroll

9.1 Introduction

In this chapter we consider several methods and techniques that fall, somehow, outside the mainstream of the other topics discussed earlier.

First we discuss elements of the Padé method, which gives rational approximations that, just as power series and asymptotic expansions, may give excellent approximations at one point, say the origin or infinity. In this way we can transform power series of functions with finite singularities, or series that become unstable for certain values of the argument or parameters, into representations which allow larger domains for numerical evaluations.

As a related topic we discuss elements of sequence transformations, with emphasis on those introduced by Levin and Weniger. With this technique slowly convergent series (or divergent asymptotic series) can be transformed into new series with better numerical properties. In particular, for certain asymptotic expansions striking improvements can be obtained.

Another form of rational approximations gives the so-called best approximation on a finite interval (or an infinite interval after a transformation), with an error estimate that holds for the entire interval, in contrast to the Padé method, which gives (in its original form) an optimal approximation in one point of the interval or domain.

Finally we consider how to use Taylor expansions for computing solutions of linear second order ordinary differential equations.

9.2 Padé approximations

As an alternative to approximating a function by a power series we consider Padé approximations. Padé (1863–1953) was a French mathematician and a student of Hermite (1822–1901) who used Padé approximants to prove that e is transcendental. From a numerical point of view, the Padé approximants are important for computing functions outside the disk of convergence of the power series of the function, as well as inside the disk (for example, near the boundary of the disk). The Padé method can also be successfully applied for locating zeros and poles of the function.

Consider the power series

$$f(z) = c_0 + c_1 z + c_2 z^2 + \cdots, \tag{9.1}$$

with $c_0 \neq 0$. This series may be convergent or just a formal power series. We introduce a rational function $N_m^n(z)/D_m^n(z)$, where $N_m^n(z)$ and $D_m^n(z)$ are polynomials of maximal degree n and m, respectively. That is,

$$N_m^n(z) = a_0 + a_1 z + \cdots + a_n z^n, \quad D_m^n(z) = b_0 + b_1 z + \cdots + b_m z^m. \tag{9.2}$$

We choose these polynomials such that the power series expansion of $N_m^n(z) - f(z) D_m^n(z)$ starts with a term $A_{n,m} z^{n+m+1}$. The ratio $N_m^n(z)/D_m^n(z)$, of which the polynomials $N_m^n(z)$ and $D_m^n(z)$ satisfy the conditions

$$\begin{aligned} \text{degree } N_m^n(z) &\leq n, \quad \text{degree } D_m^n(z) \leq m, \\ N_m^n(z) - f(z) D_m^n(z) &= A_{n,m} z^{n+m+1} + \cdots, \end{aligned} \tag{9.3}$$

is called a *Padé approximant* of type (n, m) to the power series (9.1) (the function f). The ratio $N_m^n(z)/D_m^n(z)$ is denoted by $[n/m]_f$.

For each pair (n, m) at least one rational function exists that satisfies the conditions in (9.3), and this function can be found by solving the equations

$$\begin{cases} a_0 = c_0 b_0, \\ a_1 = c_1 b_0 + c_0 b_1, \\ \vdots \\ a_n = c_n b_0 + c_{n-1} b_1 + \cdots + c_{n-m} b_m, \end{cases} \quad \begin{cases} 0 = c_{n+1} b_0 + \cdots + c_{n-m+1} b_m, \\ \vdots \\ 0 = c_{n+m} b_0 + \cdots + c_n b_m, \end{cases} \tag{9.4}$$

where $c_j = 0$ if $j < 0$. When $m = 0$ the system of equations at the right is empty. In this case $a_j = c_j (j = 0, 1, \ldots, n)$ and $b_0 = 1$, and the partial sums of (9.1) yield the Padé approximants of type $(n, 0)$. In general, first the set at the right-hand side of (9.4) is solved (a homogeneous set of m equations for the $m+1$ values b_j), which has at least one nontrivial solution. We take a normalization, for example, by taking $b_0 = 1$ (see also the discussion in [8, p. 18]), and with this choice the last m equations give $b_1, \ldots, b_m$ as the solution of a system of m linear equations. The set on the left-hand side in (9.4) then yields $a_0, \ldots, a_n$.

Lemma 9.1. *Let* $\{N_m^n(z), D_m^n(z)\}$ *and* $\{\widetilde{N}_m^n(z), \widetilde{D}_m^n(z)\}$ *satisfy* (9.3). *Then*

$$N_m^n(z) \widetilde{D}_m^n(z) = \widetilde{N}_m^n(z) D_m^n(z). \tag{9.5}$$

9.2. Padé approximations

Proof. Because

$$N_m^n(z)\widetilde{D}_m^n(z) - \widetilde{N}_m^n(z)D_m^n(z) \qquad (9.6)$$
$$= [f(z)\widetilde{D}_m^n(z) - \widetilde{N}_m^n(z)]D_m^n(z) - [f(z)D_m^n(z) - N_m^n(z)]\widetilde{D}_m^n(z),$$

and both quantities between the square brackets satisfy (9.3), it follows that

$$N_m^n(z)\widetilde{D}_m^n(z) - \widetilde{N}_m^n(z)D_m^n(z) = B_{n+m+1}z^{n+m+1} + \cdots. \qquad (9.7)$$

But $N_m^n(z)\widetilde{D}_m^n(z) - \widetilde{N}_m^n(z)D_m^n(z)$ is a polynomial of degree at most $n+m$. So $N_m^n(z)\widetilde{D}_m^n(z) = \widetilde{N}_m^n(z)D_m^n(z)$. □

It follows that the ratios $N_m^n(z)/D_m^n(z)$ and $\widetilde{N}_m^n(z)/\widetilde{D}_m^n(z)$ are the same, and that all nontrivial solutions of (9.3) give the same irreducible Padé approximant. In fact, when forming the irreducible Padé approximant $N_m^n(z)/D_m^n(z)$ a nontrivial factor (a polynomial) may be factored out. Anyhow, we have the following theorem.

Theorem 9.2. *For every nonnegative n and m, a unique Padé approximant of type (n,m) for f of (9.1) exists.*

The array of Padé approximants

$$\begin{matrix} [0/0]_f & [0/1]_f & [0/2]_f & \cdots \\ [1/0]_f & [1/1]_f & [1/2]_f & \cdots \\ [2/0]_f & [2/1]_f & [2/2]_f & \cdots \\ \vdots & \vdots & \vdots & \ddots \end{matrix} \qquad (9.8)$$

is called a *Padé table*. It is arranged here so that approximants with the same denominator degree are located in the same column. As remarked earlier, the first column corresponds to the partial sums of the power series in (9.1). The elements of the first row correspond to the partial sums of the power series of $1/f$.

It may happen that certain approximants occur several times in the table (because the polynomials $N_m^n(z)$ and $D_m^n(z)$ for certain (n,m) are not relatively prime). An approximant $[n,m]_f$ is called *normal* if it occurs only once in the Padé table.

In the literature special attention is paid to the diagonal elements $[n,n]_f$ of the table, with applications to orthogonal polynomials, quadrature formulas, moment problems, and other problems of classical analysis.

In applied mathematics and in theoretical physics, Padé approximants have become a useful tool for overcoming convergence problems with power series. The popularity of Padé approximants in theoretical physics is due to Baker [6], who also wrote a monograph on Padé approximants [7]. Of interest also is the monograph by Baker and Graves-Morris [8].

An extended bibliography on Padé approximants and related matters containing several thousand references was published by Brezinski in [20]. For an annotated bibliography focusing on computational aspects, see [244]. Luke gives many rational approximations of special functions, and usually these are Padé approximants; see [143, 144], and also §9.2.4.

9.2.1 Padé approximants and continued fractions

One of the many relations between Padé approximants and continued fractions is the following result (see [119, pp. 190–195]).

Theorem 9.3. *Let* $f(z) = 1 + c_1 z + c_2 z^2 + \cdots$ *be a (formal) power series with Padé approximants* $R_{n,m}(z) = N_m^n(z)/D_m^n(z)$ *and assume that in the staircase sequence*

$$R_{0,0}, R_{1,0}, R_{1,1}, R_{2,1}, R_{2,2}, R_{3,2}, \ldots \qquad (9.9)$$

all terms are normal. Then there exists a continued fraction of the form

$$1 + \frac{a_1 z}{1+} \frac{a_2 z}{1+} \frac{a_3 z}{1+} \cdots, \quad a_n \neq 0, \qquad (9.10)$$

with nth approximant C_n *satisfying*

$$C_{2k} = R_{k,k}, \quad C_{2k+1} = R_{k+1,k}, \quad k = 0, 1, 2, \ldots. \qquad (9.11)$$

In [8, §4.7] several examples are given of continued fractions which are Padé approximants. This reference also gives the connection to continued fractions and Gauss quadrature and convergence results of Padé approximants.

9.2.2 How to compute the Padé approximants

The approximants can be computed by *Wynn's cross rule*. Any five Padé approximants arranged in the Padé table as

$$
\begin{array}{ccc}
& N & \\
W & C & E \\
& S &
\end{array}
$$

satisfy Wynn's cross rule (see [246])

$$(N - C)^{-1} + (S - C)^{-1} = (W - C)^{-1} + (E - C)^{-1}. \qquad (9.12)$$

Starting with the first column $[n/0]_f$, $n = 0, 1, 2, \ldots$, initializing the preceding column by $[n/-1]_f = \infty$, $n = 1, 2, \ldots$, (9.12) enables us to compute the lower triangular part of the table. Likewise, the upper triangular part follows from the first row $[0/n]_f$, $n = 0, 1, 2, \ldots$, by initializing $[-1/n]_f = 0$, $n = 1, 2, \ldots$.

The elements of the Padé table can also be computed by the *epsilon algorithm* of Wynn [245]. We consider the recursions

$$\varepsilon_{-1}^{(n)} = 0, \quad \varepsilon_0^{(n)} = s_n, \quad n = 0, 1, 2, \ldots,$$

$$\varepsilon_{m+1}^{(n)} = \varepsilon_{m-1}^{(n+1)} + \frac{1}{\varepsilon_m^{(n+1)} - \varepsilon_m^{(n)}}, \quad n, m = 0, 1, 2, \ldots. \qquad (9.13)$$

9.2. Padé approximations

Table 9.1. *Padé approximants* $[n+k/k]_f$, $k = 1, 2$, $n = 0, 1, 2, 3, 4$, *for* $f(z) = z^{-1}\log(1+z)$.

n	$[n+1/1]_f$	$[n+2/2]_f$
0	$\dfrac{6+z}{2(3+2z)}$	$\dfrac{z^2+21z+30}{3(3z^2+12z+10)}$
1	$-\dfrac{-24-6z+z^2}{6(4+3z)}$	$-\dfrac{z^3-12z^2-150z-180}{12(6z^2+20z+15)}$
2	$\dfrac{60+18z-4z^2+z^3}{12(5+4z)}$	$\dfrac{2z^4-15z^3+120z^2+1170z+1260}{60(10z^2+30z+21)}$
3	$-\dfrac{-360-120z+30z^2-10z^3+3z^4}{60(6+5z)}$	$-\dfrac{z^5-6z^4+30z^3-200z^2-1680z-1680}{60(15z^2+42z+28)}$
4	$\dfrac{420+150z-40z^2+15z^3-6z^4+2z^5}{60(7+6z)}$	$\dfrac{4z^6-21z^5+84z^4-350z^3+2100z^2+15960z+15120}{420(21z^2+56z+36)}$

If s_n is the nth partial sum of a power series f, then $\varepsilon_{2k}^{(n)}$ is the Padé approximant $[n+k/k]_f$ (cf. (9.8)). We give details in the following algorithm, where we assume that N is divisible by 4. The elements $\varepsilon_{2k+1}^{(n)}$ are only auxiliary quantities which diverge if the whole transformation process converges and shouldn't be used for convergence tests or output. A recent review of the applications of the epsilon algorithm can be found in [102].

ALGORITHM 9.1. Computing Padé approximants $[n+k/k]_f$.
Input: N, a_m, $m = 0, 1, \ldots, N$.

Output: Padé approximants $[n+k/k]_f$, $k = 0, 1, \ldots, N/4$, $n = 0, 1, \ldots, N/2$.

- $M = N/2$; $m = 0$;
- $\varepsilon_{-1}^{(n)} = 0$, $\varepsilon_0^{(n)} = s_n = \sum_{m=0}^{n} a_m z^m$, $0 \leq n \leq N$
- DO WHILE $m < M+1$: $n = 0$
-     DO WHILE $n < N-m$: $\varepsilon_{m+1}^{(n)} = \varepsilon_{m-1}^{(n+1)} + \dfrac{1}{\varepsilon_m^{(n+1)} - \varepsilon_m^{(n)}}$
- $n = 0$
- DO WHILE $n < M+1$: $k = 0$
-     DO WHILE $k < N/4+1$: $[n+k/k]_f = \varepsilon_{2k}^{(n)}$

Example 9.4 (the logarithmic function). We use the epsilon algorithm for obtaining a few Padé approximants to the logarithmic function $f(z) = z^{-1}\log(1+z) = 1 - \frac{1}{2}z + \frac{1}{3}z^2 - \cdots$, whose series converges for $|z| < 1$. We take $N = 8$ and obtain the results shown in Table 9.1 (we skip $\varepsilon_0^{(n)} = s_n = \sum_{m=0}^{n}(-1)^m \frac{z^m}{m+1}$).

Next we take $N = 20$ and compute the relative errors for $z = 1$, a point on the radius of convergence. The results are given in Table 9.2. ∎

Table 9.2. *Relative errors in* $[n+k/k]_f$, *for* $z=1$, $f(z) = z^{-1}\log(1+z)$.

n	$[n+1/1]_f$	$[n+2/2]_f$	$[n+3/3]_f$	$[n+4/4]_f$	$[n+5/5]_f$
0	$0.99\,10^{-02}$	$0.27\,10^{-03}$	$0.76\,10^{-05}$	$0.22\,10^{-06}$	$0.64\,10^{-08}$
1	$0.39\,10^{-02}$	$0.83\,10^{-04}$	$0.21\,10^{-05}$	$0.55\,10^{-07}$	$0.15\,10^{-08}$
2	$0.19\,10^{-02}$	$0.32\,10^{-04}$	$0.69\,10^{-06}$	$0.16\,10^{-07}$	$0.42\,10^{-09}$
3	$0.10\,10^{-02}$	$0.14\,10^{-04}$	$0.26\,10^{-06}$	$0.56\,10^{-08}$	$0.13\,10^{-09}$
4	$0.64\,10^{-03}$	$0.71\,10^{-05}$	$0.11\,10^{-06}$	$0.22\,10^{-08}$	$0.46\,10^{-10}$
5	$0.42\,10^{-03}$	$0.39\,10^{-05}$	$0.53\,10^{-07}$	$0.90\,10^{-09}$	$0.18\,10^{-10}$
6	$0.29\,10^{-03}$	$0.22\,10^{-05}$	$0.27\,10^{-07}$	$0.41\,10^{-09}$	$0.73\,10^{-11}$
7	$0.21\,10^{-03}$	$0.14\,10^{-05}$	$0.14\,10^{-07}$	$0.20\,10^{-09}$	$0.32\,10^{-11}$
8	$0.15\,10^{-03}$	$0.87\,10^{-06}$	$0.81\,10^{-08}$	$0.99\,10^{-10}$	$0.15\,10^{-11}$
9	$0.12\,10^{-03}$	$0.58\,10^{-06}$	$0.47\,10^{-08}$	$0.53\,10^{-10}$	$0.72\,10^{-12}$
10	$0.92\,10^{-04}$	$0.39\,10^{-06}$	$0.29\,10^{-08}$	$0.29\,10^{-10}$	$0.37\,10^{-12}$

In applications one usually concentrates on obtaining diagonal elements $[n/n]_f$ and elements not far away from the diagonal; see [233], which also has an efficient modified algorithm for these elements.

9.2.3 Padé approximants to the exponential function

There are several elegant methods for obtaining the polynomials that constitute the elements of the Padé table for e^z. For this function it is not necessary to solve the linear equations in (9.4), to use Wynn's cross rule, or to use the epsilon algorithm. See, for example, [179, §75]. We derive the polynomials by using a few properties of the confluent hypergeometric function.

We start with the well-known relation (see [219, p. 173])

$$_1F_1\left(\begin{matrix}a\\c\end{matrix};z\right) = e^z \,_1F_1\left(\begin{matrix}c-a\\c\end{matrix};-z\right) \qquad (9.14)$$

and take $a = -n$, $n = 0, 1, 2, \ldots$. This gives a polynomial on the left-hand side. From the literature (see [179, §75]), we know that the two polynomials $_1F_1(-n;-n-m;z)$ and $_1F_1(-m;-n-m;-z)$ are the desired polynomials for constructing the Padé approximations to e^z. When we try to use $a = -n$ and $c = -n - m$ in (9.14) we obtain the remarkable relation

$$_1F_1\left(\begin{matrix}-n\\-m-n\end{matrix};z\right) \stackrel{?}{=} e^z \,_1F_1\left(\begin{matrix}-m\\-m-n\end{matrix};-z\right), \quad n,m = 0,1,2,\ldots, \qquad (9.15)$$

9.2. Padé approximations

in which the left-hand side is a polynomial and the right-hand side is not a polynomial. We conclude that (9.15) is false; it would give an extreme example for the Padé theory in which an entire function (e^z, not a polynomial) would be approximated by a rational function, with infinite precision.

Remark 13. This is an instructive example for explaining the danger of using relations of the hypergeometric functions in the case that upper and lower parameters assume negative integer values. For similar examples for the Gauss hypergeometric function, see [219, pp. 109 and 111].

Next we take $a = -n$ and $c = \varepsilon - n - m$ (with ε a small number) in (9.14), and we will see what happens in the valid relation

$$_1F_1\left(\begin{array}{c}-n\\ \varepsilon - m - n\end{array}; z\right) = e^z {}_1F_1\left(\begin{array}{c}\varepsilon - m\\ \varepsilon - m - n\end{array}; -z\right) \qquad (9.16)$$

when we let $\varepsilon \to 0$. Multiplying both sides by $\Gamma(\varepsilon)/\Gamma(\varepsilon - n - m)$, we obtain

$$\sum_{k=0}^{n} \frac{\Gamma(\varepsilon)(-n)_k}{\Gamma(k + \varepsilon - n - m)} \frac{z^k}{k!} = e^z \sum_{k=0}^{\infty} \frac{\Gamma(\varepsilon)(\varepsilon - m)_k}{\Gamma(k + \varepsilon - n - m)} \frac{(-z)^k}{k!}. \qquad (9.17)$$

We use the limit

$$\lim_{\varepsilon \to 0} \frac{\Gamma(\varepsilon)}{\Gamma(\varepsilon - n)} = (-1)^n n!, \quad n = 0, 1, 2, \ldots. \qquad (9.18)$$

This gives in (9.17)

$$\sum_{k=0}^{n} (-n)_k (n + m - k)! \frac{(-z)^k}{k!} = e^z \sum_{k=0}^{m} (-m)_k (n + m - k)! \frac{z^k}{k!} + S_1 + S_2, \qquad (9.19)$$

where

$$\begin{aligned} S_1 &= (-1)^{n+m} e^z \lim_{\varepsilon \to 0} \sum_{k=m+1}^{n+m} \frac{\Gamma(\varepsilon)(\varepsilon - m)_k}{\Gamma(k + \varepsilon - n - m)} \frac{(-z)^k}{k!}, \\ S_2 &= (-1)^{n+m} e^z \lim_{\varepsilon \to 0} \sum_{k=n+m+1}^{\infty} \frac{\Gamma(\varepsilon)(\varepsilon - m)_k}{\Gamma(k + \varepsilon - n - m)} \frac{(-z)^k}{k!}. \end{aligned} \qquad (9.20)$$

Each term in the sum S_1 vanishes, and for each term in S_2 we have a nonvanishing limit. We obtain

$$\begin{aligned} \sum_{k=0}^{n} (-n)_k (n + m - k)! \frac{(-z)^k}{k!} &= e^z \sum_{k=0}^{m} (-m)_k (n + m - k)! \frac{z^k}{k!} \\ &+ (-1)^n m! e^z \sum_{k=n+m+1}^{\infty} \frac{\Gamma(k - m)}{\Gamma(k - n - m)} \frac{(-z)^k}{k!}. \end{aligned} \qquad (9.21)$$

In other words,

$$\sum_{k=0}^{n}\binom{n+m-k}{m}\frac{z^k}{k!} = e^z \sum_{k=0}^{m}\binom{n+m-k}{n}\frac{(-z)^k}{k!}$$
$$+ \frac{(-1)^{m+1}z^{n+m+1}}{(n+m+1)!}{}_1F_1\left(\begin{array}{c}m+1\\n+m+2\end{array};z\right). \qquad (9.22)$$

Because
$$(n+m-k)! = (-1)^k \frac{(n+m)!}{(-n-m)_k}, \quad 0 \le k \le n+m, \qquad (9.23)$$

we also have

$${}_1F_1\left(\begin{array}{c}-n\\-m-n\end{array};z\right) - e^z {}_1F_1\left(\begin{array}{c}m\\-m-n\end{array};-z\right)$$
$$= \frac{(-1)^{m+1}z^{n+m+1}}{n+m+1}{}_1F_1\left(\begin{array}{c}m+1\\n+m+2\end{array};z\right). \qquad (9.24)$$

This repairs the invalid relation in (9.15), and we conclude that the Padé approximations to the exponential function are given by

$$N_m^n(z) = {}_1F_1\left(\begin{array}{c}-n\\-m-n\end{array};z\right), \quad D_m^n(z) = {}_1F_1\left(\begin{array}{c}-m\\-m-n\end{array};-z\right). \qquad (9.25)$$

In addition, we have an explicit form of the remainder, as shown in (9.24).

In terms of the confluent hypergeometric function $U(a,c,z)$ we have

$$\frac{(n+m)!}{m!} N_m^n(z) = z^{n+m+1} U(m+1, n+m+2, z),$$
$$\frac{(n+m)!}{n!} D_m^n(z) = z^{n+m+1} U(n+1, n+m+2, -z), \qquad (9.26)$$

with integral representations (see (2.136))

$$N_m^n(z) = \frac{1}{(n+m)!} \int_0^\infty e^{-t} t^m (t+z)^n \, dt,$$
$$D_m^n(z) = \frac{1}{(n+m)!} \int_0^\infty e^{-t} t^n (t-z)^m \, dt. \qquad (9.27)$$

Remark 14. The diagonal numerator polynomials $N_n^n(z)$ are of interest in the representation of the K-Bessel functions of half-integer order. We have (see (2.135))

$$K_{n+\frac{1}{2}}(z) = \sqrt{\frac{\pi}{2z}} 2^n \left(\frac{1}{2}\right)_n z^{-n} e^{-z} N_n^n(2z), \quad n = 0, 1, 2, \ldots. \qquad (9.28)$$

9.2.4 Analytic forms of Padé approximations

We mention a few examples given by Luke for hypergeometric functions. See also [197, §17.3] for Padé approximations for hypergeometric series of $_1F_1$, $_2F_1$, and $_2F_0$ types.

The Gauss hypergeometric function

In [144, pp. 276–278] we find results for the Gauss hypergeometric function. Let $H(z) = {}_2F_1(1, \sigma; \rho + 1; -z)$. Then we have with $a = 0$ or $a = 1$

$$H(z) = \frac{A_n(z)}{B_n(z)} + \frac{S_n(z)}{B_n(z)},$$

$$B_n(z) = {}_2F_1\left(\begin{array}{c} -n, \ -n + a - \sigma \\ -2n + a - \rho \end{array}; -z\right), \tag{9.29}$$

and A_n and B_n satisfy the same recurrence formula

$$B_{n+1}(z) = C_n B_n(z) + D_n B_{n-1}(z),$$

$$C_n = 1 + z \frac{2n^2 + 2n(\rho + 1 - a) + (\rho - a)(\sigma + 1 - a)}{(2n + \rho - a)(2n + \rho + 2 - a)}, \tag{9.30}$$

$$D_n = -\frac{n(n + \rho - \sigma)(n + \rho - a)(n + \sigma - a)z^2}{(2n + \rho - 1 - a)(2n + \rho - a)^2(2n + \rho + 1 - a)},$$

with initial values

$$A_0 = 1, \quad A_1 = 1 + \frac{(\rho + 1 - \sigma)z}{(\rho + 1)(\rho + 2)} \quad \text{if } a = 0,$$

$$A_0 = 0, \quad A_1 = 1 \quad \text{if } a = 1, \tag{9.31}$$

$$B_0 = 1, \quad B_1 = 1 + \frac{(\sigma + 1 - a)z}{\rho + 2 - a} \quad \text{if } a = 0 \text{ or } a = 1.$$

If $a = 0$, then the ratio $A_n(z)/B_n(z)$ is the diagonal element $[n/n]_{{}_2F_1}$ of the Padé table, and if $a = 1$, it is the subdiagonal element $[n - 1/n]_{{}_2F_1}$.

$S_n(z)$ of (9.29) is given by

$$S_n(z) = \frac{(-\sigma)^{1-a}\rho^a(\sigma + 1 - a)_n(\rho + 1 - \sigma)_n n! \Gamma(n + \rho - a + 1)}{(\rho - a + 1)_n(n + \rho + 1 - a)_n \Gamma(2n + \rho - a + 2)}$$

$$\times (1 + z)^{\rho - \sigma} z^{2n - a + 1} {}_2F_1\left(\begin{array}{c} n + \rho + 1 - \sigma, \ n + \rho - a + 1 \\ 2n + \rho - a + 2 \end{array}; -z\right), \tag{9.32}$$

and Luke gives asymptotic estimates of $S_n(z)/B_n(z)$, from which it follows that

$$\lim_{n \to \infty} \frac{S_n(z)}{B_n(z)} = 0, \quad |\mathrm{ph}(1 + z)| < \pi. \tag{9.33}$$

We conclude from this example that the Padé approximants extend the original domain of the power series (the unit disk) to the complete complex domain outside the cut from -1 to $-\infty$. As an example we have the case $\sigma = \rho = 1$:

$$_2F_1\left(\begin{matrix}1, 1\\2\end{matrix}; -z\right) = \frac{\ln(1+z)}{z}. \tag{9.34}$$

For $n = 10$ and $z = 1$ we obtain

$$\frac{A_n(z)}{B_n(z)} = 0.69314\ 71805\ 59945\ 40350, \tag{9.35}$$

giving an absolute error of $0.94\ 10^{-16}$ compared with $\ln(2)$.

The incomplete gamma functions

Our next case is the incomplete gamma function

$$\Gamma(v, z) = \int_z^\infty t^{v-1} e^{-t}\, dt. \tag{9.36}$$

From [144, §4.3.2] we derive

$$z^{1-v} e^z \Gamma(v, z) = \frac{G_n(v, z)}{H_n(v, z)} + \frac{S_n(v, z)}{H_n(v, z)},$$
$$H_n(z) = (2 - a - v)_n {}_1F_1(-n; 2 - a - v; -z) = n!\, L_n^{(1-a-v)}(-z), \tag{9.37}$$
$$S_n(v, z) = (v - 1)^{1-a}(2 - a - v)_n z^a n!\, U(n + 1; v + a; z),$$

where $a = 0$ or $a = 1$, $L_n^{(\alpha)}(z)$ is the Laguerre polynomial, and $U(a; c; z)$ is a confluent hypergeometric function. The Padé approximants $G_n(v, z)/H_n(v, z)$ are of the type $(n - a, m)$.

G_n and H_n satisfy the same recurrence formula

$$H_{n+1} = (z + 2n + 2 - a - v) H_n - n(n + 1 - a - v) H_{n-1}, \tag{9.38}$$

with initial values

$$\begin{aligned}G_0(v, z) &= 1, & G_1(v, z) &= 1 + z & &\text{if } a = 0,\\G_0(v, z) &= 0, & G_1(v, z) &= z & &\text{if } a = 1,\\H_0(v, z) &= 1, & H_1(v, z) &= 2 - a - v + z & &\text{if } a = 0 \text{ or } a = 1.\end{aligned} \tag{9.39}$$

The error is given by

$$\frac{S_n(v, z)}{H_n(v, z)} = \frac{2\pi(v - 1)^{1-a} z^{1-v} e^{z - 4\sqrt{kz}}}{\Gamma(2 - a - v)} \left(1 + \mathcal{O}(k^{-\frac{1}{2}})\right), \tag{9.40}$$

9.2. Padé approximations

where $k = n + 1 - \frac{1}{2}(a + v)$, $z \neq 0$, z and v are bounded, and $|\text{ph}(\sqrt{kz})| \leq \frac{1}{2}\pi - \delta$, δ being a fixed small positive number. For a detailed discussion on estimates of the error term for large k that hold uniformly with respect to z, we refer to [144, pp. 84–94]. Luke's result may be updated by using more recent results on uniform asymptotic approximations of the confluent hypergeometric functions.

It follows from (9.40) that the approximants converge everywhere in the complex z-plane except on the negative real axis and except, possibly, in the neighborhood of the zeros of $H_n(v, z)$.

For convenience the polynomials $G_n(z)$ and $H_n(z)$ are given as polynomials of z. However, the Padé approximants are connected with the asymptotic expansion (see (8.3))

$$z^{1-v} e^z \Gamma(v, z) \sim \sum_{k=0}^{\infty} (-1)^k (1 - v)_k z^{-k}, \quad z \to \infty. \tag{9.41}$$

The Padé property of the approximation in (9.37) follows from the asymptotic expansion (see [219, p. 280])

$$U(n+1; v+a; z) \sim z^{-n-1} \sum_{k=0}^{\infty} (-1)^k \frac{(n+1)_k (n - v - a + 2)_k}{k! z^k}, \quad z \to \infty, \tag{9.42}$$

from which it follows that

$$\frac{S_n(v, z)}{H_n(v, z)} = \mathcal{O}\left(z^{a-2n-1}\right), \quad z \to \infty. \tag{9.43}$$

For $a = 0$ and $v = \frac{1}{2}$ we have computed the Padé approximants for the following cases. For $n = 15$ and $z = 10$ we obtain

$$\frac{G_n(z)}{H_n(z)} = 0.95608\,66129\,30276\,73758, \tag{9.44}$$

giving an absolute error of $0.11\,10^{-16}$. For $n = 100$ and $z = 1$ the result is

$$\frac{G_n(z)}{H_n(z)} = 0.75787\,21561\,41312\,14071, \tag{9.45}$$

with absolute error $0.38\,10^{-16}$. The error estimates from (9.40) for these cases are $0.39\,10^{-16}$ and $0.35\,10^{-16}$, respectively.

For Padé approximants for the incomplete gamma function

$$\gamma(v, z) = \int_0^z t^{v-1} e^{-t}\, dt, \quad \Re v > 0, \tag{9.46}$$

we refer to [144, pp. 79–82].

9.3 Sequence transformations

When applying numerical techniques to physical problems, results are usually produced in the form of sequences. Examples are iterative methods, discretization methods, perturbation methods, and—most important in the context of special functions—series expansions. Often, the sequences that are produced in this way converge too slowly to be numerically useful. When dealing with asymptotic series, summation of the sequences may also be difficult.

Sequence transformations are tools to overcome convergence problems of that kind. A slowly convergent (or even divergent in the asymptotic sense) sequence $\{s_n\}_{n=0}^{\infty}$, whose elements may be the partial sums

$$s_n = \sum_{k=0}^{n} a_k \qquad (9.47)$$

of a convergent or formal infinite series, is converted into a new sequence $\{s'_n\}_{n=0}^{\infty}$ with hopefully better numerical properties.

We discuss sequence transformations that are useful in the context of special functions. For many special functions power series are available, which can be used for their evaluation. In addition, many special functions possess asymptotic expansions which diverge for every finite argument. Consequently, the emphasis in this section will be on sequence transformations that are able either to accelerate the convergence of slowly convergent power series effectively or to sum divergent asymptotic series.

In §9.2 we have already seen how the Padé method transforms the partial sums of a (formal) power series into a sequence of rational functions. We discuss a few other sequence transformations that, in the case of power series, produce different rational approximants, and they can also be applied to other convergence acceleration problems.

Details on the history of sequence transformations and related topics, starting from the 17th century, can be found in [21]; see also [22]. For review papers, with many references to monographs devoted to this topic, we refer the reader to [106, 233]. See also Appendix A in [19], written by Dirk Laurie, with interesting observations and opinions about sequence transformations.

9.3.1 The principles of sequence transformations

Let s_n be the partial sum of (9.47). If $\lim_{n \to \infty} s_n = s$ exists, we can write

$$s_n = s + r_n. \qquad (9.48)$$

When the limit does not exist, this relation may have a meaning in the theory of asymptotic expansions, say, when a_k depends on a large parameter z and by using an estimate of the form $r_n = \mathcal{O}(z^{-n})$, as $z \to \infty$.

When convergence is slow, we can try to improve convergence by transforming the sequence $\{s_n\}$ into a new sequence $\{s'_n\}$, and consider the relation

$$s'_n = s + r'_n. \qquad (9.49)$$

9.3. Sequence transformations

With the exception of some more or less trivial model problems, the transformed remainders $\{r'_n\}_{n=0}^{\infty}$ will be different from zero for all finite values of n. However, convergence is accelerated if the transformed remainders $\{r'_n\}_{n=0}^{\infty}$ vanish more rapidly than the original remainders $\{r_n\}_{n=0}^{\infty}$ as $n \to \infty$, that is,

$$\lim_{n \to \infty} \frac{s'_n - s}{s_n - s} = \lim_{n \to \infty} \frac{r'_n}{r_n} = 0, \tag{9.50}$$

with modifications for asymptotic sequences.

9.3.2 Examples of sequence transformations

First we mention Levin's sequence transformation [135], which is defined by

$$\mathcal{L}_k^{(n)}(s_n, \omega_n) = \frac{\sum_{j=0}^{k}(-1)^j \binom{k}{j} \frac{(n+j+1)^{k-1}}{(\zeta+n+k)^{k-1}} \frac{s_{n+j}}{\omega_{n+j}}}{\sum_{j=0}^{k}(-1)^j \binom{k}{j} \frac{(n+j+1)^{k-1}}{(\zeta+n+k)^{k-1}} \frac{1}{\omega_{n+j}}}, \tag{9.51}$$

where s_n are the partial sums of (9.47) and the quantities ω_n are remainder estimates. For example, we can simply take

$$\omega_n = s_{n+1} - s_n = a_{n+1}, \tag{9.52}$$

but more explicit remainder estimates can be used.

Another transformation is due to Weniger [233], who replaced the powers $(n+j+1)^{k-1}$ in Levin's transformation by Pochhammer symbols $(n+j+1)_{k-1}$. That is, Weniger's transformation reads

$$\mathcal{S}_k^{(n)}(s_n, \omega_n) = \frac{\sum_{j=0}^{k}(-1)^j \binom{k}{j} \frac{(\zeta+n+j)_{k-1}}{(\zeta+n+k)_{k-1}} \frac{s_{n+j}}{\omega_{n+j}}}{\sum_{j=0}^{k}(-1)^j \binom{k}{j} \frac{(\zeta+n+j)_{k-1}}{(\zeta+n+k)_{k-1}} \frac{1}{\omega_{n+j}}}. \tag{9.53}$$

Other sequence transformations can be found in [233] or in [23, §2.7]. The sequence transformations (9.51) and (9.53) differ from other sequence transformations because not only the elements of a sequence $\{s_n\}$ are required, but also explicit remainder estimates $\{\omega_n\}$. For special functions this information is usually available when divergent asymptotic expansions are considered. It was shown in several articles that the transformation (9.53) is apparently very effective, in particular if divergent asymptotic series are to be summed; see [14, 232, 235].

9.3.3 The transformation of power series

For transforming partial sums $f_n(z) = \sum_{k=0}^{n} \gamma_k z^k$ of a formal power series

$$f(z) = \sum_{k=0}^{\infty} \gamma_k z^k, \tag{9.54}$$

we can take the remainder estimates

$$\omega_n = \gamma_{n+1} z^{n+1}, \tag{9.55}$$

and we replace z by $1/z$ in the case of an asymptotic series.

With these modifications the transformations (9.51) and (9.53) become rational functions of the variable z. If the coefficients γ_n in (9.54) are all different from zero, these rational functions satisfy the asymptotic error estimates [236, eqs. (4.28)–(4.29)]

$$\begin{aligned} f(z) - \mathcal{L}_k^{(n)}(f_n(z), \gamma_{n+1} z^{n+1}) &= \mathcal{O}(z^{k+n+2}), \quad z \to 0, \\ f(z) - \mathcal{S}_k^{(n)}(f_n(z), \gamma_{n+1} z^{n+1}) &= \mathcal{O}(z^{k+n+2}), \quad z \to 0. \end{aligned} \tag{9.56}$$

These estimates imply that all terms of the formal power series, which were used for construction of the rational approximants in this way, are reproduced exactly by a Taylor expansion around $z = 0$. Thus, the transformations $\mathcal{L}_k^{(n)}(f_n(z), \gamma_{n+1} z^{n+1})$ and $\mathcal{S}_k^{(n)}(f_n(z), \gamma_{n+1} z^{n+1})$ are formally very similar to the analogous estimate (9.3) satisfied by the Padé approximants $[n/m]_f(z) = N_m^n(z)/D_m^n(z)$.

9.3.4 Numerical examples

Simple test problems, which nevertheless demonstrate convincingly the power of sequence transformations using explicit remainder estimates, are the integrals

$$\mathbf{E}^{(\nu)}(z) = \int_0^{\infty} \frac{e^{-t} \, dt}{1 + z t^{\nu}} \tag{9.57}$$

and their associated divergent asymptotic expansions

$$\mathbf{E}^{(\nu)}(z) \sim \sum_{k=0}^{\infty} (\nu k)! (-z)^k, \quad z \to 0. \tag{9.58}$$

For $\nu = 1$, $\mathbf{E}^{(\nu)}(z)$ is the exponential integral E_1 with argument $1/z$ according to $\mathbf{E}^{(1)}(z) = e^{1/z} E_1(1/z)/z$. For $\nu = 2$ or $\nu = 3$, $\mathbf{E}^{(\nu)}(z)$ cannot be expressed in terms of known special functions.

In order to demonstrate the use of sequence transformations with explicit remainder estimates, both $\mathcal{S}_k^{(n)}(f_n(z), \gamma_{n+1} z^{n+1})$ and Padé approximants are applied to the partial sums

$$E_n^{(\nu)}(z) = \sum_{k=0}^{n} (\nu k)! (-z)^k, \quad 0 \le n \le 50, \tag{9.59}$$

9.3. Sequence transformations

of the asymptotic series (9.58) for $\nu = 1, 2, 3$. The Padé approximants were computed with the help of Wynn's epsilon algorithm (see §9.2.2). All calculations were done in Maple, and the integrals $\mathbf{E}^{(\nu)}(z)$ were computed to the desired precision with the help of numerical quadrature. For the remainder estimates we took $\omega_n = (\nu(n+1))!(-z)^{n+1}$.

The results for $\mathbf{E}^{(1)}(z)$ with $z = 1$ are

$$\mathbf{E}^{(1)}(1) = 0.59634\ 73623\ 23194\ 07434\ 10785,$$
$$\mathcal{L}_0^{(50)}(E_0^{(1)}(1), \omega_{50}) = 0.59634\ 73623\ 23194\ 07434\ 10759,$$
$$\mathcal{S}_0^{(50)}(E_0^{(1)}(1), \omega_{50}) = 0.59634\ 73623\ 23194\ 07434\ 10785, \qquad (9.60)$$
$$[25/24] = 0.59634\ 7322,$$
$$[25/25] = 0.59634\ 7387.$$

The Padé approximants are not very efficient. Nevertheless, it seems that they are able to sum the divergent series (9.58) for $\nu = 1$.

The results for $\mathbf{E}^{(2)}(z)$ with $z = 1/10$ are

$$\mathbf{E}^{(2)}(1/10) = 0.88425\ 13061\ 26979,$$
$$\mathcal{L}_0^{(50)}(E_0^{(2)}(1/10), \omega_{50}) = 0.88425\ 13061\ 26980,$$
$$\mathcal{S}_0^{(50)}(E_0^{(2)}(1/10), \omega_{50}) = 0.88425\ 13061\ 26985, \qquad (9.61)$$
$$[25/24] = 0.88409,$$
$$[25/25] = 0.88437.$$

Here, the Padé approximants are certainly not very useful since they can only extract an accuracy of three places.

The results for $\mathbf{E}^{(3)}(z)$ with $z = 1/100$ are

$$\mathbf{E}^{(3)}(1/100) = 0.96206\ 71061,$$
$$\mathcal{S}_0^{(50)}(E_0^{(3)}(1/100), \omega_{50}) = 0.96206\ 71055,$$
$$\mathcal{L}_0^{(50)}(E_0^{(3)}(1/100), \omega_{50}) = 0.96206\ 71057, \qquad (9.62)$$
$$[25/24] = 0.960,$$
$$[25/25] = 0.964.$$

In [100] it is shown that an asymptotic series, whose coefficients grow more rapidly than $(2n)!$, is not Padé summable since subsequences $[n + j/n]$ in the Padé table converge to different, j-dependent limits as $n \to \infty$. The Levin and Weniger transformations are apparently able to sum the asymptotic series (9.58) even for $\nu = 3$.

Other numerical examples of sequence transformations using explicit remainder estimates can be found in [14, 234, 235].

9.4 Best rational approximations

In a previous section we considered the Padé method as a local approximation at one point. In that case the error is smallest at one point. For the exponential function this point was $z = 0$; for the incomplete gamma function $\Gamma(\nu, z)$ it was the point at infinity. It is possible to extend the method by including more points, which turns the method into the *multipoint Padé method*; see [8, Part II, §1.2]. In the case of multipoint approximants the error is minimal at several points of the interval, and the method may seem to be more attractive than the classical Padé method. However, for practical use in the evaluation of special functions the computation of coefficients of the polynomials constituting the rational function becomes rather complicated, and explicit forms of the polynomials are not available.

In the theory of best rational approximation (which includes the best polynomial approximation) the goal is to find a rational function that approximates a function f on a finite real interval as best as possible. To explain this, let $\mathcal{R}_m^n$ be the family of all rational functions that can be written in the form

$$\frac{N_m^n(x)}{D_m^n(x)} = \frac{a_0 + a_1 x + \cdots + a_n x^n}{b_0 + b_1 x + \cdots + b_m x^m}. \tag{9.63}$$

Let $[a, b]$ be a finite real interval and let f be a function defined for all $x \in [a, b]$. Let $\|\cdot\|$ be the maximum norm

$$\|f\| = \max_{x \in [a,b]} |f(x)|. \tag{9.64}$$

Then $\mathcal{R}_m^n$ contains a unique element R of best approximation to f. That is,

$$\inf_{R \in \mathcal{R}_m^n} \|R - f\| \tag{9.65}$$

is actually achieved by one and only one element R_m^n of $\mathcal{R}_m^n$. As in the Padé method, the *form* in which this rational function of best approximation is written is never unique, unless we require it to be irreducible, for example, by imposing the condition $D_m^n(a) = 1$.

The characterization of the best approximation to f may be given in terms of oscillations of the error curve.

Theorem 9.5 (Chebyshev's theorem). *Let $R = N/D$ be an irreducible element of $\mathcal{R}_m^n$. A necessary and sufficient condition that R be the best approximation to f is that the error function $R(x) - f(x)$ exhibits at least $2 + \max\{m + \partial N, n + \partial D\}$ points of alternation. (Here ∂P denotes the degree of the polynomial P.)*

Proof. For the proof see [152]. □

For the elementary and well-known higher transcendental functions the polynomials in the best rational approximations are not explicitly known, and the coefficients of these polynomials should be computed by an algorithm. This algorithm is not as simple as the one for computing the Padé approximants, which can be based on solving the set of linear equations in (9.4). For best approximation the *second algorithm of Remes* can be used [187, p. 176], and for a Fortran program see [116].

For many elementary and special functions best rational approximations have been computed. See [104] for many tables (and an explanation of the Remes algorithm). For several other special functions we refer to the survey [43]. Computer algebra packages, such as Maple, also have programs for computing best rational approximants.

In Chapter 3 we have discussed the expansion of functions in series of Chebyshev polynomials. Partial sums of these series are in fact polynomial approximations, and for practical purposes, the coefficients of Chebyshev series (and, hence, those of the partial sums) are usually much easier to obtain than the coefficients of the polynomials of best approximation; see [38]. And for each choice of n (or of n and m in rational approximations), new coefficients have to be computed.

The numerical difference between partial sums of the Chebyshev series

$$f(x) = \sum_{k=0}^{\infty} c_k T_k(x), \quad x \in [-1, 1], \tag{9.66}$$

and the polynomials of best approximation is not very great. Because the series (9.66) usually converges rapidly (for example, for functions in C^{∞} on $[-1, 1]$), we obtain a very good first approximation to the polynomial $p_n(x)$ of best approximation for $[-1, 1]$ if we truncate (9.66) at its $(n + 1)$th term. This is because

$$f(x) - \sum_{k=0}^{n} c_k T_k(x) = c_{n+1} T_{n+1}(x), \tag{9.67}$$

approximately, and the right-hand side enjoys exactly those properties concerning its maxima and minima that are required for the polynomial of best approximation.

More precisely, it is known that for the interval $[-1, 1]$, the ratio of the maximum value of the remainder

$$\left| \sum_{k=n+1}^{\infty} c_k T_k(x) \right| \tag{9.68}$$

to the maximum error of the polynomial of best approximation $p_n(x)$ is bounded by $1 + L_n$, where L_n is the nth *Lebesgue constant* for Fourier series. Since $L_0 = 1$, L_n is a monotonically increasing function of n, and, e.g., $L_{1000} = 4.07\ldots$, this means that in practice the gain in replacing a truncated Chebyshev series expansion by the corresponding minimax polynomial approximation is hardly worthwhile; see [184] and §3.4.

9.5 Numerical solution of ordinary differential equations: Taylor expansion method

The special functions of mathematical physics usually arise as special solutions of ordinary linear differential equations, which follow from certain forms of the wave equation. Separation of the variables and the use of domains such as spheres, circles, cylinders, and so on, are the standard ways of introducing Bessel functions, Legendre functions, and confluent hypergeometric functions (also called Whittaker functions). For an introduction to this topic, see [219, Chap. 10].

In numerical mathematics, computing solutions of ordinary linear differential equations is a vast research area, with popular methods such as, for example, Runge–Kutta methods. These techniques are usually not used for computing special functions, mainly because so many other efficient methods are available for these functions. However, when the differential equation has coefficients in terms of analytic functions, as is the case for the equations of special functions, a method based on Taylor expansions may be considered as an alternative method, in particular for solving the equation in the complex plane.

In this section we give the basic steps for the Taylor expansion method, and we consider linear second order equations of the form

$$\frac{d^2 w}{dz^2} + f(z)\frac{dw}{dz} + g(z)w = h(z), \tag{9.69}$$

where f, g, and h are analytic functions in a domain $D \subset \mathbb{C}$. If $h = 0$, the differential equation is *homogeneous*; otherwise it is *inhomogeneous*. For applications to special functions f, g, and h are often simple rational functions, and usually the equation is homogeneous.

For classification of singularities of (9.69) and expansions of solutions in the neighborhoods of singularities, see §2.2.

9.5.1 Taylor-series method: Initial value problems

Assume that we wish to integrate (9.69) along a finite path $\mathcal{P}$ from $z = a$ to $z = b$ in the domain D. The path is partitioned at $P + 1$ points labeled successively $z_0, z_1, \ldots, z_P$, with $z_0 = a$, $z_P = b$.

By repeated differentiation of (9.69) all derivatives of $w(z)$ can be expressed in terms of $w(z)$ and $w'(z)$ as follows. Write

$$w^{(s)}(z) = f_s(z)w(z) + g_s(z)w'(z) + h_s(z), \quad s = 0, 1, 2, \ldots, \tag{9.70}$$

with

$$\begin{aligned} f_0(z) &= 1, & g_0(z) &= 0, & h_0(z) &= 0, \\ f_1(z) &= 0, & g_1(z) &= 1, & h_1(z) &= 0. \end{aligned} \tag{9.71}$$

Then for $s = 2, 3, \ldots,$

$$\begin{aligned} f_s(z) &= f'_{s-1}(z) - g(z)g_{s-1}(z), \\ g_s(z) &= f_{s-1}(z) - f(z)g_{s-1}(z) + g'_{s-1}(z), \\ h_s(z) &= h(z)g_{s-1}(z) + h'_{s-1}(z). \end{aligned} \tag{9.72}$$

Write $\tau_j = z_{j+1} - z_j$, $j = 0, 1, \ldots, P$, expand $w(z)$ and $w'(z)$ in Taylor series centered at $z = z_j$, and apply (9.70). Then

$$\begin{bmatrix} w(z_{j+1}) \\ w'(z_{j+1}) \end{bmatrix} = \mathbf{A}(\tau_j, z_j) \begin{bmatrix} w(z_j) \\ w'(z_j) \end{bmatrix} + \mathbf{b}(\tau_j, z_j), \tag{9.73}$$

9.5 Numerical solution of ODEs: Taylor expansion method

where $\mathbf{A}(\tau, z)$ is the matrix

$$\mathbf{A}(\tau, z) = \begin{bmatrix} A_{11}(\tau, z) & A_{12}(\tau, z) \\ A_{21}(\tau, z) & A_{22}(\tau, z) \end{bmatrix}, \tag{9.74}$$

$\mathbf{b}(\tau, z)$ is the vector

$$\mathbf{b}(\tau, z) = \begin{bmatrix} b_1(\tau, z) \\ b_2(\tau, z) \end{bmatrix}, \tag{9.75}$$

and the quantities $A_{jk}(\tau, z)$ and $b_j(\tau, z)$ can be expanded as follows:

$$A_{11}(\tau, z) = \sum_{s=0}^{\infty} \frac{\tau^s}{s!} f_s(z), \qquad A_{12}(\tau, z) = \sum_{s=0}^{\infty} \frac{\tau^s}{s!} g_s(z),$$

$$A_{21}(\tau, z) = \sum_{s=0}^{\infty} \frac{\tau^s}{s!} f_{s+1}(z), \qquad A_{22}(\tau, z) = \sum_{s=0}^{\infty} \frac{\tau^s}{s!} g_{s+1}(z), \tag{9.76}$$

$$b_1(\tau, z) = \sum_{s=0}^{\infty} \frac{\tau^s}{s!} h_s(z), \qquad b_2(\tau, z) = \sum_{s=0}^{\infty} \frac{\tau^s}{s!} h_{s+1}(z).$$

If the solution $w(z)$ that we are seeking grows in magnitude at least as fast as all other solutions of (9.69) as we pass along $\mathcal{P}$ from a to b, then $w(z)$ and $w'(z)$ may be computed in a stable manner for $z = z_0, z_1, \ldots, z_P$ by successive application of (9.73) for $j = 0, 1, \ldots, \mathcal{P} - 1$, beginning with initial values $w(a)$ and $w'(a)$.

Similarly, if $w(z)$ is decaying at least as fast as all other solutions along $\mathcal{P}$, then we may reverse the labeling of the z_j along $\mathcal{P}$ and begin with initial values $w(b)$ and $w'(b)$.

9.5.2 Taylor-series method: Boundary value problem

Now suppose the path $\mathcal{P}$ is such that the rate of growth of $w(z)$ along $\mathcal{P}$ is intermediate to that of two other solutions. (This can happen only for inhomogeneous equations.) Then to compute $w(z)$ in a stable manner we solve the set of equations (9.73) simultaneously for $j = 0, 1, \ldots, P$, as follows. Let $\mathbf{A}$ be the $(2P) \times (2P + 2)$ band matrix

$$\mathbf{A} = \begin{bmatrix} -\mathbf{A}(\tau_0, z_0) & \mathbf{I} & \mathbf{0} & \cdots & \mathbf{0} & \mathbf{0} \\ \mathbf{0} & -\mathbf{A}(\tau_1, z_1) & \mathbf{I} & \cdots & \mathbf{0} & \mathbf{0} \\ \vdots & \vdots & \vdots & \ddots & \vdots & \vdots \\ \mathbf{0} & \mathbf{0} & \mathbf{0} & & -\mathbf{A}(\tau_{P-1}, z_{P-1}) & \mathbf{I} \end{bmatrix} \tag{9.77}$$

($\mathbf{I}$ and $\mathbf{0}$ being the identity and zero matrices of order 2×2). Also let $\mathbf{w}$ denote the $(2P+2) \times 1$ vector

$$\mathbf{W} = \begin{bmatrix} w(z_0), w'(z_0), w(z_1), w'(z_1), \ldots, w(z_P), w'(z_P) \end{bmatrix}^\mathrm{T}, \tag{9.78}$$

and $\mathbf{b}$ the $(2P) \times 1$ vector

$$\mathbf{b} = [b_1(\tau_0, z_0), b_2(\tau_0, z_0), b_1(\tau_1, z_1), b_2(\tau_1, z_1), \ldots, b_1(\tau_{P-1}, z_{P-1}), b_2(\tau_{P-1}, z_{P-1})]^\mathrm{T}. \tag{9.79}$$

Then
$$\mathbf{Aw} = \mathbf{b}. \tag{9.80}$$

This is a set of $2P$ equations for the $2P + 2$ unknowns, $w(z_j)$ and $w'(z_j)$, $j = 0, 1, \ldots, P$. The remaining two equations are supplied by boundary conditions of the form

$$\alpha_0 w(z_0) + \beta_0 w'(z_0) = \gamma_0, \quad \alpha_1 w(z_P) + \beta_1 w'(z_P) = \gamma_1, \tag{9.81}$$

where the α's, β's, and γ's are constants.

If, for example, $\beta_0 = \beta_1 = 0$, then on moving the contributions of $w(z_0)$ and $w(z_P)$ to the right-hand side of (9.80) the resulting system of equations is not tridiagonal, but can readily be made tridiagonal by annihilating the elements of $\mathbf{A}$ that lie below the main diagonal and its two adjacent diagonals. The equations can then be solved by the method of Gaussian elimination for a tridiagonal matrix or by Olver's algorithm (see §4.7.3). The latter would be especially useful if the endpoint b of $\mathcal{P}$ were at ∞.

For further information and examples, see [162, §7] and [139]. For an application to compute solutions in the complex plane of the Airy differential equation, see [65] with a Fortran computer program in [64].

9.6 Other quadrature methods

We briefly describe additional quadrature rules not discussed in Chapter 5. Romberg quadrature (§9.6.1) provides a scheme for computing successively refined rules with a higher degree of exactness; Fejér and Clenshaw–Curtis quadratures (§9.6.2) are interpolatory rules which behave quite similarly to Gauss–Legendre rules but which are easier to compute and provide nested rules. Other nested rules, related to Gauss quadrature but harder to compute than Clenshaw–Curtis, are Kronrod and Patterson quadratures (§9.6.3). Finally, we describe specific methods for oscillatory integrands, focusing on Filon's method (§9.6.4).

9.6.1 Romberg quadrature

As we discussed before, in general the trapezoidal rule is a rule of second order in the sense that the remainder in (5.14) satisfies $R_n = \mathcal{O}(h^2)$ as $h \to 0$. We now briefly describe how to use the expression (5.32) for building quadrature rules of order $2k + 2$ from quadrature rules of order $2k$. This method is known as *Romberg quadrature*.

Consider an initial step h (for instance, $h = b - a$) and let $R(i, k)$ be the quadrature rule of step

$$h_i = h/2^{i-1}, \quad i = 1, 2, \ldots, \tag{9.82}$$

and order $2k$. In this way, $R(1, 1)$ will be the trapezoidal rule with step h. From (5.32), we know that

$$I(f) = \int_a^b f(x)\,dx = T(f, h) + a_2 h^2 + a_4 h^4 + \cdots = R(1, 1) + a_2 h^2 + a_4 h^4 + \cdots, \tag{9.83}$$

where a_{2j} do not depend on h. We write this as formal series expansions, neglecting the remainders.

9.6. Other quadrature methods

Next consider

$$I(f) = R(1, 1) + a_2 h^2 + a_4 h^4 + \cdots,$$
$$I(f) = R(2, 1) + a_2 (h/2)^2 + a_4 (h/2)^4 + \cdots. \tag{9.84}$$

Multiplying the second equation by 4 and subtracting the first, we obtain

$$I(f) = \tfrac{1}{3}[4R(2, 1) - R(1, 1)] + \widehat{a}_4 (h/2)^4 + \cdots, \tag{9.85}$$

which is a fourth order rule with step size $h/2$. Therefore, we can write $R(2, 2) = \tfrac{1}{3}[4R(2, 1) - R(1, 1]$, or, in other words,

$$S(f, h/2) = \tfrac{1}{3}[4T(f, h/2) - T(f, h)] = \tfrac{1}{3} h(f_0 + 4f_1 + 2f_2 + 4f_3 + 2f_4 + \cdots + f_n), \tag{9.86}$$

which, indeed, is the *compound Simpson's rule*.

This process can be generalized: after $k - 1$ steps, we arrive at a quadrature rule of order $2k$,

$$I(f) = R(i, k) + b_{2k} h_i^{2k} + b_{2k+2} h_i^{2k+2} + \cdots, \tag{9.87}$$

and we also have

$$I(f) = R(i + 1, k) + b_{2k} (h_i/2)^{2k} + b_{2k+2} (h_i/2)^{2k+2} + \cdots. \tag{9.88}$$

So, similar to before,

$$I(f) = \frac{4^k R(i+1, k) - R(i, k)}{4^k - 1} + \widehat{b}_{2k+2} h_{i+1}^{2k+2}, \quad h_{i+1} = h_i/2. \tag{9.89}$$

It follows that

$$R(i+1, k+1) = \frac{4^k R(i+1, k) - R(i, k)}{4^k - 1}, \quad i = 1, 2, \ldots, \quad k = 1, 2, \ldots, i, \tag{9.90}$$

which can be also written as

$$\begin{aligned} R(i, k+1) &= R(i, k) + \frac{R(i, k) - R(i-1, k)}{4^k - 1} \\ &= R(i, k) + \frac{\nabla_i R(i, k)}{4^k - 1}, \quad i = 2, 3, \ldots, \quad k = 1, 2, \ldots, i - 1. \end{aligned} \tag{9.91}$$

It is quite natural to consider the term $\nabla_i R(i, k)/(4^k - 1)$ as an estimation of the error, because we may assume that the results are improving (which is not always true) when higher values of i and k are used.

The basic ideas of Romberg quadrature can be summarized in the following algorithm.

ALGORITHM 9.2. Romberg quadrature.
Input: $\epsilon > 0$, $f(x)$, b, a, $n \in \mathbb{N}$, $n_f \in \mathbb{N}$.
Output: $\approx \int_a^b f(x)\, dx$.

- $h = (b-a)/n; i = 1; \Delta = 1 + \epsilon;$

- $I(i) = \frac{h}{2}(f(a) + f(b));$

- IF $n > 1$ THEN $I(i) = I(i) + h \sum_{j=1}^{n-1} f(a + jh)$

- DO WHILE $(\Delta > \epsilon, i < n_f)$

    $i = i + 1; h = h/2;$

    $I(i) = I(i-1)/2 + h \sum_{j=1}^{n} f(a + (2j-1)h);$

    $n = 2n; k = 0;$

    DO WHILE $(\Delta > \epsilon, k < i - 1)$

    $\quad k = k + 1;$

    $\quad \Delta_k = (I(i-k+1) - I(i-k))/(4^k - 1); \Delta = |\Delta_k|;$

    $\quad I(i-k) = I(i-k+1) + \Delta_k;$

- $I = I(i-k).$

9.6.2 Fejér and Clenshaw–Curtis quadratures

As discussed in Chapter 5, Gauss–Legendre quadrature is the quadrature with the highest possible degree of exactness for computing integrals of the type

$$\int_{-1}^{1} f(x)\, dx. \tag{9.92}$$

However, as with all Gaussian quadratures (except the Gauss–Chebyshev quadrature), it has the inconvenient feature that the nodes and weights are not easy to compute, because no explicit formulas exist.

The idea behind Fejér [67] and Clenshaw–Curtis [39] quadratures consists in abandoning the interpolation at the zeros of Legendre polynomials $P_n^{(0,0)}(x)$ (where $P_n^{(\alpha,\beta)}(x)$ is a Jacobi polynomial) and considering instead the nodes of $P_n^{(-\frac{1}{2},-\frac{1}{2})}(x)$ or $P_n^{(\frac{1}{2},\frac{1}{2})}(x)$ (which are Chebyshev polynomials; see Chapter 3).

The degree of precision for this type of quadrature is $n-1$, which is the minimum for an interpolatory quadrature rule based on n nodes. In this sense, Gauss–Legendre quadrature is superior. However, because the distribution of the zeros of Legendre and Chebyshev polynomials is very similar, particularly when n is large, in many cases the performance is also very similar (see, for instance, [224]). The Clenshaw–Curtis rule has the advantages that the nodes and weights are explicitly known and the number of nodes can be incremented with ease, particularly when it is doubled; this is crucial for efficiency and control of the error.

9.6. Other quadrature methods

The first Fejér rule

Let us approximate $f(x)$ by the the n-point interpolatory rule taking as nodes the zeros of $T_n(x)$. This is the same construction considered in Example 5.10, except that now the weight function is $w(x) = 1$ and we are thus considering the approximation

$$\int_{-1}^{1} f(x)\,dx \approx I^{F1}(f) = \int_{-1}^{1} f_{n-1}(x)\,dx, \tag{9.93}$$

where $f_{n-1}(x)$ is the polynomial of degree at most $n-1$ which interpolates f at the nodes $x_k = \cos[(k-1/2)(\pi/n)]$, $k = 1, \ldots, n$. This gives

$$I^{F1}(f) = \sum_{j=0}^{n-1}{}' c_j \int_{-1}^{1} T_j(x)\,dx, \tag{9.94}$$

with

$$c_j = \frac{2}{n}\sum_{k=1}^{n} f(x_k) T_j(x_k), \tag{9.95}$$

which, as discussed in Chapter 3, can be interpreted as a discrete cosine transform.

Using (3.25) and $T_n(1) = 1$, $T_n(-1) = (-1)^n$, we obtain

$$\int_{-1}^{1} T_j(x)\,dx = \frac{1}{2}\left[\frac{T_{j+1}(x)}{j+1} - \frac{T_{j-1}(x)}{j-1}\right]_{-1}^{1} = \frac{1+(-1)^j}{1-j^2}, \tag{9.96}$$

which vanishes when j is odd (also for $j = 1$). With this, and using $T_j(\cos\theta) = \cos(j\theta)$, we have the *first Fejér rule*

$$I^{F1}(f) = \sum_{k=1}^{n} w_k f(x_k), \tag{9.97}$$

where

$$w_k = \frac{2}{n}\left(1 - 2\sum_{j=1}^{\lfloor n/2 \rfloor}\frac{\cos 2j\theta_k}{4j^2-1}\right), \quad \theta_k = \left(k - \frac{1}{2}\right)\frac{\pi}{n}. \tag{9.98}$$

Clenshaw–Curtis quadrature

The *Clenshaw–Curtis rule* is a variant of the Fejér rule. The integrand $f(x)$ is interpolated using as nodes the extrema of $T_n(x)$ (that is, the zeros of $U_{n-1}(x)$ and including the endpoints $x = -1, 1$). When the points $x = -1, 1$ are excluded, the resulting rule is named the *second Fejér rule*.

The polynomial interpolating $f(x)$ at the extrema of $T_n(x)$ is given by (3.60), (3.61). The Clenshaw–Curtis quadrature can be expressed as follows:

$$\int_{-1}^{1} f(x)\,dx \approx I^{CC}(f) = \int_{-1}^{1} f_n(x)\,dx = \sum_{k=0}^{n}{}'' c_k \int_{-1}^{1} T_k(x)\,dx, \tag{9.99}$$

where

$$c_k = \frac{2}{n}\sum_{j=0}^{n}{}'' f(x_j)T_k(x_j), \quad x_j = \cos(j\pi/n). \tag{9.100}$$

With this, it is immediate to check that

$$I^{CC}(f) = \sum_{k=0}^{n} w_k f(x_k), \tag{9.101}$$

where

$$w_k = \frac{g_k}{n}\left(1 - \sum_{j=1}^{\lfloor n/2 \rfloor} \frac{b_j}{4j^2 - 1}\cos(2jk\pi/n)\right). \tag{9.102}$$

The values of the constants are

$$g_k = \begin{cases} 1, & k = 0, n; \\ 2, & 1 \leq k \leq n; \end{cases} \qquad b_j = \begin{cases} 1, & j = n/2; \\ 2 & \text{otherwise}. \end{cases} \tag{9.103}$$

Because the coefficients c_k are discrete cosine transforms, their computation can be efficiently performed with fast Fourier transform techniques (see [226]). Increasing the number of nodes by doubling n can be done economically, by reusing the old nodes in the new quadrature and adding one node between each of the old nodes (equispaced in the θ variable).

Other types of nested rules (with an increasing number of interlaced nodes) are possible (Kronrod, Patterson), but they are harder to compute (see §9.6.3).

The weights w_k can also be directly computed in terms of discrete Fourier transforms (see [228]).

9.6.3 Other Gaussian quadratures

A number of variants of Gauss quadratures exist with different properties. For instance, when the interval of integration $[a, b]$ is such that a and/or b is finite, one can consider the possibility of building quadrature rules which have one of the endpoints (Gauss–Radau) or both (Gauss–Lobatto) as nodes and have the maximal degree of exactness.

The starting point for building such quadrature rules is the following result, which can be proved in the same way as Theorem 5.9.

Theorem 9.6. *Given $k \in \mathbb{N}$, $0 \leq k \leq n$, a quadrature rule $Q(f)$ with n nodes of type (5.39) has degree of exactness $d = n - 1 + k$ if and only if the following conditions are satisfied:*

1. *The quadrature formula is interpolatory.*

2. *The node polynomial $q_n(x) = (x - x_1)\cdots(x - x_n)$ satisfies*

$$\int_a^b p(x)q_n(x)w(x)\,dx = 0 \quad \forall p \in \mathbb{P}_{k-1}. \tag{9.104}$$

9.6. Other quadrature methods

Proof. For the proof see [80, pp. 21–22]. □

In fact, the degree of exactness for Gauss rules can be found as a particular case of this theorem with $k = n$.

For instance, for building a Gauss–Radau formula which has a as a node and n additional nodes $x_1, \ldots, x_n$, giving the rule

$$\int_a^b f(x)w(x)\,dx \approx w_0 f(a) + \sum_{j=1}^n w_j f(x_j), \qquad (9.105)$$

in order to reach the maximum possible degree of exactness we should select the nodes x_j in such a way that the node polynomial $q_{n+1}(x) = (x - a)(x - x_1) \cdots (x - x_n)$ satisfies the condition

$$\int_a^b p(x) q_{n+1}(x) w(x)\,dx = \int_a^b p(x) q_n(x)(x - a) w(x)\,dx = 0 \qquad (9.106)$$

for the highest possible degree k. Because

$$w_R(x) = (x - a)w(x) \qquad (9.107)$$

is a nonnegative weight function, by choosing $p_n(x)$ to be the monic orthogonal polynomial of degree n with respect to the weight $w_R(x)$, we get the optimal situation in which (9.106) is satisfied for all polynomials $p(x)$ of degree not larger than $n - 1$. Therefore, the rule has, according to Theorem 9.6 (with n replaced by $n + 1$ and k by $n - 1$), degree of exactness $d = 2n$, which is in agreement with the fact that the number of parameters to be determined is $2n + 1$ (the dimension of the space $\mathbb{P}_{2n}$).

Using the same type of arguments, it is possible to show that the Gauss–Lobatto rule with $2n + 2$ nodes (two of them being $x = a$ and $x = b$) has degree of exactness $2n + 1$.

The weights and nodes for these rules can be computed in a similar way to those for standard Gauss rules, by solving an associated eigenvalue problem (see [80]).

Gauss–Kronrod quadrature

The Gauss–Kronrod rule also consists in adding additional nodes to a Gauss rule, but with a different philosophy. The idea is to build a Gauss rule with some control of the error.

The main difficulty in the Gauss rules is that one needs to compute nodes and weights. For this one has to decide how many nodes will be considered. The method is certainly optimal with respect to the degree of exactness and, therefore, it may give a large reward. However, the method is rather rigid and does not provide an effective method for testing the accuracy.

The rule invented by Kronrod [129, 130] (see also [28, 49, 80, 133, 173]) provides a method for control of the accuracy.

The idea is, given a number of nodes n of a certain Gauss rule (Theorem 5.9), to compare the results of this rule with a finer rule which has $n + 1$ additional nodes in the interval $[a, b]$ such that it has the highest degree of exactness. We write

$$\int_a^b f(x) w(x)\,dx \approx \sum_{i=1}^n w_i^K f(x_i) + \sum_{i=1}^{n+1} \hat{w}_i f(\hat{x}_i), \qquad (9.108)$$

where x_i are the given Gaussian nodes, $\hat{x}_i$ are the new nodes (the *Kronrod nodes*), which have to be computed as well as the new weights w_i^K (which are not the Gaussian weights corresponding to the nodes x_i) and $\hat{w}_i$. We have $n + 1$ new nodes plus $2n + 1$ new weights. This suggests that this degree of freedom can be used to build a quadrature with degree of exactness $3n + 1$.

According to Theorem 9.6, in order to have degree of precision $3n + 1$ it is needed that

$$\int_a^b p(x) E_{n+1}(x) p_n(x) w(x) \, dx = 0 \quad \forall p \in \mathbb{P}_n, \tag{9.109}$$

where $E_{n+1}(x) = (x - \hat{x}_1) \cdots (x - \hat{x}_{n+1})$ and $p_n(x)$ is the monic orthogonal polynomial of degree n with respect to $w(x)$.

If this condition can be met, after determining the nodes, the weights can be computed from the interpolatory nature of the quadrature. The problem is that this condition is of a different orthogonality nature, since we require that the polynomial $E_{n+1} \in \mathbb{P}_{n+1}$, called a *Stieltjes polynomial*, be orthogonal to all polynomials in $\mathbb{P}_n$ with respect to a weight function

$$w_K(x) = p_n(x) w(x), \tag{9.110}$$

which is not a usual weight function in the sense of Definition 5.7. Indeed, w_K, just like p_n, is oscillating on $[a, b]$ and is not a nonnegative weight.

If w is a nonnegative weight function, it can be proved [80, Thm. 3.15] that the Stieltjes polynomial satisfying the property (9.109) exists. However, the nice properties for the zeros of the polynomials orthogonal with respect to nonnegative weights are not always true for Stieltjes polynomials: not all zeros are necessarily real, and some real zeros may lie outside the interval $[a, b]$. Moreover, and of course, they are not necessarily interlaced with those of $p_n(x)$ (which is not necessary but is indeed an interesting property). Therefore, a Gauss–Kronrod rule is not always possible.

Gauss–Kronrod quadrature is known to be possible for the Gegenbauer case (see Table 5.2) when $0 < \lambda \leq 2$ (that the Stieltjes polynomials have the required properties has been known since the work of Szegö (1935) [205]). On the other hand, Gauss–Kronrod is not possible for $\lambda > 3$ and large n [177] (see [178] for analogous results for the Jacobi case). For the successful Gegenbauer cases (and also for the Legendre case $\lambda = 1/2$), the degree of exactness is even larger in some cases: it is precisely $3n + 1$ for n even, but $3n + 2$ for n odd due to symmetry; in addition, when $\lambda = 1$ (Chebyshev) it is $4n + 1$. For the Laguerre and Hermite cases, the situation is even worse [121] and, for instance, for the Hermite case it is not possible unless $n = 1, 2, 4$. When Gauss–Kronrod is not possible, one can always try to require less degree of exactness or add fewer additional nodes.

When Gauss–Kronrod is possible, matrix algorithms are also available [28, 80, 133]. In this case, one can compute the integral with the Gauss n rule and then test the accuracy with $n + 1$ additional function evaluations; the previously evaluated function values do not need to be recomputed, because the Gauss nodes x_i, $i = 1, \ldots, n$, are used again in (9.108).

When this can be done, the Gauss–Kronrod quadrature (or variants) will give a warning when it fails to give a sufficiently accurate result. But when it does, we will still have to recompute a higher number of nodes and weights from scratch.

Patterson quadrature

The limited level of flexibility of Gauss–Kronrod quadrature suggests building a process with a higher degree of recursivity. Let us observe that, given n preassigned nodes, adding $n + 1$ additional nodes will lead to a rule of $3n + 1$ degree of exactness if the new nodes can be chosen in an optimal way (which is not always possible). In principle, it is therefore possible to construct nested rules with an increasing number of nodes which are chosen in an optimal way.

For instance, for integrals $\int_{-1}^{1} f(x)\,dx$, it is possible to generate quadrature rules with an increasing number of nodes chosen in an optimal way, starting with the Legendre quadrature with one node, which is the midpoint rule

$$\int_{-1}^{1} f(x)\,dx \approx 2 f(0). \tag{9.111}$$

Next, two additional nodes are included such that the resulting rule has the highest possible degree of exactness (this is the three-point Legendre rule). Then four more nodes are added, maximizing the degree of exactness (this is no longer a Gauss rule, but it is a Gauss–Kronrod rule). We can continue this process and, after the rule with $2^k + 1$ nodes has been computed, 2^k more nodes are added in such a way that the degree of exactness is optimized.

Nodes and weights for this case are given in [173] up to a 127-point rule of degree 191. See also [128, 175, 176] for improvements and algorithmic details. This is a convenient procedure in adaptive quadrature.

9.6.4 Oscillatory integrals

In §5.5 we have discussed how to deal with integrals on contours in the complex plane and how strong oscillations can be handled by choosing appropriate contours, through saddle points, for example (see also §11.5).

In this section we discuss a few other aspects of oscillatory integrals. For further discussions we refers to [16], where a number of related methods for the evaluation of oscillatory integrals over infinite ranges are compared.

We start with integrals of the form

$$I(f; p) = \int_{0}^{\infty} f(x) e^{ipx}\,dx, \tag{9.112}$$

in which p may be a complex parameter, and f is a function sufficiently smooth on $(0, \infty)$ and with sufficient decay at ∞ to ensure convergence of the integral.

First we make a few observations concerning integrals of type (9.112). To conclude we give the details of Filon's method, which can be used on a finite interval.

Asymptotic expansion

When derivatives of f are available and when p is large we can first integrate by parts to obtain the main contribution to the integral. In this way,

$$I(f; p) = \frac{i}{p} f(0) - \frac{1}{p^2} f'(0) - \frac{1}{p^2} \int_0^\infty f''(x) e^{ipx} \, dx, \tag{9.113}$$

which may be continued in order to obtain higher approximations, as far as the smoothness and growth conditions of the derivatives of f allow.

Odd or even functions

When f is an even function, all its odd derivatives at the origin vanish; hence, all terms with even powers of p^{-1} vanish when we continue the expansion in (9.113). For example, when we take $f(x) = 1/(1 + x^2)$, we have, when $p > 0$,

$$I(f; p) = \int_0^\infty \frac{e^{ipx}}{1 + x^2} \, dx = \frac{1}{2} \pi e^{-p} + i \int_0^\infty \frac{\sin(px)}{1 + x^2} \, dx, \tag{9.114}$$

in which the real part of $I(f; p)$ is exponentially small and the imaginary part is $\mathcal{O}(p^{-1})$ when p is large. So, computing $I(f; p)$ of (9.112) by using a quadrature rule when f is a real even function may give a large relative error (but a small absolute error) in the real part of $I(f; p)$. Similarly when f is an odd function.

Analytic functions

When f is slowly decreasing at ∞, convergence of a quadrature rule may be rather poor. When f is analytic in the right half-plane, and p is positive, we may investigate if turning the path of integration up into the complex plane is possible. In that case convergence of the integral and of the quadrature rule may be improved. See also §5.5.

Orthogonal polynomials

For integrals of the form

$$I(f; p) = \int_{-1}^{1} e^{ipx} f(x) \, dx, \tag{9.115}$$

we can try to expand $f(x)$ in terms of Legendre polynomials, $f(x) = \sum_{n=0}^\infty c_n P_n(x)$, and obtain

$$I(f; p) = \sqrt{\frac{2\pi}{p}} \sum_{n=0}^\infty (-i)^n c_n J_{n+\frac{1}{2}}(p) \tag{9.116}$$

in terms of spherical Bessel functions; see [219, eq. (6.64)].

In the same manner we can expand the function $f(x)$ in

$$I(f; p) = \int_{-1}^{1} e^{ipx} \frac{f(x)}{\sqrt{1 - x^2}} \, dx \tag{9.117}$$

in terms of Chebyshev polynomials, $f(x) = \sum_{n=0}^\infty c_n T_n(x)$, and obtain

$$I(f; p) = \pi \sum_{n=0}^\infty i^n c_n J_n(p) \tag{9.118}$$

9.6. Other quadrature methods

in terms of ordinary Bessel functions, which follows from [219, eq. (9.20)].

For further examples on the use of orthogonal polynomials (also on infinite intervals), see [174].

More general forms

Oscillatory integrals also occur in the form

$$I(f; p) = \int_0^\infty f(x)\Phi(xp)\,dx, \tag{9.119}$$

where Φ is an oscillatory function. For example, in the case of Bessel functions we have the class of Hankel transforms

$$I_{\mu,\nu}(f; p) = \int_0^\infty x^{\mu-1} f(x) J_\nu(px)\,dx, \tag{9.120}$$

which play an important role in applied mathematics. In [242] it is explained how quadrature rules of Gauss type can be constructed for these integrals and also for integrals of type (9.112). In the latter case Wong gives a modification of the Gauss–Laguerre rule, and the method works for functions f that are analytic in the right half-plane.

Reducing the interval

Because the exponential function in (9.112) is periodic, with interval of periodicity $[0, 2\pi/p]$, we can write $I(f; p)$ in the form

$$I(f; p) = \frac{1}{p} \int_0^{2\pi} e^{it} S_p(t)\,dt, \quad S_p(t) = \sum_{k=0}^\infty f\left(\frac{2\pi k + t}{p}\right). \tag{9.121}$$

Another summation method

When in (9.112) the exponential function is written in terms of the sine and cosine functions, the resulting integrals can be written as alternating series of positive and negative subintegrals that are computed individually (for example, when f is positive). A similar method can also be used for (9.119) and (9.120) by using subintervals with endpoints the zeros of $\Phi(x)$ or $J_\nu(px)$; see [137]. Convergence acceleration schemes, for example, Levin's or Weniger's transformations (cf. §9.3.2), can be used when evaluating the series. For further information see [35, 141, 145].

Filon's method for oscillatory integrals

Oscillatory integrals of the form

$$I(f; p) = \int_a^b f(x) e^{ipx}\,dx \tag{9.122}$$

can be evaluated using Filon's method [70]. The method, especially useful when p is large, is based on the piecewise approximation of $f(x)$ on the interval of integration by low-degree polynomials. The basic ideas of the method are as follows.

The interval of integration $[a, b]$ is divided into $2N$ equally spaced subintervals,

$$a = x_0 < x_1 < \cdots < x_{2N} = b,$$

$$x_k = a + hk, \quad k = \frac{b-a}{2N}. \tag{9.123}$$

On each subinterval $[x_{2k-2}, x_{2k}]$ the function $f(x)$ is locally approximated by a polynomial $P_k(x)$ of degree 2 at most. That is,

$$f(x) \approx P_k(x) = P_k(x_{2k-1} + ht) = \phi_k(t), \quad t \in [-1, 1], \tag{9.124}$$

and the polynomial $\phi_k(t)$ is expressed in terms of the function values at different nodes, as follows:

$$\begin{aligned}\phi_k(t) &= f(x_{2k-1}) + \tfrac{1}{2}\left(f(x_{2k}) - f(x_{2k-2})\right) t \\ &+ \tfrac{1}{2}\left(f(x_{2k}) - 2f(x_{2k-1}) + f(x_{2k-2})\right) t^2\end{aligned} \tag{9.125}$$

for $k = 1, 2, \ldots, N$. By substituting (9.124) into (9.122) we obtain

$$\begin{aligned}I(f; p) = \int_a^b f(x) e^{ipx}\, dx &\approx \sum_{k=1}^N \int_{x_{2k-2}}^{x_{2k}} f(x) e^{ipx}\, dx \\ &\approx h \sum_{k=1}^N e^{ipx_{2k-1}} \int_{-1}^1 \phi_k(t) e^{i\theta t}\, dt,\end{aligned} \tag{9.126}$$

where $\theta = ph$.

For the evaluation of the integral

$$I_{\phi_k;\theta} = \int_{-1}^1 \phi_k(t) e^{i\theta t}\, dt, \tag{9.127}$$

we consider (9.125). In this way,

$$I_{\phi_k;\theta} = A f(x_{2k-2}) + B f(x_{2k-1}) + C f(x_{2k}), \tag{9.128}$$

with

$$\begin{aligned}A &= \frac{1}{2}\int_{-1}^1 (t^2 - t) e^{i\theta t}\, dt = \frac{(\theta^2 - 2)\sin\theta + 2\theta\cos\theta}{\theta^3} + i\frac{\theta\cos\theta - \sin\theta}{\theta^2}, \\ B &= \frac{1}{2}\int_{-1}^1 (1 - t^2) e^{i\theta t}\, dt = \frac{4}{\theta^3}\left(\sin\theta - \theta\cos\theta\right), \\ C &= \frac{1}{2}\int_{-1}^1 (t^2 - t) e^{-i\theta t}\, dt = \frac{(\theta^2 - 2)\sin\theta + 2\theta\cos\theta}{\theta^3} - i\frac{\theta\cos\theta - \sin\theta}{\theta^2}.\end{aligned} \tag{9.129}$$

9.6. Other quadrature methods

Then we have

$$I(f; p) \approx h \left\{ i\alpha \left(e^{ipa} f(a) - e^{ipb} f(b) \right) + \beta E_{2N} + \gamma E_{2N-1} \right\}, \quad (9.130)$$

with

$$\begin{aligned}
\theta^3 \alpha &= \theta^2 + \theta \sin\theta \cos\theta - 2\sin^2\theta, \\
\theta^3 \beta &= 2\left[\theta(1 + \cos^2\theta) - 2\sin\theta\cos\theta\right], \\
\theta^3 \gamma &= 4\left[\sin\theta - \theta\cos\theta\right],
\end{aligned} \quad (9.131)$$

and

$$E_{2N} = \sum_{k=0}^{N}{}'' f(x_{2k}) e^{iwx_{2k}}, \quad E_{2N-1} = \sum_{k=1}^{N} f(x_{2k-1}) e^{iwx_{2k-1}}. \quad (9.132)$$

The double prime over the first summation indicates that the first and last terms are to be multiplied by $\frac{1}{2}$.

The quantities α, β, and γ defined in (9.131) need to be recomputed when we change h, or for different p. When θ is small the right-hand sides in (9.131) should be expanded in powers of θ to preserve accuracy. We have

$$\begin{aligned}
\alpha &= \tfrac{2}{45}\theta^3 - \tfrac{2}{315}\theta^5 + \tfrac{2}{4725}\theta^7 + \cdots, \\
\beta &= \tfrac{2}{3} + \tfrac{2}{15}\theta^2 - \tfrac{4}{105}\theta^4 + \tfrac{2}{567}\theta^6 + \cdots, \\
\gamma &= \tfrac{4}{3} - \tfrac{2}{15}\theta^2 + \tfrac{1}{210}\theta^4 - \tfrac{1}{11340}\theta^6 + \cdots.
\end{aligned} \quad (9.133)$$

It can be easily verified that for $p = 0$, that is, $\theta = 0$, Filon's method becomes Simpson's extended rule; see §9.6.1.

For recent investigations of Filon-type quadrature applied to highly oscillatory integrals with extensive numerical and asymptotic analysis, see [113, 114, 115].

Part III

Related Topics and Examples

Chapter 10

Inversion of Cumulative Distribution Functions

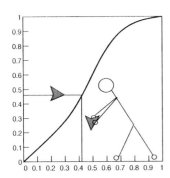

We are not certain, we are never certain. If we were we could reach some conclusions, and we could, at last, make others take us seriously.
—Albert Camus, French philosopher and writer

10.1 Introduction

The inversion of cumulative distribution functions is an important topic in statistics, probability theory, and econometrics, in particular for computing percentage points of chi-square, F, and Student's t-distributions. In the tails of these distributions the numerical inversion is not very easy, and for the standard distributions asymptotic formulas are available.

In this chapter we use the uniform asymptotic expansions of the incomplete gamma functions and incomplete beta functions, which are the basic functions for several distribution functions, for inverting these functions for large values of one or two parameters. The methods have been developed in [216, 217], and we summarize the main parts of these papers. For algorithms, including Fortran subroutines, for computing the incomplete gamma function ratios and their inverse, we refer to [55, 56].

We start with the relatively simple problem of inverting the complementary error function, again by using asymptotic methods.

In a final section we consider the inversion of the incomplete gamma function $P(a, x)$ for small values of a by using a high order Newton-like method.

10.2 Asymptotic inversion of the complementary error function

We recall the definition of the complementary error function (we use here real arguments)

$$\operatorname{erfc} x = \frac{2}{\sqrt{\pi}} \int_x^\infty e^{-t^2}\, dt \qquad (10.1)$$

and the asymptotic expansion

$$\operatorname{erfc} x \sim \frac{e^{-x^2}}{x\sqrt{\pi}} \sum_{k=0}^{\infty} \frac{(-1)^k (2k)!}{k! (2x)^{2k}} = \frac{e^{-x^2}}{x\sqrt{\pi}} \left(1 - \frac{1}{2}x^{-2} + \frac{3}{4}x^{-4} - \frac{15}{8}x^{-6} + \cdots\right), \quad (10.2)$$

as $x \to \infty$.

We derive an asymptotic expansion for the inverse $x(y)$ of the function $y(x) = \operatorname{erfc} x$ for small positive values of y.

Let t, α, and β be defined by

$$t = \frac{2}{\pi y^2}, \quad \alpha = \frac{1}{\ln t}, \quad \beta = \ln(\ln t). \quad (10.3)$$

Then we have the expansion

$$x(y) \sim \frac{1}{\sqrt{2\alpha}} \left(1 + x_1 \alpha + x_2 \alpha^2 + x_3 \alpha^3 + x_4 \alpha^4 + \cdots\right). \quad (10.4)$$

The first coefficients x_k are given by

$$\begin{aligned}
x_1 &= -\tfrac{1}{2}\beta, \\
x_2 &= -\tfrac{1}{8}\left(\beta^2 - 4\beta + 8\right), \\
x_3 &= -\tfrac{1}{16}\left(\beta^3 - 8\beta^2 + 32\beta - 56\right), \\
x_4 &= -\tfrac{1}{384}\left(15\beta^4 - 184\beta^3 + 1152\beta^2 - 4128\beta + 7040\right), \\
x_5 &= -\tfrac{1}{768}\left(21\beta^5 - 352\beta^4 + 3056\beta^3 - 16752\beta^2 + 57536\beta - 97984\right).
\end{aligned} \quad (10.5)$$

We explain how these coefficients can be obtained. The equation $y = \operatorname{erfc} x$, with y small, will be solved for x by using the asymptotic expansion in (10.2). We square the equation and write

$$\frac{1}{2}\pi y^2 = \frac{e^{-\xi}}{\xi} S^2(\xi), \quad (10.6)$$

where $\xi = 2x^2$ and $S(\xi)$ denotes the function that has the power series in (10.2) as its asymptotic expansion with $2x^2$ replaced by ξ.

We can rewrite (10.6) in the form

$$\xi e^{\xi} = t S^2(\xi), \quad (10.7)$$

where t is defined in (10.3). We solve this equation for ξ, with t large.

We observe that this equation has been discussed in detail in [50, §2.4] for the case that $S(\xi) = 1$. We can apply the same method for constructing an asymptotic expansion of the equation in (10.7). We write

$$\xi = \ln t - \ln(\ln t) + \eta \quad (10.8)$$

10.2. Asymptotic inversion of the complementary error function

Table 10.1. *Solutions $x(y)$ of the equation $y = \operatorname{erfc} x$ by using* (10.4).

y	$\tilde{x}$	$|\operatorname{erfc}(\tilde{x})/y - 1|$	$\hat{x}$	$|\tilde{x}/\hat{x} - 1|$
10^{-02}	1.820630554	$0.31\,10^{-02}$	1.821591563	$0.53\,10^{-03}$
10^{-03}	2.326648925	$0.51\,10^{-03}$	2.326782380	$0.57\,10^{-04}$
10^{-04}	2.751038248	$0.15\,10^{-03}$	2.751070914	$0.12\,10^{-04}$
10^{-05}	3.123404718	$0.56\,10^{-04}$	3.123415612	$0.35\,10^{-05}$
10^{-06}	3.458907270	$0.25\,10^{-04}$	3.458911685	$0.13\,10^{-05}$
10^{-07}	3.766560973	$0.13\,10^{-04}$	3.766563021	$0.54\,10^{-06}$
10^{-08}	4.052236421	$0.69\,10^{-05}$	4.052237469	$0.26\,10^{-06}$
10^{-09}	4.320004932	$0.41\,10^{-05}$	4.320005509	$0.14\,10^{-06}$
10^{-10}	4.572824704	$0.25\,10^{-05}$	4.572825039	$0.74\,10^{-07}$

and find that η satisfies the relation

$$(\ln t - \ln(\ln t) + \eta)\, e^{\eta} = \ln t\, S^2(\ln t - \ln(\ln t) + \eta). \tag{10.9}$$

The quantity η can be expanded in the form

$$\eta = \eta_1 \alpha + \eta_2 \alpha^2 + \eta_3 \alpha^3 + \cdots, \tag{10.10}$$

where α is defined in (10.3). By using a few terms in the expansion of $S(\xi)$ we find

$$\eta_1 = \beta - 2, \quad \eta_2 = \tfrac{1}{2}\beta^2 - 3\beta + 7, \tag{10.11}$$

where β is defined in (10.3).

The expansion for η gives an expansion for ξ (see (10.8)), and by using $x = \sqrt{\xi/2}$ we obtain the expansion given in (10.4).

In Table 10.1 we give values of the approximation $\tilde{x}$ of the solution of the equation $y = \operatorname{erfc} x$ for several values of y. We have used the asymptotic expansion (10.4) with the terms up to and including x_4. The values $\hat{x}$ are obtained by using one Newton–Raphson step. We also give relative errors. The computations are done in Maple with default 10-digit arithmetic.

For other methods of the inversion of the error functions, we refer to [203], where coefficients of the Maclaurin expansion of $x(y)$, the inverse of $y = \operatorname{erf} x$, are given, with Chebyshev coefficients for an expansion on the y-interval $[-0.8, 0.8]$. For small values of y (not smaller than 10^{-300}) high-precision coefficients of Chebyshev expansions are given for the numerical evaluation of the inverse of $y = \operatorname{erfc} x$. For rational Chebyshev (near-minimax) approximations for the inverse of the complementary error function $y = \operatorname{erfc} x$, we refer to [15], where y-values are considered in the y-interval $[10^{-10000}, 1]$, with relative errors ranging down to 10^{-23}. An asymptotic formula for the region $y \to 0$ is also given.

Remark 15. In Chapter 7, §7.6.3, we use similar asymptotic inversion methods for finding complex zeros of the error function. The present case is simpler because we want to find only one real solution with the inversion method.

Remark 16. The solution of the equation $\xi e^\xi = t$ (see (10.7)) can be expressed in terms of Lambert's W-function: $\xi = W(t)$.

10.3 Asymptotic inversion of incomplete gamma functions

We solve the equations

$$P(a,x) = p, \quad Q(a,x) = q, \quad 0 \le p \le 1, \quad 0 \le q \le 1, \qquad (10.12)$$

where $P(a, x)$ and $Q(a, x)$ are the incomplete gamma functions introduced in (8.12) in §8.3. We invert the equations for x, with a as a large positive parameter. This problem is of importance in probability theory and mathematical statistics. Several approaches are available in the (statistical) literature, where often a first approximation of x is constructed, based on asymptotic expansions, but this first approximation is not always reliable. Higher approximations may be obtained by numerical inversion techniques, which require evaluation of the incomplete gamma functions. This may be rather time consuming, especially when a is large.

In the present method we also use an asymptotic result. The approximation is quite accurate, especially when a is large. It follows from numerical results, however, that a three-term asymptotic expansion already gives an accuracy of 4 significant digits for $a = 2$, uniformly with respect to $p, q \in [0, 1]$.

The method is rather general. In a later section we mention application of the same method on a wider class of cumulative distribution functions.

The approximations are obtained by using the uniform asymptotic expansions of the incomplete gamma functions given in §8.3, in which an error function is the dominant term. The inversion problem is started by inverting this error function term.

In §10.6 we consider the inversion of $P(a, x)$ for small values of a.

10.3.1 The asymptotic inversion method

We perform the inversion of (10.12) with respect to the parameter η by using the representations (8.17), with z replaced by x throughout, and large positive values of a. Afterwards we have to compute λ and x from the relation for η in (8.18) and $\lambda = x/a$. We concentrate on the second equation in (10.12). Let us rewrite the inversion problem in the form

$$\tfrac{1}{2}\operatorname{erfc}\left(\eta\sqrt{a/2}\right) + R_a(\eta) = q, \quad q \in [0, 1], \qquad (10.13)$$

which is equivalent to the second equation in (10.12), and we denote the solution of the above equation by $\eta(q, a)$.

To start the procedure we consider $R_a(\eta)$ in (10.13) as a perturbation, and we define the number $\eta_0 = \eta_0(q, a)$ as the real number that satisfies the equation

$$\tfrac{1}{2}\operatorname{erfc}\left(\eta_0\sqrt{a/2}\right) = q. \qquad (10.14)$$

10.3. Asymptotic inversion of incomplete gamma functions

Known values are

$$\eta_0(0, a) = +\infty, \quad \eta_0\left(\tfrac{1}{2}, a\right) = 0, \quad \eta_0(1, a) = -\infty. \tag{10.15}$$

We note the symmetry $\eta_0(q, a) = -\eta_0(p, a)$. Computation of η_0 requires an inversion of the error function, but this problem has been satisfactorily solved in the literature; see §10.2. The value η defined by (10.13) is, for large values of a, approximated by the value η_0. We write

$$\eta(q, a) = \eta_0(q, a) + \varepsilon(\eta_0, a), \tag{10.16}$$

and we try to determine the function ε. It appears that we can expand this quantity in the form

$$\varepsilon(\eta_0, a) \sim \frac{\varepsilon_1}{a} + \frac{\varepsilon_2}{a^2} + \frac{\varepsilon_3}{a^3} + \cdots, \tag{10.17}$$

as $a \to \infty$. The coefficients ε_i will be written explicitly as functions of η_0.

We first remark that (10.13) yields the relation

$$\frac{dq}{d\eta} = \frac{d}{d\eta} Q(a, x) = \frac{d}{dx} Q(a, x) \frac{dx}{d\eta}. \tag{10.18}$$

Using the definition of $Q(a, x)$ and the relation for η in (8.18), we obtain after straightforward calculations

$$\frac{dq}{d\eta} = -\frac{1}{\Gamma^*(a)} \sqrt{\frac{a}{2\pi}} f(\eta) e^{-\frac{1}{2} a \eta^2}, \tag{10.19}$$

where $\Gamma^*(a)$ is defined in (8.28), and

$$f(\eta) = \frac{\eta}{\lambda - 1}, \tag{10.20}$$

the relation between η and λ being given in (8.18). For small values of η we can expand

$$f(\eta) = 1 - \tfrac{1}{3}\eta + \tfrac{1}{12}\eta^2 + \cdots. \tag{10.21}$$

From (10.14) we obtain

$$\frac{dq}{d\eta_0} = -\sqrt{\frac{a}{2\pi}} e^{-\frac{1}{2} a \eta_0^2}. \tag{10.22}$$

Upon dividing (10.19) and (10.22), we eliminate q, although it is still present in η_0. So we obtain

$$\frac{d\eta}{d\eta_0} = \frac{\Gamma^*(a)}{f(\eta)} e^{\frac{1}{2} a (\eta^2 - \eta_0^2)}, \quad -\infty < \eta_0 < \infty. \tag{10.23}$$

Substitution of (10.16) gives the differential equation

$$f(\eta_0 + \varepsilon) \left[1 + \frac{d\varepsilon}{d\eta_0}\right] = \Gamma^*(a) e^{a\varepsilon(\eta_0 + \frac{1}{2}\varepsilon)}, \tag{10.24}$$

a relation between ε and η_0, with a as (large) parameter.

It is convenient to write η in place of η_0. That is, we try to find the function $\varepsilon = \varepsilon(\eta, a)$ that satisfies the equation

$$f(\eta + \varepsilon)\left[1 + \frac{d\varepsilon}{d\eta}\right] = \Gamma^*(a) e^{a\varepsilon(\eta + \frac{1}{2}\varepsilon)}. \tag{10.25}$$

When we have obtained the solution $\varepsilon(\eta, a)$ (in fact we find an approximation of the form (10.17)), we write it as $\varepsilon(\eta_0, a)$, and the final value of η follows from (10.16). The parameters λ and x of the incomplete gamma function then follow from inversion of the relation for η in (8.18).

10.3.2 Determination of the coefficients ε_i

For large values of a we have $\Gamma^*(a) = 1 + \mathcal{O}(a^{-1})$; see (8.31). Comparing dominant terms in (10.25), we infer that the first coefficient ε_1 in (10.17) is defined by

$$f(\eta) = e^{\eta \varepsilon_1}, \tag{10.26}$$

giving

$$\varepsilon_1 = \frac{1}{\eta} \ln f(\eta). \tag{10.27}$$

It is not difficult to verify that f is positive on $\mathbb{R}$, $f(0) = 1$, and that f is analytic in a neighborhood of $\eta = 0$. It follows that $\varepsilon_1 = \varepsilon_1(\eta)$ is an analytic function on $\mathbb{R}$. For small values of η we have, using (10.21),

$$\varepsilon_1 = -\tfrac{1}{3} + \tfrac{1}{36}\eta + \tfrac{1}{1620}\eta^2 + \cdots. \tag{10.28}$$

The function $\varepsilon_1(\eta)$ is nonvanishing on $\mathbb{R}$ (and hence negative). To show this, consider the equation $f^2(\eta) = 1$. From (10.20) and the relation for η in (8.18), it follows that the corresponding λ-value should satisfy

$$-\ln \lambda = (\lambda - 1)(2\lambda - 3). \tag{10.29}$$

This equation has only one real solution, $\lambda = 1$, which gives $\eta = 0$. However, for this value ε_1 equals $-\tfrac{1}{3}$.

Further coefficients in (10.17) can be obtained by using standard perturbation methods. We need the expansion of $\Gamma^*(a)$ given in (8.29), and

$$f(\eta + \varepsilon) = f(\eta) + \varepsilon f'(\eta) + \tfrac{1}{2}\varepsilon^2 f''(\eta) + \cdots, \tag{10.30}$$

in which (10.17) is substituted to obtain an expansion in powers of a^{-1}. Putting all this into (10.25), we find by comparing terms with equal powers of a^{-1},

$$\varepsilon_2 = \frac{1}{12\eta f}(12 f \varepsilon_1' + 12 f' \varepsilon_1 - f - 6 f \varepsilon_1^2) \tag{10.31}$$

10.3. Asymptotic inversion of incomplete gamma functions

and

$$\varepsilon_3 = \frac{1}{288\eta f}(288 f\varepsilon_2' + 288 f'\varepsilon_1\varepsilon_1' - 24 f\varepsilon_1' + 288 f'\varepsilon_2 + 144 f''\varepsilon_1^2 \\ - 24 f'\varepsilon_1 + f - 288 f\varepsilon_1\varepsilon_2 - 144 f\varepsilon_2^2\eta^2 - 144 f\varepsilon_2\eta\varepsilon_1^2 - 36 f\varepsilon_1^4). \tag{10.32}$$

The derivatives f' and ε_j' are with respect to η. It will be obvious that the complexity for obtaining higher order terms is considerable. The terms shown so far have been obtained by symbolic manipulation. For numerical evaluations it is very convenient to have representations free of derivatives.

The derivatives of f can be eliminated by using

$$\begin{aligned} f' &= f(1 - f^2 - f\eta)/\eta, \\ f'' &= f^2(-3\eta - 3f + 3f^3 + 5f^2\eta + 2\eta^2 f)/\eta^2, \end{aligned} \tag{10.33}$$

and so on.

The first relation easily follows from (10.20) and the relation between η and λ. Using these relations in ε_i, and eliminating the derivatives of previous ε_j, it follows that we can write $\eta^{2j-1}\varepsilon_j$ as a polynomial in η, f, ε_1. We have

$$\begin{aligned} 12\eta^3\varepsilon_2 &= 12 - 12f^2 - 12f\eta - 12f^2\eta\varepsilon_1 - 12f\eta^2\varepsilon_1 - \eta^2 - 6\eta^2\varepsilon_1^2, \\ 12\eta^5\varepsilon_3 &= -30 + 12f^2\eta\varepsilon_1 + 12f\eta^2\varepsilon_1 + 24f^2\eta^3\varepsilon_1 + 6\varepsilon_1^3\eta^3 + 60f^3\eta^2\varepsilon_1 \\ &\quad - 12f^2 + 31f^2\eta^2 + 72f^3\eta + 42f^4 + 18f^3\eta^3\varepsilon_1^2 + 6f^2\eta^4\varepsilon_1^2 + 36f^4\eta\varepsilon_1 \\ &\quad + 12\varepsilon_1^2\eta^3 f + 12\varepsilon_1^2\eta^2 f^2 - 12\eta\varepsilon_1 + \eta^3\varepsilon_1 + f\eta^3 - 12f\eta + 12\varepsilon_1^2\eta^2 f^4. \end{aligned} \tag{10.34}$$

From these representations we can conclude that the coefficients $\varepsilon_1, \varepsilon_2, \varepsilon_3$ are bounded on $\mathbb{R}$. To show this we need

$$f(\eta) \sim -\eta, \quad \eta \to -\infty, \quad f(\eta) \sim 2\eta^{-1}, \quad \eta \to +\infty, \tag{10.35}$$

and the above representations of ε_i. We find

$$\varepsilon_1 \sim \mp\frac{\ln|\eta|}{\eta}, \quad \varepsilon_2 \sim -\frac{1}{12\eta}, \quad \varepsilon_3 \sim \frac{\varepsilon_1}{12\eta^2}, \tag{10.36}$$

as $\eta \to \pm\infty$. In deriving the behavior at $-\infty$ one should take into account (see (10.20) and the relation between η and λ) that

$$f(\eta) + \eta = \frac{\lambda\eta}{\lambda - 1} \sim -\eta e^{-\frac{1}{2}\eta^2}, \quad \eta \to -\infty. \tag{10.37}$$

10.3.3 Expansions of the coefficients ε_i

As explained in §8.3, the function f is analytic in a strip $|\Im \eta| < \sqrt{2\pi}$, and it can be expanded in a Taylor series at the origin with radius of convergence $2\sqrt{\pi}$. All ε_i have similar analytic properties. That is, the coefficients ε_i can be expanded in series form,

$$\varepsilon_i = \sum_{n=0}^{\infty} c_{i,n} \eta^n, \quad |\eta| < 2\sqrt{\pi}, \quad i = 1, 2, 3, \ldots. \tag{10.38}$$

The representations of ε_i given in the previous section are not suitable for numerical computations. To facilitate numerical evaluations of $\varepsilon_1, \ldots, \varepsilon_4$ we provide for small values of η the following Taylor expansions (more details on the coefficient ε_4 are given in [216]):

$$\begin{aligned}
\varepsilon_1 &= -\tfrac{1}{3} + \tfrac{1}{36}\eta + \tfrac{1}{1620}\eta^2 - \tfrac{7}{6480}\eta^3 + \tfrac{5}{18144}\eta^4 - \tfrac{11}{382725}\eta^5 + \cdots, \\
\varepsilon_2 &= -\tfrac{7}{405} - \tfrac{7}{2592}\eta + \tfrac{533}{204120}\eta^2 - \tfrac{1579}{2099520}\eta^3 + \tfrac{109}{1749600}\eta^4 + \cdots, \\
\varepsilon_3 &= \tfrac{449}{102060} - \tfrac{63149}{20995200}\eta + \tfrac{29233}{36741600}\eta^2 + \tfrac{346793}{5290790400}\eta^3 + \cdots, \\
\varepsilon_4 &= \tfrac{319}{183708} - \tfrac{269383}{4232632320}\eta - \tfrac{449882243}{982102968000}\eta^2 + \cdots.
\end{aligned} \tag{10.39}$$

10.3.4 Numerical examples

When $p = q = \tfrac{1}{2}$, the asymptotics are quite simple. Then η_0 of (10.14) equals zero, and from (10.39), we obtain (10.16) and (10.17) in the form

$$\eta \sim -\tfrac{1}{3}a^{-1} - \tfrac{7}{405}a^{-2} + \tfrac{449}{102060}a^{-3} + \tfrac{319}{183708}a^{-4} + \cdots. \tag{10.40}$$

In this case we give an expansion of the requested value x. Recall that $x = a\lambda$ and that λ can be obtained from the relation between η and λ in (8.18) with η given by (10.40). Inverting

$$\tfrac{1}{2}\eta^2 = \tfrac{1}{2}(\lambda - 1)^2 - \tfrac{1}{3}(\lambda - 1)^3 + \tfrac{1}{4}(\lambda - 1)^4 + \cdots, \tag{10.41}$$

we obtain

$$\lambda = 1 + \eta + \tfrac{1}{3}\eta^2 + \tfrac{1}{36}\eta^3 - \tfrac{1}{270}\eta^4 + \tfrac{1}{4320}\eta^5 + \cdots. \tag{10.42}$$

Substituting (10.40), we have

$$x \sim a\left(1 - \tfrac{1}{3}a^{-1} + \tfrac{8}{405}a^{-2} + \tfrac{184}{25515}a^{-3} + \tfrac{2248}{3444525}a^{-4} + \cdots \right). \tag{10.43}$$

When $a = 1$, $q = \tfrac{1}{2}$, the equations in (10.12) reduce to $e^{-x} = \tfrac{1}{2}$, with solution $x = \ln 2 = 0.693147\ldots$, while expansion (10.43) gives $x \sim 0.694\ldots$, an accuracy of about 3 digits. When $a = 2$, $q = \tfrac{1}{2}$, the equations in (10.12) become $(1 + x)e^{-x} = \tfrac{1}{2}$, with solution $x = 1.6783469\ldots$; in this case our expansion (10.40) gives $x \sim 1.67842\ldots$, an accuracy of 4 significant digits. This shows that (10.40) is quite accurate for small values of the (large) parameter a. Computer experiments show that for other q-values the results are of the same kind. See Table 10.2.

Table 10.2. *Relative errors* $|x_a - x|/x$ *and* $|Q(a, x_a) - q|/q$ *for several values of* q *and* a; x_a *is obtained by using asymptotic expansion* (10.17); x *is a more accurate value.*

	\multicolumn{6}{c}{a}					
q	1		5		10	
0.0001	$2.3\,10^{-4}$	$2.1\,10^{-3}$	$1.1\,10^{-6}$	$1.6\,10^{-5}$	$9.4\,10^{-8}$	$1.7\,10^{-6}$
0.1	$6.6\,10^{-4}$	$1.5\,10^{-3}$	$2.0\,10^{-6}$	$9.3\,10^{-6}$	$1.4\,10^{-7}$	$8.8\,10^{-7}$
0.3	$8.7\,10^{-4}$	$1.0\,10^{-3}$	$2.3\,10^{-6}$	$6.4\,10^{-6}$	$1.6\,10^{-7}$	$6.0\,10^{-7}$
0.5	$7.0\,10^{-4}$	$4.8\,10^{-4}$	$6.7\,10^{-7}$	$1.2\,10^{-6}$	$5.4\,10^{-8}$	$1.4\,10^{-7}$
0.7	$4.9\,10^{-4}$	$1.7\,10^{-4}$	$2.7\,10^{-6}$	$2.6\,10^{-6}$	$1.7\,10^{-7}$	$2.6\,10^{-7}$
0.9	$1.9\,10^{-3}$	$2.0\,10^{-4}$	$2.5\,10^{-6}$	$8.8\,10^{-7}$	$1.8\,10^{-7}$	$9.3\,10^{-8}$
0.9999	$5.1\,10^{-3}$	$5.1\,10^{-7}$	$3.9\,10^{-6}$	$1.8\,10^{-9}$	$6.0\,10^{-8}$	$4.8\,10^{-11}$

In a second example we take $a = 2$, $q = 0.1$; inverting (10.14) we obtain $\eta_0 = 0.9061938$. Using (10.17) we compute

$$\eta \sim \eta_0 - 0.308292/2 - 0.0180893/4 + 0.0023105/8 = 0.747814. \qquad (10.44)$$

An inversion of the relation between η and λ gives $\lambda = 1.944743$, and hence $x = 2\lambda = 3.889486$. Computing $Q(2, x)$ with this value of x gives 0.1000186, an accuracy of 4 digits. A more accurate value of x can be obtained by a Newton–Raphson method, giving $x = 3.8897202$. It follows that this value of x obtained by the asymptotic method (with $a = 2$) is accurate within 4 significant digits.

In Table 10.2 we give more results of numerical experiments. We have used (10.17) with three terms. The first column, under each a-value, gives the relative accuracy $|x_a - x|/x$, where x_a is the result of the asymptotic method, and x is a more accurate value obtained by a Newton–Raphson method. The second column, under each a-value, gives the relative errors $|Q(a, x_a) - q|/q$.

10.4 Generalizations

The method described in the previous sections can be applied to other cumulative distribution functions. Consider the function

$$F_a(\eta) = \sqrt{\frac{a}{2\pi}} \int_{-\infty}^{\eta} e^{-\frac{1}{2}a\zeta^2} f(\zeta)\, d\zeta, \qquad (10.45)$$

where $a > 0$ and $\eta \in \mathbb{R}$. We assume that f is an analytic function in a domain containing the real axis, and that f is positive on $\mathbb{R}$ with the normalization $f(0) = 1$. In [211] it is shown that several well-known distribution functions can be written in this form, including the incomplete gamma and beta functions. It is also shown that the following representation holds:

$$F_a(\eta) = \tfrac{1}{2}\mathrm{erfc}\left(-\eta\sqrt{a/2}\right) F_a(\infty) + R_a(\eta), \qquad (10.46)$$

where $R_a(\eta)$ can be expanded as in (8.19). $F_a(\infty)$ is the complete integral and can be expanded in the form

$$F_a(\infty) \sim \sum_{n=0}^{\infty} \frac{A_n}{a^n}, \quad \text{as } a \to \infty, \ A_0 = 1. \tag{10.47}$$

By dividing both sides of (10.45) by $F_a(\infty)$, we obtain a further normalization, which is typical for distribution functions.

The inversion of the equation $F_a(\eta)/F_a(\infty) = q$, with $q \in [0, 1]$ and a a given (large) number, can be performed as in the case of the incomplete gamma functions. As in (10.14), let η_0 be the real number satisfying the equation

$$\tfrac{1}{2}\mathrm{erfc}\left(-\eta_0\sqrt{a/2}\right) = q. \tag{10.48}$$

Then the requested value η is written as in (10.16), and an expansion like (10.17) can be obtained by deriving the differential equation (10.24), with f of (10.45) and $\Gamma^*(a)$ replaced by $F_a(\infty)$.

In the next section we consider the incomplete beta function, for which three different inversion cases are discussed.

10.5 Asymptotic inversion of the incomplete beta function

The incomplete beta function is defined by

$$I_x(a, b) = \frac{1}{B(a, b)} \int_0^x t^{a-1}(1-t)^{b-1}\, dt, \quad a > 0, \ b > 0, \tag{10.49}$$

where $B(a, b)$ is Euler's beta integral

$$B(a, b) = \int_0^1 t^{a-1}(1-t)^{b-1}\, dt = \frac{\Gamma(a)\Gamma(b)}{\Gamma(a+b)}. \tag{10.50}$$

The incomplete beta function is a standard probability function, with as special cases the (negative) binomial distribution, Student's t-distribution, and the F-(variance-ratio) distribution.

We consider the following inversion problem. Let $p \in [0, 1]$ be given. We are interested in the x-value that solves the equation

$$I_x(a, b) = p, \tag{10.51}$$

where a and b are fixed positive numbers. We are especially interested in solving (10.51) for large values of a and b.

The method for the asymptotic inversion is the same as that for the incomplete gamma functions. The present problem is more difficult, of course, since now two large parameters are considered. In fact we consider three asymptotic representations of the incomplete beta

10.5. Asymptotic inversion of the incomplete beta function

function with $a + b \to \infty$, valid in the following cases:

- $a = b + \beta$, where β stays fixed;
- a/b and b/a are bounded away from zero;
- at least one of the parameters a, b is large.

In the first two cases both parameters are large; in the third case we allow one parameter to be fixed or substantially smaller than the other one. In the first two cases the underlying beta distribution can be approximated by a normal (Gaussian) distribution, and we use an error function as the main approximant. In the third case the distribution may be quite skewed, and we consider an approximation in terms of the gamma distribution, with an incomplete gamma function as the main approximant. It is possible to restrict ourselves to $a \geq b$, since we have the relation

$$I_x(a, b) = 1 - I_{1-x}(b, a). \tag{10.52}$$

This relation is used in the third case, where the only condition is that the sum $a + b$ should be large.

10.5.1 The nearly symmetric case

We write $b = a + \beta$, where β is fixed. We obtain from (10.49)

$$I_x(a, a + \beta) = \frac{4^{-a}}{B(a, a + \beta)} \int_0^x [4t(1-t)]^a \frac{(1-t)^\beta \, dt}{t(1-t)}. \tag{10.53}$$

We transform this into a standard form with a Gaussian character by writing

$$-\tfrac{1}{2}\zeta^2 = \ln[4t(1-t)], \quad 0 < t < 1, \quad \operatorname{sign}(\zeta) = \operatorname{sign}\left(t - \tfrac{1}{2}\right),$$
$$-\tfrac{1}{2}\eta^2 = \ln[4x(1-x)], \quad 0 < x < 1, \quad \operatorname{sign}(\eta) = \operatorname{sign}\left(x - \tfrac{1}{2}\right). \tag{10.54}$$

This gives

$$I_x(a, a + \beta) = \frac{4^{-a}}{B(a, a+\beta)} \int_{-\infty}^\eta e^{-\tfrac{1}{2}a\zeta^2} \frac{(1-t)^\beta}{t(1-t)} \frac{dt}{d\zeta} \, d\zeta. \tag{10.55}$$

We can write t as a function of ζ,

$$t = \tfrac{1}{2}\left[1 \pm \sqrt{1 - \exp(-\tfrac{1}{2}\zeta^2)}\right] = \tfrac{1}{2}\left[1 + \zeta \sqrt{[1 - \exp(-\tfrac{1}{2}\zeta^2)]/\zeta^2}\right], \tag{10.56}$$

where the second square root is nonnegative for real values of the argument. The same relation holds for x as a function of η. It easily follows that

$$\frac{1}{t(1-t)} \frac{dt}{d\zeta} = \frac{-\zeta}{1 - 2t}, \tag{10.57}$$

and that the following standard form (in the sense of §10.4) is obtained:

$$I_x(a, a+\beta) = \sqrt{\frac{a}{2\pi}} \int_{-\infty}^{\eta} e^{-\frac{1}{2}a\zeta^2} f(\zeta)\, d\zeta, \qquad (10.58)$$

with

$$f(\zeta) = \Phi(a)\phi(\zeta), \qquad (10.59)$$

where

$$\Phi(a) = \frac{1}{\sqrt{a}} \frac{\Gamma(a + \frac{1}{2}\beta)}{\Gamma(a)} \frac{\Gamma(a + \frac{1}{2}\beta + \frac{1}{2})}{\Gamma(a+\beta)},$$

$$\phi(\zeta) = [2(1-t)]^\beta \sqrt{\frac{\sigma}{1 - \exp(-\sigma)}}, \quad \sigma = \frac{1}{2}\zeta^2. \qquad (10.60)$$

This form of $\Phi(a)$ is obtained by using the duplication formula of the gamma function:

$$\sqrt{\pi}\,\Gamma(2z) = 2^{2z-1}\Gamma(z)\Gamma\!\left(z + \tfrac{1}{2}\right). \qquad (10.61)$$

From the asymptotic expansion of the ratio of gamma functions (see, for instance, [2, eq. (6.1.47)]), we obtain

$$\Phi(a) \sim c_0 + c_1 a^{-1} + c_2 a^{-2} + \cdots, \quad a \to \infty, \qquad (10.62)$$

where

$$c_0 = 1, \quad c_1 = \tfrac{1}{8}(-2\beta^2 + 2\beta - 1), \quad c_2 = \tfrac{1}{128}(4\beta^4 + 8\beta^3 - 16\beta^2 + 4\beta + 1). \qquad (10.63)$$

The function $\phi(\zeta)$ is analytic in a strip containing $\mathbb{R}$; the singularities nearest to the origin occur at $\pm 2\sqrt{\pi}\exp(\pm i\pi/4)$. The first coefficients of the Taylor expansion

$$\phi(\zeta) = d_0 + d_1\zeta + d_2\zeta^2 + d_3\zeta^3 + \cdots \qquad (10.64)$$

are

$$d_0 = 1, \; d_1 = -\tfrac{1}{2}\beta\sqrt{2}, \; d_2 = \tfrac{1}{8}(2\beta^2 - 2\beta + 1), \; d_3 = -\tfrac{1}{24}\beta\sqrt{2}(\beta^2 - 3\beta + 2), \qquad (10.65)$$

$$\begin{aligned} d_4 &= \tfrac{1}{384}(4\beta^4 - 24\beta^3 + 32\beta^2 - 12\beta + 1), \\ d_5 &= -\tfrac{1}{960}\beta\sqrt{2}(\beta^4 - 10\beta^3 + 25\beta^2 - 20\beta + 4). \end{aligned} \qquad (10.66)$$

From results in [211] it follows that the standard form (10.58) can be written in the form

$$I_x(a, a+\beta) = \tfrac{1}{2}\mathrm{erfc}(-\eta\sqrt{a/2}) - R_a(\eta), \qquad (10.67)$$

where η is defined in (10.54). We try to solve (10.51) with the above representation of the incomplete beta function. First we solve (10.51) in terms of η; afterwards, we determine x from the inverse relation of the second line in (10.54) (that is, (10.56) with t, ζ replaced by x, η, respectively). When a is large, we consider the error function in the equation

$$I_x(a, a+\beta) = \tfrac{1}{2}\mathrm{erfc}(-\eta\sqrt{a/2}) - R_a(\eta) = p \qquad (10.68)$$

10.5. Asymptotic inversion of the incomplete beta function

as the dominant term, and a first approximation η_0 of η is defined by the solution of the equation

$$\tfrac{1}{2}\operatorname{erfc}(-\eta_0\sqrt{a/2}) = p. \tag{10.69}$$

The exact solution of (10.51) (in terms of η) is written as

$$\eta = \eta_0 + \varepsilon, \tag{10.70}$$

and we try to determine ε. It appears that we can expand this quantity in the form

$$\varepsilon \sim \frac{\varepsilon_1}{a} + \frac{\varepsilon_2}{a^2} + \frac{\varepsilon_3}{a^3} + \cdots, \tag{10.71}$$

as $a \to \infty$. The coefficients ε_i can be expressed in terms of η_0 and β.

From (10.58), (10.68), and (10.69) we obtain

$$\frac{dp}{d\eta_0} = \sqrt{\frac{a}{2\pi}}e^{-\frac{1}{2}a\eta_0^2}, \quad \frac{dp}{d\eta} = \sqrt{\frac{a}{2\pi}}f(\eta)e^{-\frac{1}{2}a\eta^2}, \tag{10.72}$$

where f is given in (10.59). Upon dividing, we obtain

$$f(\eta)\frac{d\eta}{d\eta_0} = e^{\frac{1}{2}a(\eta^2-\eta_0^2)}. \tag{10.73}$$

Substitution of (10.70) gives the differential equation

$$f(\eta_0 + \varepsilon)\left(1 + \frac{d\varepsilon}{d\eta_0}\right) = e^{a\varepsilon(\eta_0 + \frac{1}{2}\varepsilon)}. \tag{10.74}$$

We write η in place of η_0; that is, we try to find ε as a function of η that satisfies (see also (10.59))

$$\phi(\eta + \varepsilon)\Phi(a)\left(1 + \frac{d\varepsilon}{d\eta}\right) = e^{a\varepsilon(\eta + \frac{1}{2}\varepsilon)}. \tag{10.75}$$

When we have obtained ε from this equation (or an approximation), we use it in (10.70) to obtain the final value η.

The first coefficient ε_1 of (10.71) is obtained by comparing dominant terms in (10.75). Since $\Phi(a) = 1 + \mathcal{O}(a^{-1})$, we obtain

$$\varepsilon_1 = \frac{1}{\eta}\ln \phi(\eta). \tag{10.76}$$

This quantity is analytic (as a function of η) on $\mathbb{R}$; $\phi(\eta)$ is positive on $\mathbb{R}$, and $\phi(0) = 1$. Using (10.64) we obtain for small values of η

$$\varepsilon_1 = -\tfrac{1}{2}\beta\sqrt{2} + \tfrac{1}{8}(1-2\beta)\eta - \tfrac{1}{48}\beta\sqrt{2}\eta^2 - \tfrac{1}{192}\eta^3 - \tfrac{1}{3840}\beta\sqrt{2}\eta^4 + \cdots. \tag{10.77}$$

Further terms ε_i can be obtained by using more terms in (10.62) and by expanding

$$\phi(\eta + \varepsilon) = \phi(\eta) + \varepsilon\phi'(\eta) + \tfrac{1}{2}\varepsilon^2\phi''(\eta) + \cdots, \tag{10.78}$$

in which (10.71) is substituted to obtain an expansion in powers of a^{-1}. In this way we find

$$\varepsilon_2 = \frac{1}{2\eta\phi}(2\phi\varepsilon_1' + 2\phi'\varepsilon_1 + 2c_1\phi - \phi\varepsilon_1^2), \tag{10.79}$$

$$\varepsilon_3 = \frac{1}{8\eta\phi}(8\phi\varepsilon_2' + 8\phi'\varepsilon_1\varepsilon_1' + 8c_1\phi\varepsilon_1' + 8\phi'\varepsilon_2 + 4\phi''\varepsilon_1^2 \\ + 8c_1\phi'\varepsilon_1 + 8c_2\phi - 8\phi\varepsilon_1\varepsilon_2 - 4\phi\varepsilon_2^2\eta^2 - 4\phi\varepsilon_2\eta\varepsilon_1^2 - \phi\varepsilon_1^4). \tag{10.80}$$

The derivatives ϕ', ε', etc., are with respect to η, and all functions are evaluated at η. For small values of η we can use the Maclaurin series

$$\varepsilon_2 = \frac{\beta\sqrt{2}}{12}(3\beta - 2) + \frac{1}{128}(20\beta^2 - 12\beta + 1)\eta + \frac{\beta\sqrt{2}}{960}(20\beta - 1)\eta^2 + \cdots, \tag{10.81}$$

$$\varepsilon_3 = \frac{\beta\sqrt{2}}{480}(-75\beta^2 + 80\beta - 16) + \cdots.$$

10.5.2 The general error function case

Let us write
$$a = r\sin^2\theta, \quad b = r\cos^2\theta, \quad 0 \le \theta \le \tfrac{1}{2}\pi. \tag{10.82}$$

Then (10.49) can be written as

$$I_x(a,b) = \frac{1}{B(a,b)} \int_0^x e^{r[\sin^2\theta \ln t + \cos^2\theta \ln(1-t)]} \frac{dt}{t(1-t)}. \tag{10.83}$$

We consider r as a large parameter, and θ bounded away from 0 and $\tfrac{1}{2}\pi$. The maximum of the exponential function occurs at $t = \sin^2\theta$. Hence, the following transformation brings the exponential part of the integrand into a Gaussian form:

$$-\frac{1}{2}\zeta^2 = \sin^2\theta \ln\frac{t}{\sin^2\theta} + \cos^2\theta \ln\frac{1-t}{\cos^2\theta}, \tag{10.84}$$

where the sign of ζ equals the sign of $t - \sin^2\theta$. The same transformation holds for $x \mapsto \eta$ if t and ζ are replaced with x and η, respectively. From (10.84) we obtain

$$-\zeta\frac{d\zeta}{dt} = \frac{\sin^2\theta - t}{t(1-t)}, \tag{10.85}$$

and we can write (10.83) in the standard form (cf. (10.58)–(10.60))

$$I_x(a,b) = \sqrt{\frac{r}{2\pi}} \int_{-\infty}^{\eta} e^{-\frac{1}{2}r\zeta^2} f(\zeta)\, d\zeta, \tag{10.86}$$

with
$$f(\zeta) = \Phi(r)\phi(\zeta), \tag{10.87}$$

10.5. Asymptotic inversion of the incomplete beta function

where

$$\Phi(r) = \frac{\Gamma^*(r)}{\Gamma^*(a)\Gamma^*(b)}, \quad \phi(\zeta) = \frac{\zeta \sin\theta \cos\theta}{t - \sin^2\theta}. \quad (10.88)$$

The function $\Gamma^*(z)$ is the slowly varying part of the Euler gamma function, introduced in (8.28).

The analogue of the expansion (10.62) is now in terms of the large parameter r:

$$\Phi(r) \sim c_0 + c_1 r^{-1} + c_2 r^{-2} + \cdots, \quad r \to \infty, \quad (10.89)$$

where

$$c_0 = 1, \quad c_1 = \frac{\sin^2\theta \cos^2\theta - 1}{3\sin^2 2\theta}, \quad c_2 = \frac{(\sin^2\theta \cos^2\theta - 1)^2}{18\sin^4 2\theta}, \quad (10.90)$$

$$c_3 = -\frac{139(\sin^6\theta \cos^6\theta - \cos^6\theta - \sin^6\theta) + 15\sin^4\theta \cos^4\theta}{810\sin^6 2\theta}. \quad (10.91)$$

The first coefficients of the Taylor expansion

$$\phi(\zeta) = d_0 + d_1\zeta + d_2\zeta^2 + d_3\zeta^3 + \cdots \quad (10.92)$$

are

$$d_0 = 1, \quad d_1 = -\frac{2}{3}\cot 2\theta, \quad d_2 = \frac{\sin^4\theta + \cos^4\theta + 1}{6\sin^2 2\theta}. \quad (10.93)$$

To solve (10.51) for large values of r, we use the method of the previous section. We write as in (10.67)

$$I_x(a, b) = \tfrac{1}{2}\mathrm{erfc}(-\eta\sqrt{r/2}) - R_a(\eta), \quad (10.94)$$

where the relation between x and η follows from (10.84) when t and ζ are replaced with x and η, respectively. A first approximation η_0 follows from the equation

$$\tfrac{1}{2}\mathrm{erfc}(-\eta_0\sqrt{r/2}) = p, \quad (10.95)$$

and the terms ε_i in the expansion

$$\varepsilon \sim \frac{\varepsilon_1}{r} + \frac{\varepsilon_2}{r^2} + \frac{\varepsilon_3}{r^3} + \cdots \quad (10.96)$$

are the same as in (10.76), (10.77), and (10.80), but with ϕ, c_1, c_2 of the present section. For small values of η we can expand

$$\varepsilon_1 = \frac{2s^2 - 1}{3sc} - \frac{5s^4 - 5s^2 - 1}{36s^2c^2}\eta + \frac{46s^6 - 69s^4 + 21s^2 + 1}{1620s^3c^3}\eta^2 + \cdots,$$

$$\varepsilon_2 = -\frac{52s^6 - 78s^4 + 12s^2 + 7}{405s^3c^3} + \frac{2s^2 - 370s^6 + 185s^8 + 183s^4 - 7}{2592s^4c^4}\eta + \cdots,$$

$$\varepsilon_3 = \frac{3704s^{10} - 9260s^8 + 6686s^6 - 769s^4 - 1259s^2 + 449}{102060s^5c^5} + \cdots, \quad (10.97)$$

where $s = \sin\theta$, $c = \cos\theta$.

The functions ε_i are now considered as functions of η_0 (instead of η), and we write

$$\eta \sim \eta_0 + \frac{\varepsilon_1}{r} + \frac{\varepsilon_2}{r^2} + \frac{\varepsilon_3}{r^3} + \cdots. \tag{10.98}$$

This approximation is substituted on the left-hand side of (10.84), and we invert this equation to obtain t or, equivalently, x.

10.5.3 The incomplete gamma function case

In this section we consider the asymptotic condition that the sum $a + b$ should be large. We concentrate on the case $a \geq b$. In the other case we can solve (10.51) by using (10.52). From [214, eq. (9.16)] it follows that we can write

$$I_x(a, b) = Q(b, \eta a) + R_{a,b}(\eta), \tag{10.99}$$

where η is given by a mapping $x \mapsto \eta$, which is defined by

$$\eta - \mu \ln \eta + A(\mu) = -\ln x - \mu \ln(1 - x) \tag{10.100}$$

and

$$\mu = \frac{b}{a}, \quad A(\mu) = (1 + \mu)\ln(1 + \mu) - \mu. \tag{10.101}$$

Q is the incomplete gamma function defined by (8.12). Corresponding points in the mapping defined in (10.100) are

$$x = 0 \leftrightarrow \eta = +\infty, \quad x = \frac{1}{1 + \mu} \leftrightarrow \eta = \mu, \quad x = 1 \leftrightarrow \eta = 0. \tag{10.102}$$

From (10.100) it follows that

$$\frac{dx}{d\eta} = \frac{\eta - \mu}{\eta} \frac{x(1 - x)}{(1 + \mu)x - 1}. \tag{10.103}$$

In [214] an asymptotic expansion of $R_{a,b}(\eta)$ in (10.99) is derived, which holds for $a \to \infty$, uniformly with respect to $x \in [0, 1]$ and $b \in [0, \infty)$.

We obtain the solution of (10.51) for large values of a by first determining η_0, the solution of the reduced equation

$$Q(b, \eta_0 a) = p. \tag{10.104}$$

This involves an inversion of the incomplete gamma function, whose problem is considered in §10.3, especially for large values of b. As in the previous sections, the exact solution of (10.51) is written as $\eta = \eta_0 + \varepsilon$, and we expand ε as in (10.71). We have (cf. (10.72))

$$\begin{aligned}\frac{dp}{d\eta_0} &= -\frac{a^b}{\eta_0 \Gamma(b)} e^{a(-\eta_0 + \mu \ln \eta_0)}, \\ \frac{dp}{d\eta} &= \frac{1}{B(a,b)x(1-x)} \frac{dx}{d\eta} e^{a[-\eta + \mu \ln \eta - A(\mu)]}.\end{aligned} \tag{10.105}$$

10.5. Asymptotic inversion of the incomplete beta function

Upon dividing these equations and using (10.103), we obtain

$$f(\eta)\frac{d\eta}{d\eta_0} = \frac{\eta}{\eta_0} e^{a[\eta-\eta_0-\mu\ln(\eta/\eta_0)]}, \qquad (10.106)$$

with $f(\eta) = \phi(\eta)\Phi(a)$, and

$$\phi(\eta) = \frac{\eta-\mu}{1-x(1+\mu)}\frac{1}{\sqrt{1+\mu}}, \quad \Phi(a) = \frac{\Gamma^*(a+b)}{\Gamma^*(a)}, \qquad (10.107)$$

where Γ^* is introduced in (8.28). By writing $\eta = \eta_0 + \varepsilon$, and writing η in place of η_0 (for the time being), (10.106) can be written as

$$\phi(\eta+\varepsilon)\Phi(a)\left(1+\frac{d\varepsilon}{d\eta}\right) = e^{a[\varepsilon-\mu\ln(1+\varepsilon/\eta)]}. \qquad (10.108)$$

The analogue of the expansion (10.62) has the coefficients

$$c_0 = 1, \quad c_1 = -\frac{\mu}{12(1+\mu)}, \quad c_2 = \frac{\mu^2}{288(1+\mu)^2}, \qquad (10.109)$$

$$c_3 = \frac{\mu(432 + 432\mu + 139\mu^2)}{51840(1+\mu)^3}. \qquad (10.110)$$

The analogue of (10.64) reads

$$\phi(\eta) = d_0 + d_1(\eta-\mu) + d_2(\eta-\mu)^2 + \cdots, \qquad (10.111)$$

with coefficients

$$d_0 = 1, \quad d_1 = \frac{w+2}{3(w+1)w}, \quad d_2 = \frac{1}{12w^2}, \quad d_3 = \frac{8w^3 + 9w^2 - 9w - 8}{540w(w+1)^3}, \qquad (10.112)$$

where

$$w = \sqrt{1+\mu}. \qquad (10.113)$$

Substituting

$$\varepsilon \sim \frac{\varepsilon_1}{a} + \frac{\varepsilon_2}{a^2} + \frac{\varepsilon_3}{a^3} + \cdots \qquad (10.114)$$

into (10.108), we find the first coefficient,

$$\varepsilon_1 = \frac{\ln\phi(\eta)}{1-\mu/\eta}, \qquad (10.115)$$

a regular function at $\eta = \mu$, as follows from the expansion (10.111) of $\phi(\eta)$ at this point. The next term is

$$\varepsilon_2 = \frac{1}{2\phi\eta(\eta-\mu)}(2\phi\varepsilon_1'\eta^2 + 2\phi'\varepsilon_1\eta^2 + 2c_1\phi\eta^2 - \phi\mu\varepsilon_1^2 - 2\varepsilon_1\phi\eta), \qquad (10.116)$$

where the derivatives are with respect to η. For small values of $|\eta-\mu|$ we can expand

$$\varepsilon_1 = \frac{(w+2)(w-1)}{3w} + \frac{w^3 + 9w^2 + 21w + 5}{36w^2(w+1)}(\eta-\mu) + \cdots, \qquad (10.117)$$

$$\varepsilon_2 = \frac{(28w^4 + 131w^3 + 402w^2 + 581w + 208)(w - 1)}{1620(w + 1)w^3} + \cdots, \qquad (10.118)$$

where w is given by (10.113).

Considering the functions ε_i as functions of η_0, we obtain using (10.114)

$$\eta \sim \eta_0 + \frac{\varepsilon_1}{a} + \frac{\varepsilon_2}{a^2} + \frac{\varepsilon_3}{a^3} + \cdots, \qquad (10.119)$$

which is substituted on the left-hand side of (10.100). Solving for x, we finally obtain the desired approximation of the solution of (10.51).

In this section, the functions $\Phi(a)$, $\phi(\eta)$, ε have expansions with coefficients c_i, d_i, ε_i in which the parameter $\mu = b/a$ may assume any value in $[0, \infty)$. This aspect demonstrates the uniform character (with respect to μ) of the present approach. In §10.5.1 large values of β are not allowed, and in §10.5.2 the value of θ should be bounded away from 0 and $\frac{1}{2}\pi$. Of course, the transformations and expansions of this section are more complicated than those in the previous sections. Moreover, to start the inversion procedure, first (10.104) including an incomplete gamma function should be solved, whereas in the foregoing cases only an error function has to be inverted; see (10.69) and (10.95).

10.5.4 Numerical aspects

In numerical applications one needs the inversion of the mappings given in (10.54), (10.84), and (10.100). Only (10.54) can be inverted directly, as shown in (10.56). For small values of $|\zeta|$ we have

$$t = \tfrac{1}{2} + \tfrac{1}{4}\sqrt{2}\zeta - \tfrac{1}{32}\sqrt{2}\zeta^3 + \tfrac{5}{1536}\sqrt{2}\zeta^5 + \cdots. \qquad (10.120)$$

The inversion of (10.100) can be based on that of (10.84), with other parameters. We give some details on the inversion of (10.84).

For small values of $|\zeta|$ we have

$$t = s^2 + sc\zeta + \frac{1 - 2s^2}{3}\zeta^2 + \frac{13s^4 - 13s^2 + 1}{36sc}\zeta^3 + \cdots, \qquad (10.121)$$

where $s = \sin\theta$, $c = \cos\theta$. For larger values of $|\zeta|$, with $\zeta < 0$, we rewrite (10.84) in the form

$$t(1 - t)^\alpha = u, \quad \alpha = \cot^2\theta, \quad u = \exp\left[\left(-\tfrac{1}{2}\zeta^2 + s^2 \ln s^2 + c^2 \ln c^2\right)/s^2\right], \qquad (10.122)$$

and for small values of u we expand

$$t = u + \alpha u^2 + \frac{3\alpha(3\alpha + 1)}{3!}u^3 + \frac{4\alpha(4\alpha + 1)(4\alpha + 2)}{4!}u^4 + \cdots. \qquad (10.123)$$

A similar approach is possible for positive values of ζ, giving an expansion for t near unity. The approximations obtained in this way may be used for starting a Newton–Raphson method for obtaining more accurate values of t.

10.6. High order Newton-like methods

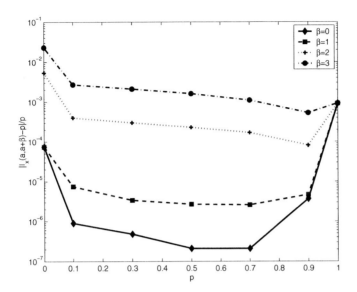

Figure 10.1. *Relative errors* $|I_x(a, a+\beta) - p|/p$ *for* $a = 10$ *and several values of* p *and* β. *The asymptotic inversion is based on the method of* §10.5.1.

We have tested the inversion process of the incomplete beta function for several values of the parameters. We describe the testing for the method of §10.5.1. After obtaining the first approximation η_0 by inverting (10.69), we compute the values of ε_1, ε_2, ε_3, (10.76)–(10.80) (with η replaced with η_0). The coefficient ε_3 is included only when η_0 is small enough for using the Maclaurin series given in §10.5.1. Next, (10.70) gives the final approximation of η, which is used in the second line in (10.54), to obtain the approximation of x. Finally we verified (10.51) by computing the incomplete beta function with this value x. We used the continued fraction given in (6.78).

In Figures 10.1, 10.2, and 10.3, we show the relative errors $|(I_x(a,b) - p)/p|$, where x is obtained by the asymptotic inversion methods of the previous sections. As is expected, it follows that the larger values of β give less accuracy in the results in Figure 10.1. The same holds for smaller values of θ in Figure 10.2. From Figure 10.3 it follows that the results are not influenced by large or small values of μ. This shows the uniform character of the method of §10.5.3. In fact, this method can be used in extreme situations: the ratio a/b may be very small and very large, and p may assume values quite close to zero or to unity.

10.6 High order Newton-like methods

Special functions usually satisfy a simple ordinary differential equation, and this equation can be used to construct Newton-like methods of high order.

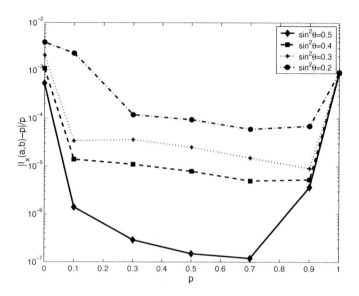

Figure 10.2. *Relative errors* $|I_x(a,b) - p|/p$ *for* $r = a + b = 10$ *and several values of p and* $\sin^2 \theta = a/r$. *The asymptotic inversion is based on the method of* §10.5.2.

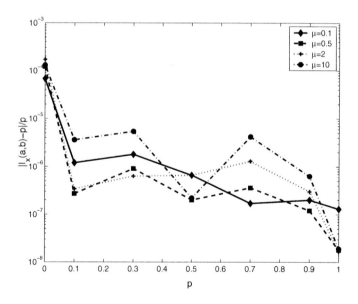

Figure 10.3. *Relative errors* $|I_x(a,b) - p|/p$ *for* $a = 10$ *and several values of p and* $\mu = b/a$. *The asymptotic inversion is based on the method of* §10.5.3.

10.6. High order Newton-like methods

Let $f(z)$ be the function, the zero ζ of which has to be computed. We put $\zeta = \zeta_0 + h$, where ζ_0 is an approximation of this zero, and we assume that we can expand in a neighborhood of this point

$$f(\zeta) = f(\zeta_0 + h) = f(\zeta_0) + hf_1 + \frac{1}{2!}h^2 f_2 + \frac{1}{3!}h^3 f_3 + \cdots, \tag{10.124}$$

where f_k denotes the kth derivative of f at ζ_0. We assume that $f(\zeta_0)$ is small and we expand

$$h = c_1 f(\zeta_0) + c_2 f^2(\zeta_0) + c_3 f^3(\zeta_0) + \cdots. \tag{10.125}$$

Substituting this expansion into (10.124), using that $f(\zeta) = 0$, and comparing equal powers of $f(\zeta_0)$, we find, when $f_1 \neq 0$,

$$c_1 = -\frac{1}{f_1}, \quad c_2 = -\frac{f_2}{2f_1^3},$$
$$c_3 = \frac{-3f_2^2 + f_3^2 f_1}{6f_1^5}, \quad c_4 = -\frac{f_4 f_1^2 + 15 f_2^3 - 10 f_2 f_3 f_1}{24 f_1^7}. \tag{10.126}$$

When we neglect in (10.125) the coefficients c_k with $k \geq 2$, we obtain Newton–Raphson, with $\zeta \doteq \zeta_0 - f(\zeta_0)/f'(\zeta_0)$.

When $f(z)$ satisfies a simple ordinary differential equation, the higher derivatives can be replaced by combinations of lower derivatives.

Example 10.1 (the inversion of the incomplete gamma function). In §10.3 we have considered the inversion of the equations

$$P(a, x) = p, \quad Q(a, x) = q, \tag{10.127}$$

where $0 < p < 1$, $0 < q < 1$, for large positive values of a. When a is small the asymptotic methods cannot be applied, although for $a = 1$ the results can be used as a first approximation. Now we take $a \in (0, 1]$, $f(x) = P(a, x) - p$, and an initial value $x_0 > 0$. We derive from (8.12) the values

$$c_1 = -x_0^{1-a} e^{x_0} \Gamma(a),$$
$$c_2 = \frac{x_0 + 1 - a}{2x_0} c_1^2,$$
$$c_3 = \frac{2x_0^2 + 4x_0(1-a) + 2a^2 - 3a + 1}{6x_0^2} c_1^3, \tag{10.128}$$
$$c_4 = \frac{6x_0^3 + 18x_0^2(1-a) + x_0(18a^2 - 29a + 11) - 6a^3 + 11a^2 - 6a + 1}{24x_0^3} c_1^4.$$

For $a = 1$ the equation $f(x) = 0$ is simple, because $P(1, x) = 1 - e^{-x}$ and the solution of $f(x) = 0$ is $x = -\ln(1 - p)$. The values c_k are in this case $c_k = (-1)^k e^{kx_0}/k$, $k = 1, 2, 3, \ldots$, and h of (10.125) becomes $h = \sum_{k=1}^{\infty} c_k f^k = -\ln(1 + e^{x_0} f(x_0))$. Using this value of h we obtain

$$x = x_0 + h = x_0 - \ln(1 + e^{x_0} f(x_0)) = -\ln(1 - p), \tag{10.129}$$

which gives the exact solution of $f(x) = 0$ for any $x_0 > 0$.

For general $a \in (0, 1]$ we derive a convenient starting value x_0. We observe that

$$P(a, x) = \frac{1}{\Gamma(a)} \int_0^x t^{a-1} e^{-t} \, dt < \frac{1}{\Gamma(a)} \int_0^x t^{a-1} \, dt = \frac{x^a}{\Gamma(a+1)}. \tag{10.130}$$

Hence, the solution x_0 of the equation $x^a = p\Gamma(a+1)$ satisfies $0 < x_0 < x$, where x is the exact solution of $f(x) = 0$.

The case $a = \frac{1}{2}$ is of special interest, because $P(\frac{1}{2}, x) = \text{erf} \sqrt{x}$, the error function. For a numerical example for that case we take $p = 0.5$. We have $x_0 = \pi/16 = 0.196349540849362$ and $f(x_0) = \text{erf} \sqrt{x_0} - \frac{1}{2} = -0.030884051069941$. Using the values c_1, c_2, c_3, c_4 from (10.128), we have $h = 0.0311185517296367$. This gives the new approximation $x \doteq x_0 + h = 0.227468092579000$ and with this value we have $f(x) = -1.12 \ldots 10^{-7}$. It is easy to iterate and to obtain much higher accuracy. ∎

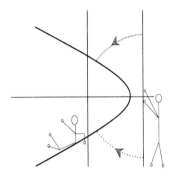

Chapter 11
Further Examples

There is nothing more frustrating than a good example.
—Mark Twain, American writer

11.1 Introduction

In this chapter we consider a few topics in which certain methods discussed in earlier chapters are considered in more detail.

We start with Euler's summation formula, an elegant technique for approximating finite or infinite series, by using an integral as a first approximation.

The chapter continues with information on Stirling numbers, in particular how to handle numerically certain expressions used in the asymptotic approximations.

In §11.4 we pay attention to Carlson's symmetric elliptic integrals and give an algorithm for one of these integrals.

The chapter concludes with a discussion of numerical inversion of Laplace transforms.

11.2 The Euler summation formula

Euler's summation formula can be used to estimate finite or infinite series by means of an integral; it includes correction terms and a remainder. With this formula it is possible to evaluate in an efficient way slowly convergent series. Turning it round, by this method also an integral can be approximated by discretization, which leads to the *trapezoidal quadrature rule*; see §5.2.

We give a few theorems, starting with finite sums, and extend the results to infinite series. First we mention properties of Bernoulli numbers and Bernoulli polynomials, which play a crucial role in the summation formula.

The Bernoulli polynomials $B_n(x)$ are defined by the generating function

$$\frac{z e^{xz}}{e^z - 1} = \sum_{n=0}^{\infty} \frac{B_n(x)}{n!} z^n, \quad |z| < 2\pi. \tag{11.1}$$

The Bernoulli numbers B_n are the values of these polynomials at $x = 0$. Because $f(z) = z/(e^z - 1) - 1 + \frac{1}{2}z$ is an even function, we have

$$B_{2n+1} = 0, \quad n = 1, 2, 3, \ldots. \tag{11.2}$$

The first nonvanishing numbers are

$$B_0 = 1, \quad B_1 = -\tfrac{1}{2}, \quad B_2 = \tfrac{1}{6}, \quad B_4 = -\tfrac{1}{30}, \quad B_6 = \tfrac{1}{42}, \quad B_8 = -\tfrac{1}{30},$$
$$B_{10} = \tfrac{5}{66}, \quad B_{12} = -\tfrac{691}{2730}, \quad B_{14} = \tfrac{7}{6}, \quad B_{16} = -\tfrac{3617}{510}. \tag{11.3}$$

By differentiating and integrating (11.1) with respect to x, it follows that

$$B_{n-1}(x) = \frac{1}{n} B'_n(x), \quad \int_0^1 B_n(x)\, dx = 0, \quad n = 1, 2, 3, \ldots, \tag{11.4}$$

and more generally

$$\int_a^x B_{n-1}(x)\, dx = \frac{1}{n} [B_n(x) - B_n(a)], \quad n = 1, 2, 3, \ldots. \tag{11.5}$$

Now we have the following theorem.

Theorem 11.1. *Let the function $f: [0, 1] \to \mathbb{C}$ have k continuous derivatives ($k = 0, 1, 2, \ldots$). Then for $k \geq 1$*

$$f(1) = \int_0^1 f(x)\, dx + \sum_{m=1}^{k} \frac{(-1)^m B_m}{m!} \left[f^{(m-1)}(1) - f^{(m-1)}(0) \right] + R_k, \tag{11.6}$$

with

$$R_k = \frac{(-1)^{k+1}}{k!} \int_0^1 f^{(k)}(x) B_k(x)\, dx. \tag{11.7}$$

Proof. The proof runs by induction with respect to k. For $k = 1$ the claim is true, which follows from integrating by parts. Then the first property of (11.4) is used to go from $k = m \geq 1$ to $k = m + 1$. □

With similar conditions for f on the interval $[j - 1, j]$ we have

$$f(j) = \int_{j-1}^{j} f(x)\, dx + \sum_{m=1}^{k} \frac{(-1)^m B_m}{m!} \left[f^{(m-1)}(j) - f^{(m-1)}(j - 1) \right] + R_k, \tag{11.8}$$

11.2. The Euler summation formula

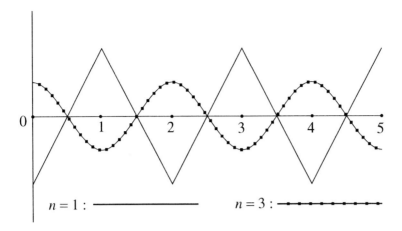

Figure 11.1. *Periodic Bernoulli functions* $\widetilde{B}_n(x)$, (11.10), $n = 1$ *and* $n = 3$.

with

$$R_k = \frac{(-1)^{k+1}}{k!} \int_{j-1}^{j} f^{(k)}(x) \widetilde{B}_k(x) \, dx, \tag{11.9}$$

where $\widetilde{B}_k(x)$ is a periodic function, the extension of the Bernoulli polynomial $B_k(x)$ from the interval [0, 1]. That is,

$$\widetilde{B}_n(x) = \begin{cases} B_n(x) & \text{if } 0 \leq x < 1, \\ \widetilde{B}_n(x-1) & \text{if } x \in \mathbb{R}. \end{cases} \tag{11.10}$$

The periodic Bernoulli functions $\widetilde{B}_n(x)$, $n = 1$ and $n = 3$, are shown in Figure 11.1 and those for $n = 2$ and $n = 4$ in Figure 11.2.

The next step joins a number of these intervals:

$$\sum_{m=1}^{n} f(m) = \int_{0}^{n} f(x) \, dx + \sum_{m=1}^{k} \frac{(-1)^i B_m}{m!} \left[f^{(m-1)}(n) - f^{(m-1)}(0) \right] + R_k, \tag{11.11}$$

with

$$R_k = \frac{(-1)^{k+1}}{k!} \int_{0}^{n} f^{(k)}(x) \widetilde{B}_k(x) \, dx. \tag{11.12}$$

For $k = 1$, this gives the formula

$$f(1) + f(2) + \cdots + f(n) = \int_{0}^{n} f(x) \, dx + \tfrac{1}{2}[f(n) - f(0)] + \int_{0}^{n} \widetilde{B}_1(x) f'(x) \, dx, \tag{11.13}$$

with $\widetilde{B}_1(x)$ a sawtooth function on $[0, n]$. This is Euler's summation formula in its simplest form.

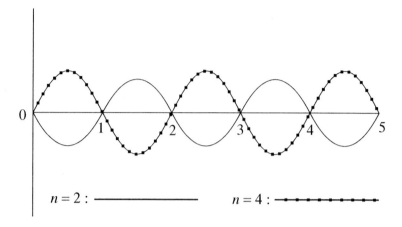

Figure 11.2. *Periodic Bernoulli functions $\widetilde{B}_n(x)$, (11.10), $n=2$ and $n=4$.*

Example 11.2 (Euler summation formula). Take $f(x) = 1/(1+x)$ and in the above formula replace n with $n-1$. Then we obtain the classical example

$$1 + \frac{1}{2} + \frac{1}{3} + \frac{1}{4} + \cdots + \frac{1}{n} = \ln n + \frac{1}{2n} + \frac{1}{2} - \int_0^{n-1} \widetilde{B}_1(x) \frac{dx}{(1+x)^2}. \tag{11.14}$$

The integral is convergent when $n \to \infty$. From this we infer that

$$\gamma = \lim_{n \to \infty} \left(1 + \tfrac{1}{2} + \tfrac{1}{3} + \tfrac{1}{4} + \cdots + \tfrac{1}{n} - \ln n\right) \tag{11.15}$$

also exists. The limit $\gamma = 0.5772\,15664\,90153\ldots$ is called *Euler's constant*. ∎

Because of (11.2) all terms with odd index can be deleted in the summation formula, except the term with index $m = 1$. And at both sides we can add the term $f(0)$. Then the result is as follows.

Theorem 11.3. *Let the function $f:[0,n] \to \mathbb{C}$ have $(2k+1)$ continuous derivatives ($k \geq 0, n \geq 1$). Then*

$$\sum_{j=0}^{n} f(j) = \int_0^n f(x)\,dx + \frac{1}{2}[f(n) + f(0)]$$
$$+ \sum_{j=1}^{k} \frac{B_{2j}}{(2j)!}\left[f^{(2j-1)}(n) - f^{(2j-1)}(0)\right] + R_k, \tag{11.16}$$

where

$$R_k = \frac{1}{(2k+1)!} \int_0^n f^{(2k+1)}(x)\widetilde{B}_{2k+1}(x)\,dx. \tag{11.17}$$

11.2. The Euler summation formula

An alternative summation formula for infinite series arises through the intermediate form

$$\sum_{j=m}^{n} f(j) = \int_{m}^{n} f(x)\,dx + \frac{1}{2}[f(n) + f(m)]$$
$$+ \sum_{j=1}^{k} \frac{B_{2j}}{(2j)!} \left[f^{(2j-1)}(n) - f^{(2j-1)}(m) \right] + R_k, \qquad (11.18)$$

where

$$R_k = \frac{1}{(2k+1)!} \int_{m}^{n} f^{(2k+1)}(x) \widetilde{B}_{2k+1}(x)\,dx. \qquad (11.19)$$

In this formula we replace n with ∞, which is allowed when the infinite series and the indefinite integrals

$$\sum_{j=m}^{\infty} f(j), \quad \int_{m}^{\infty} f(x)\,dx, \quad \int_{m}^{\infty} f^{(2k+1)}(x) \widetilde{B}_{2k+1}(x)\,dx \qquad (11.20)$$

exist. In addition, we assume that f and the derivatives occurring in the formula tend to zero when their arguments tend to infinity. The result is

$$\sum_{j=m}^{\infty} f(j) = \int_{m}^{\infty} f(x)\,dx + \frac{1}{2}f(m) - \sum_{j=1}^{k} \frac{B_{2j}}{(2j)!} f^{(2j-1)}(m) + R_k, \qquad (11.21)$$

with

$$R_k = \frac{1}{(2k+1)!} \int_{m}^{\infty} f^{(2k+1)}(x) \widetilde{B}_{2k+1}(x)\,dx. \qquad (11.22)$$

This form of Euler's summation formula can be fruitfully applied in summing infinite series. It is important to have information on the remainder R_k. It is not always necessary to know the integral in R_k exactly. Also, it is not necessary to know whether

$$\lim_{k \to \infty} R_k = 0.$$

In many cases this condition is not fulfilled, or the limit does not even exist. An estimate of the remainder can be obtained through the following theorem.

Theorem 11.4. *Let f and all its derivatives be defined on the interval $[0, \infty)$ on which they should be monotonic and tend to zero when $x \to \infty$. Then R_k of (11.16) satisfies*

$$R_k = \theta_k \frac{B_{2k+2}}{(2k+2)!} \left[f^{(2k+1)}(n) - f^{(2k+1)}(0) \right], \qquad (11.23)$$

where $0 \leq \theta_k \leq 1$.

Proof. For the proof see [219, pp. 12–13]. □

A similar result holds for formula (11.21). In this case we have

$$R_k = -\theta_k \frac{B_{2k+2}}{(2k+2)!} f^{(2k+1)}(m), \quad 0 \leq \theta_k \leq 1. \qquad (11.24)$$

We see that with proper conditions on f, the error in taking k terms of the series on the right-hand side of (11.21) is smaller than the first neglected $(k+1)$th term. In practical problems one tries to find this $(k+1)$th term that falls below the requested accuracy, and one sums the series on the right-hand side of (11.21) as far as the kth term. In other words, one may sum the series until a particular term falls below the accuracy. This naive criterion, which is very popular in summing infinite series, is fully legitimate here.

Example 11.5 (to sum the series $\sum_{j=1}^{\infty} 1/j^3$). We sum the series $\sum_{j=1}^{\infty} 1/j^3$ with an error less than 10^{-9}. First we compute $\sum_{j=1}^{9} 1/j^3 = 1.19653198567\ldots$. Next we apply (11.21) with $f(x) = 1/x^3$ and $m = 10$. Note that we do not use (11.21) with the low value $m = 1$, but with one that makes R_k small enough (for an acceptable value of k). Our f fulfills the conditions of Theorem 11.4 (extended to an infinite interval). We verify that the third term in the series on the right-hand side of (11.21) equals

$$-\tfrac{1}{42}\tfrac{1}{6!} 2520 \, 10^{-8} = -\tfrac{1}{12} 10^{-8} = -0.83 \, 10^{-9}. \tag{11.25}$$

Hence, we apply (11.21) with $k = 2$ and obtain

$$\sum_{j=1}^{\infty} \frac{1}{j^3} = \sum_{j=1}^{9} \frac{1}{j^3} + \sum_{j=10}^{\infty} \frac{1}{j^3} = 1.19653198567\ldots$$
$$+ \int_{10}^{\infty} \frac{dx}{x^3} + \frac{1}{2000} + \frac{1}{40000} - \frac{1}{12000000} \tag{11.26}$$
$$= 1.20205690234,$$

with an error that is smaller than $0.83 \, 10^{-9}$. The actual error is $0.82 \, 10^{-9}$.

From this example we see that the error estimate can be very sharp. Another point is that Euler's summation formula may produce a quite accurate result, with almost no effort. To obtain the same accuracy, straightforward numerical summation of the series $\sum j^{-3}$ requires about 22360 terms. ∎

Not all series can be evaluated by Euler's formula in this favorable way. Although the class of series for which the formula is applicable is quite interesting, Euler's method has its limitations. Alternating series should be tackled through Boole's summation method, which is based on the Euler polynomials (see [219, §1.2.2]).

To obtain information on how many terms one needs using (11.24), one may use estimates of the Bernoulli numbers. Since the radius of convergence of the series in (11.1) equals 2π, one can use the rough estimate

$$\frac{B_{2k}}{(2k)!} = \mathcal{O}\left[(2\pi)^{-2k}\right], \quad k \to \infty. \tag{11.27}$$

11.3 Approximations of Stirling numbers

The Stirling numbers play an important role in difference calculus, combinatorics, and probability theory. Many properties of Stirling numbers can be found in, for example, [120, Chap. 8], [44], and in the chapter on combinatorial analysis in [2, Chap. 24]. Recent interest with an interesting treatment of Stirling numbers can be found in [25, 26, 101, 125].

11.3. Approximations of Stirling numbers

In [218] asymptotic estimates for the Stirling numbers have been derived which hold for large values of n and which are uniformly valid for $0 \leq m \leq n$. In this section we summarize the results of [218] and we give additional information on the evaluation of the asymptotic estimates for the Stirling numbers.

11.3.1 Definitions

The Stirling numbers of the first and second kinds, denoted by $S_n^{(m)}$ and $\mathcal{S}_n^{(m)}$, respectively, are defined through the generating functions

$$x(x-1)\cdots(x-n+1) = \sum_{m=0}^{n} S_n^{(m)} x^m, \tag{11.28}$$

$$x^n = \sum_{m=0}^{n} \mathcal{S}_n^{(m)} x(x-1)\cdots(x-m+1), \tag{11.29}$$

where the left-hand side of (11.28) has the value 1 if $n = 0$; similarly for the factors in the right-hand side of (11.29) if $m = 0$. This gives the "boundary values"

$$S_n^{(n)} = \mathcal{S}_n^{(n)} = 1,\ n \geq 0,\quad \text{and}\quad S_n^{(0)} = \mathcal{S}_n^{(0)} = 0,\ n \geq 1. \tag{11.30}$$

Furthermore it is convenient to agree on $S_n^{(m)} = \mathcal{S}_n^{(m)} = 0$ if $m > n$.

The Stirling numbers are integers; apart from the above-mentioned zero values, the numbers of the second kind are positive; those of the first kind have the sign of $(-1)^{n+m}$.

Alternative generating functions are

$$\frac{[\ln(x+1)]^m}{m!} = \sum_{n=m}^{\infty} S_n^{(m)} \frac{x^n}{n!}, \tag{11.31}$$

$$\frac{(e^x-1)^m}{m!} = \sum_{n=m}^{\infty} \mathcal{S}_n^{(m)} \frac{x^n}{n!}. \tag{11.32}$$

An explicit representation of the numbers of the second kind is available as a finite sum:

$$\mathcal{S}_n^{(m)} = \frac{1}{m!} \sum_{k=0}^{m} (-1)^{m-k} \binom{m}{k} k^n. \tag{11.33}$$

Such a simple closed form for the numbers of the first kind is not available. In terms of the numbers of the second kind we have

$$S_n^{(m)} = \sum_{k=0}^{n-m} (-1)^k \binom{n-1+k}{n-m+k} \binom{2n-m}{n-m-k} \mathcal{S}_{n-m+k}^{(k)}. \tag{11.34}$$

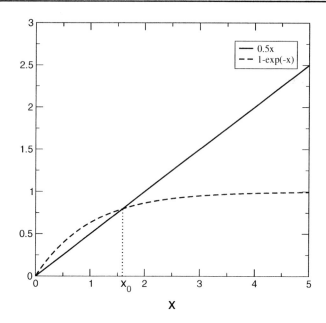

Figure 11.3. *Graphs of μx ($\mu = 0.5$) and $1 - e^{-x}$. The positive solution x_0 of* (11.39) *is also shown.*

11.3.2 Asymptotics for Stirling numbers of the second kind

The starting point in [218] is the Cauchy-type integral, which follows from (11.32),

$$S_n^{(m)} = \frac{n!}{m!} \frac{1}{2\pi i} \int (e^z - 1)^m \frac{dz}{z^{n+1}} = \frac{n!}{m!} \frac{1}{2\pi i} \int e^{\phi(z)} \frac{dz}{z}, \qquad (11.35)$$

where, initially, the contour is a circle around the origin with radius less than 2π and

$$\phi(z) = -n \ln z + m \ln(e^z - 1). \qquad (11.36)$$

The equation for the saddle points follows from $\phi'(z) = 0$, that is,

$$\mu z = 1 - e^{-z}, \quad \mu = \frac{m}{n}. \qquad (11.37)$$

The estimate given in [218, eq. (2.9)] reads

$$S_n^{(m)} \sim e^A m^{n-m} f(t_0) \binom{n}{m}, \quad n \to \infty, \qquad (11.38)$$

where $t_0 = (n - m)/m$, x_0 is the positive solution of the equation (see Figure 11.3)

$$\mu x = 1 - e^{-x}, \qquad (11.39)$$

and

$$A = n[-\ln(x_0) + \mu(x_0 - t_0 + \ln(\mu x_0)) + (1 - \mu)\ln(t_0)], \qquad (11.40)$$

11.3. Approximations of Stirling numbers

$$f(t_0) = \sqrt{\frac{t_0}{(1+t_0)(x_0 - t_0)}}. \tag{11.41}$$

Equation (11.38) can be written in the form

$$\mathcal{S}_n^{(m)} \sim \frac{n!}{m!} e^{\phi(x_0)} \frac{e^{-(n-m)}(n-m)^{n-m}}{(n-m)!} \sqrt{\frac{1-\mu}{x_0 - t_0}}. \tag{11.42}$$

Expansions of the number x_0

In Lemma 11.6 we derive simple bounds for the positive solution x_0 of (11.39), and it is rather easy to use a numerical algorithm for computing x_0. However, it is also convenient to have convergent expansions for the cases when μ is small or when μ is close to unity.

It should be observed that an explicit form of the positive solution x_0 of (11.39) can be written in terms of the Lambert W-function, that is, the function $w(z)$ that solves the equation $we^w = z$. We can write (11.39) in this form by putting

$$w = x - \frac{1}{\mu}, \quad z = -\frac{1}{\mu}e^{-\frac{1}{\mu}}, \tag{11.43}$$

where w should belong to the interval $(-1, 0)$.

First we give simple bounds of x_0.

Lemma 11.6. *The positive solution x_0 of (11.39), where $0 < \mu < 1$, satisfies $1/\mu - 1 < x_0 < 1/\mu$.*

Proof. For $x > 0$ we have
$$x/(1+x) < 1 - e^{-x} < 1, \tag{11.44}$$
which gives the bounds for x_0. □

The following lemma gives an expansion for x_0 that is very useful for small values of μ.

Lemma 11.7. *The positive solution x_0 of (11.39) has the representation*

$$x_0 = \frac{1}{\mu} - T(\mu), \quad T(\mu) = \sum_{k=1}^{\infty} \frac{k^{k-1}}{k!} \delta^k, \quad \delta = \frac{1}{\mu}e^{-\frac{1}{\mu}}, \tag{11.45}$$

where the series for $T(\mu)$ is convergent for $\mu \in (0, 1)$.

Proof. Let $x = 1/\mu + y$; then the equation for y reads $ye^y = -\delta$. One solution is $y = -1/\mu$ which corresponds to the trivial solution $x = 0$ of (11.39). We need the other y-solution that tends to zero if $\delta \to 0$. The equation for y is considered in [50, p. 23], where the convergent expansion of this lemma is given for $|\delta| < 1/e$. □

The convergence of the expansion in this lemma becomes worse if μ is close to unity. For this case we have a different expansion.

Lemma 11.8. *For small values of $|\mu - 1|$ the positive solution x_0 of (11.39) has the convergent expansion*

$$x_0 = \sum_{k=1}^{\infty} c_k (1 - \mu)^k, \qquad (11.46)$$

where the first few coefficients are given by

$$c_1 = 2, \quad c_2 = \tfrac{4}{3}, \quad c_3 = \tfrac{10}{9}, \quad c_4 = \tfrac{136}{135}, \quad c_5 = \tfrac{386}{405}, \quad c_6 = \tfrac{524}{567}. \qquad (11.47)$$

Proof. Equation (11.39) can be written in the form

$$1 - \mu = \frac{x}{f(x)}, \quad f(x) = \frac{x^2}{e^{-x} - 1 + x}. \qquad (11.48)$$

The function f is analytic in a neighborhood of the point $x = 0$ with $f(0) = 2$. According to [50, p. 22], there exist positive numbers a and b such that for $|1 - \mu| < a$ the equation has just one solution x_0 in the domain $|x| < b$, and x_0 has, for $|1 - \mu| < a$, the convergent expansion as shown in the lemma, with

$$c_k = \frac{1}{k!} \left\{ \left(\frac{d}{dx}\right)^{k-1} [f(x)]^k \right\}_{x=0}, \quad k = 1, 2, 3, \ldots. \qquad \square \qquad (11.49)$$

Evaluating the estimate (11.42)

For large values of n and m or $n - m$, the numerical evaluation of the estimate in (11.42) should be done with some care. We write

$$\mathcal{S}_n^{(m)} \sim C(m, n) D(m, n), \qquad (11.50)$$

where

$$C(m, n) = \frac{e^{-(n-m)}(n-m)^{n-m}}{(n-m)!} \sqrt{\frac{(n-m)}{n(x_0 - t_0)}}, \quad D(m, n) = \frac{n!}{m!} e^{\phi(x_0)}, \qquad (11.51)$$

in which $D(m, n)$ is the dominant part, and roughly gives the size of $\mathcal{S}_n^{(m)}$.

We use the function $\Gamma^*(x)$ defined in (8.28), that is,

$$\Gamma^*(x) = \sqrt{\frac{x}{2\pi}} e^x x^{-x} \Gamma(x), \quad x > 0, \qquad (11.52)$$

and we assume that this function is available for numerical computations for $x > 0$. We have $\Gamma^*(x) \sim 1$, as $x \to \infty$.

When $n - m > 0$ the quantity $C(m, n)$ can be written as

$$C(m, n) = \frac{1}{\Gamma^*(n-m)\sqrt{2\pi n(x_0 - t_0)}}. \qquad (11.53)$$

11.3. Approximations of Stirling numbers

Recalling that $t_0 = \frac{1}{\mu} - 1$ and using (11.45), we see that

$$x_0 - t_0 = 1 - T(\mu). \tag{11.54}$$

When μ is small, that is, when $m \ll n$, both x_0 and t_0 are of order $1/\mu$, and the quantity $x_0 - t_0$ should be computed from (11.45) and (11.54), where $T(\mu)$ is exponentially small.

For the quantity $D(m, n)$ we can write, using (11.36) and (11.39),

$$D(m, n) = \frac{n!}{m!} \frac{(e^{x_0} - 1)^m}{x_0^n} = \frac{n!}{m!} e^{mx_0} \frac{(1 - e^{-x_0})^m}{x_0^n} = \frac{m^m \, n!}{n^m \, m!} x_0^{m-n} e^{mx_0}. \tag{11.55}$$

By using (11.52), we obtain

$$D(m, n) = x_0^{m-n} e^{mx_0} e^{m-n} n^{n-m} \sqrt{\frac{n}{m}} \frac{\Gamma^*(n)}{\Gamma^*(m)}. \tag{11.56}$$

First we assume that μ is small. Because $x_0 = \frac{1}{\mu} - T(\mu)$, we have

$$D(m, n) = e^m m^{n-m} e^{n(\chi_1 + \chi_2)} \sqrt{\frac{n}{m}} \frac{\Gamma^*(n)}{\Gamma^*(m)}, \tag{11.57}$$

where

$$\chi_1 = \mu \ln(1 - \mu T(\mu)), \quad \chi_2 = -\mu T(\mu) - \ln(1 - \mu T(\mu)). \tag{11.58}$$

Because $T(\mu)$ is exponentially small, both χ_1 and χ_2 should be calculated by using, for example, Maclaurin expansions of the logarithm.

When $\mu \sim 1$ we have $x_0 \sim 2(1 - \mu)$ (see (11.46)), and no further steps seem to be needed in the evaluation of $D(m, n)$ when using (11.56).

Numerical verification of the approximation

The Stirling numbers of the second kind satisfy the following relation:

$$\mathcal{S}_{n+1}^{(m)} = m \mathcal{S}_n^{(m)} + \mathcal{S}_n^{(m-1)}, \quad n \geq m \geq 1. \tag{11.59}$$

This relation follows from the generating function in (11.32). We use it to verify numerically the asymptotic approximation. Let

$$\delta_n^{(m)} = \left| \frac{\mathcal{S}_{n+1}^{(m)}}{m \mathcal{S}_n^{(m)} + \mathcal{S}_n^{(m-1)}} - 1 \right|. \tag{11.60}$$

We compute this quantity by using the approximation given in (11.42) for a fixed value of n and all $m \in [1, n]$. In Table 11.1 we give the numbers $\delta_n^{(m^*)}$ defined by

$$\delta_n^{(m^*)} = \max_{m \in [1, n]} \delta_n^{(m)}, \tag{11.61}$$

the largest relative error for fixed n. We also give the numbers m^* for which this largest relative error is attained and the ratio m^*/n. We observe that m^* is not near 1 or near n, which shows the uniform character of the approximation. To see the growth of these Stirling numbers we give the maximal value of all $\mathcal{S}_n^{(m)}$ for fixed n. That is, we give

$$\mathcal{S}_n^{(\widehat{m})} = \max_{m \in [1, n]} \mathcal{S}_n^{(m)}. \tag{11.62}$$

Table 11.1. *Relative error $\delta_n^{(m^*)}$ and m^* for fixed n; see (11.60) and (11.61); $\widehat{m}$ is the value of m that gives the maximal value of all $S_n^{(m)}$ denoted by $S_n^{(\widehat{m})}$.*

n	m^*	m^*/n	$\delta_n^{(m^*)}$	$\widehat{m}$	$\widehat{m}/n$	$S_n^{(\widehat{m})}$
10	2	0.20	$1.221\ 10^{-3}$	5	0.50	$4.275\ 10^{+04}$
20	4	0.20	$2.982\ 10^{-4}$	8	0.40	$1.522\ 10^{+13}$
30	6	0.20	$1.313\ 10^{-4}$	11	0.37	$2.155\ 10^{+23}$
40	8	0.20	$7.352\ 10^{-5}$	14	0.35	$3.591\ 10^{+34}$
50	10	0.20	$4.692\ 10^{-5}$	16	0.32	$3.845\ 10^{+46}$

Table 11.2. *Relative error $\varepsilon_n^{(m^*)}$ and m^* for fixed large n; see (11.64); $\widehat{m}$ is the value of m that gives the maximal value of all $\log S_n^{(m)}$ denoted by $\log S_n^{(\widehat{m})}$.*

n	m^*	m^*/n	$\varepsilon_n^{(m^*)}$	$\widehat{m}$	$\widehat{m}/n$	$\log S_n^{(\widehat{m})}$
50	9	0.18	$4.726\ 10^{-07}$	16	0.32	$1.073\ 10^{+2}$
100	19	0.19	$4.575\ 10^{-08}$	28	0.28	$2.645\ 10^{+2}$
500	96	0.19	$2.406\ 10^{-10}$	106	0.21	$1.939\ 10^{+3}$
1000	193	0.19	$2.625\ 10^{-11}$	189	0.19	$4.436\ 10^{+3}$
5000	976	0.20	$1.623\ 10^{-13}$	754	0.15	$2.888\ 10^{+4}$
10000	1961	0.20	$1.850\ 10^{-14}$	1382	0.14	$6.370\ 10^{+4}$

We also give the number $\widehat{m}$ for which this largest value is attained and the ratio $\widehat{m}/n$.

For larger values of n, we use the approximation in the form of (11.50), with $C(m, n)$ and $D(m, n)$ given in (11.53) and (11.57), respectively, from which we compute the approximation of $\log S_n^{(m)}$. By taking logarithms of both sides of the relation in (11.59), we write this relation in the form

$$P = Q + \log\left(m + e^{R-Q}\right), \qquad (11.63)$$

where $P = \log S_{n+1}^{(m)}$, $Q = \log S_n^{(m)}$, and $R = \log S_n^{(m-1)}$. Next we compute

$$\varepsilon_n^{(m)} = \left|\frac{P - Q - \log\left(m + e^{R-Q}\right)}{Q}\right| \quad \text{and} \quad \varepsilon_n^{(m^*)} = \max_{m\in[2,n-1]} \varepsilon_n^{(m)}. \qquad (11.64)$$

This information is given in Table 11.2, together with the maximal value of all $\log S_n^{(m)}$ for fixed n, the value $\widehat{m}$ for which this largest value is attained, the ratios $\widehat{m}/n$, and the largest value $\log S_n^{(\widehat{m})}$.

11.3.3 Stirling numbers of the first kind

The estimate for the Stirling numbers of the first kind reads (see [218, eq. (3.5)])

$$S_{n+1}^{(m+1)} \sim (-1)^{n-m} \frac{1}{x_0} e^{\phi(x_0)} m^m n^{-n} (n-m)^{n-m} \binom{n}{m} \sqrt{\frac{m(n-m)}{n\phi''(x_0)}}, \qquad (11.65)$$

where

$$\phi(x) = \ln[(x+1)(x+2)\cdots(x+n)] - m \ln x, \qquad (11.66)$$

and x_0 is the positive solution of the equation $\phi'(x) = 0$, that is, of

$$\frac{1}{x+1} + \frac{1}{x+2} + \cdots + \frac{1}{x+n} = \frac{m}{x}. \qquad (11.67)$$

We have the following lemma.

Lemma 11.9. *The positive solution x_0 of* (11.67) *satisfies*

$$\frac{m}{n-m} < x_0 < \frac{mn}{n-m}. \qquad (11.68)$$

Proof. For $x > 0$ we have

$$\frac{n}{x+1} > \frac{1}{x+1} + \frac{1}{x+2} + \cdots + \frac{1}{x+n} > \frac{n}{x+n}, \qquad (11.69)$$

which gives the bounds for x_0. □

The upper and lower bounds in the inequalities in (11.68) differ by a factor n; compare this with the bounds in Lemma 11.6. When n and x are large, we may write $\phi(x)$ of (11.66) and $\phi'(x)$ in terms of the logarithm of the gamma function and the derivatives thereof. That is,

$$\begin{aligned}\phi(x) &= \ln \Gamma(x+n+1) - \ln \Gamma(x+1) - m \ln x, \\ \phi'(x) &= \psi(x+n+1) - \psi(x+1) - m/x.\end{aligned} \qquad (11.70)$$

Asymptotic expansions of these functions are given in [2, Chap. 6], and we have

$$\psi(x) = \ln x - \frac{1}{2x} + \mathcal{O}\left(x^{-2}\right), \quad x \to \infty. \qquad (11.71)$$

When we replace the equation for the saddle point x_0 in (11.67) by the equation

$$\ln \frac{\xi+n+1}{\xi+1} = \frac{m}{\xi}, \qquad (11.72)$$

then the root ξ_0 of this equation is a good approximation of x_0, in particular when n is large. By using the first term in the expansion

$$\ln z = 2\left(w + \frac{1}{3}w^3 + \frac{1}{5}w^5 + \cdots\right), \quad w = \frac{z-1}{z+1}, \qquad (11.73)$$

whose expansion converges for $\Re z \geq 0$, $z \neq 0$, we obtain for the solution of (11.72) the approximation

$$\xi_1 = \frac{m(n+2)}{2(n-m)}. \tag{11.74}$$

This approximation is useful when n and m are both large. For example, when $n = 1000$, $m = 900$, then $x_0 = 4349.4176\ldots$ and $\xi_1 = 4509$. The root of (11.72) is $\xi_0 = 4354.08\ldots$.

When m is small compared with n, we write (11.72) in the form

$$\xi = m / \ln \frac{\xi + n + 1}{\xi + 1} \tag{11.75}$$

and take as a first approximation

$$\xi_1 = \frac{m}{\ln(n+1)}. \tag{11.76}$$

A second iteration gives

$$\xi_2 = m / \ln \frac{\xi_1 + n + 1}{\xi_1 + 1}. \tag{11.77}$$

When $n = 1000$, $m = 100$, then $x_0 = 27.8461\ldots$ and $\xi_2 = 23.9\ldots$. The root of (11.75) is $\xi_0 = 28.0\ldots$.

For the calculation of the approximation (11.65) we write

$$S_{n+1}^{(m+1)} \sim (-1)^{n-m} \frac{1}{x_0} e^{\phi(x_0)} D(m,n) \sqrt{\frac{m(n-m)}{n\phi''(x_0)}}, \tag{11.78}$$

with

$$D(m,n) = m^m n^{-n} (n-m)^{n-m} \binom{n}{m} = \sqrt{\frac{n}{2\pi m(n-m)}} \frac{\Gamma^*(n)}{\Gamma^*(m)\Gamma^*(n-m)}, \tag{11.79}$$

where $\Gamma^*(x)$ is defined in (11.52). This gives

$$S_{n+1}^{(m+1)} \sim (-1)^{n-m} \frac{e^{\phi(x_0)}}{x_0 \sqrt{2\pi \phi''(x_0)}} \frac{\Gamma^*(n)}{\Gamma^*(m)\Gamma^*(n-m)}. \tag{11.80}$$

11.4 Symmetric elliptic integrals

Legendre's standard elliptic integrals are the *incomplete elliptic integral of the first kind*,

$$K(k) = \int_0^\phi \frac{d\theta}{\sqrt{1 - k^2 \sin^2 \theta}}, \tag{11.81}$$

the *incomplete integral of the second kind*,

$$E(k) = \int_0^\phi \sqrt{1 - k^2 \sin^2 \theta}\, d\theta, \tag{11.82}$$

11.4. Symmetric elliptic integrals

and the *incomplete elliptic integral of the third kind*,

$$\Pi(n; \phi, k) = \int_0^\phi \frac{1}{1 - n\sin^2\theta} \frac{d\theta}{\sqrt{1 - k^2\sin^2\theta}}. \tag{11.83}$$

It is assumed here that $k \in [0, 1]$, $\phi \in [0, \frac{1}{2}\pi]$, although the functions can also be defined for complex values of the parameters. Also, n is real, and if $n > 1$, the integral of the third kind should be interpreted as a Cauchy principal value integral. When $\phi = \frac{1}{2}\pi$ the integrals are called *complete elliptic integrals*.

11.4.1 The standard forms in terms of symmetric integrals

The computational problem for the elliptic integrals has received much attention in the literature, and the algorithms are usually based on successive *Landen transformations* or *Gauss transformations*, or by infinite series.

By considering a new set of integrals it is possible to compute the elliptic integrals, also by using successive transformations, by very efficient algorithms. The integrals are introduced in [29], where the following standard forms are chosen.

The standard form of the first kind:

$$R_F(x, y, z) = \frac{1}{2} \int_0^\infty \frac{dt}{\sqrt{(t+x)(t+y)(t+z)}}. \tag{11.84}$$

This function is symmetric and homogeneous of degree $-\frac{1}{2}$ in x, y, z and is normalized so that $R_F(x, x, x) = x^{-\frac{1}{2}}$.

The standard form of the third kind:

$$R_J(x, y, z, \rho) = \frac{3}{2} \int_0^\infty \frac{dt}{(t+\rho)\sqrt{(t+x)(t+y)(t+z)}}, \tag{11.85}$$

where $\rho \neq 0$. This function is symmetric in x, y, z and homogeneous of degree $-\frac{3}{2}$ in x, y, z, ρ, and normalized so that $R_J(x, x, x, x) = x^{-\frac{3}{2}}$. If $\rho < 0$, the Cauchy principal value is taken in (11.85). If ρ equals one of the other variables, R_J degenerates to an integral of the second kind,

$$R_D(x, y, z) = \frac{3}{2} \int_0^\infty \frac{dt}{(t+z)\sqrt{(t+x)(t+y)(t+z)}}. \tag{11.86}$$

If $x = 0$, the integrals are said to be complete.

Many elementary functions can be expressed in terms of these integrals. For example, when two of the variables are equal, R_F degenerates to

$$R_C(x, y) = R_F(x, y, y) = \frac{1}{2} \int_0^\infty \frac{dt}{\sqrt{(t+x)}\,(t+y)}, \tag{11.87}$$

which is a logarithm if $0 < y < x$ and an inverse circular function if $0 \le x < y$.

The three standard elliptic integrals can be written in terms of the above integrals as follows:

$$F(\phi, k) = \sin\phi R_F \left(\cos^2\phi, 1 - k^2 \sin^2\phi, 1\right), \tag{11.88}$$

$$E(\phi, k) = F(\phi, k) - \tfrac{1}{3}k^2 \sin^3\phi R_D \left(\cos^2\phi, 1 - k^2 \sin^2\phi, 1\right), \tag{11.89}$$

$$\Pi(\phi, k, n) = F(\phi, k) - \tfrac{1}{3}n \sin^3\phi R_J \left(\cos^2\phi, 1 - k^2 \sin^2\phi, 1, 1 + n \sin^2\phi\right). \tag{11.90}$$

11.4.2 An algorithm

We describe the algorithm for computing R_F of (11.84) as given in [30] for real variables. In [31] the algorithm is extended to complex variables. The algorithm is based on the following transformation:

$$R_F(x, y, z) = 2R_F(x + \lambda, y + \lambda, z + \lambda) = R_F\left(\tfrac{1}{4}(x + \lambda), \tfrac{1}{4}(y + \lambda), \tfrac{1}{4}(z + \lambda)\right), \tag{11.91}$$

where $\lambda = \sqrt{xy} + \sqrt{yz} + \sqrt{zx}$. This transformation easily follows by changing the variable of integration in the integral in (11.84). Transformation (11.91) will be repeated until the arguments of R_F are nearly equal. In that case a Taylor expansion of at most five terms is used to finish off the calculations. Observe that when indeed the arguments are equal, we have $R_F(x, x, x) = 1/\sqrt{x}$.

ALGORITHM 11.1. Carlson's algorithm for computing elliptic integrals.

Input: $x_0 \geq 0$, $y_0 > 0$, and $z_0 > 0$.

Output: $R_F(x_0, y_0, z_0)$.

For $n = 0, 1, 2, \ldots$, let

$$\lambda_n = \sqrt{x_n y_n} + \sqrt{x_n z_n} + \sqrt{y_n z_n},$$
$$\mu_n = \tfrac{1}{3}(x_n + y_n + z_n),$$
$$x_{n+1} = \tfrac{1}{4}(x_n + \lambda_n), \quad y_{n+1} = \tfrac{1}{4}(y_n + \lambda_n), \quad z_{n+1} = \tfrac{1}{4}(z_n + \lambda_n),$$
$$X_n = 1 - (x_n/\mu_n), \quad Y_n = 1 - (y_n/\mu_n), \quad Z_n = 1 - (z_n/\mu_n),$$
$$\varepsilon_n = \max\{|X_n|, |Y_n|, |Z_n|\},$$
$$s_n^{(m)} = (X_n^m + Y_n^m + Z_n^m)/(2m), \quad (m = 2, 3).$$

Then $\varepsilon_n = \mathcal{O}(4^{-n})$ as $n \to \infty$. If $\varepsilon_n < 1$, then

$$R_F(x_0, y_0, z_0) = \mu_n^{-\frac{1}{2}} \left(1 + \tfrac{1}{5}s_n^{(2)} + \tfrac{1}{7}s_n^{(3)} + \tfrac{1}{6}(s_n^{(2)})^2 + \tfrac{3}{11}s_n^{(2)}s_n^{(3)} + r_n\right),$$

$$|r_n| < \frac{\varepsilon_n^6}{4(1 - \varepsilon)} \quad \text{and} \quad r_n \sim \tfrac{5}{26}(s_n^{(2)})^3 + \tfrac{3}{26}(s_n^{(3)})^2, \quad n \to \infty.$$

11.5. Numerical inversion of Laplace transforms

Carlson [30] remarks that although this algorithm has linear rather than quadratic convergence, it is quite fast because r_n is of order $4^{-6n} = (4096)^{-n}$. For example, for the computation of the lemniscate constant

$$R_F(0, 1, 2) = \int_0^1 \frac{ds}{\sqrt{1-s^4}}, \tag{11.92}$$

we have after three iterations

$$|r_3| < 1.0 \times 10^{-10} < (4096)^{-3} = 0.145\ldots \times 10^{-10}. \tag{11.93}$$

The algorithm for R_C of (11.87) is somewhat simpler, because of the reduction to two variables, and those for R_J and R_D are slightly more complicated. For details we refer to [30], or to [31] when the variables are complex.

11.4.3 Other elliptic integrals

Carlson's method is not restricted to the integrals of the form given above. Many elliptic integrals of the form

$$\int_y^x \prod_{j=1}^n (a_j + b_j t)^{\frac{1}{2} p_j}\, dt, \tag{11.94}$$

where $p_1, \ldots, p_n$ are integers and the integrand is real, can be reduced to integrals for R_F, R_J, R_D, and R_C. For example (see [31]), consider

$$I = \int_y^x \frac{dt}{\sqrt{(f_1 + 2g_1 t + h_1 t^2)(f_2 + 2g_2 t + h_2 t^2)}}, \tag{11.95}$$

where all quantities are real, $x > y$, the two quadratic (or, if $h_j = 0$, linear) polynomials are positive on the open interval (x, y), and their product has at most simple zeros on the closed interval. Let

$$\begin{aligned}
q_j(t) &= f_j + 2g_j t + h_j t^2, \quad j = 1, 2, \\
(x-y)U &= \sqrt{q_1(x)q_2(y)} + \sqrt{q_1(y)q_2(x)}, \\
T &= 2g_1 g_2 - f_1 h_2 - f_2 h_1, \\
V &= 2\sqrt{(g_1^2 - f_1 h_1)(g_2^2 - f_2 h_2)}.
\end{aligned} \tag{11.96}$$

Then

$$I = 2R_F\left(U^2 + T + V, U^2 + T - U, U^2\right). \tag{11.97}$$

If $x = +\infty$, U can be obtained by taking a limit, that is, $U = \sqrt{h_1 q_2(y)} + \sqrt{q_1(y) h_2}$.

11.5 Numerical inversion of Laplace transforms

We consider the pair of Laplace transforms

$$F(s) = \int_0^\infty e^{-st} f(t)\, dt, \quad f(t) = \frac{1}{2\pi i} \int_{c-i\infty}^{c+i\infty} e^{st} F(s)\, ds, \tag{11.98}$$

where f should be absolutely integrable on any finite interval $[0, a]$ and the number c is chosen such that all singularities of $F(s)$ are at the left of the vertical line $\Re s = c$.

The inversion problem is to find $f(t)$ when $F(s)$ is given. To solve this problem numerically, an essential condition is whether function values of $F(s)$ are only available for real s or for complex values of s. The first case is quite difficult and requires completely different techniques compared with those for the second case. In this section we consider a few methods for the case that $F(s)$ is available as an analytic function in part of the complex s-plane. We describe two methods: one is based on complex Gauss quadrature, whereas the other uses deformations of the contour of integration.

11.5.1 Complex Gauss quadrature

First we consider a more general setting, and we give an example for Laplace inversion.

We start with the integral

$$I(G) = \frac{1}{2\pi i} \int_{c-i\infty}^{c+i\infty} e^z z^{-\nu-1} G(z)\, dz, \quad c > c_0 \geq 0, \quad (11.99)$$

and assume that $G(z)$ is analytic in the half-plane $\Re z > c_0$ and bounded as $z \to \infty$ in $|\mathrm{ph}\, z| \leq \frac{1}{2}\pi$. For this type of integral a *complex Gauss quadrature formula* is available of the form

$$I(G) = \sum_{j=1}^{n} \lambda_{j,n} G(z_{j,n}) + E_n(G), \quad (11.100)$$

where $E_n(G) = 0$ if $G(z)$ is a polynomial of degree $\leq 2n-1$ in $1/z$. The nodes $z_{j,n}$ are the zeros of the hypergeometric polynomial

$$P_n(\nu, z^{-1}) = {}_2F_0\left(\begin{matrix}-n,\, n+\nu \\ -\end{matrix};\, \frac{1}{z}\right), \quad P(\nu, z_{j,n}^{-1}) = 0, \quad (11.101)$$

which are related to *Bessel polynomials* (see [103]). The weights are given by

$$\lambda_{j,n} = \frac{(-1)^{n+1}\, n!\, (2n+\nu-1)^2}{n^2 \Gamma(n+\nu) z_{j,n}^2 [P_{n-1}(\nu, z_{j,n}^{-1})]^2}. \quad (11.102)$$

The complex orthogonal polynomials $P_n(\nu, z^{-1})$, $n = 0, 1, 2, \ldots$, satisfy the orthogonality condition

$$\int_{c-i\infty}^{c+i\infty} e^z z^{-\nu-1} P_n(\nu, z^{-1}) P_m(\nu, z^{-1})\, dz = 0, \quad m \neq n. \quad (11.103)$$

To apply this rule to the pair of transforms in (11.98) we first write

$$t^{-\nu} f(t) = \frac{1}{2\pi i} \int_{c-i\infty}^{c+i\infty} e^z z^{-\nu-1} G(z)\, dz, \quad G(z) = (z/t)^{\nu+1} F(z/t), \quad (11.104)$$

and then the quadrature rule (11.100) without the error term reads

$$f(t) \approx t^\nu \sum_{j=1}^{n} \lambda_{j,n} G(z_{j,n}). \quad (11.105)$$

The complex Gauss nodes $z_{j,n}$ have positive real part for all $\nu > -1$. The nodes and weights of the 5-point complex Gauss quadrature formula (11.100) for $\nu = 0$ are shown in Table 11.3. Extensive tables of the quadrature nodes and weights can be found in [131].

11.5. Numerical inversion of Laplace transforms

Table 11.3. *Nodes and weights for the 5-point complex Gauss quadrature formula with $\nu = 0$.*

j	$\lambda_{j,n}$		w_j	
1	3.65569 4325 +	6.54373 6899 i	3.83966 1630 −	0.27357 03863 i
2	3.65569 4325 −	6.54373 6899 i	3.83966 1630 +	0.27357 03863 i
3	5.70095 3299 +	3.21026 5600 i	−25.07945 2209 +	2.18725 22937 i
4	5.70095 3299 −	3.21026 5600 i	−25.07945 2209 −	2.18725 22937 i
5	6.28670 4752 +	0.00000 0000 i	43.47958 1157 +	0.00000 00000 i

Table 11.4. *Laplace transform inversion for the pair* (11.106).

t	$f(t)$	$J_0(t)$
0.0	1.00000 00000	1.00000 00000
0.5	0.93846 98072	0.93846 98072
1.0	0.76519 76865	0.76519 76866
2.0	0.22389 10326	0.22389 07791
5.0	−0.17902 54097	−0.17759 67713
10.0	−0.07540 53543	−0.24593 57645

Example 11.10 (Laplace transform inversion). We have the Laplace transform pair

$$f(t) = J_0(t), \quad F(s) = \frac{1}{\sqrt{s^2 + 1}}, \tag{11.106}$$

where $J_0(t)$ is the Bessel function. According to (11.104) we have $G(z) = z/\sqrt{z^2 + t^2}$ when we choose $\nu = 0$, so that $G(z) = \mathcal{O}(1)$ at infinity. Equation (11.105) gives

$$f(t) = \sum_{k=1}^{n} \frac{\lambda_{j,n} z_j}{\sqrt{z_j^2 + t^2}} \tag{11.107}$$

as an approximation of the Bessel function $J_0(t)$.

Using Table 11.3 we compute the sum in (11.107) for $f(t)$ with $n = 5$. The results are given in the middle column of Table 11.4, accompanied by values of $J_0(t)$ in the last column. ∎

11.5.2 Deforming the contour

In the previous example we have seen that the results of the complex Gauss quadrature are quite good for small values of t; see Table 11.4. For larger values, say $t \geq 2$, the method gives poor results with this number of nodes and weights ($n = 5$).

We give an example in which a much better representation of the integral is obtained by deforming the contour of integration and by using a proper value of c in the complex integral in (11.98). After selecting this new contour, the trapezoidal rule can be used for numerical quadrature. As explained in Chapter 5 this method may be very efficient for evaluating a class of integrals with analytic integrands.

Example 11.11 (the complementary error function). We have the Laplace transform pair (see [2, eq. (29.3.83)])

$$F(s) = \frac{1}{s} e^{-k\sqrt{s}} = \int_0^\infty e^{-st} \operatorname{erfc} \frac{k}{2\sqrt{t}} \, dt,$$
$$\operatorname{erfc} \frac{k}{2\sqrt{t}} = \frac{1}{2\pi i} \int_{c-i\infty}^{c+i\infty} e^{st-k\sqrt{s}} \frac{ds}{s}, \tag{11.108}$$

where in this case $c > 0$. We take $k = 2\lambda$ and $t = 1$, which gives

$$\operatorname{erfc} \lambda = \frac{1}{2\pi i} \int_{c-i\infty}^{c+i\infty} e^{s-2\lambda\sqrt{s}} \frac{ds}{s}, \tag{11.109}$$

and we assume that $\lambda > 0$. When λ is large the integral becomes exponentially small, and application of the quadrature rule of §11.5.1 is useless.

With the transformation $s = \lambda^2 t$, (11.109) becomes

$$\operatorname{erfc} \lambda = \frac{1}{2\pi i} \int_{c-i\infty}^{c+i\infty} e^{\lambda^2(t-2\sqrt{t})} \frac{dt}{t}. \tag{11.110}$$

When we take $c = 1$ the path runs through the saddle point at $t = 1$, where the exponential function of the integrand has the value $e^{-\lambda^2}$, which corresponds to the main term in the asymptotic estimate

$$\operatorname{erfc} \lambda \sim \frac{e^{-\lambda^2}}{\sqrt{\pi}\lambda}, \quad \lambda \to \infty. \tag{11.111}$$

Because the convergence at $\pm i\infty$ along the vertical through $t = 1$ is rather poor, the next step is to deform the contour into a new contour that terminates in the left half-plane, with $\Re t \to -\infty$.

In fact many contours are suitable, but there is only one contour through $t = 1$ on which no oscillations occur. That contour, the steepest descent path, is given by $\Im(t - 2\sqrt{t}) = 0$, or in polar coordinates $t = re^{i\theta}$ we have $r = \sec^2(\frac{1}{2}\theta)$. See Figure 11.4. This gives, by integrating with respect to $\theta \in [-\pi, \pi]$,

$$\operatorname{erfc} \lambda = \frac{e^{-\lambda^2}}{2\pi} \int_{-\pi}^{\pi} e^{-\lambda^2 \tan^2(\frac{1}{2}\theta)} \, d\theta. \tag{11.112}$$

As discussed in §5.2.3, the trapezoidal rule is exceptionally accurate in this case.

Table 11.5 gives the results of applying the composite trapezoidal rule with step size h; n indicates the number of function values in the rule that are larger than 10^{-15} (we exploit the fact that the integrand is even). All digits shown in the approximation in the final row are correct. ∎

11.5. Numerical inversion of Laplace transforms

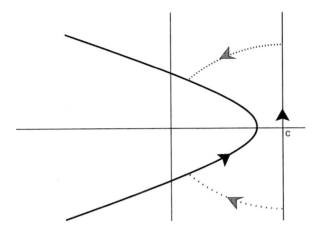

Figure 11.4. *The new contour of integration for* (11.110) *has the shape of a parabola.*

Table 11.5. *Composite trapezoidal rule for the integral in* (11.112) *with* $\lambda = 10$.

h	erfc λ	n
0.25	0.20949 49432 96679 10^{-44}	5
0.20	0.20886 11645 34559 10^{-44}	6
0.15	0.20884 87588 72946 10^{-44}	8
0.10	0.20884 87583 76254 10^{-44}	11

When $F(s)$ in (11.98) has singularities or poles, a straightforward and optimal choice of the path of integration, as in the above example, might not be easy to find. For example, for the Laplace transform pair in (11.106), a suitable path can be chosen not by saddle point analysis, but by deforming the original contour in the s-plane around the branch cuts from $\pm i$ to $-\infty$. This requires extra analysis when choosing the proper values of the function $F(s) = 1/\sqrt{s^2 + 1}$ around the branch cuts.

We omit the details of this exercise and give the result

$$J_0(t) = \frac{1}{\pi}\left[e^{i(t-\frac{1}{4}\pi)}\int_0^\infty \frac{e^{-st}\,ds}{\sqrt{s(2+is)}} + e^{-i(t-\frac{1}{4}\pi)}\int_0^\infty \frac{e^{-st}\,ds}{\sqrt{s(2-is)}}\right], \qquad (11.113)$$

whose representation has the proper oscillatory factors that describe the behavior for large values of t of this Bessel function.

When less information is available on the function $F(s)$ a less optimal contour may be chosen. For example, we can take a parabola or a hyperbola that terminates in the left half-plane at $-\infty$—for example, when we write $s = u + iv$, and consider the parabola defined by $u = p - qv^2$, and integrate with respect to v. When we choose p and q properly all singularities of $F(s)$ may remain inside the contour (unless $F(s)$ has an infinite number of singularities up to $+i\infty$ and/or $-i\infty$).

Further details can be found in [155, 190, 208]. Recent investigations are discussed in [231].

11.5.3 Using Padé approximations

In [143, §16.4] several examples are given for Laplace transform inversion by using rational approximations $R(s)$ of F in (11.98). Then $f(t)$ is approximated by

$$f(t) \approx \frac{1}{2\pi i} \int_{c-i\infty}^{c+i\infty} e^{st} R(s)\, ds. \tag{11.114}$$

Assuming no double roots in the denominator of $R(s)$ and expanding $R(s)$ in partial fractions

$$R(s) = \sum_j \frac{A_j}{s+s_j}, \tag{11.115}$$

the right-hand side can be evaluated as a sum of exponential functions. A possible choice of $R(s)$ is a Padé approximant (see §9.2) to $F(s)$ if $F(s)$ has a known power series expansion, for example, for large s.

In many applications the function $F(s)$ is known analytically in terms of special functions of which several forms of rational approximations are known. Luke gives a detailed treatment of the solution of a partial differential equation in which (see [143, p. 259])

$$F(s) = \frac{e^{rs} K_m(rs)}{se^s K'_m(s)}, \tag{11.116}$$

where $K_\nu(z)$ denotes the modified Bessel function. Luke [143, p. 229] also gives details on rational approximations of $K_\nu(z)$ (which are based on the large z behavior of this function), and in this way the inversion problem is numerically solved.

Part IV

Software

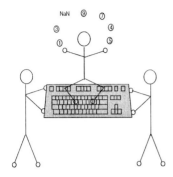

Chapter 12
Associated Algorithms

Computers are useless. They can only give you answers.
—Pablo Picasso

You can't always get what you want.
—Mick Jagger and Keith Richards, *Let It Bleed*, 1969

12.1 Introduction

As an illustration of the different methods for computing special functions, we now describe algorithms for the computation of a number of selected functions. The corresponding routines can be downloaded from the web page

http://functions.unican.es

These routines are based on algorithms which have been published by the authors in *ACM Transactions on Mathematical Software* or *Computer Physics Communications*. A description of the available modules for computing special functions is shown in Table 12.1.

A complete survey of the available software (up to 1993) for the computation of special functions can be found in [140]. Among the mathematical libraries including software for computing special functions, we could mention CALGO [27], SLATEC [199], CERN [32], IMSL [109], and NAG [156]. On the other hand, *Mathematica* [149], MATLAB [150], and Maple [147] are examples of interactive systems also including software for such purposes.

In the following sections, we provide a brief description of the algorithms. But before that, we make a few remarks on basic terminology and about the design and testing of software for special functions.

Table 12.1. *Description of the modules for computing functions and associated values.*

Module	Description
AiryScorerC	Airy and Scorer functions of complex arguments and their derivatives
AsocLegendre	Legendre functions of integer and half-integer degrees
BesselIK	Modified Bessel functions of integer and half-integer orders
KiaLia	Modified Bessel functions of purely imaginary orders
Parabolic	Parabolic cylinder functions
ZerosBessel	Zeros of the first kind Bessel function $J_\nu(x)$

12.1.1 Errors and stability: Basic terminology

Although we assume that the reader has a basic familiarity with computing and/or programming concepts, it is convenient to state here some standard principles.

Floating-point arithmetic

For representing numbers and doing manipulations with them, most computers use the binary number system. In connection with this point, having a standard for representing numbers, agreed upon and accepted by most computer manufacturers, was a key step for improving the portability between computers. This was the case of the IEEE-754 standard [112] for base-2 floating-point arithmetic.

A computer works with *floating-point numbers* in terms of a sign, an exponent (E), and a mantissa (M):

$$x = \pm M \times 2^E, \quad \begin{array}{ll} M = 0 \text{ or } 1 \leq M < 2, & \text{for } E > E_{\min}, \\ 0 < M < 1, & \text{for } E = E_{\min}. \end{array} \quad (12.1)$$

Values with $E = E_{\min}$ are called *subnormal*. Table 12.2 shows the main characteristics of the *IEEE floating-point formats*. In particular, we mention the *overflow* and the *underflow* thresholds, which are the largest and smallest, respectively, positive numbers which can be represented in floating-point format, and the *machine-ϵ*, which is the difference between 1 and the next larger floating-point number.

Errors. Stability and condition

If x_A is an approximation to the true value x_T, then a measure of the accuracy of this approximation is provided by the *absolute error* and the *relative error*, which are defined as follows:

$$\begin{aligned} E_{\text{abs}} &= |x_T - x_A|, \\ E_{\text{rel}} &= \left|1 - \frac{x_A}{x_T}\right| \quad \text{if } x_T \neq 0. \end{aligned} \quad (12.2)$$

Among the possible sources of errors arising in a numerical computation, we mention the following relevant possibilities:

12.1. Introduction

Table 12.2. *Characteristics of IEEE floating-point formats.* (∗) *Hidden bit convention.* (∗∗) *Minimum subnormal positive number.*

	Single precision	Double precision
Number of bits in sign	1	1
Number of bits in exponent	8	11
Number of bits in mantissa	$23 + 1^{(*)}$	$52 + 1^{(*)}$
Range for the exponent	-126 to 127	-1022 to 1023
Overflow threshold	$\sim 10^{38}$	$\sim 10^{308}$
Underflow threshold$^{(**)}$	$\sim 10^{-45}$	$\sim 10^{-324}$
machine-ϵ	$\sim 10^{-8}$	$\sim 10^{-16}$

- Errors related to the fixed-length representation of floating-point numbers: roundoff errors, loss of significant digits due to cancellations (when computing the difference of nearly identical numbers), overflow or underflow problems.

- Discretization errors, which arise in the approximation of a continuous problem by a discrete sequence.

- Truncation errors, which are those made in truncating an infinite process.

Two important words in numerical analysis are *condition* and *stability*. These are related concepts, although the first refers to problems and the second to methods.

A problem which is inherently sensitive to roundoff errors and other uncertainties is said to be *ill conditioned* or *poorly conditioned*. As an example, in Chapter 4, the condition of recursive processes for computing special functions was analyzed and illustrated with examples.

On the other hand, the *stability of a numerical method* reflects the effect of roundoff errors in the implementation of the method: a method will be *numerically unstable* if the effects of the roundoff errors are magnified by the method.

A numerical method for an *ill-conditioned* problem will be *numerically unstable*. However, the opposite implication is not necessarily true.

12.1.2 Design and testing of software for computing functions: General philosophy

The rules for designing software for computing special functions are, of course, the same as those that apply to any kind of mathematical software: portability, efficiency, and accuracy must be the guidelines for building the algorithms. Portability can be guaranteed by using the programming standards. CPU times spent by the routines will acknowledge an efficient implementation of the algorithms. Related to this point, when two or more methods providing the same accuracy are available for a certain range of parameters of the function, the fastest method will be the reasonable choice.

The accuracy can be controlled a priori by using rigorous error bounds (when available) of the implemented methods of computation. Alternative strategies are as follows:

- Comparison with existing algorithms. For this kind of test, the verified accuracy is limited by the accuracy of the algorithm used for comparison.

- Consistency between different methods of evaluation of the function (with overlapping regions of validity) within a single algorithm.

- Verification of functional relations such as Wronskians, special values of the function (zeros, for example), sum relations, and so on. Of course, a crucial point is to be sure about the conditioning of the functional relations used for verification. For example, for testing the accuracy of an algorithm for computing complex Airy functions $\mathrm{Ai}(z)$ and $\mathrm{Bi}(z)$, the Wronskian

$$\mathcal{W}[\mathrm{Ai}(z), \mathrm{Bi}(z)] = \pi^{-1} \qquad (12.3)$$

is only well conditioned in the sector $|\mathrm{ph}\, z| \leq \pi/3$, where $\mathrm{Ai}(z)$ is recessive when $|z| \to \infty$. Outside this sector, both functions are dominant and two large quantities are subtracted in the Wronskian to give $1/\pi$.

- Consideration of extended precision algorithms. This type of comparison is only a test of consistency of the particular methods used in the algorithm under consideration. This type of test can be used to estimate the accuracy of the algorithm assuming that the methods work well.

- Use of local Taylor series. Assuming the value of a function and its first derivative are known at a given point x, if the function satisfies a "simple" second order ordinary differential equation, the successive derivatives of the function can be obtained by differentiating the ordinary differential equation repeatedly. In this way, it is possible to build a Taylor series which can be used to verify the values of the function and its first derivative in a neighborhood of the point x in question. See §9.5.

12.1.3 Scaling the functions

Last but not least, an important issue in the design of algorithms for computing functions is the possibility of evaluating *scaled functions*. If, for example, a function $f(z)$ increases exponentially for large $|z|$, the factorization of the exponential term and the computation of a scaled function $\tilde{f}(z)$ (without the exponential term) can be used to avoid degradations in the accuracy and overflow problems as z increases.

Then, the appropriate scaling of functions can be used to enlarge the range of computation. In addition, the accuracy of the computed scaled function is usually higher than that of the original function. For an earlier example of the use of the scaled gamma function for computing Stirling numbers, we refer to §11.3.2. See also Chapter 1, where the scaling of Airy functions is discussed.

Table 12.3. *Definition of scaled homogeneous* ($\widetilde{\mathrm{Ai}}(z)$, $\widetilde{\mathrm{Bi}}(z)$) *and inhomogeneous* ($\widetilde{\mathrm{Gi}}(z)$, $\widetilde{\mathrm{Hi}}(z)$) *Airy functions. We denote* $\zeta = \frac{2}{3}z^{\frac{3}{2}}$.

Function	Definition	Range
$\widetilde{\mathrm{Ai}}(z)$	$e^{\zeta}\mathrm{Ai}(z)$	for all z
$\widetilde{\mathrm{Bi}}(z)$	$e^{-\zeta}\mathrm{Bi}(z)$ $e^{\zeta}\mathrm{Bi}(z)$	if $|\mathrm{ph}\,z| < \pi/3$ if $|\mathrm{ph}\,z| \geq \pi/3$
$\widetilde{\mathrm{Hi}}(z)$	$e^{-\zeta}\mathrm{Hi}(z)$	if $\mathrm{ph}\,z \in [-\frac{1}{3}\pi, \frac{1}{3}\pi]$
$\widetilde{\mathrm{Gi}}(z)$	$e^{\zeta}\mathrm{Gi}(z)$	if $\mathrm{ph}(-z) \in [-\frac{2}{3}\pi, \frac{2}{3}\pi]$

12.2 Airy and Scorer functions of complex arguments

12.2.1 Purpose

The module **AiryScorerC** includes routines for the computation of the homogeneous (Ai(z), Bi(z)) and inhomogeneous (Gi(z), Hi(z)) *Airy functions* of complex arguments and their first derivatives. The functions Gi(z) and Hi(z) are also called *Scorer functions*. These algorithms are based on the references [88] (Airy functions) and [89] (Scorer functions).

The Airy functions, Ai(z) and Bi(z), are linearly independent solutions of the differential equation

$$w'' - zw = 0. \tag{12.4}$$

The Scorer functions, Gi(z) and Hi(z), are solutions of the inhomogeneous Airy equations

$$\begin{aligned} w'' - zw &= -\tfrac{1}{\pi}, & w(z) &= \mathrm{Gi}(z), \\ w'' - zw &= \tfrac{1}{\pi}, & w(z) &= \mathrm{Hi}(z). \end{aligned} \tag{12.5}$$

Because of the exponential behavior for large $|z|$ of the Airy functions (and only in particular regions of the complex plane in the case of Scorer functions), the routines have the option of computing scaled functions in order to enlarge the range of computation.

In Table 12.3, we show the definition used for scaled Airy functions and Scorer functions. We denote $\zeta = \frac{2}{3}z^{\frac{3}{2}}$.

12.2.2 Algorithms

ALGORITHM 12.1. Homogeneous complex Airy functions.

The algorithm for the computation of the function Ai(z) has the following structure.

Scheme for Ai(z)

- For z such that $-2.6 < x < 1.3$, $-3 \leq y \leq 3$ ($z = x + iy$), compute Maclaurin series (see [2, eq. (10.4.2)]).

- For z such that $|\text{ph } z| \leq 2\pi/3$, $|z| < 15$, and where Maclaurin series are not used, compute the integral representation

$$\text{Ai}(z) = a(z) \int_0^\infty \left(2 + \frac{t}{\zeta}\right)^{-\frac{1}{6}} t^{-\frac{1}{6}} e^{-t} dt, \quad a(z) = \frac{1}{\sqrt{\pi}(48)^{\frac{1}{6}} \Gamma(5/6)} e^{-\zeta} \zeta^{-\frac{1}{6}} \quad (12.6)$$

by using Gauss–Laguerre quadrature with parameter $\alpha = -1/6$ (40-point Gauss for double precision).

As noted in Chapter 5, near the Stokes line $\text{ph } z = 2\pi/3$ this integral becomes singular. This problem can be avoided by turning the path of integration over a given angle τ; namely, by considering the replacement $t \to t(1 + i \tan \tau)$ in (12.6),

$$\text{Ai}(z) = a(z) \left(\frac{e^{i\tau}}{\cos \tau}\right)^{\frac{5}{6}} \int_0^\infty \left(2 + \frac{t}{\tilde{\zeta}}\right)^{-\frac{1}{6}} e^{-it \tan(\tau)} t^{-\frac{1}{6}} e^{-t} dt, \quad (12.7)$$

where $\tilde{\zeta} = \cos \tau e^{-i\tau} \zeta$ and a convenient choice of τ is $\tau = 3(\text{ph } z - \pi/2)/2$. At $\text{ph } z = \pi/2$ the expressions (12.6) and (12.7) coincide, and for $\pi/2 < \text{ph } z < 2\pi/3$ the new integrands are free of singularities.

After turning the path of integration we can safely use the integral (12.6) for $0 \leq \text{ph } z \leq \pi/2$ and (12.7) for $\pi/2 < \text{ph } z \leq 2\pi/3$ for not too small $|z|$. Of course, as z becomes large the integrands become smoother and the Gauss–Laguerre quadrature tends to work better.

- For z such that $|\text{ph } z| \leq 2\pi/3$ and $|z| > 15$, compute the asymptotic expansion

$$\text{Ai}(z) \sim \frac{\pi^{-\frac{1}{2}} z^{-\frac{1}{4}}}{2} e^{-\zeta} \sum_{k=0}^\infty (-1)^k c_k \zeta^{-k}, \quad c_0 = 1, \quad c_k = \frac{\Gamma(3k + 1/2)}{54^k k! \Gamma(k + 1/2)}. \quad (12.8)$$

- For z such that $2\pi/3 < |\text{ph } z| \leq \pi$ and where Maclaurin series are not used, compute the connection formula

$$\text{Ai}(z) = -e^{-2\pi i/3} \text{Ai}(e^{-2\pi i/3} z) - e^{2\pi i/3} \text{Ai}(e^{2\pi i/3} z). \quad (12.9)$$

This connection formula must be modified for the computation of $\widetilde{\text{Ai}}(z)$,

$$\widetilde{\text{Ai}}(z) = -e^{-2\pi i/3} e^{\frac{4}{3} z^{\frac{3}{2}}} \widetilde{\text{Ai}}(ze^{-2\pi i/3}) - e^{2\pi i/3} \widetilde{\text{Ai}}(ze^{2\pi i/3}). \quad (12.10)$$

The conjugation property $\text{Ai}(x + iy) = \overline{\text{Ai}(x - iy)}$ is considered for $\Im z < 0$.

12.2. Airy and Scorer functions of complex arguments

Scheme for Bi(z)

The function Bi(z) is computed through the following two connection formulas:

$$\begin{aligned} \text{Bi}(z) &= i\text{Ai}(z) + 2e^{-i\pi/6}\text{Ai}(e^{-2\pi i/3}z), \\ \text{Bi}(z) &= e^{i\pi/6}\text{Ai}(e^{2\pi i/3}z) + e^{-i\pi/6}\text{Ai}(e^{-2\pi i/3}z). \end{aligned} \quad (12.11)$$

We apply the first connection formula for $0 < |\text{ph } z| \leq 2\pi/3$ and the second one for the rest of the complex plane. In this way, both relations are numerically satisfactory. The connection formulas in (12.11) must be modified when computing scaled functions. In that case, we use the following expressions:

$$\begin{aligned} \widetilde{\text{Bi}}(z) &= ie^{-2\zeta}\widetilde{\text{Ai}}(z) + 2e^{-i\pi/6}\widetilde{\text{Ai}}(e^{-2\pi i/3}z) & \text{when } 0 \leq \text{ph } z < \pi/3, \\ \widetilde{\text{Bi}}(z) &= i\widetilde{\text{Ai}}(z) + 2e^{-i\pi/6}e^{2\zeta}\widetilde{\text{Ai}}(e^{-2\pi i/3}z) & \text{when } \pi/3 \leq \text{ph } z \leq 2\pi/3, \quad (12.12) \\ \widetilde{\text{Bi}}(z) &= e^{i\pi/6}\widetilde{\text{Ai}}(e^{2\pi i/3}z) + e^{-i\pi/6}e^{2\zeta}\widetilde{\text{Ai}}(e^{-2\pi i/3}z) & \text{when } 2\pi/3 < \text{ph } z \leq \pi, \end{aligned}$$

where $\zeta = \frac{2}{3}z^{\frac{3}{2}}$ as in Table 12.3.

ALGORITHM 12.2. Scorer functions.

Scheme for Hi(z)

- For $|z| \leq 1.5$, compute the power series

$$\text{Hi}(z) = \frac{1}{\pi}\sum_{k=0}^{\infty} h_k \frac{z^k}{k!}, \quad h_k = \int_0^{\infty} t^k e^{-\frac{1}{3}t^3} dt = 3^{\frac{1}{3}(k-2)}\Gamma\left(\frac{k+1}{3}\right). \quad (12.13)$$

- For z such that $0 \leq \text{ph } z < \frac{2}{3}\pi$ and where power series are not used, compute the connection formula

$$\text{Hi}(z) = e^{2\pi i/3}\text{Hi}\left(ze^{2\pi i/3}\right) + 2e^{-\pi i/6}\text{Ai}\left(ze^{-2\pi i/3}\right). \quad (12.14)$$

- For z such that $\frac{2}{3}\pi \leq \text{ph } z \leq \pi$, $|z| < 20$, and where power series are not used, compute the integral representation

$$\text{Hi}(z) = \frac{1}{\pi}\int_0^{\infty} e^{zt - \frac{1}{3}t^3} dt. \quad (12.15)$$

This integral representation needs to be modified in order to avoid oscillations in the integrand. We do this by writing

$$z = x + iy, \quad t = u + iv, \quad \phi(t) = \frac{1}{3}t^3 - zt = \phi_r(u, v) + i\phi_i(u, v), \quad (12.16)$$

where

$$\phi_r(u, v) = \frac{1}{3}u^3 - uv^2 - xu + yv, \quad \phi_i(u, v) = u^2 v - \frac{1}{3}v^3 - xv - yu. \quad (12.17)$$

Then, we integrate along the contour defined by $\phi_i(u, v) = \phi(0, 0) = 0$. We obtain

$$\operatorname{Hi}(z) = \frac{1}{\pi} \int_0^\infty e^{-\phi_r(u, v(u))} h(u) \, du, \qquad (12.18)$$

where $v(u)$ is the solution of $\phi_i(u, v) = 0$, and

$$h(u) = 1 + i\frac{dv(u)}{du} = 1 + i\frac{2uv - y}{v^2 - u^2 + x}. \qquad (12.19)$$

In this way the integral becomes nonoscillating. Near the upper boundary of the sector ph $z \in [\frac{2}{3}\pi, \pi]$, the relation between v and u becomes singular; in this case, it is better to use a different relation. We use a simple relation that fits the exact solution of $\phi_i(u, v) = 0$ at $u = 0$ and at $u = \infty$ by writing

$$v(u) = -\frac{y}{x}\frac{u}{u^2 + 1}, \quad \frac{dv(u)}{du} = -\frac{y}{x}\frac{1 - u^2}{(u^2 + 1)^2}. \qquad (12.20)$$

This gives

$$\operatorname{Hi}(z) = \frac{1}{\pi} \int_0^\infty e^{-\phi_r(u, v(u)) - i\phi_i(u, v(u))} h(u) \, du, \qquad (12.21)$$

where again $h(u) = 1 + i \, dv(u)/du$ with the new expression for the derivative.

These representations are efficiently computed using the trapezoidal rule.

- For z such that $\frac{2}{3}\pi \leq \operatorname{ph} z \leq \pi$, $|z| \geq 20$, compute the asymptotic expansion

$$\operatorname{Hi}(z) \sim -\frac{1}{\pi z}\left[1 + \frac{1}{z^3}\sum_{s=0}^\infty \frac{(3s + 2)!}{s!(3z^3)^s}\right]. \qquad (12.22)$$

Scheme for **Gi(z)**

- For $|z| \leq 1.5$, compute the power series

$$\operatorname{Gi}(z) = \frac{1}{\pi}\sum_{k=0}^\infty g_k \frac{z^k}{k!}, \quad g_k = -h_k \cos\frac{2}{3}\pi(k + 1), \qquad (12.23)$$

h_k being the coefficients given in (12.13).

- For z such that $\frac{2}{3}\pi \leq \operatorname{ph} z \leq \pi$ and where power series are not used, compute the connection formula

$$\operatorname{Gi}(z) = \operatorname{Bi}(z) - \operatorname{Hi}(z). \qquad (12.24)$$

- For z such that $|\operatorname{ph} z| < \frac{1}{3}\pi - 0.3$, $|z| > 30$, compute the asymptotic expansion

$$\operatorname{Gi}(z) \sim -\frac{1}{\pi z}\left[1 + \frac{1}{z^3}\sum_{s=0}^\infty \frac{(3s + 2)!}{s!(3z^3)^s}\right]. \qquad (12.25)$$

- For z such that $0 \leq \mathrm{ph}\, z < \frac{2}{3}\pi$ and where power series or asymptotic expansions are not used, compute the connection formula

$$\mathrm{Gi}(z) = -e^{2\pi i/3}\mathrm{Hi}\left(ze^{2\pi i/3}\right) + i\mathrm{Ai}(z). \qquad (12.26)$$

The scaling of the Scorer function $\mathrm{Hi}(z)$ is relevant only in the sector $\mathrm{ph}\, z \in [-\frac{1}{3}\pi, \frac{1}{3}\pi]$, where the function increases exponentially for large $|z|$. In the case of the Scorer function $\mathrm{Gi}(z)$, scaling is important in the sector $\mathrm{ph}(-z) \in [-\frac{2}{3}\pi, \frac{2}{3}\pi]$. See Table 12.3.

For computing the scaled function $\widetilde{\mathrm{Hi}}(z)$, we use

$$\widetilde{\mathrm{Hi}}(z) = e^{2\pi i/3}e^{-(2/3)z^{\frac{3}{2}}}\mathrm{Hi}(ze^{2\pi i/3}) + 2e^{-\pi i/6}\widetilde{\mathrm{Ai}}(ze^{-2\pi i/3}), \qquad (12.27)$$

where $\widetilde{\mathrm{Ai}}$ is the scaled Airy function $\mathrm{Ai}(z)$. In the remaining part of the plane, $\mathrm{Hi}(z)$ is of order $\mathcal{O}(1/z)$. For $\widetilde{\mathrm{Gi}}(z)$ we use

$$\begin{aligned}\widetilde{\mathrm{Gi}}(z) &= -e^{2\pi i/3}e^{(2/3)z^{\frac{3}{2}}}\mathrm{Hi}(ze^{2\pi i/3}) + i\widetilde{\mathrm{Ai}}(z) & \text{when } \pi/3 \leq |\mathrm{ph}\, z| \leq 2\pi/3, \\ \widetilde{\mathrm{Gi}}(z) &= -e^{(2/3)z^{\frac{3}{2}}}\mathrm{Hi}(z) + \widetilde{\mathrm{Bi}}(z) & \text{when } 2\pi/3 < |\mathrm{ph}\, z| \leq \pi,\end{aligned} \qquad (12.28)$$

where $\widetilde{\mathrm{Bi}}$ is the scaled Airy function $\mathrm{Bi}(z)$.

12.3 Associated Legendre functions of integer and half-integer degrees

12.3.1 Purpose

The module **AsocLegendre** computes *associated Legendre functions* (of first and second kinds) of integer and half-integer degrees. Associated Legendre functions of integer degrees are also known as *prolate spheroidal harmonics* when the argument of the functions is real (x, $x > 1$) and *oblate spheroidal harmonics* when the argument is purely imaginary (ix, $x > 0$). On the other hand, associated Legendre functions of half-integer degrees and arguments larger than one are known as *toroidal harmonics*. These functions are the natural basis to solve the Dirichlet problem for domains bounded by spheroids and tori, respectively.

This module is based on the references [82] (prolate and oblate harmonics) and [196, 83] (toroidal harmonics).

Associated Legendre functions $P_\nu^m(z)$, $Q_\nu^m(z)$ are solutions of the Legendre equation

$$(1-z^2)\frac{d^2u}{dz^2} - 2z\frac{du}{dz} + \left[\nu(\nu+1) - \frac{m^2}{1-z^2}\right]u = 0. \qquad (12.29)$$

We will concentrate on the evaluation of $\{P_n^m(x), Q_n^m(x)\}$, x real and $x > 1$, n, m integers and $n \geq m$ (prolate spheroidal harmonics), $\{P_n^m(ix), Q_n^m(ix)\}$, x real and $x > 0$, n, m integers and $n \geq m$ (oblate spheroidal harmonics), and $\{P_{n-1/2}^m(x), Q_{n-1/2}^m(x)\}$, x real and $x > 1$, n, m integers (toroidal harmonics).

In the case of oblate spheroidal harmonics, the algorithm computes an equivalent set of functions $R_n^m(x)$, $T_n^m(x)$, which are defined as follows:

$$\begin{aligned} R_n^m(x) &= \exp\left(-i\frac{\pi n}{2}\right) P_n^m(ix), \\ T_n^m(x) &= i \exp\left(i\frac{\pi n}{2}\right) Q_n^m(ix). \end{aligned} \qquad (12.30)$$

These functions have the benefit of being real valued.

The algorithm also has the option of computing scaled functions in order to enlarge the range of computation. The scaled functions are defined as follows.

Scaled prolate spheroidal harmonics

$$\{\widetilde{P}_n^m(x),\ \widetilde{Q}_n^m(x)\} = \left\{\frac{P_n^m(x)}{(2m-1)!!},\ \frac{Q_n^m(x)}{(2m)!!}\right\}. \qquad (12.31)$$

The double factorial of n ($n!!$) is defined as follows:

$$n!! = \begin{cases} n(n-2)\cdots 5\cdot 3\cdot 1, & n > 0 \text{ odd}, \\ n(n-2)\cdots 6\cdot 4\cdot 2, & n > 0 \text{ even}, \\ 1, & n = -1, 0. \end{cases} \qquad (12.32)$$

Scaled oblate spheroidal harmonics

$$\{\widetilde{R}_n^m(x),\ \widetilde{T}_n^m(x)\} = \left\{\frac{R_n^m(x)}{(2m-1)!!},\ \frac{T_n^m(x)}{(2m)!!}\right\}. \qquad (12.33)$$

Scaled toroidal harmonics

$$\{\widetilde{P}_{n-1/2}^m(x),\ \widetilde{Q}_{n-1/2}^m(x)\} = \left\{\frac{P_{n-1/2}^m(x)}{\Gamma(m+1/2)},\ \frac{Q_{n-1/2}^m(x)}{\Gamma(m+1/2)}\right\}. \qquad (12.34)$$

12.3.2 Algorithms

ALGORITHM 12.3. Prolate spheroidal harmonics.
The algorithm evaluates prolate spheroidal harmonics for a given order m, from the lowest degree $n = m$ to a maximum degree $n = m + N$ in a single run.

Scheme for $\{P_n^m(x),\ Q_n^m(x)\},\ n \geq m,\ x > 1$

- Compute the initial values $P_m^m(x) = (2m-1)!!(x^2-1)^{m/2}$ and $P_{m+1}^m(x) = x(2m+1)P_m^m(x)$.

- Compute $P_n^m(x)$ for $n = m, \ldots, m+N$ by forward application of the recurrence relation

$$(n-m+1)P_{n+1}^m(x) - (2n+1)xP_n^m(x) + (n+m)P_{n-1}^m(x) = 0. \qquad (12.35)$$

12.3. Associated Legendre functions of integer and half-integer degrees

The starting values of this TTRR are $P_m^m(x)$ and $P_{m+1}^m(x)$.

The same TTRR is also satisfied by $Q_n^m(x)$. As discussed in Chapter 4, the functions $Q_n^m(x)$ are the minimal solution of the TTRR, while the functions $P_n^m(x)$ are a dominant solution. Hence, (12.35) can be applied forward for the computation of $P_n^m(x)$.

- Compute the ratio $Q_{m+N}^m(x)/Q_{m+N-1}^m(x)$ by means of the continued fraction representation, which is obtained by iterating the expression (for $\nu = m + N$)

$$H_Q(\nu, m, x) = \frac{Q_\nu^m(x)}{Q_{\nu-1}^m(x)} = \cfrac{1}{\cfrac{2\nu+1}{\nu+m} x - \cfrac{\nu-m+1}{\nu+m} \cfrac{Q_{\nu+1}^m}{Q_\nu^m}}. \quad (12.36)$$

- Evaluate $Q_{n+N-1}^m(x)$ in terms of the already computed values $P_{m+N}^m(x)$, $P_{m+N-1}^m(x)$, $H_Q(m+N, m, x)$ and using the relation (for $\nu = m + N$)

$$P_\nu^m(x) Q_{\nu-1}^m(x) - P_{\nu-1}^m(x) Q_\nu^m(x) = \frac{\Gamma(\nu+m)}{\Gamma(\nu-m+1)}(-1)^m. \quad (12.37)$$

The resulting expression for $Q_{m+N-1}^m(x)$ is given by

$$Q_{m+N-1}^m(x) = \frac{(2m+N-1)!}{N!} \frac{(-1)^m}{P_{m+N}^m(x) - P_{m+N-1}^m(x) H_Q(m+N, m, x)}. \quad (12.38)$$

And then $Q_{m+N}^m(x) = Q_{m+N-1}^m(x) H_Q(m+N, m, x)$.

- Compute $Q_n^m(x)$ from $n = m + N - 2$ to $n = m$ by backward application of the recurrence relation

$$(n - m + 1) Q_{n+1}^m(x) - (2n+1) x Q_n^m(x) + (n+m) Q_{n-1}^m(x) = 0. \quad (12.39)$$

The starting values of this recurrence relation are $Q_{m+N}^m(x)$ and $Q_{m+N-1}^m(x)$.

ALGORITHM 12.4. Oblate spheroidal harmonics.

Specifically, the algorithm computes the equivalent set of functions given in (12.30). Oblate spheroidal harmonics for a given order m, from the lowest degree $n = m$ to a maximum degree $n = m + N$, are evaluated in a single run. From (12.35), the recurrence relations for $R_n^m(x)$ and $T_n^m(x)$ are written as

$$(n - m + 1) R_{n+1}^m(x) = (2n+1) x R_n^m(x) + (n+m) R_{n-1}^m(x), \quad (12.40)$$

$$(n + m) T_{n-1}^m(x) = (n - m + 1) T_{n+1}^m(x) + (2n+1) x T_n^m(x). \quad (12.41)$$

The function $R_n^m(x)$ is a dominant solution of the TTRR (12.40), while the function $T_n^m(x)$ is the minimal solution of the TTRR (12.41).

For simplicity, let us define

$$r_s = R_{m+s}^m(x), \quad t_s = T_{m+s}^m(x). \quad (12.42)$$

Scheme for $\{r_s(x), t_s(x)\}$, $s = 0, 1, 2, \ldots, x > 0$

- Compute the initial values $r_0(x) = (2m-1)!!(x^2+1)^{m/2}$ and $r_1(x) = x(2m+1)r_0(x)$.

- Compute $r_s(x)$ for $s = 0, \ldots, N$ by forward application of the recurrence relation

$$(s+1)r_{s+1}(x) = (2s+2m+1)xr_s(x) + (s+2m)r_{s-1}(x). \tag{12.43}$$

The starting values of this recurrence relation are $r_0(x)$ and $r_1(x)$.

- Compute the ratio $t_N(x)/t_{N-1}(x)$ by means of a continued fraction representation, which is obtained by iterating the expression (for $\nu = m + N$)

$$H_T(\nu, m, x) = \frac{t_N(x)}{t_{N-1}(x)} = \frac{T_\nu^m(x)}{T_{\nu-1}^m(x)} = \cfrac{1}{\frac{2\nu+1}{\nu+m}x + \frac{\nu-m+1}{\nu+m}\frac{T_{\nu+1}^m}{T_\nu^m}}. \tag{12.44}$$

- Evaluate $t_{N-1}(x)$ in terms of the already computed values of $r_N(x)$, $r_{N-1}(x)$, $H_T(m+N, m, x)$ by means of the relation

$$t_{N-1}(x) = \frac{(2m+N-1)!}{N!}(-1)^m \frac{1}{r_N(x) + r_{N-1}(x)H_T(m+N, m, x)}. \tag{12.45}$$

And then $t_N = t_{N-1}H_T(m+N, m, x)$.

- Compute $t_s(x)$ from $s = N-2$ to $s = 0$ by backward application of the recurrence relation

$$(s+2m)t_{s-1}(x) = (s+1)t_{s+1}(x) + (2s+2m+1)xt_s(x). \tag{12.46}$$

The starting values of this recurrence relation are $t_N(x)$ and $t_{N-1}(x)$.

ALGORITHM 12.5. Toroidal harmonics.

The algorithm evaluates toroidal harmonics $\{P_{n-1/2}^m(x), Q_{n-1/2}^m(x)\}$ for $m = 0, \ldots, M$ and $n = 0, \ldots, N$ in a single run.

Scheme for $\{P_{n-1/2}^m(x), Q_{n-1/2}^m(x)\}$, $n, m \geq 0$, $x > 1$

○ For $1 < x < \sqrt{2}$ the functions are computed using the so-called *Dual Algorithm*.

○ For $x \geq \sqrt{2}$ the so-called *Primal Algorithm* is applied.

Primal Algorithm for $\{P_{n-1/2}^m(x), Q_{n-1/2}^m(x)\}$

- Compute the set $\{P_{\pm 1/2}^m(x), m = 0, \ldots, M\}$ by considering the following steps.

12.3. Associated Legendre functions of integer and half-integer degrees

(i) Evaluation of $Q^0_{-1/2}(x)$ and $Q^1_{-1/2}(x)$ through their relation with *elliptic integrals*,

$$Q^0_{-1/2}(x) = \sqrt{\frac{2}{x+1}} K\left(\sqrt{\frac{2}{x+1}}\right), \quad (12.47)$$

$$Q^1_{-1/2}(x) = \frac{-1}{\sqrt{2(x-1)}} E\left(\sqrt{\frac{2}{x+1}}\right). \quad (12.48)$$

These are computed by using algorithms from SLATEC [199], which are based on Carlson's algorithms for symmetric elliptic integrals (see also §11.4).

(ii) If $M > 1$, compute $Q^m_\nu(x)$ for $m = 2, \ldots, M$ by using the TTRR (in the particular case of $\nu = -1/2$) in the forward direction,

$$Q^{m+2}_\nu(x) + \frac{2(m+1)x}{(x^2-1)^{\frac{1}{2}}} Q^{m+1}_\nu(x) - (\nu - m)(\nu + m + 1) Q^m_\nu(x) = 0, \quad (12.49)$$

which is also satisfied by $P^m_\nu(x)$. For this TTRR, the function $Q^m_\nu(x)$ is a dominant solution, while $P^m_\nu(x)$ is the minimal one.

(iii) Compute $P^M_{-1/2}(x)$ by considering different methods depending on the values of x and M.

□ If $x \leq 20$ and $M/x > 3$, use the continued fraction representation obtained by iterating the expression (for $\nu = -1/2$)

$$H_P(\nu, m, x) = \frac{P^m_\nu(x)}{P^{m-1}_\nu(x)} = \frac{(\nu - m + 1)(\nu + m)}{\frac{2mx}{\sqrt{x^2-1}} + \frac{P^{m+1}_\nu(x)}{P^m_\nu(x)}}, \quad (12.50)$$

together with the relation

$$P^m_\nu(x) Q^{m+1}_\nu(x) - P^{m+1}_\nu(x) Q^m_\nu(x) = \frac{\Gamma(\nu + m + 1)}{\Gamma(\nu - m + 1)} \frac{(-1)^{m+1}}{\sqrt{x^2-1}}. \quad (12.51)$$

Both expressions are evaluated for $m = M$ and $\nu = -1/2$. This leads to

$$P^M_{-1/2}(x) = -\frac{\Gamma^2(M - 1/2)}{\pi \sqrt{x^2 - 1}} \frac{1}{\frac{Q^M_{-1/2}(x)}{H_P(-1/2, M, x)} - Q^{M-1}_{-1/2}(x)}. \quad (12.52)$$

□ If $x > 5$ and $M/x \leq 3$, use the series expansion

$$P^M_{-1/2}(x) = \frac{2(-1)^M}{\pi^{\frac{3}{2}}} \left(1 - \frac{1}{x^2}\right)^{M/2}$$

$$\times \sum_{r=0}^{\infty} \{\log(2x) - \psi(M + 2r + 1/2) + \psi(r + 1)\} \quad (12.53)$$

$$\times \frac{\Gamma(1/2 + M + 2r)}{\Gamma^2(r+1)} \frac{1}{(2x)^{2r+1/2}},$$

where ψ is the logarithmic derivative of the Γ function.

□ Where the previous approaches are not used, compute the asymptotic expansion

$$P^M_{-1/2}(x) \sim (-1)^M \frac{\Gamma(M+1/2)}{\pi^{\frac{3}{2}}} \xi^{-\frac{1}{2}} \sum_{l=0}^{1} K_l(M\alpha/2) \sum_{k=0}^{\infty} \frac{a_{2k+l}}{M^{2k+l}}, \quad (12.54)$$

with $\xi = (x-1)/2$ and $\alpha = \ln[(x+1)/(x-1)]$. $K_0(x)$ and $K_1(x)$ are modified Bessel functions and the first coefficients a_k are

$$\begin{aligned}
a_0 &= \sqrt{\alpha/(e^\alpha - 1)}, \\
a_1/a_0 &= -\tfrac{1}{48}\alpha + \tfrac{1}{2880}\alpha^3 - \tfrac{1}{120960}\alpha^5 + \cdots, \\
a_2/a_0 &= \tfrac{7}{7680}\alpha^2 - \tfrac{13}{322560}\alpha^4 + \cdots, \\
a_3/a_0 &= \tfrac{7}{1920}\alpha - \tfrac{571}{2580480}\alpha^3 + \tfrac{1697}{154828800}\alpha^5 + \cdots.
\end{aligned} \quad (12.55)$$

(iv) Compute $P^{M-1}_{-1/2}(x)$. The method of computation depends on the method used for $P^M_{-1/2}(x)$ in (iii).

□ If $P^M_{-1/2}(x)$ was computed using the continued fraction for $H_P(-1/2, M, x)$, then use this expression and the calculated value of $P^M_{-1/2}(x)$ (see (12.52)) to obtain

$$P^{M-1}_{-1/2} = \frac{P^M_{-1/2}(x)}{H_P(-1/2, M, x)}. \quad (12.56)$$

□ If $P^M_{-1/2}(x)$ was computed using a series or asymptotic expansion, then use the relation of (12.51) (for $m = M-1$, $\nu = -1/2$) and the calculated values of $Q^M_{-1/2}(x)$, $Q^{M-1}_{-1/2}(x)$, and $P^M_{-1/2}(x)$:

$$P^{M-1}_{-1/2} = \frac{1}{Q^M_{-1/2}(x)} \left[P^M_{-1/2}(x) Q^{M-1}_{-1/2}(x) - \frac{\Gamma^2(M-1/2)}{\pi\sqrt{x^2-1}} \right]. \quad (12.57)$$

(v) Compute $P^M_{1/2}(x)$ by using the continued fraction $H_Q(1/2, M, x)$ of (12.36), together with the relation given by (12.37) (for $\nu = 1/2$). It follows that

$$P^M_{1/2}(x) = P^M_{-1/2}(x) H_Q(1/2, M, x) \\
+ \frac{1}{Q^M_{-1/2}(x)} \frac{\Gamma(3/2+M)}{\Gamma(3/2-M)} \frac{(-1)^M}{(M+1/2)}. \quad (12.58)$$

(vi) Compute $P^{M-1}_{1/2}(x)$ by using the continued fraction representation for H_P in (12.50) (for $\nu = 1/2$) and the computed value of $P^M_{1/2}(x)$. That is,

$$P^{M-1}_{1/2}(x) = \frac{P^M_{1/2}(x)}{H_P(1/2, M, x)}. \quad (12.59)$$

(vii) Compute the set $\{P^m_{-1/2}, m = 0, \ldots, M\}$ by backward application of the recurrence relation of (12.49) for $\nu = -1/2$. The starting values of the recurrence are $P^M_{-1/2}(x)$ and $P^{M-1}_{-1/2}(x)$.

(viii) Compute the set $\{P^m_{1/2}, m = 0, \ldots, M\}$ by backward application of the recurrence relation of (12.49) for $\nu = 1/2$. The starting values of the recurrence are $P^M_{1/2}(x)$ and $P^{M-1}_{1/2}(x)$.

- Compute the set $\{P^m_{n-1/2}, m = 0, \ldots, M, n = 0, \ldots, N\}$ by forward application for each m of the TTRR in (12.35).

- Compute the set $\{Q^m_{n-1/2}, m = 0, \ldots, M, n = 0, \ldots, N\}$. The following sequence of steps is considered.

    (i) Compute $Q^0_{N-1/2}(x)$ and $Q^0_{N-3/2}(x)$ by using the calculated values of $P^0_{N-1/2}(x)$, $P^0_{N-3/2}(x)$, the continued fraction representation for $H_Q(N-1/2, 0, x)$, and the relation of (12.37) for $m = 0$ and $\nu = N - 1/2$.

    (ii) Compute $Q^1_{N-1/2}(x)$ and $Q^1_{N-3/2}(x)$ by using the calculated values of $P^1_{N-1/2}(x)$, $P^1_{N-3/2}(x)$, the continued fraction representation for $H_Q(N-1/2, 1, x)$, and the relation of (12.37) for $m = 1$ and $\nu = N - 1/2$.

    (iii) Compute the set $\{Q^m_{N-1/2}, m = 0, \ldots, M\}$ by forward application of the TTRR of (12.49). The starting values of the recurrence are $Q^0_{N-1/2}(x)$ and $Q^1_{N-1/2}(x)$.

    (iv) Compute the set $\{Q^m_{N-3/2}, m = 0, \ldots, M\}$ by forward application of the TTRR of (12.49). The starting values of the recurrence are $Q^0_{N-3/2}(x)$ and $Q^1_{N-3/2}(x)$.

    (v) Compute the set $\{Q^m_{n-1/2}, m = 0, \ldots, M, n = 0, \ldots, N\}$ by backward application for each $m = 0, 1, \ldots, M$ of the TTRR of (12.35). The starting values for each m of the recurrence relations are $Q^m_{N-1/2}(x)$ and $Q^m_{N-3/2}(x)$.

Dual Algorithm for $\{P^m_{n-1/2}(x), Q^m_{n-1/2}(x)\}$

Replace P by Q, the recursion over n by recursion over m, and the Wronskian relating consecutive orders (m) by the Wronskian relating consecutive degrees (n). In this way, the starting point would be the values $P^0_{-1/2}(x)$ and $P^0_{1/2}(x)$, and the forward recurrence would be applied together with a continued fraction (CF) for $Q^0_{N-1/2}(x)/Q^0_{N-3/2}(x)$ and a Wronskian in order to obtain $Q^0_{N-1/2}(x)$, and so on. In this case, the four continued fractions that would come into play correspond to $Q^0_{N-1/2}(x)/Q^0_{N-3/2}(x)$, $P^1_{N-1/2}(x)/P^0_{N-1/2}(x)$, $P^1_{k-1/2}(x)/P^0_{k-1/2}(x), k = 0, 1$.

12.4 Bessel functions

We describe algorithms for computing the modified Bessel functions $K_\nu(x)$, $I_\nu(x)$ for integer orders and the *modified spherical Bessel functions* (modified Bessel functions of half-integer orders with a factor). Additionally, we also consider the case of purely imaginary orders. In this case, instead of computing $I_{ia}(x)$ (which is not real), we consider a related function which is a numerically satisfactory companion of $K_{ia}(x)$: the function $L_{ia}(x)$.

12.4.1 Modified Bessel functions of integer and half-integer orders

Purpose

The module **BesselIK** computes the modified Bessel functions $K_\nu(x)$, $I_\nu(x)$ for integer orders. The functions of integer orders $K_n(x)$ and $I_n(x)$ are linearly independent solutions of the differential equation

$$x^2 w'' + x w' + (x^2 + n^2) w = 0. \tag{12.60}$$

The module **BesselIK** also computes the *modified spherical Bessel functions* of the *first, second,* and *third* kinds (see [2, eqs. (10.2.2)–(10.2.4)]).

These algorithms are an improved version of the programs in [194].

We use the notation

$$i_n(x) = \sqrt{\frac{\pi}{2x}} I_{n+1/2}(x),$$

$$i_{-n}(x) = \sqrt{\frac{\pi}{2x}} I_{-n-1/2}(x), \tag{12.61}$$

$$k_n(x) = \sqrt{\frac{\pi}{2x}} K_{n+1/2}(x) = \frac{1}{2}\pi(-1)^{n+1} \left(i_n(x) - i_{-n}(x)\right)$$

for the Bessel functions of first, second, and third kinds, respectively.

Algorithms

ALGORITHM 12.6. Modified Bessel functions of integer orders.

The algorithm calculates the set of functions $\{I_n(x), K_n(x) : n = 0, 1, 2, \ldots, N\}$ in a single run.

Scheme for $\{I_n(x), K_n(x)\}$

- Compute $K_0(x)$ and $K_1(x)$ by using Chebyshev expansions (see Tables 15–17 of [38]).

- Compute $K_n(x)$ for $n = 0, \ldots, N$ by forward application of the recurrence relation

$$K_{n+1}(x) = \frac{2n}{x} K_n(x) + K_{n-1}(x), \quad n = 1, \ldots, N-1. \tag{12.62}$$

The starting values of this recurrence relation are $K_0(x)$ and $K_1(x)$.

- Compute the ratio $I_N(x)/I_{N-1}(x)$ by means of a continued fraction representation, which is obtained by iterating the following expression (for $\nu = N$):

$$H_I(\nu, x) = \frac{I_\nu(x)}{I_{\nu-1}(x)} = \frac{1}{\dfrac{2\nu}{x} + \dfrac{I_{\nu+1}(x)}{I_\nu(x)}}. \tag{12.63}$$

12.4. Bessel functions

- Evaluate $I_N(x)$ in terms of the already computed values of $K_N(x)$, $K_{N-1}(x)$, $H_I(N, x)$ by means of the relation

$$I_N(x) = \frac{1}{x\left(K_{N-1}(x) + \frac{K_N(x)}{H_I(N, x)}\right)}. \quad (12.64)$$

And then $I_{N-1}(x) = \frac{I_N(x)}{H_I(N,x)}$.

- Compute $I_n(x)$ from $n = N - 2$ to $n = 0$ by backward application of the recurrence relation

$$I_{n-1}(x) = \frac{2n}{x} I_n(x) + I_{n+1}(x). \quad (12.65)$$

The starting values of this TTRR are $I_N(x)$ and $I_{N-1}(x)$.

ALGORITHM 12.7. Modified spherical Bessel functions.

The algorithm calculates the set of functions $\{i_n(x), k_n(x) : n = 0, 1, 2, \ldots, N\}$ in a single run.

Scheme for $\{i_n(x), k_n(x)\}$

- Compute the initial values $k_0(x) = \frac{1}{2}\pi \frac{e^{-x}}{x}$ and $k_1(x) = \frac{1}{2}\pi e^{-x}(\frac{1}{x} + \frac{1}{x^2})$.

- Compute $k_n(x)$ for $n = 0, \ldots, N$ by forward application of the recurrence relation

$$k_{n+1}(x) = \frac{(2n+1)}{x} k_n(x) + k_{n-1}(x). \quad (12.66)$$

The starting values of this recurrence relation are $k_0(x)$ and $k_1(x)$.

- Compute the ratio $i_N(x)/i_{N-1}(x)$ by means of a continued fraction representation, which is obtained by iterating the expression given in (12.63) for $\nu = N + 1/2$.

- Evaluate $i_N(x)$ in terms of the already computed values of $k_N(x)$, $k_{N-1}(x)$, and $H_I(N + 1/2, x)$ by means of the relation

$$i_N(x) = \frac{\pi/2}{x^2 \left(H_I(N + 1/2, x)k_{N-1}(x) + k_N(x)\right)}. \quad (12.67)$$

And then $i_{N-1}(x) = \frac{i_N(x)}{H_I(x)}$.

- Compute $i_n(x)$ from $n = N - 2$ to $n = 0$ by backward application of the recurrence relation

$$i_{n-1}(x) = \frac{(2n+1)}{x} i_n(x) + i_{n+1}(x). \quad (12.68)$$

The starting values of this TTRR are $i_N(x)$ and $i_{N-1}(x)$.

12.4.2 Modified Bessel functions of purely imaginary orders

Purpose

The module **KiaLia** computes both of the *modified Bessel functions of purely imaginary orders* $K_{ia}(x)$ and $L_{ia}(x)$, which constitute a numerically satisfactory pair of independent solutions of the modified Bessel equation of imaginary order,

$$x^2 w'' + x w' + (a^2 - x^2) w = 0. \tag{12.69}$$

This module is based on the references [94, 95].

In terms of the modified Bessel function of the first kind $I_\nu(x)$, the solutions are defined as

$$K_{ia}(x) = \frac{\pi}{2i \sinh(\pi a)} [I_{-ia}(x) - I_{ia}(x)], \quad L_{ia}(x) = \frac{1}{2} [I_{-ia}(x) + I_{ia}(x)], \tag{12.70}$$

with Wronskian

$$\mathcal{W}[K_{ia}(x), L_{ia}(x)] = \frac{1}{x}. \tag{12.71}$$

Both $K_{ia}(x)$ and $L_{ia}(x)$ are real solutions for real $x > 0$ and $a \in \mathbb{R}$.

The algorithm computes either the functions $K_{ia}(x)$, $L_{ia}(x)$ and their derivatives or scaled functions, which are defined as follows:

$$\widetilde{K}_{ia}(x) = \begin{cases} e^{\lambda(x,a)} K_{ia}(x), & x \geq a, \\ e^{a\pi/2} K_{ia}(x), & x < a, \end{cases} \quad \widetilde{K}'_{ia}(x) = \begin{cases} e^{\lambda(x,a)} K'_{ia}(x), & x \geq a, \\ e^{a\pi/2} K'_{ia}(x), & x < a, \end{cases} \tag{12.72}$$

and

$$\widetilde{L}_{ia}(x) = \begin{cases} e^{-\lambda(x,a)} L_{ia}(x), & x \geq a, \\ e^{-a\pi/2} L_{ia}(x), & x < a, \end{cases} \quad \widetilde{L}'_{ia}(x) = \begin{cases} e^{-\lambda(x,a)} L'_{ia}(x), & x \geq a, \\ e^{-a\pi/2} L'_{ia}(x), & x < a, \end{cases} \tag{12.73}$$

where $\lambda(x, a) = x \cos\theta + a\theta$, $a = x \sin\theta$.

In the description of the algorithms we will consider a positive because $K_{ia}(x)$ and $L_{ia}(x)$ are even functions of a. Of course, when applying the scaling factors for negative a, one should replace a by $|a|$ in the exponential scaling factors of (12.72), (12.73). In this way, the scaled functions are also even functions of a.

Algorithm

ALGORITHM 12.8. Modified Bessel functions of purely imaginary orders.

The algorithm combines different methods of evaluation in different regions. Figures 12.1 and 12.2 show the regions in the plane (x, a) where the different methods are used in the

12.4. Bessel functions

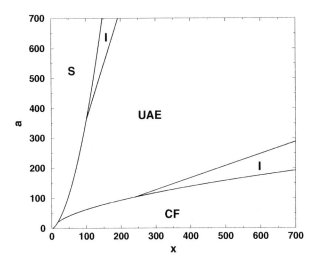

Figure 12.1. *Regions of applicability of each method for the $K_{ia}(x)$, $K'_{ia}(x)$ functions.* UAE: *uniform Airy-type asymptotic expansion*; I: *integral representations*; S: *power series*; *and* CF: *continued fraction.*

algorithm for computing the functions $K_{ia}(x)$ and $L_{ia}(x)$, respectively. More specifically, the methods of computation are as follows.

1. *Series expansions.*

$$K_{ia}(x) = \frac{1}{n(a)} \sum_{k=0}^{\infty} f_k c_k, \quad K'_{ia}(x) = \frac{1}{n(a)} \frac{2}{x} \sum_{k=0}^{\infty} \left[k f_k - \frac{r_k}{2} \right] c_k,$$

$$L_{ia}(x) = n(a) \sum_{k=0}^{\infty} r_k c_k, \quad L'_{ia}(x) = n(a) \frac{2}{x} \sum_{k=0}^{\infty} \left[k r_k + a^2 \frac{f_k}{2} \right] c_k, \quad (12.74)$$

where

$$n(a) = e^{\pi a/2} \sqrt{\frac{1 - e^{-2\pi a}}{2\pi a}}, \quad c_k = (x/2)^{2k}/k!, \quad (12.75)$$

and

$$f_k = \frac{\sin(\phi_{a,k} - a \ln(x/2))}{(a^2(1 + a^2) \cdots (k^2 + a^2))^{\frac{1}{2}}},$$

$$f_k/r_k = \frac{1}{a} \tan(\phi_{a,k} - a \ln(x/2)), \quad \text{with } \phi_{a,k} = \text{ph}(\Gamma(1 + k + ia)). \quad (12.76)$$

2. *Continued fraction for $K_{ia}(x)$ ($K'_{ia}(x)$).*

The Bessel function $K_\nu(x)$ can be expressed in terms of the functions $z_n(x) = U(\nu + \frac{1}{2} + n, 2\nu + 1, 2x)$:

$$K_\nu(x) = \pi^{\frac{1}{2}}(2x)^\nu e^{-x} z_0(x). \quad (12.77)$$

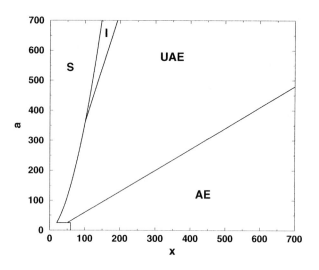

Figure 12.2. *Regions of applicability of each method for the $L_{ia}(x)$, $L'_{ia}(x)$ functions.* UAE: *uniform Airy-type asymptotic expansion*; I: *integral representations*; S: *power series; and* AE: *asymptotic expansions.*

For the calculation of z_0, a continued fraction representation for the ratio z_1/z_0 is considered,

$$\frac{z_1}{z_0} = \frac{1}{b_1 +} \frac{a_2}{b_2 + \cdots}, \qquad (12.78)$$

where

$$a_{n+1} = -[(n+1/2)^2 - \nu^2], \quad b_n = 2(n+x), \qquad (12.79)$$

together with a normalization condition

$$\sum_{n=0}^{\infty} C_n z_n = \left(\frac{1}{2x}\right)^{\nu+1/2}. \qquad (12.80)$$

The coefficients C_n are given by

$$C_n = \frac{(-1)^n}{n!} \frac{\Gamma(\nu+1/2+n)}{\Gamma(\nu+1/2-n)}. \qquad (12.81)$$

3. *Asymptotic expansions.*

$$K_{ia}(x) = \left(\frac{\pi}{2x}\right)^{\frac{1}{2}} e^{-x} \left\{ \sum_{k=0}^{n-1} \frac{(ia,k)}{(2x)^k} + \gamma_n \right\},$$

$$L_{ia}(x) = \frac{1}{\sqrt{2\pi x}} e^x \left\{ \sum_{k=0}^{n-1} (-1)^k \frac{(ia,k)}{(2x)^k} + \delta_n \right\}, \qquad (12.82)$$

12.4. Bessel functions

where (ia, m) is the Hankel symbol, which satisfies

$$(ia, k+1) = -\frac{(k+1/2)^2 + a^2}{k+1}(ia, k), \quad (ia, 0) = 1. \tag{12.83}$$

Bounds for the error terms (γ_n, δ_n) can be found in [168, p. 269].

4. *Airy-type uniform asymptotic expansions.*

The expansions for $K_{ia}(x)$ and $L_{ia}(x)$ in terms of Airy functions (Ai(x), Bi(x) and their derivatives) reads

$$K_{ia}(az) = \frac{\pi e^{-a\pi/2}\phi(\zeta)}{a^{\frac{1}{3}}}\left[\text{Ai}(-a^{\frac{2}{3}}\zeta)F_a(\zeta) + \frac{1}{a^{\frac{4}{3}}}\text{Ai}'(-a^{\frac{2}{3}}\zeta)G_a(\zeta)\right],$$

$$L_{ia}(az) = \frac{e^{a\pi/2}\phi(\zeta)}{2a^{\frac{1}{3}}}\left[\text{Bi}(-a^{\frac{2}{3}}\zeta)F_a(\zeta) + \frac{1}{a^{\frac{4}{3}}}\text{Bi}'(-a^{\frac{2}{3}}\zeta)G_a(\zeta)\right], \tag{12.84}$$

where

$$F_a(\zeta) \sim \sum_{s=0}^{\infty}(-)^s\frac{a_s(\zeta)}{a^{2s}}, \quad G_a(\zeta) \sim \sum_{s=0}^{\infty}(-)^s\frac{b_s(\zeta)}{a^{2s}}, \tag{12.85}$$

as $a \to \infty$ uniformly with respect to $z \in [0, \infty)$. Error bounds for the asymptotic expansion of the $K_{ia}(x)$ and $L_{ia}(x)$ are given in [58]. Details on the evaluation of the coefficients $a_s(\zeta)$ and $b_s(\zeta)$ can be found in [92, 220].

The quantity ζ is given by

$$\frac{2}{3}\zeta^{\frac{3}{2}} = \log\frac{1+\sqrt{1-z^2}}{z} - \sqrt{1-z^2}, \quad 0 \le z \le 1,$$

$$\frac{2}{3}(-\zeta)^{\frac{3}{2}} = \sqrt{z^2-1} - \arccos\frac{1}{z}, \quad z \ge 1, \tag{12.86}$$

and

$$\phi(\zeta) = \left(\frac{4\zeta}{1-z^2}\right)^{\frac{1}{4}}, \quad \phi(0) = 2^{\frac{1}{3}}. \tag{12.87}$$

For computing $\phi(\zeta)$ for small values, a series expansion around $z = 1$ is used.

5. *Nonoscillating integral representations.*

(a) Monotonic case $(x > a)$.

We use the following integral representations for $K_{ia}(x)$ and $K'_{ia}(x)$ in a portion of the monotonic region:

$$K_{ia}(x) = e^{-\lambda}\int_0^\infty e^{-x\Phi(\tau)}d\tau,$$

$$K'_{ia}(x) = -e^{-\lambda}\int_0^\infty \left[\cos\theta + \frac{\cosh\tau - 1 + 2\sin^2\frac{1}{2}(\theta-\sigma)}{\cos\sigma}\right]e^{-x\Phi(\tau)}d\tau, \tag{12.88}$$

where
$$\lambda = x\cos\theta + a\theta, \quad a = x\sin\theta, \quad \sin\sigma = \left(\sin\theta\frac{\tau}{\sinh\tau}\right), \quad (12.89)$$
and $\theta \in [0, \pi/2)$, $\sigma \in (0, \theta]$.

The argument of the exponential in the integrand is
$$\Phi(\tau) = (\cosh\tau - 1)\cos\sigma + 2\sin\left(\frac{\theta-\sigma}{2}\right)\sin\left(\frac{\theta+\sigma}{2}\right) + (\sigma-\theta)\sin\theta. \quad (12.90)$$

The difference $\theta - \sigma$ can be computed in a stable way for small values of τ by using the expression
$$\sin(\theta - \sigma) = \frac{\sin\theta}{\cos\theta\dfrac{\tau}{\sinh\tau} + \cos\sigma}\left[1 - \frac{\tau^2}{\sinh^2\tau}\right], \quad (12.91)$$

together with the definition of σ in (12.89) and specific algorithms to compute $\cosh(\tau) - 1$ and $1 - \sinh(\tau)^2/\tau^2$ for small τ.

(b) Oscillatory case ($x < a$).

We use the following approximations for moderately large a:
$$K_{ia}(x) \approx e^{-\pi a/2}\left[\int_{\tau_0}^{\infty} e^{-\Psi(\tau)}\left(\cos\chi + \sin\chi\frac{d\sigma}{d\tau}\right)d\tau + \mathcal{O}(e^{-\frac{\pi a}{2}})\right],$$
$$K'_{ia}(x) \approx e^{-\pi a/2}\left[\int_{\tau_0}^{\infty} e^{-\Psi(\tau)}(\cos\chi\, A(\tau) + \sin\chi\, C(\tau))d\tau + \mathcal{O}(e^{-\frac{\pi a}{2}})\right],$$
$$L_{ia}(x) \approx \frac{e^{\pi a/2}}{2\pi}\left[\int_{\tau_0}^{\infty} e^{-\Psi(\tau)}\left(\sin\chi - \cos\chi\frac{d\sigma}{d\tau}\right)d\tau + \mathcal{O}(e^{-\frac{\pi a}{2}})\right], \quad (12.92)$$
$$L'_{ia}(x) \approx \frac{e^{\pi a/2}}{2\pi}\left[\int_{\tau_0}^{\infty} e^{-\Psi(\tau)}(\sin\chi\, A(\tau) - \cos\chi\, C(\tau))d\tau + \mathcal{O}(e^{-\frac{\pi a}{2}})\right],$$

where
$$\chi = x\sinh\mu - a\mu, \quad \cosh\mu = \frac{a}{x}\,(\mu > 0), \quad \tau_0 = \mu - \tanh\mu, \quad (12.93)$$
$$\sin\sigma = \frac{(\tau-\mu)\cosh\mu + \sinh\mu}{\sinh\tau}. \quad (12.94)$$

The argument of the exponential in the integrand is
$$\Psi(\tau) = x\cosh\tau\cos\sigma + a\left(\sigma - \frac{1}{2}\pi\right), \quad (12.95)$$

and
$$\begin{aligned}A(\tau) &= -\cosh\tau\cos\sigma + \sinh\tau\sin\sigma\frac{d\sigma}{d\tau}, & B(\tau) &= A(\tau)\frac{d\tau}{d\sigma}, \\ C(\tau) &= -\sinh\tau\sin\sigma - \cosh\tau\cos\sigma\frac{d\sigma}{d\tau}, & D(\tau) &= C(\tau)\frac{d\tau}{d\sigma}.\end{aligned} \quad (12.96)$$

12.5 Parabolic cylinder functions

12.5.1 Purpose

The module **Parabolic** computes both *parabolic cylinder functions* (PCFs) $U(a, x)$ and $V(a, x)$, which constitute a satisfactory pair of independent solutions of the differential equation

$$w'' - \left(\frac{1}{4}x^2 + a\right)w = 0, \tag{12.97}$$

and their derivatives. Also scaled functions can be computed which can be used for unrestricted values of real a and x.

This module is based on the references [85, 86].

The scaled PCFs for $x \geq 0$ are defined as follows:

$$\widetilde{U}(a, x) = F(a, x)U(a, x), \quad \widetilde{U}'(a, x) = F(a, x)U'(a, x),$$
$$\widetilde{V}(a, x) = \frac{V(a, x)}{F(a, x)}, \quad \widetilde{V}'(a, x) = \frac{V'(a, x)}{F(a, x)}, \tag{12.98}$$

where the scaling factor is given by

$$F(a, x) = |f(a, x)|, \tag{12.99}$$

and the function $f(a, x)$ is defined as follows:

$$f(a, x) = \left(\frac{1}{2}x + \sqrt{\frac{1}{4}x^2 + a}\right)^a \exp\left(\frac{1}{2}x\sqrt{\frac{1}{4}x^2 + a} - \frac{1}{2}a\right). \tag{12.100}$$

We introduce the variable

$$t = \frac{x}{2\sqrt{|a|}} \tag{12.101}$$

and the notation $f(a)$ such that $f(a, x)$ can be written as

$$f(a, x) = f(a)e^{2|a|\Theta}, \tag{12.102}$$

where

$$\Theta = \begin{cases} \tilde{\xi} = \frac{1}{2}\left[t\sqrt{t^2+1} + \log(t + \sqrt{t^2+1})\right], & a > 0, \\ \xi = \frac{1}{2}\left[t\sqrt{t^2-1} - \log(t + \sqrt{t^2-1})\right], & a < 0, \quad t > 1, \\ i\eta = \frac{1}{2}i\left[\arccos t - t\sqrt{1-t^2}\right], & a < 0, \quad 0 \leq t \leq 1. \end{cases} \tag{12.103}$$

In the oscillatory region $x^2/4 + a < 0$ (that is, $a < 0$ and $0 \leq t < 1$), the values for the square roots in (12.100) are taken in the principal branch. In this region, the imaginary part indicates that the PCFs oscillate with a phase function $\phi(a, x) \sim 2a\eta$ as $a \to -\infty$.

For the oscillatory domain, the scaling factor will also be denoted by

$$f(a) = |f(a, x)| = |a|^{a/2}e^{-a/2}. \tag{12.104}$$

12.5.2 Algorithm

ALGORITHM 12.9. Parabolic cylinder functions.

Different methods of computation are considered in the algorithm for computing PCFs depending on the range of the variable and the order a.

The regions in the (a, x)-plane where different methods of computation for PCFs are considered are shown in Figure 12.3.

The curves f_i, $i = 1, \ldots, 12$, appearing in Figure 12.3, together with the equations of two additional curves (f_{13} and f_{14}) which are outside of the represented domain, are the following (we use the notation $x_{i,j}$ to denote the x-value for the intersection of the curves f_i and f_j which appear in the figure):

$$
\begin{aligned}
&f_1 : a = -0.23x^2 + 1.2x + 18.72, && 0 \le x \le x_{1,9} = 30, \\
&f_2 : a = \tfrac{3.75}{x} - 1.25, && x_{1,2} \le x \le x_{2,3} = 3, \\
&f_3 : a = -30/(x - 0.3) + 100/9, && x_{3,4} \le x \le x_{2,3} = 3, \\
&f_4 : a = -0.21x^2 - 4.5x - 40, && 0 \le x \le x_{4,10} = 30, \\
&f_5 : a = x - 14, && x_{5,12} = 4 \le x \le x_{5,6}, \\
&f_6 : a = -7 - 0.14(x - 4.8)^2, && x_{5,6} \le x \le x_{1,6}, \\
&f_7 : a = 2.5x - 30, && x_{1,7} = 12 \le x \le x_{7,13} = 72, \\
&f_8 : a = -2.5x + 30, && x_{1,8} = 12 \le x \le x_{8,14} = 72, \\
&f_9 : a = -0.1692x^2, && x > x_{1,9} = 30, \\
&f_{10} : a = -0.295x^2 + 0.3x - 107.5, && x > x_{4,10} = 30, \\
&f_{11} : a = 0, && x_{2,3} = 3 \le x \le x_{1,7} = x_{1,8} = 12, \\
&f_{12} : x = 4, && x_{4,12} = -61.26 \le a \le -10 = x_{5,12}, \\
&f_{13} : a = 150, && x > x_{7,13} = 72, \\
&f_{14} : a = -150, && x > x_{8,14} = 72.
\end{aligned} \qquad (12.105)
$$

The values of $x_{1,2}$, $x_{3,4}$, $x_{5,6}$, and $x_{1,6}$ are (approximately)

$$x_{1,2} = 0.18, \quad x_{3,4} = 0.84, \quad x_{5,6} = 6.56, \quad x_{1,6} = 17.15. \qquad (12.106)$$

With three exceptions, the curves shown in Figure 12.3 are all the curves separating the regions where different methods are applied. The first exception is related to the validity of Poincaré asymptotic expansions, which is not used for $|a| > 150$ (curves f_{13} and f_{14}, which are outside the region depicted in Figure 12.3). The second exception is the rectangular region $|a| < 0.7$ and $x \in [2.5, 12.5]$, which includes the vertices A and B of Figure 12.3 (left). The third exception is for positive a and small x (close to zeros of the V functions for positive a) where uniform asymptotic expansions are replaced by Maclaurin series for the V-functions.

12.5. Parabolic cylinder functions

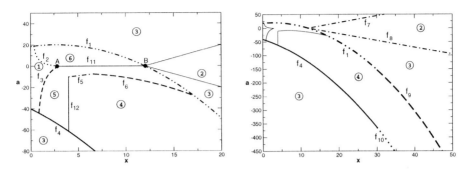

Figure 12.3. *Regions in the (a, x)-plane where different methods of computation for PCFs are considered.*

Let us identify which methods are used in each of the 8 regions in which the (a, x)-plane is divided by the 14 curves f_i. If not stated otherwise, the same method is applied for both $U(a, x)$ and $V(a, x)$ and their derivatives. Notice that the numbering of the methods in the following list corresponds to the labeling of the regions in Figure 12.3.

1. *Maclaurin series.*

They are applied in the bounded region delimited by the curves f_1, f_2, f_3, and f_4.
We use

$$y_1(a, x) = 1 + a\frac{x^2}{2!} + (a^2 + 1/2)\frac{x^4}{4!} + \cdots,$$
$$y_2(a, x) = x + a\frac{x^3}{3!} + (a^2 + 3/2)\frac{x^5}{5!} + \cdots$$
(12.107)

and

$$U(a, x) = U(a, 0)y_1(a, x) + U'(a, 0)y_2(a, x),$$
$$V(a, x) = V(a, 0)y_1(a, x) + V'(a, 0)y_2(a, x),$$
(12.108)

where

$$U(a, 0) = \frac{\sqrt{\pi}}{2^{a/2+1/4}\Gamma(\frac{3}{4} + \frac{1}{2}a)}, \qquad U'(a, 0) = -\frac{\sqrt{\pi}}{2^{a/2-1/4}\Gamma(\frac{1}{4} + \frac{1}{2}a)},$$
$$V(a, 0) = \frac{2^{a/2+1/4}\sin\pi(\frac{3}{4} - \frac{1}{2}a)}{\Gamma(\frac{3}{4} - \frac{1}{2}a)}, \qquad V'(a, 0) = \frac{2^{a/2+3/4}\sin\pi(\frac{1}{4} - \frac{1}{2}a)}{\Gamma(\frac{1}{4} - \frac{1}{2}a)}.$$
(12.109)

For scaled functions, we compute

$$\widetilde{U}(a, x) = u_1 y_1(a, x) + u_2 y_2(a, x),$$
$$\widetilde{V}(a, x) = v_1 y_1(a, x) + v_2 y_2(a, x),$$
(12.110)

where the coefficients u_i, v_i are given by

$$u_1 = F(a,x)U(a,0) = \frac{1}{\sqrt{2}}H(a)e^{2a\tilde{\xi}},$$

$$u_2 = F(a,x)U'(a,0) = -\frac{\beta(a)}{\sqrt{2}H(a)}e^{2a\tilde{\xi}},$$

$$v_1 = \frac{1}{F(a,x)}V(a,0) = \frac{1+\sin\pi a}{\sqrt{\pi}}\frac{H(a)}{\beta(a)}e^{-2a\tilde{\xi}},$$

$$v_2 = \frac{1}{F(a,x)}V'(a,0) = \frac{1-\sin\pi a}{\sqrt{\pi}}\frac{1}{H(a)}e^{-2a\tilde{\xi}};$$
(12.111)

$\tilde{\xi}$ is defined in (12.103). In addition,

$$H(a) = \beta\left(\frac{1}{4}+\frac{1}{2}a\right)\left(a+\frac{1}{2}\right)^{-\frac{1}{4}}\left[S\left(\frac{1}{2a}\right)\right]^{-\frac{1}{4}},$$
(12.112)

$$\beta(\lambda) = \frac{\sqrt{2\pi}\lambda^\lambda e^{-\lambda}}{\Gamma\left(\lambda+\frac{1}{2}\right)} \sim 1 \quad \text{as } \lambda \to +\infty,$$
(12.113)

and

$$S(y) = \frac{1}{e}(1+y)^{1/y} \sim 1 \quad \text{as } y \to 0.$$
(12.114)

2. *Asymptotic expansions for large x (Poincaré type).*

They are applied in the unbounded region between $f_7 \cup f_{13}$ and $f_8 \cup f_{14}$ ($x > x_{7,8} = 12$). We use

$$U(a,x) \sim \frac{1}{\sqrt{x}\psi(a,x)}\sum_{k=0}^{\infty}\frac{a_k}{x^{2k}}, \quad U'(a,x) \sim -\frac{\sqrt{x}}{2\psi(a,x)}\sum_{k=0}^{\infty}\frac{b_k}{x^{2k}},$$

$$V(a,x) \sim \sqrt{\frac{2}{\pi x}}\psi(a,x)\sum_{k=0}^{\infty}\frac{c_k}{x^{2k}}, \quad V'(a,x) \sim \sqrt{\frac{x}{2\pi}}\psi(a,x)\sum_{k=0}^{\infty}\frac{d_k}{x^{2k}},$$
(12.115)

where

$$\psi(a,x) = x^a e^{x^2/4}.$$
(12.116)

For scaled functions, we use

$$\tilde{U}(a,x) \sim \frac{1}{\sqrt{x}\phi(a,x)}\sum_{k=0}^{\infty}\frac{a_k}{x^{2k}}, \quad \tilde{U}'(a,x) \sim -\frac{\sqrt{x}}{2\phi(a,x)}\sum_{k=0}^{\infty}\frac{b_k}{x^{2k}},$$

$$\tilde{V}(a,x) \sim \sqrt{\frac{2}{\pi x}}\phi(a,x)\sum_{k=0}^{\infty}\frac{c_k}{x^{2k}}, \quad \tilde{V}'(a,x) \sim \sqrt{\frac{x}{2\pi}}\phi(a,x)\sum_{k=0}^{\infty}\frac{d_k}{x^{2k}},$$
(12.117)

where $\phi(a,x) = \psi(a,x)/F(a,x)$.

12.5. Parabolic cylinder functions

The coefficients can be computed recursively in the following way:

$$a_k = -\frac{(a+2k-\frac{3}{2})(a+2k-\frac{1}{2})}{2k}a_{k-1}, \quad a_0 = 1,$$

$$b_k = a_k + (2a+4k-3)a_{k-1}, \quad b_0 = 1,$$

$$c_k = \frac{(a-2k+\frac{3}{2})(a-2k+\frac{1}{2})}{2k}c_{k-1}, \quad c_0 = 1,$$

$$d_k = c_k + (2a-4k+3)c_{k-1}, \quad d_0 = 1.$$

(12.118)

3. *Uniform asymptotic expansions in terms of elementary functions.*

Three different sets of expansions are used in three different regions.

(a) Unbounded region for positive a above the curves f_1, f_7, and f_{13} (f_{13} is not shown in Figure 12.3).

We use

$$\widetilde{U}(a,x) = \frac{1}{\sqrt{2}}\left(\frac{1}{4}x^2+a\right)^{-\frac{1}{4}} F_a,$$

$$\widetilde{U}'(a,x) = -\frac{1}{\sqrt{2}}\left(\frac{1}{4}x^2+a\right)^{\frac{1}{4}} G_a,$$

(12.119)

$$\widetilde{V}(a,x) = \frac{1}{\sqrt{\pi}}\left(\frac{1}{4}x^2+a\right)^{-\frac{1}{4}} [P_a + \sin\pi a M(a,x)F_a],$$

$$\widetilde{V}'(a,x) = \frac{1}{\sqrt{\pi}}\left(\frac{1}{4}x^2+a\right)^{\frac{1}{4}} [Q_a - \sin\pi a M(a,x)G_a],$$

(12.120)

where F_a, G_a, P_a, and Q_a have the expansions

$$F_a \sim \sum_{s=0}^{\infty} \frac{\phi_s(\tau)}{(-2a)^s}, \quad G_a \sim \sum_{s=0}^{\infty} \frac{\psi_s(\tau)}{(-2a)^s},$$

(12.121)

$$P_a \sim \sum_{s=0}^{\infty} \frac{\phi_s(\tau)}{(2a)^s}, \quad Q_a \sim \sum_{s=0}^{\infty} \frac{\psi_s(\tau)}{(2a)^s},$$

(12.122)

and

$$\tau = \frac{1}{2}\left[\frac{x/2}{\sqrt{x^2/4+a}}-1\right] = \frac{-a/2}{x^2/4+a+\frac{x}{2}\sqrt{x^2/4+a}},$$

(12.123)

$$M(a,x) = \frac{\Gamma(a+1/2)}{\sqrt{2\pi}F(a,x)^2},$$

(12.124)

where $F(a,x)$ is given in (12.99).

The coefficients $\phi_s(\tau)$ are polynomials in τ which follow from the recursion

$$\phi_{s+1}(\tau) = -4\tau^2(\tau+1)^2\frac{d\phi_s(\tau)}{d\tau} - \frac{1}{4}\int_0^\tau (20\tau'^2 + 20\tau' + 3)\phi_s(\tau')d\tau', \quad (12.125)$$

with $\phi_0(\tau) = 1, \phi_{-1}(\tau) = 0$.

The coefficients $\psi_s(\tau)$ can be obtained from the relation

$$\psi_s(\tau) = \phi_s(\tau) + 2\tau(\tau+1)(2\tau+1)\phi_{s-1}(\tau) + 8\tau^2(\tau+1)^2\frac{d\phi_{s-1}(\tau)}{d\tau}. \quad (12.126)$$

The unscaled functions can be obtained by multiplying (12.119) or dividing (12.120) by $F(a, x)$.

(b) Unbounded region for negative a between $f_8 \cup f_{14}$ and $f_1 \cup f_9$.

For $\widetilde{V}(a, x), \widetilde{V}'(a, x)$, we consider

$$\begin{aligned}\widetilde{V}(a,x) &= \frac{1}{\sqrt{\pi}}\left(\frac{1}{4}x^2 + a\right)^{-\frac{1}{4}} P_a, \\ \widetilde{V}'(a,x) &= \frac{1}{\sqrt{\pi}}\left(\frac{1}{4}x^2 + a\right)^{\frac{1}{4}} Q_a. \end{aligned} \quad (12.127)$$

Equation (12.119) is used for computing $\widetilde{U}(a, x), \widetilde{U}'(a, x)$. These expansions are valid for $a < 0$.

(c) Unbounded region for negative values of a below $f_4 \cup f_{10}$.

We use

$$\begin{aligned}\widetilde{U}(a,x) &= \frac{\sqrt{2}}{\lambda(x,a)} G(\mu)[C_1(a,x)\cos\phi(a,x) + S_1(a,x)\sin\phi(a,x)], \\ \widetilde{U}'(a,x) &= \sqrt{2}\lambda(x,a) G(\mu)[-S_2(a,x)\sin\phi(a,x) + C_2(a,x)\cos\phi(a,x)], \\ \widetilde{V}(a,x) &= \frac{\beta(|a|)}{\sqrt{\pi}\lambda(x,a)} G(\mu)[C_1(a,x)\sin\phi(a,x) - S_1(a,x)\cos\phi(a,x)], \\ \widetilde{V}'(a,x) &= \frac{\lambda(x,a)\beta(|a|)}{\sqrt{\pi}} G(\mu)[S_2(a,x)\cos\phi(a,x) + C_2(a,x)\sin\phi(a,x)], \end{aligned}$$
$$(12.128)$$

where

$$\lambda(x,a) = \left|\frac{1}{4}x^2 + a\right|^{\frac{1}{4}}, \quad (12.129)$$

$$\phi(a,x) = -\mu^2\eta + \frac{1}{4}\pi = 2a\eta + \frac{1}{4}\pi; \quad (12.130)$$

η is defined in (12.103), $\beta(|a|)$ is given by (12.113), $G(\mu)$ is given by

$$G(\mu) \sim \left(\sum_{s=0}^{\infty} \frac{g_s}{\mu^{2s}}\right)^{-1}, \quad \mu = \sqrt{2|a|}, \quad (12.131)$$

12.5. Parabolic cylinder functions

and

$$g_s = \lim_{t \to \infty} \frac{u_s(t)}{(t^2 - 1)^{3s/2}}. \tag{12.132}$$

The quantities $S_i(a, x)$ and $C_i(a, x)$, $i = 1, 2$, are available in the form of asymptotic expansions:

$$C_1(a, x) \sim \sum_{s=0}^{\infty} \frac{(-1)^s u_{2s}(t)}{(1 - t^2)^{3s} \mu^{4s}}, \qquad S_1(a, x) \sim \sum_{s=0}^{\infty} \frac{(-1)^s u_{2s+1}(t)}{(1 - t^2)^{3s+3/2} \mu^{4s+2}},$$

$$C_2(a, x) \sim \sum_{s=0}^{\infty} \frac{(-1)^s v_{2s+1}(t)}{(1 - t^2)^{3s+3/2} \mu^{4s+2}}, \qquad S_2(a, x) \sim \sum_{s=0}^{\infty} \frac{(-1)^s v_{2s}(t)}{(1 - t^2)^{3s} \mu^{4s}}. \tag{12.133}$$

The coefficients $u_s(t)$ satisfy the relation

$$(t^2 - 1)u'_s(t) - 3stu_s(t) = r_{s-1}(t), \quad s = 1, 2, 3, \ldots, \tag{12.134}$$

where $u_0(t) = 1$, $r_{-1} = 0$, and for $s = 0, 1, 2, \ldots$,

$$8r_s(t) = (3t^2 + 2)u_s(t) - 12(s + 1)tr_{s-1}(t) + 4(t^2 - 1)r'_{s-1}(t). \tag{12.135}$$

The coefficients $v_s(t)$ are given by $v_0(t) = 1$ and

$$v_s(t) = u_s(t) + \frac{1}{2}tu_{s-1}(t) - r_{s-2}(t), \quad s = 1, 2, 3, \ldots. \tag{12.136}$$

4. *Uniform Airy-type asymptotic expansion.*

These expansions are applied in the unbounded region for negative a delimited by the curves f_5, f_6, f_1 ($x \geq x_{1,6}$), f_9, f_{12}, f_4 ($x \geq x_{4,12}$), and f_{10}.
We use

$$\tilde{U}(a, x) = 2^{\frac{3}{4}} \pi^{\frac{1}{2}} \mu^{-\frac{1}{6}} G(\mu) \phi(\zeta) \left[\widehat{\mathrm{Ai}}(\mu^{\frac{4}{3}} \zeta) A_\mu(\zeta) + \frac{\widehat{\mathrm{Ai}'}(\mu^{\frac{4}{3}} \zeta)}{\mu^{8/3}} B_\mu(\zeta) \right], \tag{12.137}$$

$$\tilde{U}'(a, x) = 2^{\frac{1}{4}} \pi^{\frac{1}{2}} \mu^{\frac{1}{6}} \frac{G(\mu)}{\phi(\zeta)} \left[\frac{\widehat{\mathrm{Ai}}(\mu^{\frac{4}{3}} \zeta)}{\mu^{\frac{4}{3}}} C_\mu(\zeta) + \widehat{\mathrm{Ai}'}(\mu^{\frac{4}{3}} \zeta) D_\mu(\zeta) \right], \tag{12.138}$$

$$\tilde{V}(a, x) = 2^{\frac{1}{4}} \mu^{-\frac{1}{6}} \frac{\phi(\zeta)}{G(\mu) S(\mu)} \left[\widehat{\mathrm{Bi}}(\mu^{\frac{4}{3}} \zeta) A_\mu(\zeta) + \frac{\widehat{\mathrm{Bi}'}(\mu^{\frac{4}{3}} \zeta)}{\mu^{8/3}} B_\mu(\zeta) \right], \tag{12.139}$$

$$\tilde{V}'(a, x) = 2^{-\frac{1}{4}} \mu^{\frac{1}{6}} \frac{1}{G(\mu) S(\mu) \phi(\zeta)} \left[\frac{\widehat{\mathrm{Bi}}(\mu^{\frac{4}{3}} \zeta)}{\mu^{\frac{4}{3}}} C_\mu(\zeta) + \widehat{\mathrm{Bi}'}(\mu^{\frac{4}{3}} \zeta) D_\mu(\zeta) \right]. \tag{12.140}$$

$\widehat{\mathrm{Ai}}$, $\widehat{\mathrm{Ai}'}$, $\widehat{\mathrm{Bi}}$, and $\widehat{\mathrm{Bi}'}$ are scaled Airy functions for positive x and plain Airy functions for negative x.

The parameter ζ is defined as $\zeta = \frac{3}{2}\Theta^{\frac{2}{3}}$ with Θ as given in (12.103) and the values are taken in the principal branch. $G(\mu)$ is given in (12.131), and for $S(\mu)$ we have

$$S(\mu) \sim \sum_{k=0}^{\infty} \frac{w_k}{\mu^{4k}}. \qquad (12.141)$$

The first four values of w_k are $w_0 = 1$, $w_1 = -1/576$, $w_2 = 2021/2488320$, $w_3 = -337566547/300987187200$. In addition,

$$\phi(\zeta) = \left(\frac{\zeta}{t^2 - 1}\right)^{\frac{1}{4}}, \qquad (12.142)$$

$$A_\mu(\zeta) \sim \sum_{s=0}^{\infty} \frac{a_s(\zeta)}{\mu^{4s}}, \quad B_\mu(\zeta) \sim \sum_{s=0}^{\infty} \frac{b_s(\zeta)}{\mu^{4s}},$$

$$C_\mu(\zeta) \sim \sum_{s=0}^{\infty} \frac{c_s(\zeta)}{\mu^{4s}}, \quad D_\mu(\zeta) \sim \sum_{s=0}^{\infty} \frac{d_s(\zeta)}{\mu^{4s}}. \qquad (12.143)$$

5. *Integral representations.*

Integral representations are considered in the bounded region for negative a encircled by the curves f_{11}, f_3, f_4 ($x_{3,4} \leq x \leq x_{4,12}$), f_{12}, f_5, f_6, and f_1 ($12 \leq x \leq x_{1,6}$). Different representations are used for the monotonic and oscillatory regions:

(a) When $\frac{1}{4}x^2 + a < 0$ we use

$$\tilde{U}(a, x) = \sqrt{\frac{2}{\pi}} |a|^{\frac{1}{4}} [G_1(a, x) \cos \phi(a, x) + G_2(a, x) \sin \phi(a, x)], \qquad (12.144)$$

$$\tilde{U}'(a, x) = \sqrt{\frac{2}{\pi}} |a|^{\frac{3}{4}} [H_1(a, x) \cos \phi(a, x) + H_2(a, x) \sin \phi(a, x)], \qquad (12.145)$$

$$\tilde{V}(a, x) = \frac{1}{\pi} |a|^{\frac{1}{4}} \beta(|a|) [G_1(a, x) \sin \phi(a, x) - G_2(a, x) \cos \phi(a, x)], \qquad (12.146)$$

$$\tilde{V}'(a, x) = \frac{1}{\pi} |a|^{\frac{3}{4}} \beta(|a|) [H_1(a, x) \sin \phi(a, x) - H_2(a, x) \cos \phi(a, x)], \qquad (12.147)$$

with $\phi(a, x)$ as given in (12.130) and $\beta(|a|)$ as defined in (12.113). $G_j(a, x)$ and $H_j(a, x)$ are computed by means of stable integral representations based on steepest descent methods. See [86] for further details.

(b) When $x^2/4 + a > 0$ we use

$$\tilde{U}(a, x) = \sqrt{\frac{2}{\pi}} |a|^{\frac{1}{4}} \hat{G}_1, \qquad (12.148)$$

$$\tilde{U}'(a, x) = \sqrt{\frac{2}{\pi}} |a|^{\frac{3}{4}} \hat{H}_1, \qquad (12.149)$$

$$\tilde{V}(a,x) = \frac{1}{\pi}|a|^{\frac{1}{4}}\left(\beta(|a|)\hat{G}_3 + N(a,x)\hat{G}_2\right), \qquad (12.150)$$

$$\tilde{V}'(a,x) = \frac{1}{\pi}|a|^{\frac{3}{4}}\left(\beta(|a|)\hat{H}_3 + N(a,x)\hat{H}_2\right), \qquad (12.151)$$

where

$$N(a,x) = \frac{\sqrt{2\pi}}{\Gamma(\frac{1}{2}-a)F(a,x)^2}; \qquad (12.152)$$

$\beta(|a|)$ is given by (12.113). The functions H_j and G_j are computed by means of stable integral representations based on the steepest descent method. See [86] for further details.

6. *Series for V and recurrence relations for U.*

These methods are applied in the bounded region for positive values of a encircled by the curves f_1, f_2, and f_{11}. Maclaurin series are used for V (see (12.108)) and recurrence relations for U

$$U(a-1,x) = xU(a,x) + \left(a+\frac{1}{2}\right)U(a+1,x), \qquad (12.153)$$

$$U'(a,x) = -\frac{1}{2}xU(a,x) - \left(a+\frac{1}{2}\right)U(a+1,x), \qquad (12.154)$$

with starting values $a \in [21, 23)$ (computed with the uniform asymptotic expansions corresponding to region 3).

12.6 Zeros of Bessel functions

12.6.1 Purpose

The module **ZerosBessel** computes the zeros of Bessel functions of the first kind for real orders ν. This module is based on the reference [195].

12.6.2 Algorithm

ALGORITHM 12.10. Zeros of the first kind Bessel function $J_\nu(x)$.

The algorithm is based on the evaluation of the ratio $H_\nu(x) = J_\nu(x)/J_{\nu-1}(x)$ and the monotonicity properties of the function $f_\nu(x) = x^{2\nu-1}H_\nu(x)$. We then apply a Newton–Raphson iteration on $f_\nu(x)$ which is convergent for all real values ν and starting values $x_0 > 0$.

For the computation of the ratio $H_\nu(x)$, the algorithm computes a continued fraction representation for this ratio:

$$H_\nu(x) = \frac{J_\nu(x)}{J_{\nu-1}(x)} = \cfrac{1}{\frac{2\nu}{x} - \cfrac{1}{\frac{2(\nu+1)}{x} - \cfrac{1}{\frac{2(\nu+2)}{x} - \cdots}}}. \qquad (12.155)$$

We use the modified Lentz algorithm (see §6.6.2) to evaluate the continued fraction $H_\nu(x)$.

The algorithm (ELF) finds the real zero $j_{\nu,s}$ in the neighborhood of a given value x_0. We will say that a value x_0 is in the neighborhood of $j_{\nu,s}$ (and vice versa) when both numbers lie in the same branch of $H_\nu(x) = J_\nu(x)/J_{\nu-1}(x)$.

Since the resulting algorithm is quite short, we give it explicitly coded in a Fortran-like syntax.

    SUBROUTINE ELF(x,ν,ϵ,xc)

Inputs: x (starting value), ν (order of the Bessel function), ϵ (demanded accuracy).

Output: xc (zero of $J_\nu(x)$ in the neighborhood of x).

Parameters: *dwarf* (underflow limit), N_{max} (maximum number of iterations).

(1) *dwarfsq*=SQRT(*dwarf*); $dev = 1$; $xc = x$; iter=0
(2) IF (INT(ν)=ν) ν=ABS(ν)
(3) DO WHILE((dev> ϵ).AND.($xc \leq \nu$))
(4)   iter = iter + 1
(5)   IF(*iter* > N_{max}) THEN STOP !Convergence failure
(6) $m = 0$; $b = 2*\nu/xc$; $a = 1$; $fc = $ *dwarfsq*; $c0 = fc$; $d0 = 0$; $delta = 0$
(7)   DO WHILE (ABS($delta - 1$)> ϵ)
(8) $d0 = b + a*d0$
(9)   IF(ABS($d0$)< *dwarfsq*) $d0 = $ *dwarfsq*
(10) $c0 = b + a/c0$
(11)   IF(ABS($c0$)< *dwarfsq*) $c0 = $ *dwarfsq*
(12) $d0 = 1/d0$; $delta = c0*d0$; $fc = fc*delta$; $m = m+1$; $a = -1$; $b = 2*(\nu+m)/xc$
(13)   END WHILE
(14)   IF (ABS(fc)> 1) $fc = fc/$ABS(fc)
(15) $xa = xc$; $xc = xc - fc/(1 + fc*fc)$
(16)   IF($xc \leq 0$) $xc = xa*0.5$
(17) dev=ABS(fc)/xa
(18) END WHILE
(19) IF($xc < \nu$) $xc = 0$
(20) RETURN
(21) END

List of Algorithms

Algorithm 3.1	Clenshaw's method for a Chebyshev sum	75
Algorithm 3.2	Modified Clenshaw's method for a Chebyshev sum	78
Algorithm 4.1	Miller's method for nonvanishing minimal solutions	106
Algorithm 4.2	Miller's algorithm with a normalizing sum	108
Algorithm 4.3	Anti-Miller algorithm	112
Algorithm 5.1	Recursive trapezoidal rule	130
Algorithm 5.2	Gram–Schmidt orthogonalization	134
Algorithm 5.3	Stieltjes procedure	139
Algorithm 5.4	Golub–Welsch algorithm	145
Algorithm 6.1	Steed's algorithm	182
Algorithm 6.2	Modified Lentz algorithm	185
Algorithm 7.1	Zeros of cylinder functions of order $\nu > 1/2$	216
Algorithm 7.2	Zeros of cylinder functions of order $\nu < 1/2$	217
Algorithm 7.3	Zeros of special functions from a fixed point method	221
Algorithm 9.1	Computing Padé approximants $[n + k/k]_f$	279
Algorithm 9.2	Romberg quadrature	295
Algorithm 11.1	Carlson's algorithm for computing elliptic integrals	346
Algorithm 12.1	Homogeneous complex Airy functions	359
Algorithm 12.2	Scorer functions	361
Algorithm 12.3	Prolate spheroidal harmonics	364
Algorithm 12.4	Oblate spheroidal harmonics	365
Algorithm 12.5	Toroidal harmonics	366
Algorithm 12.6	Modified Bessel functions of integer orders	370
Algorithm 12.7	Modified spherical Bessel functions	371
Algorithm 12.8	Modified Bessel functions of purely imaginary orders	372
Algorithm 12.9	Parabolic cylinder functions	378
Algorithm 12.10	Zeros of the first kind Bessel function $J_\nu(x)$	385

Bibliography

[1] M. ABRAMOWITZ, *Evaluation of the integral $\int_0^\infty e^{-u^2-x/u}\,du$*, J. Math. Phys., 32 (1953), pp. 188–192.

[2] M. ABRAMOWITZ AND I. A. STEGUN, *Handbook of Mathematical Functions with Formulas, Graphs, and Mathematical Tables*, vol. 55 of National Bureau of Standards Applied Mathematics Series, U.S. Government Printing Office, Washington, D.C., 1964.

[3] D. E. AMOS, *Algorithm 644: A portable package for Bessel functions of a complex argument and nonnegative order*, ACM Trans. Math. Software, 12 (1986), pp. 265–273.

[4] G. E. ANDREWS, R. ASKEY, AND R. ROY, *Special Functions*, vol. 71 of Encyclopedia of Mathematics and Its Applications, Cambridge University Press, Cambridge, UK, 1999.

[5] M. BAIN, *On the uniform convergence of generalized Fourier series*, J. Inst. Math. Appl., 21 (1978), pp. 379–386.

[6] G. A. BAKER, JR., *The theory and application of the Padé approximant method*, in Advances in Theoretical Physics, Vol. 1, Academic Press, New York, 1965, pp. 1–58.

[7] ———, *Essentials of Padé Approximants*, Academic Press, New York, London, 1975.

[8] G. A. BAKER, JR. AND P. GRAVES-MORRIS, *Padé Approximants*, vol. 59 of Encyclopedia of Mathematics and Its Applications, 2nd ed., Cambridge University Press, Cambridge, UK, 1996.

[9] J. S. BALL, *Automatic computation of zeros of Bessel functions and other special functions*, SIAM J. Sci. Comput., 21 (2000), pp. 1458–1464.

[10] ———, *Half-range generalized Hermite polynomials and the related Gaussian quadratures*, SIAM J. Numer. Anal., 40 (2003), pp. 2311–2317.

[11] A. R. BARNETT, D. H. FENG, J. W. STEED, AND L. J. B. GOLDFARB, *Coulomb wave functions for all real η and ρ*, Comput. Phys. Comm., 8 (1974), pp. 377–395.

[12] W. BECKEN AND P. SCHMELCHER, *The analytic continuation of the Gaussian hypergeometric function $_2F_1(a, b; c; z)$ for arbitrary parameters*, J. Comput. Appl. Math., 126 (2000), pp. 449–478.

[13] M. V. BERRY AND C. J. HOWLS, *Hyperasymptotics for integrals with saddles*, Proc. Roy. Soc. London Ser. A, 434 (1991), pp. 657–675.

[14] R. BHATTACHARYA, D. ROY, AND S. BHOWMICK, *Rational interpolation using Levin-Weniger transforms*, Comput. Phys. Comm., 101 (1997), pp. 213–222.

[15] J. M. BLAIR, C. A. EDWARDS, AND J. H. JOHNSON, *Rational Chebyshev approximations for the inverse of the error function*, Math. Comp., 30 (1976), pp. 7–68.

[16] M. BLAKEMORE, G. A. EVANS, AND J. HYSLOP, *Comparison of some methods for evaluating infinite range oscillatory integrals*, J. Comput. Phys., 22 (1976), pp. 352–376.

[17] G. BLANCH, *Numerical evaluation of continued fractions*, SIAM Rev., 6 (1964), pp. 383–421.

[18] N. BLEISTEIN, *Uniform asymptotic expansions of integrals with stationary point near algebraic singularity*, Comm. Pure Appl. Math., 19 (1966), pp. 353–370.

[19] F. BORNEMANN, D. LAURIE, S. WAGON, AND J. WALDVOGEL, *The SIAM 100-Digit Challenge: A Study in High-Accuracy Numerical Computing*, SIAM, Philadelphia, 2004.

[20] C. BREZINSKI, *A Bibliography on Continued Fractions, Padé Approximation, Sequence Transformation and Related Subjects*, Prensas Universitarias de Zaragoza, Zaragoza, 1991.

[21] ——, *History of Continued Fractions and Padé Approximants*, Springer-Verlag, Berlin, 1991.

[22] ——, *Convergence acceleration during the 20th century*, J. Comput. Appl. Math., 122 (2000), pp. 1–21.

[23] C. BREZINSKI AND M. REDIVO-ZAGLIA, *Extrapolation Methods. Theory and Practice*, vol. 2 of Studies in Computational Mathematics, North–Holland, Amsterdam, 1991.

[24] W. BÜHRING, *An analytic continuation of the hypergeometric series*, SIAM J. Math. Anal., 18 (1987), pp. 884–889.

[25] P. L. BUTZER AND M. HAUSS, *Stirling functions of first and second kind; some new applications*, in Approximation Interpolation and Summability (Ramat Aviv, 1990/Ramat Gan, 1990), vol. 4 of Israel Math. Conf. Proc., Bar-Ilan Univ., Ramat Gan, 1991, pp. 89–108.

[26] P. L. BUTZER, M. HAUSS, AND M. SCHMIDT, *Factorial functions and Stirling numbers of fractional orders*, Results Math., 16 (1989), pp. 16–48.

[27] CALGO, TOMS Collected Algorithms. See http://www.acm.org/calgo.

[28] D. Calvetti, G. H. Golub, W. B. Gragg, and L. Reichel, *Computation of Gauss-Kronrod quadrature rules*, Math. Comp., 69 (2000), pp. 1035–1052.

[29] B. C. Carlson, *Special Functions of Applied Mathematics*, Academic Press, New York, 1977.

[30] ———, *Computing elliptic integrals by duplication*, Numer. Math., 33 (1979), pp. 1–16.

[31] ———, *Numerical computation of real or complex elliptic integrals*, Numer. Algorithms, 10 (1995), pp. 13–26.

[32] CERNLIB. See http://cernlib.web.cern.ch.

[33] C. Chester, B. Friedman, and F. Ursell, *An extension of the method of steepest descents*, Proc. Cambridge Philos. Soc., 53 (1957), pp. 599–611.

[34] P. A. Clarkson and E. L. Mansfield, *The second Painlevé equation, its hierarchy and associated special polynomials*, Nonlinearity, 16 (2003), pp. R1–R26.

[35] W. W. Clendenin, *A method for numerical calculation of Fourier integrals*, Numer. Math., 8 (1966), pp. 422–436.

[36] C. W. Clenshaw, *A note on the summation of Chebyshev series*, Math. Tables Aids Comput., 9 (1955), pp. 118–120.

[37] ———, *The numerical solution of linear differential equations in Chebyshev series*, Proc. Cambridge Philos. Soc., 53 (1957), pp. 134–149.

[38] ———, *Chebyshev Series for Mathematical Functions*, National Physical Laboratory Mathematical Tables, Vol. 5, Department of Scientific and Industrial Research, Her Majesty's Stationery Office, London, 1962.

[39] C. W. Clenshaw and A. R. Curtis, *A method for numerical integration on an automatic computer*, Numer. Math., 2 (1960), pp. 197–205.

[40] J. A. Cochran, *The zeros of Hankel functions as functions of their order*, Numer. Math., 7 (1965), pp. 238–250.

[41] J. A. Cochran and J. N. Hoffspiegel, *Numerical techniques for finding v-zeros of Hankel functions*, Math. Comp., 24 (1970), pp. 413–422.

[42] W. J. Cody, *Rational Chebyshev approximations for the error function*, Math. Comp., 23 (1969), pp. 631–637.

[43] ———, *A survey of practical rational and polynomial approximation of functions*, SIAM Rev., 12 (1970), pp. 400–423.

[44] L. Comtet, *Advanced Combinatorics*, D. Reidel, Dordrecht, 1974.

[45] S. D. CONTE AND C. DE BOOR, *Elementary Numerical Analysis. An Algorithmic Approach*, 3rd ed., International Series in Pure and Applied Mathematics, McGraw-Hill, New York, 1980.

[46] E. T. COPSON, *Asymptotic Expansions*, vol. 55 of Cambridge Tracts in Mathematics, Cambridge University Press, Cambridge, UK, 2004.

[47] A. CUYT, V. PETERSEN, B. VERDONK, H. WAADELAND, W. B. JONES, AND C. BONAN-HAMADA, *Handbook of Continued Fractions for Special Functions*, Kluwer, Dordrecht, 2007.

[48] P. J. DAVIS, *Interpolation and Approximation*, Dover, New York, 1975. Republication, with minor corrections, of the 1963 original, with a new preface and bibliography.

[49] P. J. DAVIS AND P. RABINOWITZ, *Methods of Numerical Integration*, Academic Press, Orlando, FL, 1984.

[50] N. G. DE BRUIJN, *Asymptotic Methods in Analysis*, 3rd ed., Dover, New York, 1981.

[51] A. DEAÑO AND J. SEGURA, *Transitory minimal solutions of hypergeometric recursions and pseudoconvergence of associated continued fractions*, Math. Comp., 76 (2007), pp. 879–901.

[52] A. DEAÑO, A. GIL, AND J. SEGURA, *New inequalities from classical Sturm theorems*, J. Approx. Theory, 131 (2004), pp. 208–230.

[53] A. DEAÑO AND J. SEGURA, *Global Sturm inequalities for the real zeros of the solutions of the Gauss hypergeometric equation*, J. Approx. Theory, 148 (2007), pp. 92–110.

[54] R. L. DEVANEY, *An Introduction to Chaotic Dynamical Systems*, Benjamin/Cummings, Menlo Park, CA, 1986.

[55] A. R. DIDONATO AND A. H. MORRIS, JR., *Computation of the incomplete gamma function ratios and their inverse*, ACM Trans. Math. Software, 12 (1986), pp. 377–393.

[56] ———, *Algorithm 654: FORTRAN subroutines for computing the incomplete gamma function ratios and their inverse*, ACM Trans. Math. Software, 13 (1987), pp. 318–319.

[57] T. M. DUNSTER, *Uniform asymptotic expansions for Whittaker's confluent hypergeometric functions*, SIAM J. Math. Anal., 20 (1989), pp. 744–760.

[58] ———, *Bessel functions of purely imaginary order, with an application to second-order linear differential equations having a large parameter*, SIAM J. Math. Anal., 21 (1990), pp. 995–1018.

[59] T. M. DUNSTER AND D. A. LUTZ, *Convergent factorial series expansions for Bessel functions*, SIAM J. Math. Anal., 22 (1991), pp. 1156–1172.

[60] S. ELAYDI, *An Introduction to Difference Equations*, 3rd ed., Undergraduate Texts in Mathematics, Springer, New York, 2005.

[61] Á. ELBERT AND A. LAFORGIA, *Further results on McMahon's asymptotic approximations*, J. Phys. A, 33 (2000), pp. 6333–6341.

[62] D. ELLIOTT, *Error analysis of an algorithm for summing certain finite series*, J. Aust. Math. Soc., 8 (1968), pp. 213–221.

[63] A. ERDÉLYI, W. MAGNUS, F. OBERHETTINGER, AND F. G. TRICOMI, eds., *Higher Transcendental Functions*, McGraw–Hill, New York, 1953–1955. Based on notes left by Harry Bateman. Reprinted in 1981 by Robert E. Krieger, Melbourne, FL, 3 volumes.

[64] B. R. FABIJONAS, *Algorithm 838: Airy functions*, ACM Trans. Math. Software, 30 (2004), pp. 491–501.

[65] B. R. FABIJONAS, D. W. LOZIER, AND F. W. J. OLVER, *Computation of complex Airy functions and their zeros using asymptotics and the differential equation*, ACM Trans. Math. Software, 30 (2004), pp. 471–490.

[66] B. R. FABIJONAS AND F. W. J. OLVER, *On the reversion of an asymptotic expansion and the zeros of the Airy functions*, SIAM Rev., 41 (1999), pp. 762–773.

[67] L. FEJÉR, *Mechanische Quadraturen mit positiven Cotesschen Zahlen*, Math. Z., 37 (1933), pp. 287–309.

[68] C. FERREIRA, J. L. LÓPEZ, AND E. PÉREZ-SINUSÍA, *Incomplete gamma functions for large values of their variables*, Adv. in Appl. Math., 34 (2005), pp. 467–485.

[69] H. E. FETTIS, J. C. CASLIN, AND K. R. CRAMER, *Complex zeros of the error function and of the complementary error function*, Math. Comp., 27 (1973), pp. 401–407.

[70] L. N. G. FILON, *On a quadrature formula for trigonometric integrals*, Proc. Roy. Soc. Edinburgh, 49 (1928), pp. 38–47.

[71] R. C. FORREY, *Computing the hypergeometric function*, J. Comput. Phys., 137 (1997), pp. 79–100.

[72] J. FRANKLIN AND B. FRIEDMAN, *A convergent asymptotic representation for integrals*, Proc. Cambridge Philos. Soc., 53 (1957), pp. 612–619.

[73] I. GARGANTINI AND P. HENRICI, *A continued fraction algorithm for the computation of higher transcendental functions in the complex plane*, Math. Comp., 21 (1967), pp. 18–29.

[74] L. GATTESCHI AND C. GIORDANO, *Error bounds for McMahon's asymptotic approximations of the zeros of the Bessel functions*, Integral Transforms Spec. Funct., 10 (2000), pp. 41–56.

[75] W. GAUTSCHI, *Computational aspects of three-term recurrence relations*, SIAM Rev., 9 (1967), pp. 24–82.

[76] ———, *Algorithm 471—Exponential integrals*, Comm. ACM, 16 (1973), pp. 761–763.

[77] ———, *Anomalous convergence of a continued fraction for ratios of Kummer functions*, Math. Comp., 31 (1977), pp. 994–999.

[78] ———, *A computational procedure for incomplete gamma functions*, ACM Trans. Math. Software, 5 (1979), pp. 466–481.

[79] ———, *Computation of Bessel and Airy functions and of related Gaussian quadrature formulae*, BIT, 42 (2002), pp. 110–118.

[80] ———, *Orthogonal Polynomials: Computation and Approximation*, Numerical Mathematics and Scientific Computation, Oxford University Press, New York, 2004.

[81] A. GIL, W. KOEPF, AND J. SEGURA, *Computing the real zeros of hypergeometric functions*, Numer. Algorithms, 36 (2004), pp. 113–134.

[82] A. GIL AND J. SEGURA, *A code to evaluate prolate and oblate spheroidal harmonics*, Comput. Phys. Comm., 108 (1998), pp. 267–278.

[83] ———, *DTORH3 2.0: A new version of a computer program for the evaluation of toroidal harmonics*, Comput. Phys. Comm., 139 (2001), pp. 186–191.

[84] ———, *Computing the zeros and turning points of solutions of second order homogeneous linear ODEs*, SIAM J. Numer. Anal., 41 (2003), pp. 827–855.

[85] A. GIL, J. SEGURA, AND N. M. TEMME, *Algorithm 850: Real parabolic cylinder functions $U(a, x)$, $V(a, x)$*, ACM Trans. Math. Software, 32 (2006), pp. 102–112.

[86] ———, *Computing the real parabolic cylinder functions $U(a, x)$, $V(a, x)$*, ACM Trans. Math. Software, 32 (2006), pp. 70–101.

[87] A. GIL, J. SEGURA, AND N. M. TEMME, *On nonoscillating integrals for computing inhomogeneous Airy functions*, Math. Comp., 70 (2001), pp. 1183–1194.

[88] ———, *Algorithm 819: AIZ, BIZ: Two Fortran 77 routines for the computation of complex Airy functions*, ACM Trans. Math. Software, 28 (2002), pp. 325–336.

[89] ———, *Algorithm 822: GIZ, HIZ: Two Fortran 77 routines for the computation of complex Scorer functions*, ACM Trans. Math. Software, 28 (2002), pp. 436–447.

[90] ———, *Computing complex Airy functions by numerical quadrature*, Numer. Algorithms, 30 (2002), pp. 11–23.

[91] ———, *Evaluation of the modified Bessel function of the third kind of imaginary orders*, J. Comput. Phys., 175 (2002), pp. 398–411.

[92] ——, *Computation of the modified Bessel function of the third kind of imaginary orders: Uniform Airy-type asymptotic expansion*, J. Comput. Appl. Math., 153 (2003), pp. 225–234.

[93] ——, *On the zeros of the Scorer functions*, J. Approx. Theory, 120 (2003), pp. 253–266.

[94] ——, *Algorithm 831: Modified Bessel functions of imaginary order and positive argument*, ACM Trans. Math. Software, 30 (2004), pp. 159–164.

[95] ——, *Computing solutions of the modified Bessel differential equation for imaginary orders and positive arguments*, ACM Trans. Math. Software, 30 (2004), pp. 145–158.

[96] ——, *Numerically satisfactory solutions of hypergeometric recursions*, Math. Comp., 76 (2007), pp. 1449–1468.

[97] G. H. GOLUB AND J. H. WELSCH, *Calculation of Gauss quadrature rules*, Math. Comp., 23 (1969), pp. 221–230.

[98] E. T. GOODWIN, *The evaluation of integrals of the form $\int_{-\infty}^{\infty} f(x)e^{-x^2}dx$*, Proc. Cambridge Philos. Soc., 45 (1949), pp. 241–245.

[99] J. GRAD AND E. ZAKRAJŠEK, *Method for evaluation of zeros of Bessel functions*, J. Inst. Math. Appl., 11 (1973), pp. 57–72.

[100] S. GRAFFI AND V. GRECCHI, *Borel summability and indeterminacy of the Stieltjes moment problem: Application to the anharmonic oscillators*, J. Math. Phys., 19 (1978), pp. 1002–1006.

[101] R. L. GRAHAM, D. E. KNUTH, AND O. PATASHNIK, *Concrete mathematics*, 2nd ed., Addison–Wesley, Reading, MA, 1994.

[102] P. R. GRAVES-MORRIS, D. E. ROBERTS, AND A. SALAM, *The epsilon algorithm and related topics*, J. Comput. Appl. Math., 122 (2000), pp. 51–80.

[103] E. GROSSWALD, *Bessel Polynomials*, vol. 698 of Lecture Notes in Mathematics, Springer, Berlin, 1978.

[104] J. F. HART, E. W. CHENEY, C. L. LAWSON, H. J. MAEHLY, C. K. MESZTENYI, J. R. RICE, H. C. THACHER, JR., AND C. WITZGALL, *Computer Approximations*, SIAM Ser. in Appl. Math., John Wiley and Sons, New York, 1968.

[105] H. W. HETHCOTE, *Error bounds for asymptotic approximations of zeros of transcendental functions*, SIAM J. Math. Anal., 1 (1970), pp. 147–152.

[106] H. H. H. HOMEIER, *Scalar Levin-type sequence transformations*, J. Comput. Appl. Math., 122 (2000), pp. 81–147.

[107] Y. IKEBE, *The zeros of regular Coulomb wave functions and of their derivatives*, Math. Comp., 29 (1975), pp. 878–887.

[108] Y. IKEBE, Y. KIKUCHI, I. FUJISHIRO, N. ASAI, K. TAKANASHI, AND M. HARADA, *The eigenvalue problem for infinite compact complex symmetric matrices with application to the numerical computation of complex zeros of $J_0(z) - iJ_1(z)$ and of Bessel functions $J_m(z)$ of any real order m*, Linear Algebra Appl., 194 (1993), pp. 35–70.

[109] IMSL. See http://www.vni.com/products/imsl.

[110] E. L. INCE, *On the continued fractions associated with the hypergeometric equation*, Edinb. M. S. Proc., 34 (1915), pp. 146–154.

[111] ———, *On the continued fractions connected with the hypergeometric equation*, Lond. M. S. Proc., 18 (1919), pp. 236–248.

[112] AMERICAN NATIONAL STANDARDS INSTITUTE, INSTITUTE OF ELECTRICAL, AND ELECTRONIC ENGINEERS, *IEEE Standard for Binary Floating-Point Arithmetic*, ANSI/IEEE Standard Std 754-1985, New York, 1985.

[113] A. ISERLES, *On the numerical quadrature of highly-oscillating integrals. I. Fourier transforms*, IMA J. Numer. Anal., 24 (2004), pp. 365–391.

[114] ———, *On the numerical quadrature of highly-oscillating integrals. II. Irregular oscillators*, IMA J. Numer. Anal., 25 (2005), pp. 25–44.

[115] A. ISERLES AND S. P. NØRSETT, *Efficient quadrature of highly oscillatory integrals using derivatives*, Proc. R. Soc. Lond. Ser. A Math. Phys. Eng. Sci., 461 (2005), pp. 1383–1399.

[116] J. H. JOHNSON AND J. M. BLAIR, *REMES2—A Fortran Program to Calculate Rational Minimax Approximations to a Given Function*, Technical Report AECL-4210, Atomic Energy of Canada Limited, Chalk River Nuclear Laboratories, Chalk River, ON, Canada, 1973.

[117] N. L. JOHNSON, S. KOTZ, AND N. BALAKRISHNAN, *Continuous univariate distributions. Vol. 1*, 2nd ed., Wiley Series in Probability and Mathematical Statistics: Applied Probability and Statistics, John Wiley and Sons, New York, 1994.

[118] W. B. JONES AND W. J. THRON, *Numerical stability in evaluating continued fractions*, Math. Comp., 28 (1974), pp. 795–810.

[119] ———, *Continued Fractions. Analytic Theory and Applications*, vol. 11 of Encyclopedia of Mathematics and Its Applications, Addison–Wesley, Reading, MA, 1980.

[120] C. JORDAN, *The Calculus of Finite Differences*, 2nd ed., Chelsea, New York, 1947.

[121] D. K. KAHANER AND G. MONEGATO, *Nonexistence of extended Gauss-Laguerre and Gauss-Hermite quadrature rules with positive weights*, Z. Angew. Math. Phys., 29 (1978), pp. 983–986.

[122] E. H. KAUFMAN, JR. AND TERRY D. LENKER, *Linear convergence and the bisection algorithm*, Amer. Math. Monthly, 93 (1986), pp. 48–51.

[123] A. N. KHOVANSKII, *The Application of Continued Fractions and Their Generalizations to Problems in Approximation Theory*, P. Noordhoff N. V., Groningen, 1963.

[124] K. KNOPP, *Theorie and Anwendung der unendlichen Reihen*, Fünfte berichtigte Auflage. Die Grundlehren der Mathematischen Wissen schen Wissenschaften, Band 2, Springer, Berlin, 1964. English translation: *Theory and Application of Infinite Series*, 2nd ed., Blackie & Son, London, 1951.

[125] D. E. KNUTH, *Two notes on notation*, Amer. Math. Monthly, 99 (1992), pp. 403–422.

[126] P. KRAVANJA, O. RAGOS, M. N. VRAHATIS, AND F. A. ZAFIROPOULOS, *Zebec: A mathematical software package for computing simple zeros of Bessel functions of real order and complex argument*, Comput. Phys. Comm., 113 (1998), pp. 220–238.

[127] P. KRAVANJA, M. VAN BAREL, O. RAGOS, M. N. VRAHATIS, AND F. A. ZAFIROPOULOS, *Zeal: A mathematical software package for computing zeros of analytic functions*, Comput. Phys. Comm., 124 (2000), pp. 212–232.

[128] F. T. KROGH AND W. VAN SNYDER, *Algorithm 699: A new representation of Patterson's quadrature formulae*, ACM Trans. Math. Software, 37 (2000), pp. 457–461.

[129] A. S. KRONROD, *Integration with control of accuracy*, Soviet Phys. Dokl., 9 (1964), pp. 17–19.

[130] ———, *Nodes and Weights of Quadrature Formulas. Sixteen-Place Tables*, Consultants Bureau, New York, 1965.

[131] V. I. KRYLOV AND N. S. SKOBLYA, *A Handbook of Methods of Approximate Fourier Transformation and Inversion of the Laplace Transformation*, Mir, Moscow, 1985.

[132] LAPACK95. See http://www.netlib.org/lapack95.

[133] D. P. LAURIE, *Calculation of Gauss-Kronrod quadrature rules*, Math. Comp., 66 (1997), pp. 1133–1145.

[134] N. N. LEBEDEV, *Special functions and their applications*, revised ed., Dover, New York, 1972.

[135] D. LEVIN, *Development of non-linear transformations of improving convergence of sequences*, Int. J. Comput. Math., 3 (1973), pp. 371–388.

[136] X. LI AND R. WONG, *On the asymptotics of the Meixner-Pollaczek polynomials and their zeros*, Constr. Approx., 17 (2001), pp. 59–90.

[137] I. M. LONGMAN, *Note on a method for computing infinite integrals of oscillatory functions*, Proc. Cambridge Philos. Soc., 52 (1956), pp. 764–768.

[138] L. LORENTZEN AND H. WAADELAND, *Continued Fractions with Applications*, vol. 3 of Studies in Computational Mathematics, North–Holland, Amsterdam, 1992.

[139] D. W. LOZIER AND F. W. J. OLVER, *Airy and Bessel functions by parallel integration of ODEs*, in Proceedings of the Sixth SIAM Conference on Parallel Processing for Scientific Computing, Vol. II, R. F. Sincovec, D. E. Keyes, M. R. Leuze, L. R. Petzold, and D. A. Reed, eds., SIAM, Philadelphia, 1993, pp. 530–538.

[140] ———, *Numerical evaluation of special functions*, in Mathematics of Computation 1943–1993: A Half-Century of Computational Mathematics (Vancouver, BC, 1993), vol. 48 of Proc. Sympos. Appl. Math., Amer. Math. Soc., Providence, RI, 1994, pp. 79–125. Updates are available at http://math.nist.gov/mcsd/Reports/2001/nesf/.

[141] S. K. LUCAS AND H. A. STONE, *Evaluating infinite integrals involving Bessel functions of arbitrary order*, J. Comput. Appl. Math., 64 (1995), pp. 217–231.

[142] Y. L. LUKE, *The Special Functions and Their Approximations, Vol. I*, vol. 53 of Mathematics in Science and Engineering, Academic Press, New York, 1969.

[143] ———, *The Special Functions and Their Approximations. Vol. II*, vol. 53 of Mathematics in Science and Engineering, Academic Press, New York, 1969.

[144] ———, *Mathematical Functions and Their Approximations*, Academic Press, New York, 1975.

[145] J. N. LYNESS, *Integrating some infinite oscillating tails*, in Proceedings of the International Conference on Computational and Applied Mathematics (Leuven, 1984), Vol. 12/13, 1985, pp. 109–117.

[146] A. J. MACLEOD, *Chebyshev expansions for Abramowitz functions*, Appl. Numer. Math., 10 (1992), pp. 129–137.

[147] MAPLE. See http://www.maplesoft.com.

[148] J. C. MASON AND D. C. HANDSCOMB, *Chebyshev Polynomials*, Chapman & Hall/CRC, Boca Raton, FL, 2003.

[149] MATHEMATICA. See http://www.mathworks.com/.

[150] MATLAB. See http://www.mathworks.com/.

[151] G. MATVIYENKO, *On the evaluation of Bessel functions*, Appl. Comput. Harmon. Anal., 1 (1993), pp. 116–135.

[152] G. MEINARDUS, *Approximation of Functions: Theory and Numerical Methods*, vol. 13 of Springer Tracts in Natural Philosophy, Springer, New York, 1967.

[153] Y. MIYAZAKI, Y. ASAI, N. KIKUCHI, D. CAI, AND Y. IKEBE, *Computation of multiple eigenvalues of infinite tridiagonal matrices*, Math. Comp., 73 (2004), pp. 719–730.

[154] M. MORI AND M. SUGIHARA, *The double-exponential transformation in numerical analysis*, J. Comput. Appl. Math., 127 (2001), pp. 287–296.

[155] A. MURLI AND M. RIZZARDI, *Algorithm 682: Talbot's method for the Laplace inversion problem*, ACM Trans. Math. Software, 16 (1990), pp. 158–168.

[156] NAGLIB. See http://www.nag.co.uk.

[157] N. NIELSEN, *Handbuch der Theorie der Gammafunktion*, B. G. Teubner, Leipzig, Germany, 1906.

[158] N. E. NÖRLUND, *Vorlesungen über Differenzenrechnung.*, Springer, Berlin, 1924.

[159] A. B. OLDE DAALHUIS, *Hyperasymptotic expansions of confluent hypergeometric functions*, IMA J. Appl. Math., 49 (1992), pp. 203–216.

[160] ———, *Hyperasymptotic solutions of second-order linear differential equations. II*, Methods Appl. Anal., 2 (1995), pp. 198–211.

[161] A. B. OLDE DAALHUIS AND F. W. J. OLVER, *Hyperasymptotic solutions of second-order linear differential equations. I*, Methods Appl. Anal., 2 (1995), pp. 173–197.

[162] ———, *On the asymptotic and numerical solution of linear ordinary differential equations*, SIAM Rev., 40 (1998), pp. 463–495.

[163] A. B. OLDE DAALHUIS AND N. M. TEMME, *Uniform Airy-type expansions of integrals*, SIAM J. Math. Anal., 25 (1994), pp. 304–321.

[164] J. OLIVER, *An error analysis of the modified Clenshaw method for evaluating Chebyshev and Fourier series*, J. Inst. Math. Appl., 20 (1977), pp. 379–391.

[165] ———, *Rounding error propagation in polynomial evaluation schemes*, J. Comput. Appl. Math., 5 (1979), pp. 85–97.

[166] F. W. J. OLVER, *Uniform asymptotic expansions for Weber parabolic cylinder functions of large orders*, J. Res. Nat. Bur. Standards Sect. B, 63B (1959), pp. 131–169.

[167] ———, *Numerical solution of second-order linear difference equations*, J. Res. Nat. Bur. Standards Sect. B, 71B (1967), pp. 111–129.

[168] ———, *Asymptotics and Special Functions*, AKP Classics, A K Peters, Wellesley, MA, 1997.

[169] R. B. PARIS, *On the use of Hadamard expansions in hyperasymptotic evaluation of Laplace-type integrals. I. Real variable*, J. Comput. Appl. Math., 167 (2004), pp. 293–319.

[170] ———, *On the use of Hadamard expansions in hyperasymptotic evaluation of Laplace-type integrals. II. Complex variable*, J. Comput. Appl. Math., 167 (2004), pp. 321–343.

[171] R. B. PARIS AND D. KAMINSKI, *Asymptotics and Mellin-Barnes integrals*, vol. 85 of Encyclopedia of Mathematics and Its Applications, Cambridge University Press, Cambridge, UK, 2001.

[172] L. PASQUINI, *Accurate computation of the zeros of the generalized Bessel polynomials*, Numer. Math., 86 (2000), pp. 507–538.

[173] T. N. L. PATTERSON, *The optimum addition of points to quadrature formulae*, Math. Comp. 22 (1968), 847–856; addendum, ibid., 22 (1968), pp. C1–C11.

[174] ———, *On high precision methods for the evaluation of Fourier integrals with finite and infinite limits*, Numer. Math., 27 (1976/77), pp. 41–52.

[175] ———, *Algorithm 672: Generation of interpolatory quadrature rules of the highest degree of precision with preassigned nodes for general weight functions*, ACM Trans. Math. Software, 15 (1989), pp. 137–143.

[176] ———, *Stratified nested and related quadrature rules*, J. Comput. Appl. Math., 112 (1999), pp. 243–251.

[177] F. PEHERSTORFER AND K. PETRAS, *Ultraspherical Gauss–Kronrod quadrature is not possible for $\lambda > 3$*, SIAM J. Numer. Anal., 37 (2000), pp. 927–948.

[178] ———, *Stieltjes polynomials and Gauss-Kronrod quadrature for Jacobi weight functions*, Numer. Math., 95 (2003), pp. 689–706.

[179] O. PERRON, *Die Lehre von den Kettenbrüchen. Dritte, verbesserte und erweiterte Aufl. Bd.* II. Analytisch-funktionentheoretische Kettenbrüche, B. G. Teubner Verlagsgesellschaft, Stuttgart, 1957.

[180] R. PIESSENS, *Chebyshev series approximations for the zeros of the Bessel functions*, J. Comput. Phys., 53 (1984), pp. 188–192.

[181] ———, *A series expansion for the first positive zero of the Bessel functions*, Math. Comp., 42 (1984), pp. 195–197.

[182] ———, *On the computation of zeros and turning points of Bessel functions*, Bull. Soc. Math. Grèce (N.S.), 31 (1990), pp. 117–122.

[183] H. POINCARÉ, *Sur les équations linéaires aux différentielles ordinaires et aux différences finies*, Amer. J. Math., 7 (1885), pp. 203–258.

[184] M. J. D. POWELL, *On the maximum errors of polynomial approximations defined by interpolation and by least squares criteria*, Comput. J., 9 (1967), pp. 404–407.

[185] W. H. PRESS, S. A. TEUKOLSKY, W. T. VETTERLING, AND B. P. FLANNERY, *Numerical recipes in Fortran: The art of scientific computing*, 2nd ed., Cambridge University Press, Cambridge, UK, 1992. Versions in other languages are available: C language (1988), (1992); Pascal (1989); Fortran 90 (1996).

[186] P. J. PRINCE, *Algorithm 498—Airy functions using Chebyshev series approximations*, ACM Trans. Math. Software, 1 (1975), pp. 372–379.

[187] J. R. RICE, *The approximation of functions. Vol. I: Linear theory*, Addison–Wesley, Reading, MA, London, 1964.

[188] È. YA. RIEKSTYN'SH, *Asymptotics and Estimates of the Roots of Equations*, "Zinatne," Riga, 1991.

[189] T. J. RIVLIN, *An Introduction to the Approximation of Functions*, Dover, New York, 1981.

[190] M. RIZZARDI, *A modification of Talbot's method for the simultaneous approximation of several values of the inverse Laplace transform*, ACM Trans. Math. Software, 21 (1995), pp. 347–371.

[191] H. RUTISHAUSER, *Der Quotienten-Differenzen-Algorithmus*, Mitt. Inst. Angew. Math. Zürich, 1957 (1957), p. 74.

[192] C. SCHWARTZ, *Numerical integration of analytic functions*, J. Comput. Phys., 4 (1969), pp. 19–29.

[193] J. SEGURA, *The zeros of special functions from a fixed point method*, SIAM J. Numer. Anal., 40 (2002), pp. 114–133.

[194] J. SEGURA, P. FERNÁNDEZ DE CÓRDOBA, AND YU. L. RATIS, *A code to evaluate modified Bessel functions based on the continued fraction method.*, Comput. Phys. Comm., 105 (1997), pp. 263–272.

[195] J. SEGURA AND A. GIL, *ELF and GNOME: Two tiny codes to evaluate the real zeros of the Bessel functions of the first kind for real orders*, Comput. Phys. Comm., 117 (1999), pp. 250–262.

[196] ——, *Evaluation of toroidal harmonics*, Comput. Phys. Comm., 124 (2000), pp. 104–122.

[197] A. SIDI, *Practical Extrapolation Methods*, vol. 10 of Cambridge Monographs on Applied and Computational Mathematics, Cambridge University Press, Cambridge, UK, 2003.

[198] K. SIKORSKI, *Bisection is optimal*, Numer. Math., 40 (1982), pp. 111–117.

[199] SLATEC. See http://www.netlib.org/slatec/.

[200] L. J. SLATER, *Confluent Hypergeometric Functions*, Cambridge University Press, New York, 1960.

[201] M. A. SNYDER, *Chebyshev Methods in Numerical Approximation*, Prentice–Hall, Englewood Cliffs, NJ, 1966.

[202] A. N. STOKES, *A stable quotient-difference algorithm*, Math. Comp., 34 (1980), pp. 515–519.

[203] A. STRECOK, *On the calculation of the inverse of the error function*, Math. Comp., 22 (1968), pp. 144–158.

[204] P. N. SWARZTRAUBER, *On computing the points and weights for Gauss–Legendre quadrature*, SIAM J. Sci. Comput., 24 (2002), pp. 945–954.

[205] G. SZEGÖ, *Über gewisse orthogonale Polynome, die zu einer oszillierenden Belegungsfunktion gehören*, Math. Ann., 110 (1935), pp. 501–513.

[206] H. TAKAHASI AND M. MORI, *Quadrature formulas obtained by variable transformation*, Numer. Math., 21 (1973/74), pp. 206–219.

[207] ———, *Double exponential formulas for numerical integration.*, Publ. Res. Inst. Math. Sci., 9 (1974), pp. 721–741.

[208] A. TALBOT, *The accurate numerical inversion of Laplace transforms*, J. Inst. Math. Appl., 23 (1979), pp. 97–120.

[209] N. M. TEMME, *On the numerical evaluation of the modified Bessel function of the third kind*, J. Comput. Phys., 19 (1975), pp. 324–337.

[210] ———, *An algorithm with Algol 60 program for the computation of the zeros of ordinary Bessel functions and those of their derivatives*, J. Comput. Phys., 32 (1979), pp. 270–279.

[211] ———, *The uniform asymptotic expansion of a class of integrals related to cumulative distribution functions*, SIAM J. Math. Anal., 13 (1982), pp. 239–253.

[212] ———, *The numerical computation of the confluent hypergeometric function $U(a, b, z)$*, Numer. Math., 41 (1983), pp. 63–82.

[213] ———, *Uniform asymptotic expansions of Laplace integrals*, Analysis, 3 (1983), pp. 221–249.

[214] ———, *Incomplete Laplace integrals: Uniform asymptotic expansion with application to the incomplete beta function*, SIAM J. Math. Anal., 18 (1987), pp. 1638–1663.

[215] ———, *Asymptotic estimates for Laguerre polynomials*, Z. Angew. Math. Phys., 41 (1990), pp. 114–126.

[216] ———, *Asymptotic inversion of incomplete gamma functions*, Math. Comp., 58 (1992), pp. 755–764.

[217] ———, *Asymptotic inversion of the incomplete beta function*, J. Comput. Appl. Math., 41 (1992), pp. 145–157.

[218] ———, *Asymptotic estimates of Stirling numbers*, Stud. Appl. Math., 89 (1993), pp. 233–243.

[219] ———, *Special Functions: An Introduction to the Classical Functions of Mathematical Physics*, A Wiley-Interscience Publication, John Wiley and Sons, New York, 1996.

[220] ———, *Numerical algorithms for uniform Airy-type asymptotic expansions*, Numer. Algorithms, 15 (1997), pp. 207–225.

[221] ———, *Numerical and asymptotic aspects of parabolic cylinder functions*, J. Comput. Appl. Math., 121 (2000), pp. 221–246.

[222] ———, *Numerical aspects of special functions*, Acta Numer., 16 (2007), pp. 379–478.

[223] I. J. THOMPSON AND A. R. BARNETT, *Coulomb and Bessel functions of complex arguments and order*, J. Comput. Phys., 64 (1986), pp. 490–509.

[224] L. TREFETHEN, *Is Gauss quadrature better than Clenshaw–Curtis?*, SIAM Rev. (to appear).

[225] F. G. TRICOMI, *Asymptotische Eigenschaften der unvollständigen Gammafunktion*, Math. Z., 53 (1950), pp. 136–148.

[226] C. VAN LOAN, *Computational Frameworks for the Fast Fourier Transform*, vol. 10 of Frontiers in Applied Mathematics, SIAM, Philadelphia, 1992.

[227] H.-J. WAGNER, *An algorithm for asymptotic approximation of Laplace integrals*, Computing, 18 (1977), pp. 51–58.

[228] J. WALDVOGEL, *Fast construction of the Fejér and Clenshaw-Curtis quadrature rules*, BIT, 46 (2006), pp. 195–202.

[229] H. S. WALL, *Analytic Theory of Continued Fractions*, D. Van Nostrand, New York, 1948.

[230] G. N. WATSON, *A Treatise on the Theory of Bessel Functions*, 2nd ed., Cambridge Mathematical Library, Cambridge University Press, Cambridge, UK, 1944.

[231] J. A. C. WEIDEMAN, *Optimizing Talbot's Contours for the Inversion of the Laplace Transform*, Technical Report NA 05/05, Oxford University Computing Lab, 2005.

[232] E. J. WENIGER, *Computation of the Whittaker function of the second kind by summing its divergent asymptotic series with the help of nonlinear sequence transformations*, Comput. Phys., 10 (1996), pp. 496–503.

[233] ——, *Nonlinear sequence transformations for the acceleration of convergence and the summation of divergent series*, Comput. Phys. Reports, 10 (1989), pp. 189–371.

[234] ——, *On the summation of some divergent hypergeometric series and related perturbation expansions*, J. Comput. Appl. Math., 32 (1990), pp. 291–300.

[235] E. J. WENIGER AND J. ČÍŽEK, *Rational approximations for the modified Bessel function of the second kind*, Comput. Phys. Comm., 59 (1990), pp. 471–493.

[236] E. J. WENIGER, J. ČÍŽEK, AND F. VINETTE, *The summation of the ordinary and renormalized perturbation series for the ground state energy of the quartic, sextic, and octic anharmonic oscillators using nonlinear sequence transformations*, J. Math. Phys., 34 (1993), pp. 571–609.

[237] E. WHITTAKER AND G. ROBINSON, *The Calculus of Observations: An Introduction to Numerical Analysis*, 4th ed., Dover, New York, 1967.

[238] E. T. WHITTAKER AND G. N. WATSON, *A Course of Modern Analysis*, Cambridge Mathematical Library, Cambridge University Press, Cambridge, UK, 1996. An introduction to the general theory of infinite processes and of analytic functions; with an account of the principal transcendental functions.

[239] H. S. WILF, *Mathematics for the Physical Sciences*, Dover Books in Advanced Mathematics, John Wiley and Sons, New York, 1962.

[240] ———, *A global bisection algorithm for computing the zeros of polynomials in the complex plane*, J. Assoc. Comput. Mach., 25 (1978), pp. 415–420.

[241] J. WIMP, *Computation with Recurrence Relations*, Applicable Mathematics Series, Pitman (Advanced Publishing Program), Boston, MA, 1984.

[242] R. WONG, *Quadrature formulas for oscillatory integral transforms*, Numer. Math., 39 (1982), pp. 351–360.

[243] ———, *Asymptotic Approximations of Integrals*, vol. 34 of Classics in Applied Mathematics, SIAM, Philadelphia, 2001.

[244] L. WUYTACK, *Commented bibliography on techniques for computing Padé approximants*, in Padé Approximation and Its Applications (Proc. Conf., Univ. Antwerp, Antwerp, 1979), vol. 765 of Lecture Notes in Math., Springer, Berlin, 1979, pp. 375–392.

[245] P. WYNN, *On a device for computing the $e_m(S_n)$ tranformation*, Math. Tables Aids Comput., 10 (1956), pp. 91–96.

[246] ———, *Upon systems of recursions which obtain among the quotients of the Padé table*, Numer. Math., 8 (1966), pp. 264–269.

Index

Abramowitz function
    computed by Clenshaw's method, 74
absolute error, 356
Airy function
    contour integral for, 166
Airy functions
    algorithm, 359
    asymptotic estimate of, 18
    asymptotic expansions, 81, 360
    Chebyshev expansions, 80, 85
    computing
        complex arguments, 359
        Gauss quadrature, 145
        scaled functions, 359
        zeros, 224
    connection formulas, 360, 361
    contour integral for, 264
    differential equation, 249, 359
    relation with hypergeometric function, 28
    used in uniform asymptotic expansion, 250
Airy-type asymptotic expansion
    for modified Bessel functions of purely imaginary order, 375
    for parabolic cylinder functions, 383
    obtained from integrals, 249, 264
algorithm
    for Airy functions, 359
    for computing zeros of Bessel functions, 385
    for modified Bessel functions, 370
    for oblate spheroidal harmonics, 365
    for parabolic cylinder functions, 378
    for prolate spheroidal harmonics, 364
    for Scorer functions, 361
    for toroidal harmonics, 366
    of Remes, 290
analytic continuation of generalized hypergeometric function, 27
anomalous behavior of recursions, 118
    a warning, 122
    confluent hypergeometric functions, 120
    exponential integrals, 121
    first order inhomogeneous equation, 121
    modified Bessel functions, 118
anti-Miller algorithm, 110, 112
associated Legendre functions
    computation for $\Re z > 0$, 363
asymptotic expansion
    uniform, 237
asymptotic expansions
    alternative asymptotic representation for $\Gamma(z)$, 49
    alternative expansion
        for $\Gamma(z)$, 49
        for $\Gamma(a, z)$, 47
    convergent asymptotic representation, 46
    converging factor, 40
    exponentially improved, 39
        for $\Gamma(a, z)$, 39
    exponentially small remainders, 38
    hyperasymptotics, 40
    of exponential integral, 37, 38

of incomplete gamma function
  $\Gamma(a, z)$, 37
of modified Bessel function $K_\nu(z)$,
  43
of Poincaré type, 34
of the exponential integral, 34
Stokes phenomenon, 40
to compute zeros, 199, 200
  of Airy functions, 224
  of Bessel functions, 233
  of Bessel functions with
    McMahon expansions, 200
  of error functions, 229
  of orthogonal polynomials, 234
  of parabolic cylinder functions,
    233
  of Scorer functions, 227
transforming into factorial series,
  44
uniform, 239
  for the incomplete gamma
    functions, 240
upper bound for remainder, 39
  for log $\Gamma(z)$, 39
Wagner's modification, 48
Watson's lemma, 36
asymptotic inversion
  of distribution functions, 317
  of incomplete beta functions, 318
  of incomplete gamma functions,
    312
  of the incomplete beta function
    error function case, 322
    incomplete gamma function
      case, 324
    symmetric case, 319

backsubstitution in Olver's method, 117
backward recurrence algorithm, *see also*
  Miller algorithm
  for computing continued fractions,
    181
backward sweep, 215
base-2 floating-point arithmetic, 356
Bernoulli numbers and polynomials,
  131, 331

order estimate, 336
Bessel functions
  Airy-type expansions, 250
  algorithms for computing, 369
  computing zeros, 197, 204, 385
    asymptotic expansions, 200, 233
    asymptotic expansions of Airy
      type, 204
    eigenvalue problems, 208, 212
    McMahon expansions, 200, 204
  differential equation, 19, 24
  $J_0(x)$ computation
    Chebyshev expansion, 83
    numerical inversion of Laplace
      transform, 349
    the trapezoidal rule, 128
  $J_\nu(z)$ as hypergeometric function,
    28
  Neumann function $Y_\nu(z)$, 25
  recurrence relations, 96
  recursion for $J_\nu(z)$ and $Y_\nu(z)$, 87
  series expansion for $J_\nu(z)$, 24
  Wronskian, 255
Bessel polynomials, 348
best approximation, 51
  Jackson's theorem, 63
  polynomial, 290
    versus Chebyshev series, 291
  rational, 290
    oscillations of the error curve,
      290
binomial coefficient
  gamma functions, 27
  Pochhammer symbol, 27
bisection method, 191, 193, 195
  order of convergence, 194
Bolzano's theorem, 193
Boole's summation method, 336
boundary value problem
  for differential equations in the
    complex plane
    Taylor-series method, 293
Bühring's analytic continuation formula
  for hypergeometric functions,
    31

Carlson's symmetric elliptic integrals, 345
Casorati determinant, 89
    its use in anti-Miller algorithm, 110
Cauchy's form for the remainder of Taylor's formula, 16
Cauchy's inequality, 16
Cauchy–Riemann equations, 162
chaotic behavior in the complex plane, 197
characteristic equation, 92
Chebyshev equioscillation theorem, 63
Chebyshev expansion
    computing coefficients, 69
    convergence properties, 68
        analytic functions, 68
    for Airy functions, 80, 85
    for error function, 83
    for J-Bessel functions, 83
    for Kummer U-function, 84
    of a function, 66
Chebyshev interpolation, 62
    computing the polynomial, 64
    of the second kind, 65
Chebyshev polynomial, 56, 140
Chebyshev polynomials
    as particular case of Jacobi polynomials, 62
    discrete orthogonality relation, 59
    economization of power series, 80
    equidistant zeros and extrema, 61
    expansion of a function, 66
    minimax approximation, 58
    of the first kind, 56
    of the second, third, and fourth kinds, 60
    orthogonality relation, 59
    polynomial representation, 59
    shifted polynomial $T_n^*(x)$, 60
Chebyshev sum
    evaluated by Clenshaw's method, 75
Christoffel numbers for Gauss quadrature, 136
classical orthogonal polynomials, 140

Clenshaw's method
    for evaluating a Chebyshev sum, 65, 75
    error analysis, 76
    modification, 78
    for solving differential equations, 70
    for the Abramowitz function, 74
    for the J-Bessel function, 72
Clenshaw–Curtis quadratures, 62, 296, 297
compact operator in a Hilbert space, 209
complementary error function
    as normal distribution function, 242
    computed by numerical inversion of Laplace transform, 350
    contour integral, 350
    in uniform asymptotic approximations, 242
complex Gauss quadrature formula, 348
    nodes and weights, 349
complex orthogonal polynomials, 348
compound trapezoidal rule, 126
condition of TTRRs, 88
confluent hypergeometric functions
    anomalous behavior of recursion, 120
    Chebyshev expansion for U-function, 84
    differential equation, 19
    integral representation for $U(a, c, z)$, 43
    M in terms of hypergeometric function, 28
    recurrence relations, 96, 99
conical functions
    computing zeros, 211, 223
    recurrence relation, 103, 211
conjugate harmonic functions, 162
continued fraction, 173
    computing, 181
        backward recurrence algorithm, 181
        forward recurrence algorithm, 181

forward series recurrence
    algorithm, 181
  modified Lentz algorithm, 183
  Steed's algorithm, 181
 contractions, 175
 convergence, 175, 179
 equivalence transformations, 175
 even and odd part, 175
 for incomplete beta function, 189
 for incomplete gamma function,
    176
 for incomplete gamma function
    $\Gamma(a, z)$, 186
 for ratios of Gauss hypergeometric
    function, 187
 for special functions, 185
 Jacobi fraction, J-fraction, 179
 linear transformations, 174
 nth convergent, nth approximant,
    174
 numerical evaluation, 181
 of Gauss, 188
 recursion for convergents, 174
 relation with
    ascending power series, 178, 179
    Padé approximant, 278
    Padé approximants, 179
    three-term recurrence relation,
      95
 Stieltjes fraction, S-fraction, 178
 theorems on convergence, 180
 value of the, 174
contour integrals in the complex plane
 quadrature for, 157
convergence properties
 Chebyshev expansion, 68
    analytic functions, 68
 continued fraction, 175
convergent power series, 15
converging factor for asymptotic
    expansion, 40
Coulomb wave functions
 recurrence relations, 98
cylinder functions, 233, *see also* Bessel
    functions

degree of exactness, 124, 132
difference equation
 first order inhomogeneous, 112
 second order homogeneous, 87
differential equation
 Frobenius method, 22
 fundamental system of solutions,
    21
 homogeneous linear second order,
    292
 in the complex plane
    Taylor-series method, 292
    Taylor-series method for
      boundary value problem, 293
    Taylor-series method for initial
      value problem, 292
 inhomogeneous linear second
    order, 292
 of Airy functions, 359
 of Bessel functions, 19
 of confluent hypergeometric
    functions, 19
 of Gauss hypergeometric functions,
    18
 of Hermite polynomials, 19
 of Legendre functions, 19, 363
 of modified Bessel functions, 370
    of purely imaginary order, 372
 of parabolic cylinder functions, 19,
    377
 of Whittaker functions, 19
 singular point, 19
    irregular, 19
    regular, 19
 Taylor expansion method, 291
Dini–Lipschitz continuity, 64
discrete cosine transform, 66
dominant solution of a recurrence
    relation, 90
double factorial, 364
dual algorithm for computing toroidal
    harmonics, 369

economization of power series, 80
eigenvalue problem
 for Bessel functions, 208, 212, 213

for compact infinite matrix, 210
for conical functions, 211
for minimal solutions of three-term
    recurrence relations, 207
for orthogonal polynomials, 205
elliptic integral
    other forms, 347
elliptic integrals
    Carlson's symmetric forms, 345
    incomplete of the first kind, 344
    incomplete of the second kind, 344
    of Legendre, 344
    of the third kind, 345
epsilon algorithm of Wynn, 278
equidistant interpolation
    Runge phenomenon, 54
equioscillation property, 67
error
    absolute, relative, 356
    bound for fixed point method, 194
error functions
    Chebyshev expansion, 83
    computing zeros by using
        asymptotic expansions, 229
    inversion, 330
Euler–Maclaurin formula, 131
    relation with the trapezoidal rule,
        130
Euler's summation formula, 331
    limitations, 336
exponential function
    Padé approximants, 280
exponential integral
    anomalous behavior of recursion,
        121
    as solution of an inhomogeneous
        linear first order difference
        equations, 115
    asymptotic expansions, 37, 38
    expansion as factorial series, 45
    sequence transformations, 288
exponentially improved asymptotic
    expansions, 39

factorial series, 44
    condition for convergence, 44
    for exponential integral, 45
    for incomplete gamma function
        $\Gamma(a, z)$, 45
Fadeeva function, 229
fast cosine transforms, 67
fast Fourier transform, 69, 298
Fejér quadrature, 296
    first rule, 297
    second rule, 297
Filon's method for oscillatory integrals,
    303
first order linear inhomogeneous
    difference equations, 87
fixed point method, 192, 193, 196
    based on global strategies, 213
    error bound, 194
    Newton–Raphson method, 195
    order of convergence, 194
fixed point theorem, 193
floating-point
    IEEE formats, 356
    IEEE-754 standard for base-2
        arithmetic, 356
    numbers, 356
forward differences, 53
forward elimination in Olver's method,
    117
forward recurrence algorithm
    for computing continued fractions,
        181
forward series recurrence algorithm
    for computing continued fractions,
        181
fractal, 197
Frobenius method, 22
fundamental Lagrange interpolation
    polynomial, 52
fundamental system of solutions, 21

gamma function
    alternative asymptotic
        representation, 49
    asymptotic expansion, 243
    numerical algorithm based on
        recursion, 246

Gauss hypergeometric functions, 28
　　Bühring's analytic continuation
　　　　formula, 31
　　convergence domains of power
　　　　series, 31
　　deriving the continued fraction for
　　　　a ratio of, 187
　　differential equation, 18
　　Norlünd's continued fraction, 105
　　other power series, 30
　　Padé approximants, 283
　　recurrence relations in all
　　　　directions, 104
　　recursion for power series
　　　　coefficients, 29
　　removable singularities, 33
　　special cases, 29
　　value at $z = 1$, 29
Gauss quadrature, 132, 135, 191
　　Christoffel numbers, 136
　　computing zeros and weights, 133,
　　　　141, 145
　　example for Legendre polynomials,
　　　　145
　　for computing Airy functions, 360
　　for computing the Airy function in
　　　　the complex plane, 145
　　Gauss–Kronrod, 299
　　Gauss–Lobatto, 299
　　Gauss–Radau, 299
　　generalized Hermite polynomials,
　　　　141
　　Golub–Welsch algorithm, 133, 141,
　　　　145
　　Hermite polynomials, 145
　　Jacobi matrix
　　　　nonorthonormal case, 144
　　　　orthonormal case, 142
　　Jacobi polynomials, 145
　　Kronrod nodes, 300
　　Laguerre polynomials, 145
　　Meixner–Pollaczek polynomials,
　　　　141
　　orthonormal polynomials, 134
　　other rules, 298
　　Patterson, 301
　　recursion for orthogonal
　　　　polynomials, 143
　　Stieltjes procedure, 139
Gegenbauer polynomial, 140
　　and Gauss–Kronrod quadrature,
　　　　300
generalized
　　hypergeometric function, 27
　　　　analytic continuation, 27
　　　　terminating series, 27
　　Laguerre polynomial, 140
global fixed point methods, 213
global strategies for finding zeros, 213
Golub–Welsch algorithm for Gauss
　　quadrature, 133, 141, 145
Gram–Schmidt orthogonalization, 134

Hadamard-type expansions
　　for modified Bessel function $I_\nu(z)$,
　　　　41
Hankel transforms, 303
Hermite
　　interpolation, 53, 54, 136
Hermite polynomial
　　and Gauss–Kronrod quadrature,
　　　　300
Hermite polynomials
　　differential equation, 19
　　Gauss quadrature, 145
　　generalized for Gauss quadrature,
　　　　141
　　special case of parabolic cylinder
　　　　functions, 102
　　zeros, 102
hyperasymptotics, 40
　　for the gamma function, 40
hypergeometric functions, *see* Gauss
　　　　hypergeometric functions
hypergeometric series, 26

IEEE floating-point, *see* floating-point
ill-conditioned problem, 357
incomplete beta function
　　asymptotic inversion, 318
　　　　error function case, 322

incomplete gamma function
        case, 324
    symmetric case, 319
    deriving the continued fraction, 189
    in terms of the Gauss
        hypergeometric function, 189
incomplete gamma functions
    as solution of inhomogeneous
        linear first order difference
        equation, 114
    asymptotic expansions, 237
        alternative representation, 47
        for $\Gamma(a, z)$, 37, 238
        simpler uniform expansions, 247
        uniform, 242
    asymptotic inversion, 312, 329
    continued fraction, 176
        computing $\Gamma(a, z)$, 177, 181, 182
        for $\Gamma(a, z)$, 186
    expansion as factorial series, 45
    normalized functions $P(a, z)$ and
        $Q(a, z)$, 241
    numerical algorithm based on
        uniform expansion, 245
    Padé approximant, 284
indicial equation, 22
inhomogeneous
    Airy functions, 359
    linear difference equations, 112
    linear first order difference equation
        condition of the recursion, 113
        for exponential integrals, 115
        for incomplete gamma function,
            114
        minimal and dominant solutions,
            113
    linear first order difference
        equations, 112
    second order difference equations,
        115
        example, 115
        Olver's method, 116
        subdominant solution, 115
        superminimal solution, 115

initial value problem
    for differential equations in the
        complex plane
        Taylor-series method, 292
inner product of polynomials, 133
interpolation
    by orthogonal polynomials, 65
    Chebyshev, 62
        of the second kind, 65
    Hermite, 53, 54, 136
    Lagrange, 52, 54
    Runge phenomenon, 54
interpolation polynomial
    fundamental Lagrange, 52
inversion
    of complementary error function,
        309
    of error function, 330
    of incomplete beta functions, 318
    of incomplete gamma functions,
        312, 329
irregular singular point of a differential
    equation, 19

Jackson's theorem, 63
Jacobi continued fraction, 179
Jacobi matrix, 205
    Gauss quadrature
        nonorthonormal case, 144
    used in Gauss quadrature, 142
Jacobi polynomial, 60, 140
    as hypergeometric series, 60
    Gauss quadrature, 145
    zeros, 191
Julia set, 197

Kummer functions, *see* confluent
    hypergeometric functions

Lagrange
    interpolation, 52
        formula for the error, 125
        fundamental polynomials, 124
        polynomial, 54
        remainder of Taylor's formula, 16

Laguerre polynomial, 140
    computing zeros, 221
Laguerre polynomials
    Gauss quadrature, 145
Lambert's W-function, 312
Laplace transform
    inversion by using Padé
        approximations, 352
    numerical inversion, 347, 349
Lebesgue constants for Fourier series, 291
Legendre functions
    associated functions, 363
    associated functions $P_\nu^\mu(z)$, $Q_\nu^\mu(z)$, 363
    differential equation, 19, 363
    oblate spheroidal harmonics, 363
    prolate spheroidal harmonics, 363
    recurrence relations, 103
        for conical functions, 103
        with respect to the degree, 104
        with respect to the order, 103
    toroidal harmonics, 363
Legendre polynomial, 140
    example for Gauss quadrature, 145
Legendre's elliptic integrals, 344
Levin's sequence transformation, 287
linear
    differential equations
        regular and singular points, 19
        solved by Taylor expansion, 291
    homogeneous three-term
        recurrence relation, 87
    independent solutions of a
        recurrence relation, 89
    inhomogeneous first order
        difference equations, 87
Liouville–Green approximation, 26
Liouville transformation, 25
local strategies for finding zeros, 197
logarithmic derivative of the gamma
    function, 33
Longman's method for computing
    oscillatory integrals, 303

machine-ϵ, 356

Maclaurin series, 16
mathematical libraries for computing
    special functions, 355
McMahon expansions for zeros of
    Bessel functions, 200, 204
Meixner–Pollaczek polynomials, 141
method of Taylor series
    for differential equations in the
        complex plane, 292
Miller algorithm
    condition for convergence, 108
    estimating starting value N, 110
    for computing modified Bessel
        functions $I_{n+1/2}(x)$, 106
    numerical stability, 109
    numerical stability of the
        normalizing sum, 109
    when a function value is known, 105
    with a normalizing sum, 107
minimal solution of a recurrence
    relation, 90
    how to compute by backward
        recursion, 105
minimax approximation, 51
    Jackson's theorem, 63
modified Bessel functions
    algorithm, 370
    anomalous behavior of recursion, 118
    asymptotic expansion for $K_\nu(z)$, 43
    Chebyshev expansions for $K_0(x)$
        and $K_1(x)$, 370
    differential equation, 370
    expansion for $K_\nu(z)$ in terms of
        confluent hypergeometric
        functions, 43
    of integer and half-integer orders, 370
    of purely imaginary order, 372
        Airy-type asymptotic
            expansions, 375
        algorithms for $K_{ia}(x)$, $L_{ia}(x)$, 372
        asymptotic expansions, 374

continued fraction for $K_{ia}(x)$, 373
differential equation, 372
nonoscillating integral representations, 375
scaled functions, 372
series expansions, 373
Wronskian relation, 372
Padé approximants to $K_\nu(z)$, 352
recurrence relation, 97, 370
spherical, 370
  algorithm, 371
  notation, 370
  recurrence relation, 371
trapezoidal rule for $K_0(x)$, 153
modified Lentz algorithm
  for computing continued fractions, 183
modulus of continuity, 63
monic
  orthogonal polynomials, 134
  orthonormal polynomials, 134

Newton's binomial formula, 27
Newton's divided difference formula, 53
Newton–Raphson method, 191, 193, 195
  high order inversion, 196, 327
  order of convergence, 195
nodes of a quadrature rule, 124
nonlinear differential equations, 25
Norlünd's continued fraction for Gauss hypergeometric functions, 105
normalized incomplete gamma function
  asymptotic estimate, 42
normalized incomplete gamma functions
  asymptotic estimate for $P(a, z)$, 41
  Hadamard-type expansions, 41
  relation with chi-square probability functions, 240
  uniform asymptotic expansions, 242
numerical condition, 357
numerical inversion of Laplace transforms, 347, 349
  by deforming the contour, 350

complex Gauss quadrature formula, 348
  to compute Bessel function $J_0(x)$, 349
  to compute the complementary error function, 350
numerical stability, 357
numerically unstable method, 357

oblate spheroidal harmonics, 363
  algorithm, 365
  recurrence relations, 365
  scaled functions, 364
Olver's method for inhomogeneous second order difference equations, 116
order of convergence
  asymptotic error constant, 194
  fixed point methods, 194
ordinary differential equation, see differential equation
orthogonal basis with respect to inner product, 134
orthogonal polynomials
  computing zeros by using asymptotic expansions, 234
  on complex contour, 348
  zeros, 135
orthogonality with respect to inner product, 134
oscillatory integrals, 301
  asymptotic expansion, 301
  convergence acceleration schemes, 303
  Filon's method, 303
  general forms, 303
  Hankel transforms, 303
  Longman's method for computing, 303
overflow threshold, 356

Padé approximants, 276
  continued fractions, 278
  diagonal elements in the table, 277
  generating the lower triangular part in the table, 278

how to compute, 278
  by Wynn's cross rule, 278
  Luke's examples for special
    functions, 283
  normality, 277
  relation with continued fractions,
    179
  table, 277
  to the exponential function, 280
  to the Gauss hypergeometric
    function, 283
  to the incomplete gamma functions,
    284
  to the modified Bessel function
    $K_\nu(z)$, 352
  Wynn's cross rule for, 278
parabolic cylinder functions, 377
  algorithm, 378
    Maclaurin series, 379
      regions in (a, x)−plane, 378
  asymptotic expansions for large x,
    380
  computing zeros by using
    asymptotic expansions, 233
  contour integral for, 168
  definition, 101
  differential equation, 19, 377
  integral representations, 384
  oscillatory behavior, 102
  recurrence relation for $U(a, x)$, 385
  relation with Hermite polynomials,
    102
  scaled functions, 377
  three-term recurrence relations, 101
  uniform Airy-type asymptotic
    expansion, 383
  uniform asymptotic expansions in
    elementary functions, 381
  Wronskian relation, 101
Perron's theorem, 92, 93
  intuitive form, 92
Pincherle's theorem, 95
plasma-dispersion function, 229
Pochhammer symbol, 27
polynomial
  Stieltjes, 300

polynomial approximation
  minimax, 51
    Jackson's theorem, 63
poorly conditioned problem, 357
power series
  of Bessel function $J_\nu(z)$, 24, 28
  of confluent hypergeometric
    M-function, 28
  of Gauss hypergeometric functions,
    28
  of hypergeometric type, 26
  of the Airy functions, 18
  of the exponential function, 17
primal algorithm
  for computing toroidal harmonics,
    366
prolate spheroidal harmonics, 363
  algorithm, 364
  recurrence relation for $P_n^m(x)$, 365
  recurrence relation for $Q_n^m(x)$, 365
  scaled functions, 364

quadrature
  characteristic function for the error,
    149
  Clenshaw–Curtis, 296, 297
  degree of exactness, 124
  double exponential formulas, 156
  erf-rule, 155
  Fejér, 296
    first rule, 297
    second rule, 297
  for contour integrals in the complex
    plane, 157
  Gauss–Kronrod, 299
  Gauss–Lobatto, 299
  Gauss quadrature, 132
  Gauss–Radau, 299
  other Gauss rules, 298
  Patterson, 301
  Romberg quadrature, 294
  simple trapezoidal rule, 124
  Simpson's rule, 295
  tanh-rule, 154
  the trapezoidal rule on $\mathbb{R}$, 147
  transforming the variable, 153

Index

weight function, 132
weights, 124
quotient-difference algorithm, 178

recurrence relation
    for Bessel functions, 96
    for computing modified Bessel
        function $K_\nu(z)$, 100
    for confluent hypergeometric
        functions, 99
        in all directions, 99
        in the $(++)$ direction, 100
        in the $(+0)$ direction, 99
        in the $(0+)$ direction, 100
    for Coulomb wave functions, 98
    for Legendre functions, 103
    for modified Bessel functions, 97, 370
    for modified spherical Bessel
        functions, 371
    for parabolic cylinder functions, 101, 385
    for prolate spheroidal harmonics
        $P_n^m(x)$, 365
    for prolate spheroidal harmonics
        $Q_n^m(x)$, 365
    for toroidal harmonics, 367
recurrent trapezoidal rule, 129
regular point a differential equation, 19
regular singular point of a differential
    equation, 19
relative error, 356
Remes' second algorithm of, 290
repeated nodes in Hermite interpolation, 54
reverting asymptotic series, 226
Riccati–Bessel functions
    difference-differential system, 213
    zeros, 213
Romberg quadrature, 294
Runge phenomenon, 54

saddle point method, 158
    saddle point, 158
scalar product of polynomials, 133
scaling functions, 358
    to enlarge the domain of
        computation, 358
    to obtain higher accuracy, 358
Schwarzian derivative, 26
Scorer functions
    algorithm for Hi(z), 361
    asymptotic expansion for Gi(z), 362
    asymptotic expansion for Hi(z), 362
    computation for complex
        arguments, 359
    computing scaled functions, 359, 363
    computing zeros by using
        asymptotic expansions, 227
    connection formulas for Gi(z), 362
    integral representation for Hi(z), 361
    power series for Gi(z), 362
secant method, 191
second algorithm of Remes, 290
second order homogeneous linear
    difference equation, 87
sequence transformations, 286
    for asymptotic expansion of
        exponential integral, 288
    Levin's transformation, 287
    numerical examples, 288
    of asymptotic series, 288
    of power series, 288
    Weniger's transformation, 287
    with remainder estimates, 287
Simpson's rule, 125, 295
software survey for computing special
    functions, 355
sources of errors
    due to discretization, 357
    due to fixed-length representations, 357
    due to truncation, 357
    in computations, 357
special functions computing
    Airy functions of complex
        arguments, 359
    mathematical libraries, 355

ratios of Bessel functions, 218
Scorer functions of complex
arguments, 359
software survey, 355
stability of a numerical method, 357
Steed's algorithm
for computing continued fractions,
181
steepest descent path, 158
Stieltjes
continued fraction, 178
procedure for recurrence relations,
139
Stirling numbers
definitions, 337
explicit representations, 337
generating functions, 337
of the first kind
uniform asymptotic expansion,
343
of the second kind, 44
uniform asymptotic expansion,
338
Stokes phenomenon, 40
subdominant solution
of inhomogeneous second order
difference equation, 115
superminimal solution
of inhomogeneous second order
difference equations, 115
symmetric elliptic integrals, 345

Taylor series, 16
Cauchy's formula for remainder, 16
Lagrange's formula for remainder,
16
Taylor's formula for remainder, 16
Taylor-series method
for boundary value problem in the
complex plane, 293
for initial value problem in the
complex plane, 292
testing of software
for computing functions, 358
by comparison with existing
algorithms, 358

by extended precision
algorithms, 358
by verification of functional
relations, 358
consistency between different
methods, 358
three-term recurrence relation, *see also*
recurrence relation
anomalous behavior, 118
confluent hypergeometric
functions, 120
exponential integrals, 121
modified Bessel functions, 118
backward recursion, 91
condition of, 88
dominant solution, 90
forward recursion, 91
linear homogeneous, 87
linearly independent solutions, 89
minimal solution, 89, 90
relation with continued fractions,
95
scaled form, 94
with constant coefficients, 92
toroidal harmonics, 363
algorithm, 366
asymptotic expansion for $P^M_{-1/2}(x)$,
368
dual algorithm, 369
primal algorithm, 366
recurrence relation, 367
relation with elliptic integrals, 367
scaled functions, 364
series expansion for $P^M_{-1/2}(x)$, 367
trapezoidal rule, 350
simple rule, 124
compound rule, 126
Euler's summation formula, 130
for computing Scorer functions,
362
for computing the Bessel function
$J_0(x)$, 128
for computing the Bessel function
$K_0(x)$, 153
for computing the complementary
error function, 350

on $\mathbb{R}$, 147
recursive computation, 129
with exponentially small error, 151
TTRR, *see* three-term recurrence relation
turning point of a differential equation, 249

underflow threshold, 356

Wagner's modification of asymptotic expansions, 48
Watson's lemma, 36
weight function for numerical quadrature, 132
weights of a quadrature rule, 124
Weniger's sequence transformation, 287
Whittaker functions
    differential equation, 19
WKB approximation, 26
Wronskian, 21
    for Airy functions, 254
    for Bessel functions, 255
    for modified Bessel functions of purely imaginary order, 372
    for parabolic cylinder functions, 101
Wynn's cross rule
    for Padé approximants, 278
Wynn's epsilon algorithm, 278

zeros of functions, 191
    Airy functions, 224
    asymptotic approximations, 197, 200
    Bessel functions, 204, 233, 385
        complex zeros, 197
        eigenvalue problem, 208, 212
        from Airy-type asymptotic expansions, 204
        McMahon expansions, 200, 204
    bisection method, 191, 193
    complex zeros, 197
    computation based on asymptotic approximations, 199
    conical functions, 211, 223
    cylinder functions, 233
    eigenvalue problem for orthogonal polynomials, 205
    error functions, 229
    fixed point method, 193
    fixed point methods and asymptotics, 199
    global strategies, 204, 213
    Jacobi polynomials, 191
    Laguerre polynomials, 221
    local strategies, 197
    matrix methods, 204
    Newton–Raphson method, 191, 193
    orthogonal polynomials, 135, 234
    parabolic cylinder functions, 233
    Riccati–Bessel functions, 213
    Scorer functions, 227
    secant method, 191